SYSTEMS
ANALYSIS

HARRY J. WHITE / **SELMO TAUBER**

PROFESSOR AND HEAD, PROFESSOR OF
DEPARTMENT OF APPLIED SCIENCE MATHEMATICS

PORTLAND STATE UNIVERSITY, PORTLAND, OREGON

W. B. SAUNDERS COMPANY
PHILADELPHIA • *LONDON* • *TORONTO* • *1969*

W. B. Saunders Company: West Washington Square
Philadelphia, Pa. 19105

12 Dyott Street
London W.C. 1

1835 Yonge Street
Toronto 7, Ontario

Systems Analysis

PREFACE

Development of the field of systems science has been forced by the increasing urgency of dealing with the many modern technological, scientific, and organizational problems which, because of their complex structure and the large number of variables involved, defy the fragmented approach of traditional disciplines. The central feature of the systems approach, and the one which distinguishes it from earlier methods and requires a truly multidisciplinary attack, is the concept of treating problems as a whole rather than piecemeal.

Underlying reasons for the rapid development of complex systems in our era are found in the explosion of knowledge in science, engineering, and allied fields, the demands of society for more effective and efficient solutions of complex environmental, production, transportation, and other problems, the rapid development of military and defense technology, and the concurrent advances in computer science and computer applications.

Forces generated by these changes have led to revision and updating of educational methods and curricula in engineering, business administration, and the social and life sciences. As an example, engineering requires a mathematical base at least an order of magnitude higher than that of a generation ago. As a result curricula and courses featuring the systems philosophy are receiving increased attention in colleges and universities, many of which now offer systems courses and programs in engineering, business, and other fields. However, one of the major roadblocks in developing systems programs has been that the scope of systems science is such as to necessitate interdisciplinary treatment. In addition, there is the problem that as yet no generally accepted range of subject matter or patterns of treatment exist for texts in the systems field.

This book is an outgrowth of an interdisciplinary systems program begun at Portland State College in 1962. The material has been used as the text for courses given to seniors and graduate students in applied science, engineering, and mathematics, as well as to professional engineers from industry and government. In addition to use as a text for senior and graduate-level courses, the book is sufficiently complete and self-contained to serve the requirements of practicing engineers who wish,

iii

by self study, to broaden and update their knowledge in the rapidly growing field of systems and systems analysis. In writing the book, we have followed an interdisciplinary approach based on mathematics and applied science. The fundamental concept followed is to provide, first, the mathematical and physical foundations common to a broad class of engineering and physical systems, and second, to demonstrate the wide scope of application made possible by this approach. This philosophy of treatment is consonant with today's requirements for broadening students' horizons yet simultaneously achieving economy of time and effort within the confines of college and university curricula.

The book covers discrete systems, i.e., those that can be treated primarily by methods based on ordinary differential equations and algebra. These represent a vast array of practical applications including, for example, electric networks, structural mechanics, space mechanics, vibrations and acoustics, linear programming, growth and decay problems, feedback control systems, communications, power, and electromechanical systems. In addition, management-decision problems such as those involving urban planning, transportation, and traffic control fit into this framework. Other fertile areas of systems applications are those in which the methods of mathematics and physical science are being applied to such fields as economics, biology, and psychology. In many cases new names have been coined to describe these interdisciplinary fields.

The Introduction establishes the basic concepts, definitions, characteristics, and usefulness of systems.

Chapters 1, 2, 3, and 4 cover the mathematical concepts. Chapter 1 deals with the linear algebra essential to the quantitative treatment of multivariable systems. Since linear algebra is rapidly being integrated into undergraduate curricula, it seemed advisable simply to review major topics and required results, rather than give a more extensive treatment. Chapter 2 covers the theory of extrema of functions, while Chapter 3 gives an introduction to the calculus of variations. Both of these are vital to the optimization of systems, and, in addition, the calculus of variations is fundamental to the variational principles which are so effective in dealing with systems. Chapter 4 uses linear algebra to unify the general theory of differential equations and to cope effectively with the systems of simultaneous differential equations which occur frequently in systems problems.

Chapter 5 traces the origins of the systems concept and provides sufficient historical orientation to give the reader some sense of perspective in viewing the systems field as a whole. It is well known that different disciplines often give different names to similar concepts; by giving due attention to the historical aspects of the subjects treated we hope to avoid this disjunction and to provide motivation for further study. The mathematical-physical heritage of *systems* as we know the subject today seems sufficient justification for locating the chapter between the mathematical and physical parts of the book.

Chapters 6 and 7 cover the physical foundations. The purpose of Chapter 6 is to formulate a generalized theory of mechanics which provides means for dealing effectively and economically with a broad range of physical systems. Hamilton's principle is taken as an independent basic postulate, applicable to nonmechanical as well as to mechanical systems. It is felt that this approach is compatible with systems philosophy and the modern trend toward the clustering of knowledge around central principles. Chapter 7 covers the essential aspects of electromagnetic fields and energy relations needed to deal at a fundamental level with electrical and electromechanical systems. Maxwell's equations and the Lorentz law of force are taken as basic postulates.

The final six chapters are devoted to applications. Chapter 8 deals with linear dynamical systems, giving first an abstract general theory of such systems including the state variable approach, followed by applications to electric networks, electromechanical systems, and structural mechanics. Separate attention to linear systems is justified because of their great importance in a broad spectrum of practical applications and the existence of mathematical methods for the general solution of linear equations. Chapter 9 is devoted to the entirely different field of nonlinear satellite and orbit problems of space science. Chapter 10 deals with generalized energy conversion, using Kron's pioneering concepts, the deep significance and value of which are finally being recognized. Chapter 11 treats the broad area of vibrations which, although considered unglamorous by some, actually pervades most areas of science and engineering and has led to advances in many fields as, for example, in stability and oscillation problems, and electroacoustics. Chapter 12 introduces the reader to the broad field of feedback systems, and shows how the fundamental systems theory and methods covered in the earlier chapters can be applied to the feedback area. The purpose of the closing chapter is to provide additional perspective and to give some vista of the present and potential usefulness in applying systems methods, especially in the social and life sciences. Chapter bibliographies are included to provide convenient references and to indicate further readings.

The book provides sufficient material for a one-year sequence, to be given either at the Senior or first-year Graduate level. In our own teaching, we have covered Part One on the mathematical aspects during the Fall Quarter, Part Two on the physical foundations during the Winter Quarter, and Part Three on applications in the Spring Quarter. Since more material is provided on applications than can be covered in one quarter, selected portions of Part Three have been used, based on class and instructor interest. Suggested prerequisites include differential equations, vector analysis, intermediate mechanics and electricity, and exposure to computer methods. Since computers are now standard equipment, it is considered unnecessary to include numerical methods in a book of this kind.

Notation is a difficult problem in this kind of book. We have adhered to conventional notation acceptable to both the mathematician and the engineer where such notation exists, and have particularly tried to avoid surprising and confusing the reader with a flood of special and unexplained notations.

We are indebted to Mr. W. E. Zablocki of the W. B. Saunders Company for helpful discussions and suggestions on the format of the book, and to our colleague Professor H. Erzurumlu, who provided the section on structures in Chapter 8. Acknowledgment is made to our former students, Mr. Walter Peterson and Mr. John Casti, who contributed many cogent suggestions. The difficult task of typing the manuscript in effective form was intelligently and efficiently carried out by Miss Patricia Mayhew, to whom we are especially grateful.

HARRY J. WHITE

SELMO TAUBER

CONTENTS

PART ONE. MATHEMATICAL CONCEPTS

Chapter 1

Chapter 2

PART THREE. APPLICATIONS

Chapter 8

LINEAR SYSTEMS 237

Chapter 9

SATELLITE ORBITS 307

Chapter 10

GENERALIZED ENERGY CONVERTERS 326

INTRODUCTION

"All sciences which have for their end investigations concerning order and measure are related to mathematics, it being of small import whether this measure be sought in numbers, forms, stars, sounds, or any other object; there ought therefore to be a general science, namely mathematics, which should explain all that can be known about order and measure, considered independently of any application to a particular subject . . . . A proof that it far surpasses in utility and importance the sciences which depend on it is that it embraces at once all the objects to which these are devoted and a great many besides."

(Descartes: "Rules for Direction of the Mind," 1628)

The power and utility of systems concepts and methods in technology, management, economics, and other fields, although clearly traceable to classical mathematics and science and more dramatically to the complex technical and scientific developments of World War II, have become generally recognized only since the mid 1950's. There has been a growing realization of the existence of an identifiable *science of systems*, comprising a body of concepts, methods, and, above all, a philosophy of treating the whole rather than bits and pieces. The new field of *systems science* is as yet only loosely defined and has different meanings in different contexts. Its ultimate domain is still seen only dimly as compared to traditional disciplines such as physics, mathematics, and engineering.

Nevertheless, it is desirable and indeed necessary to integrate knowledge in the systems field into useful patterns. The pattern developed in this book stresses analysis, and uses mathematical, physical, engineering, and other related resources. Systems analysis, by its very meaning, cuts across academic departmental barriers; an interdisciplinary approach is therefore necessary, both in the marshaling of varied resources and in the manifold applications. Breadth of treatment is implied by the generic use of the term *system*. For the student, whose interests properly include both theory and applications, the most logical and fruitful approach to systems analysis is, first, to deal with the mathematical and physical foundations, and second, to apply these principles and methods to representative major areas. Systems theory, in the sense that it proceeds from concepts to theory to application, is science-oriented. The converse, or problem-oriented approach, necessary as it may be in certain specialized fields, is less fruitful in systems analysis because of the diversity of systems encountered in practice and the large number of variables often present.

Systems science aims at greatest generality and an ordering of knowledge into manageable and interesting patterns. Every age and generation has the responsibility for discovering new knowledge, and for updating and reorganizing old knowledge into new and more meaningful forms. In this sense, systems science is the outgrowth of the search for a body of theory and methods which gives greater unity to previously compartmentalized scientific disciplines. Nowhere has the need for such unification been greater than in the broad field of engineering. Not only have major engineering areas such as electrical and mechanical engineering maintained a parochial separation from each other, starting often with the freshman or sophomore year in college, but they also have a long tradition of separation from the so-called pure sciences and mathematics.

One result of this isolation is the multiplicity of duplicating and overlapping concepts, theories, and methods which have grown up in the different fields. For example, the theory of vibrations, which is fundamental to physics, mechanical engineering, and electrical engineering, as well as to several other fields, is generally treated separately by each discipline and taught in separate courses by the respective departments in universities and colleges. The general theory of vibrations of a dynamical system, which is developed in physics and mathematics using the concepts and methods of Lagrange and Hamilton, becomes electric circuit theory in the lexicon of electrical engineering, mechanical vibrations in mechanical engineering, and a ready example for illustrating differential equations in mathematics. Thus the student, scientist, or engineer whose interests encompass several disciplines is commonly faced with unnecessary repetitions of essentially similar material. The global or systems approach seeks to avoid these repetitions and inefficiencies by unification of underlying principles.

It is self-evident that the central theme of quantitative systems theory is abstraction and generalization in the mathematical sense. In this sense a system is specified before the elements are identified. Identification of the elements will then place the system in the locale or setting, for example, of engineering, biology, physics, economics, management, or a combination of these. As new fields evolve, or old fields change in content and method, the basic system theory is not invalidated or rendered obsolete. Rather, new identifications are given to the elements of a system, but the theoretical framework remains essentially unchanged.

The common mathematical basis for many fields of application has long been recognized. In science and technology, the concept of similar mathematical systems for dissimilar physical systems is generally presented under the banner of analogs or analogous systems. Analogs are important knowledge multipliers because they enable specialized knowledge in one field to be transferred more or less bodily to another field. In a broad sense system science is seen to be a generalization of this process. The exponential growth of knowledge, the breakdown of traditional barriers between academic fields, and the frustrations of attempting to cope with complex systems piecemeal are forcing the use of more comprehensive philosophy and methods.

1 FUNDAMENTAL CONCEPT OF A SYSTEM

Although *system* is a general term used in many senses, it does convey a very important meaning not readily described in any other way. The word derives from a Greek verb meaning to place or to set together, and Webster's New World Dictionary gives the definition: "A set or arrangement of things so related or connected as to form a unity or whole: as, a solar system, irrigation system, supply system." It is apparent that the idea of a system is embedded in the general culture and antedates modern science and technology by many centuries and even millennia. Discussion of the fundamental concept of a system in its modern scientific sense is therefore necessary to clarify meaning and to better define the broad area of systems.

Many definitions of *system* and related terms are current in science and engineering. Representative examples are:

(1) *Systems science* (Ref. 1): "The science that is common to all large collections of interacting functional units that are combined to achieve purposeful behavior."

(2) *Systems engineering* (Ref. 2): "A process in which complex systems are idealized, designed and manipulated by conscious rational processes based upon the scientific method."

(3) *System* (Ref. 3): "A set of objects with relationships between the objects and their attributes."

(4) *System* (Ref. 4): "An integrated assembly of interacting elements designed to carry out cooperatively a predetermined function."

Definitions of *systems science* and *systems engineering* generally include requirements for utility or at least directed behavior. By contrast, definitions of *system* tend to be more abstract.

In order to achieve greatest generality, and therefore greatest applicability, we define a system abstractly as *a set of interacting elements*. When dealing with real systems, the elements and interactions may be chosen concretely, as for example the solar system, in which the elements are the sun, planets, satellites or moons of the planets, and comets. The interaction among these elements is the force of gravitation as described and quantified by Newton. Note that a collection of isolated elements does *not* constitute a system because the elements do not interact.

Systems may be classified as natural or man-made. Natural systems are those which occur in nature, as for example polyatomic molecules, biological structures, and solar systems. Description of natural systems is the task of the physicist, chemist, astronomer, biologist, and physiologist. Man-made systems are those derived by man, presumably for some useful purpose. Familiar examples are communication systems, transportation systems, electric power systems, chemical process systems, and management-control-decision systems.

The concept of a man-made system usually includes the idea of optimizing certain parameters such as cost, efficiency, size, or reliability, in terms of criteria derived from externally imposed value systems. The value systems are subjective and are based on a variety of factors such as economic, social, or even political. Adjustments or trade-off values between such considerations as cost, reliability, and prestige are frequently necessary. Note that natural systems do not involve performance criteria, although the scientific description of natural phenomena may be based on certain maxima and minima principles, as for example the principle of least action in mechanics.

Systems may be further classified as closed or open. If all elements and interactions of a system are contained within a given closed boundary, the system is called closed, as for example the solar system or certain mathematical systems. If on the other hand exchanges do occur across a given closed boundary, the system is called open. However, it is evident that by sufficiently extending the boundary any system can be made closed.

Actual choice of a boundary for a given case may be based on convenience or on the generality of treatment, or both. For instance, the loads on an electric power system may be considered either as a part of the system or as an output of the system.

In most cases we shall deal with open systems, i.e., those which exchange energy, matter, information, or other entities with their environments; expressed in another way, systems with inputs and outputs. As an example, a data-processing system accepts input data, transforms

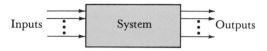

FIGURE 1 Schematic representation of a system.

or processes the data in accord with the directions or program supplied to the computer, and finally generates an output of processed data. Such a system may be represented schematically as shown in Figure 1, sometimes referred to as a "black box." Symbolically, the input may be represented by a quantity x, the output by a quantity y, and the system by an operator T. The system relations may then be expressed in the form

$$y = Tx \qquad (1)$$

which may be interpreted "the operator T transforms an input x into an output y." This interpretation can be given precise mathematical form.

Most open systems have network or flow characteristics, e.g., the flow of matter in a transport system, the flow of energy in a power system, and the flow of information in a computer or in a communication system. Physical systems are usually characterized by the flow of matter or energy.

Systems for which a portion of the output is reintroduced or fed back to the input (usually for the purpose of affecting succeeding outputs) are called feedback systems, and may be represented graphically as shown in Figure 2. Examples of feedback systems are servomechanisms, automated production systems, management systems, and cybernetic systems.

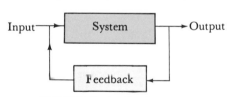

FIGURE 2 Feedback system.

2 WHAT IS SYSTEMS ANALYSIS?

Before essaying a definition, it may be helpful to state what systems analysis *is not*. It is not an inventory of data-processing techniques for automating business operations. It is not anything "made easy" for the life scientist or the economist. It is neither applied physics nor an exhibition of pure science for the engineer. Finally, it is not selected topics in applied mathematics for nonmathematicians.

With these provisos in mind, we may define systems analysis as follows:

Systems analysis: "The analytic study of systems, where analytic is taken in its most general sense."

The analysis relies primarily on mathematical and physical methods, but is not limited to these. For instance, biological and economic concepts and methods may be incorporated in the body of systems theory with enrichment of the whole fabric of systems science.

Systems analysis clearly is an area of applied science, based as it is primarily on scientific and mathematical disciplines rather than on empirical or trial and error methods.

3 EXAMPLES OF SYSTEMS

Systems may be classified in many ways, as for example natural and man-made, physical and engineering, discrete and continuous, deterministic and stochastic, and optimum and adaptive. However, rather than devoting space to elaborate classification schemes, it is regarded as more profitable here to concentrate instead on examples of systems which are both interesting and instructive.

Rational mechanics in the sense of Lagrange and Hamilton is the almost perfect example of an analytical system, containing as it does the elements of abstraction, generality, and applicability to a wide spectrum of physical phenomena and technical problems. Electromagnetic theory, quantum mechanics, and Mendeleev's periodic table of the chemical elements are other examples of systems in the field of physical science.

In technology, an outstanding example is that of automatic control systems. These systems range in complexity from the elementary with only a few variables to the very sophisticated with many variables and multiple inputs and outputs. A well-delineated body of theory, design methods, and experience exists for effective application of control systems to diverse problems in many industries. Other examples in technology are power systems, communication systems, production processes, space systems, and the almost uncountable sub-systems of these.

Many examples of the application of systems analysis to biological processes exist but are still by and large in the experimental and exploratory stages. The problems of adapting mathematical lore devised primarily for dealing with the physical sciences to the incomparably more complex biological systems are uncertain and difficult. Nevertheless, significant progress is being made in this direction. Beyond the fairly mature field of biostatistics lie numerous mathematical studies of such regulatory processes as homeostasis, i.e., regulation at a constant value of a physiological function such as body temperature and such voluntary operations as the focusing of the eye.

The increasing complexity of business and management problems has given rise to a great number of systems applications in these areas. Examples are the simulation of business operations, e.g., inventory control, production, transportation logistics, sales prediction, and advertising schedules; how to make optimal use of resources and facilities; and how to schedule complex projects. The simpler problems involve no more than methods for economical processing of huge volumes of detailed figures. More complicated problems are approached through linear programming, often involving several hundred variables, systematic scheduling techniques such as PERT (program evaluation and review technique), and so-called game and decision theory methods. A somewhat different systems approach in the business field, using basically methods long applied to science and engineering, is being developed by some investigators under the name of industrial dynamics (Ref. 5).

Finally, it may be observed that, according to the definition given earlier, mathematics itself is a powerful system.

A representative, although by no means exhaustive, listing of major areas of application of systems methods is given in Table 1 for orientation and reference purposes.

Table I. Major Areas of Application

A. Kinds of systems:	B. Fields included under systems analysis:
1. Physical	1. Automatic feedback control systems
2. Mathematical	2. Automatic process control
3. Biological	3. Linear programming and dynamic
4. Management and corporate	programming
5. Chemical process	4. Electric network analysis
6. Data processing	5. Mechanical vibration systems
7. Engineering	6. Process dynamics
8. Management-control-decision	7. Business dynamics
9. Military	8. Analog simulation and computation
10. Communication	9. Stochastic service systems
11. Electric power	10. PERT and CPM scheduling methods
12. Transportation	

4 MATHEMATICAL ASPECTS

Mathematical methods and mathematical thinking are central to the quantitative treatment of systems. The use of mathematics may be viewed from several different levels. It is not uncommon in dealing with applications to regard mathematics as no more than a set of tools or techniques to be selected and used as required for the job at hand. While this position may be defended on pragmatic grounds for narrow fields of application, somewhat akin to a craftsman's tools, a much broader perspective is necessary in systems analysis and theory. Here mathematics, with its abstractions, generalizations, and infinite variety of identifications, forms much of the body as well as the philosophical approach to systems theory. True, mathematics is by no means the whole of systems science, but no systems theory worthy of the name could exist without a strong mathematical foundation.

Traditionally, mathematical topics in scientific and technical textbooks are handled in several different ways. One approach, common in engineering works, is to introduce the mathematics piecemeal as needed to solve problems. This may be called the problem-oriented approach. A second method, characteristic of books on applied mathematics, is to develop a mathematical topic and then exhibit a variety of applications to show its importance. This may be called the topical approach. Attitudes toward mathematical rigor also vary, ranging from the oppressive rigor of some treatments of pure mathematics to the unconvincing plausibility arguments not uncommon in technical writings.

Since the essence of systems methods is generality of application and economy of treatment, the problem-oriented approach is immediately ruled out, while the topical approach usually lacks structure. To best fulfill the goals of this book, the mathematics is developed separately from, but structured with, the applications.

Systems analysis includes the idea of mathematical modeling, that is, the representation of real systems by mathematical systems. The mathematical image of even a relatively simple real system, as, for example, the oscillations of a plane pendulum, involves many simplifications and idealizations. Additional effects may be taken into account in order to obtain a more precise mathematical description of the motion, but only at the cost of added complexity, time, and effort in the analysis. Therefore the mathematical model of a system should not be made unnecessarily complicated by inclusion of trivial or irrelevant factors.

On the other hand, all significant features must be retained. Accuracy of the representation needed obviously varies with the system or problem at hand. For example, an accuracy of only a few percent is quite satisfactory for many applications and therefore only a crude model is required. By contrast, a highly refined model may be needed, as when dealing with an atomic system or the motion of a guided missile where an accuracy of one part in 10 million or better is required for a useful result. Clearly the development of mathematical models requires as much skill in knowing what to neglect as in knowing what to include.

Mathematical models provide several major gains in addition to obtaining solutions of specific problems. First is the generalizing power, since mathematical models (theories) encompass entire fields of phenomena, as, for example, Newton's laws of motion in mechanics and Maxwell's equations in electromagnetic theory. Second, a mathematical model or theory acts as a powerful knowledge multiplier; the innumerable scientific and technical advances stemming from Maxwell's equations stand in evidence of this benefit. Third, mathematical theories always suggest novel experiments which open the way to new results, often of an unexpected nature, and thereby contribute tellingly to the advancement of pure as well as applied science. A dramatic example of the power of this process is Einstein's famous equation of relativity theory, $E = mc^2$, which revealed the possibility of obtaining enormous power from the atom. Lesser examples of all three kinds of benefits of mathematical models or theories are commonplace.

5 APPLICATION ASPECTS

It is evident that the use of quantitative systems methods requires a strong foundation in mathematics and in the methods of physical science. Traditionally, education in applied mathematics and applied physics has stressed these areas. More recently the broadening of science and its associated technologies during the 1940's and 1950's set in motion forces which caused the center of gravity of engineering education to move much closer to science and mathematics. Thus it is not surprising that the most advanced use of quantitative systems methods has been in the fields of applied mathematics, applied physics, and engineering.

Because of the extent and magnitude of systems methods in engineering it is of interest to consider this field more closely. Engineering, by its very nature, is deeply concerned with systems. Engineering structures and engineering processes characteristically involve many interrelated elements, optimization requirements with respect to cost, performance, reliability, and the like, and design methods to meet these requirements. The notable shift in engineering education since about 1950 toward science and mathematics-oriented curricula has undoubtedly led to the use of more rigorous and more powerful quantitative methods in the design and operation of engineering systems, and to the ever wider use of systems methods generally in engineering.

In addition, many other important fields of application are emerging—business administration, management, economics, biological sciences, and psychology, for example. Application of systems methods to these latter fields has, until very recently, been hindered by the lack of adequate mathematical background, but fortunately, with the upgrading of mathematical requirements in the curricula for these fields, this situation is now changing.

BIBLIOGRAPHY

References Mentioned in Text

1. Institute of Electrical and Electronics Engineers, Systems Science Committee, Charter.
2. M. E. Salveson, "Suggestions for Graduate Study in Operations Research and Systems Engineering." *Journal of Engineering Education, 47* 211, 1956.

3. A. Hall and R. Fagan, "Definition of a System." *General Systems* (Yearbook of the Society for General Systems), I 18, 1956. (This Society was organized in the mid-1950's to counteract the trend toward compartmentalization of the sciences, and to develop a unity of concepts and principles.)

4. R. E. Gibson, "A Systems Approach to Research Management." Part 1 *Research Management*, 5 215, 1962. (Deals primarily with the application of the systems concept to industrial research and development.)

5. J. W. Forrester, *Industrial Dynamics*. New York: Wiley, 1961.

Philosophy of Systems

6. N. Wiener, *Cybernetics*, Second Edition. New York: Wiley, 1961.

7. D. D. Ellis and F. J. Ludwig, *Systems Philosophy*. Englewood Cliffs, N.J.: Prentice-Hall, 1962.

8. M. D. Mesarovic, ed., *Views on General Systems Theory* (Proceedings of Second Systems Symposium at Case Institute of Technology, Apr. 1963). New York: Wiley, 1964.

See also References 3, 9, 16, and 37.

Systems Engineering

9. A. D. Hall, *A Methodology for Systems Engineering*. Princeton: Van Nostrand, 1962.

10. W. A. Lynch and J. G. Truxal, *Introductory Systems Analysis*. New York: McGraw-Hill, 1961.

11. H. M. Paynter, *Analysis and Design of Engineering Systems*. Cambridge, Mass.: Massachusetts Institute of Technology Press, 1960.

12. D. O. Domasch and C. W. Laudeman, *Principles Underlying Systems Engineering*. New York: Pitman, 1962.

13. Institute of Radio Engineers, *Proceedings of the Workshop on Systems Engineering*. IRE Transactions on Education, Vol. E-5, June 1962.

14. Institute of Electrical and Electronics Engineers, *Systems Science Issue*. IEEE Transactions on Military Electronics, Vol. MIL-8, April 1964.

15. G. Murphy, *Similitude in Engineering*. New York: Wiley, 1965.

16. H. Chestnut, *Systems Engineering Tools*. New York: Wiley, 1965.

17. K. L. Nielson, "Systems Analysis and its Place in Technology." *Battelle Technical Review*, 13 9, 1964. (A view of systems analysis from the standpoint of the engineer and operations researcher.)

18. H. A. Affel, Jr., "System Engineering." *International Science and Technology*, p. 18 (Nov. 1964). (Gives a general discussion of systems engineering, its philosophy, methods, and techniques at a semi-popular level.)

19. *Systems Engineering Conference, 1964 Proceedings*. New York: Clapp and Poliak, 1964.

20. G. W. Gilman, "Systems Engineering in Bell Telephone Laboratories." *Bell Laboratories Record*, 31 8, 1953.

21. I. G. Wilson and M. E. Wilson, *Information, Computers, and System Design*. New York: Wiley, 1965.

22. H. E. Koenig et al., *Analysis of Discrete Physical Systems*. New York: McGraw-Hill, 1967.

23. A. G. J. MacFarlane, *Engineering Systems Analysis*. London: George C. Harrap and Co., Ltd., 1964. (U.S. edition distributed by Addison Wesley.)

24. J. Peschon, ed., *Disciplines and Technique of Systems Control*. New York: Blaisdel, 1965.

25. W. E. Wilson, *Concepts of Engineering System Design*. New York: McGraw-Hill, 1965.

26. R. E. Machol, ed., *Systems Engineering Handbook*. New York: McGraw-Hill, 1965.

27. W. A. Porter, *Modern Foundations of Systems Engineering*. New York: Macmillan, 1966.

28. H. Chestnut, *Systems Engineering Methods*. New York: Wiley, 1967.

29. S. M. Shinners, *Techniques of System Engineering*. New York: McGraw-Hill, 1967.

Mathematics of Systems

30. J. E. Alexander and J. M. Bailey, *Systems Engineering Mathematics*. Englewood Cliffs, N.J.: Prentice-Hall, 1962.

31. P. M. Morse and G. E. Kimball, *Methods of Operations Research*, 1st Edition Revised. New York: Wiley, 1951.
32. S. Crandall, *Engineering Analysis*. New York: McGraw-Hill, 1956.
33. G. B. Dantzig, *Linear Programming and Extensions*. Princeton: Princeton University Press, 1963.
34. D. P. Campbell, *Process Dynamics*. New York: Wiley, 1958.
35. R. Oldenburger, *Mathematical Engineering Analysis*. New York: Dover, 1961.
36. H. T. Davis, *Introduction to Nonlinear Differential and Integral Equations*. United States Atomic Energy Commission. U.S. Government Printing Office, Washington, D.C.: 1960.
37. T. L. Saaty and J. Bram, *Nonlinear Mathematics*. New York: McGraw-Hill, 1964.
38. *Mathematical Systems Theory*, published quarterly by Springer-Verlag.

General References

39. D. N. Chorafas, *Systems and Simulation*. New York: Academic Press, 1965.
40. *Operations Research*, Journal of the Operations Research Society of America, published bi-monthly.
41. *IEEE Transactions on Systems Science and Cybernetics*, published quarterly by the Institute of Electrical and Electronics Engineers, New York.
42. Proceedings of Symposium on System Theory, Polytechnic Institute of Brooklyn (April, 1965).
43. Proceedings 1965 Systems Science Conference, sponsored by the Institute of Electrical and Electronics Engineers, Systems Science and Cybernetics Group, Case Institute of Technology (October, 1965).
44. Program of 27th Annual Meeting, Operations Research Society of America, Bulletin, Supplement 1, *13* 8–31, 1965.
45. *Management Science*, Journal of the Institute of Management Sciences.
46. R. W. Jones and J. S. Gray, "System Theory and Physiological Processes." *Science*, *140* 461, 1963.
47. *Journal of Industrial Engineering*, published monthly by the American Institute of Industrial Engineers.
48. J. P. Eberhard, "Technology for the City." *International Science and Technology*, p. 18 (Sept. 1966). (Discusses systems approach to urban problems.)
49. *Journal of Computer and Computer Sciences*, published quarterly by Academic Press.

Part I

Mathematical Concepts

1
LINEAR ALGEBRA

Linear algebra is a basic mathematical subject for dealing with multivariable systems, and is used throughout this book, both in the theoretical developments and in the solution of problems. Combination of the efficient mathematical methods of linear algebra with the numerical capability of computers makes practical the numerical solution of complicated systems problems not otherwise possible. These methods are being applied with conspicuous success in most areas of engineering, in the physical, biological, and social sciences, and to many management and business problems.

Inasmuch as linear algebra is rapidly being integrated into undergraduate curricula, the treatment in this book is limited to a summary of the major topics and required results. These are amplified by examples that are worked out and a bibliography that gives references for further reading. For those readers acquainted with the subject, the material given will serve as a review. For those with lesser familiarity, the material given will serve as an introduction and covers all that is necessary for use of this book.

1.1 VECTOR SPACES

Definition 1. *A vector space is a set V of elements called vectors satisfying axioms A-1 to A-10:*

A-1 : To every pair $\mathbf{x} \in V$ and $\mathbf{y} \in V$ corresponds a vector $\mathbf{z} \in V$ called the sum of $\mathbf{x}$ and $\mathbf{y}$ and written $\mathbf{x} + \mathbf{y} = \mathbf{z}$.

A-2 : The sum is commutative, i.e., $\mathbf{x} + \mathbf{y} = \mathbf{y} + \mathbf{x}$.

A-3 : The sum is associative, i.e., $\mathbf{x} + (\mathbf{y} + \mathbf{z}) = (\mathbf{x} + \mathbf{y}) + \mathbf{z}$.

A-4: There exists a unique zero vector $\mathbf{0}$ such that $\mathbf{x} + \mathbf{0} = \mathbf{x}$. Since there is no danger of misunderstanding, we can write $\mathbf{0} = 0$.

A-5: To every vector $\mathbf{x}$ corresponds a unique vector $(-\mathbf{x})$ such that $\mathbf{x} + (-\mathbf{x}) = \mathbf{0} = 0$.

A-6: To any number α, called a scalar in opposition to a vector, and any vector $\mathbf{x} \in V$ corresponds a vector $\mathbf{w} \in V$ such that $\alpha\mathbf{x} = \mathbf{w}$. The vector $\mathbf{w}$ is called the *product* of the vector $\mathbf{x}$ by the scalar α.

A-7: The product defined by A-6 is associative, i.e.,

$$\alpha(\beta\mathbf{x}) = (\alpha\beta)\mathbf{x}$$

A-8: For $\alpha = 1$ in A-6 we clearly have $1 \cdot \mathbf{x} = \mathbf{x}$ for all $\mathbf{x} \in V$.

A-9: The product defined by A-6 is distributive with respect to vector addition defined by A-1, i.e., $\alpha(\mathbf{x} + \mathbf{y}) = \alpha\mathbf{x} + \alpha\mathbf{y}$.

A-10: The product defined by A-6 is distributive with respect to scalar addition, i.e.,

$$(\alpha + \beta)\mathbf{x} = \alpha\mathbf{x} + \beta\mathbf{x}$$

The preceding axioms are not necessarily a minimum set of axioms but rather are a convenient characterization of a vector space.

Definition 2.   *If a_m and $\mathbf{x}_m$, $m = 1, 2, \ldots, n$, are respectively scalars and vectors, we call a linear combination of the vectors $\mathbf{x}_m$ a relation of the form,*

$$\mathbf{y} = \sum_{m=1}^{n} a_m \mathbf{x}_m \tag{1.1}$$

Definition 3.   *A set of n vectors $\mathbf{x}_1, \mathbf{x}_2, \ldots, \mathbf{x}_n$ are said to be* linearly dependent *if there exists between them a relation of the form*

$$\sum_{m=1}^{n} c_m \mathbf{x}_m = 0 \tag{1.2}$$

where not all *of the c_m are equal to zero. If such a relation does not exist, the vectors are said to be* linearly independent. *This is equivalent to saying that if the vectors are linearly independent then in* (1.2) *all c_m are zero.*

Definition 4.   *A basis in a vector space V is a set X of linearly independent vectors, and such that any vector $\mathbf{x} \in V$ can be expressed as a linear combination of the elements of X.*

Definition 5.   *A vector is said to be* finite dimensional *if it has a finite basis i.e., a basis with a finite number of elements. The number of elements is called the* dimension *of V.*

1.2   INNER PRODUCT SPACES

Definition 1.   *The inner product of two vectors $\mathbf{x}, \mathbf{y} \in V$ is a function of the two vectors $\mathbf{x}, \mathbf{y}$, written $(\mathbf{x}, \mathbf{y})$, and satisfying the following postulates:*

B-1: $(\mathbf{x}, \mathbf{y}) = \overline{(\mathbf{y}, \mathbf{x})}$

(where $\overline{(\mathbf{x}, \mathbf{y})}$ is the complex conjugate of $(\mathbf{x}, \mathbf{y})$)

B-2: $$(\alpha_1 \mathbf{x}_1 + \alpha_2 \mathbf{x}_2, \mathbf{y}) = \alpha_1(\mathbf{x}_1, \mathbf{y}) + \alpha_2(\mathbf{x}_2, \mathbf{y})$$

B-3: $$(\mathbf{x}, \mathbf{x}) \geq 0$$

B-4: $$(\mathbf{x}, \mathbf{x}) = 0 \text{ if and only if } \mathbf{x} = 0$$

Definition 2. *An inner product space is a vector space with an inner product.*

Definition 3. *The quantity* $(\mathbf{x}, \mathbf{x})^{1/2} = \|\mathbf{x}\|$ *is called the norm, or the length, or the modulus of the vector. A vector for which* $\|\mathbf{x}\| = 1$*, is said to be normalized.*

Definition 4. *A real inner product space is called* Euclidean; *a complex inner product space is called* Hermitian *or* Unitary.

Definition 5. *Two vectors* $\mathbf{x}$*, and* $\mathbf{y}$ *are said to be orthogonal if* $(\mathbf{x}, \mathbf{y}) = 0$*.*

Definition 6. *A set* $\mathbf{x}_1, \mathbf{x}_2, \ldots, \mathbf{x}_n$ *of vectors is said to be an orthonormal set if*

$$(\mathbf{x}_i, \mathbf{x}_j) = \delta_{ij} = \begin{cases} 0 \text{ if } i \neq j \\ 1 \text{ if } i = j \end{cases} \tag{1.3}$$

δ_{ij} *is called the Kronecker delta.*

Definition 7. *An orthonormal set is said to be complete if it is not contained in any larger orthogonal set in* V*.*

Schwarz's Inequality: If V is an inner product vector space then for $\mathbf{x}, \mathbf{y} \in V$

$$|(\mathbf{x}, \mathbf{y})| \leq \|\mathbf{x}\| \cdot \|\mathbf{y}\|$$

Triangular Inequality: If V is a vector space and $\mathbf{x}, \mathbf{y} \in V$ then

$$\|\mathbf{x} + \mathbf{y}\| \leq \|\mathbf{x}\| + \|\mathbf{y}\| \tag{1.4}$$

and more generally

$$\left\| \sum_{m=1}^{n} \mathbf{x}_m \right\| \leq \sum_{m=1}^{n} \|\mathbf{x}_m\| \tag{1.5}$$

Orthogonalization of vectors: Given a set of n vectors $\mathbf{x}_1, \mathbf{x}_2, \ldots, \mathbf{x}_n$, that are linearly independent, by a procedure due to Schmidt, one can construct a set of vectors $\mathbf{y}_1, \mathbf{y}_2, \ldots, \mathbf{y}_n$ that is orthogonal. The procedure is shown by the following relations.

$$\mathbf{y}_1 = \mathbf{x}_1$$
$$\mathbf{y}_2 = \mathbf{x}_2 + a_{21}\mathbf{y}_1 \qquad a_{21} = -(\mathbf{y}_1, \mathbf{x}_2)/(\mathbf{y}_1, \mathbf{y}_1)$$
$$\mathbf{y}_3 = \mathbf{x}_3 + a_{32}\mathbf{y}_2 + a_{31}\mathbf{y}_1 \qquad a_{32} = -(\mathbf{y}_2, \mathbf{x}_3)/(\mathbf{y}_2, \mathbf{y}_2)$$
$$a_{31} = -(\mathbf{y}_1, \mathbf{x}_3)/(\mathbf{y}_1, \mathbf{y}_1) \tag{1.6}$$

$$\mathbf{y}_n = \mathbf{x}_n - \sum_{m=1}^{n-1} \mathbf{y}_{m-1} \, (\mathbf{y}_{m-1}, \mathbf{x}_m)/(\mathbf{y}_{m-1}, \mathbf{y}_{m-1})$$

1.3 *n*-DIMENSIONAL VECTORS

The *n*-dimensional vector we consider here is a direct generalization of the ordinary three-dimensional vector used in geometry and physics, and is defined as a sequence of *n* numbers, i.e., an *n*-plet of complex numbers written

$$\mathbf{x} = [x_1, x_2, \ldots, x_n]$$

Let V_n be the set of all such vectors; we write, by definition that if $\mathbf{x} \in V_n$ and $\mathbf{y} \in V_n$, and α is a scalar, then

$$\mathbf{x} + \mathbf{y} = [x_1 + y_1, x_2 + y_2, \ldots, x_n + y_n] = \mathbf{z} \tag{1.7}$$

$$\alpha\mathbf{x} = [\alpha x_1, \alpha x_2, \ldots, \alpha x_n] \tag{1.8}$$

$$-\mathbf{x} = [-x_1, -x_2, \ldots, -x_n] \tag{1.9}$$

$$0 = [0, 0, \ldots, 0] \tag{1.10}$$

$$\mathbf{x} \cdot \mathbf{y} = (\mathbf{x}, \mathbf{y}) = \sum_{m=1}^{n} x_m \bar{y}_m \tag{1.11}$$

$$\|\mathbf{x}\| = \sqrt{(\mathbf{x}, \mathbf{x})} = \left[\sum_{m=1}^{n} |x_m|^2 \right]^{1/2} \tag{1.12}$$

It is easily checked that the vectors so defined satisfy the postulates A-1 through A-10 and B-1 through B-4 and, therefore, all properties given in the preceding sections.

In particular, the set of vectors

$$\begin{aligned} \mathbf{e}_1 &= [1, 0, 0, \ldots, 0] \\ \mathbf{e}_2 &= [0, 1, 0, \ldots, 0] \\ & \cdot \quad \cdot \quad \cdot \quad \cdot \quad \cdot \quad \cdot \quad \cdot \\ \mathbf{e}_n &= [0, 0, 0, \ldots, 1] \end{aligned} \tag{1.13}$$

are:
(1) linearly independent
(2) orthonormal, i.e., $\mathbf{e}_k \cdot \mathbf{e}_j = \delta_{kj}$

Any vector $\mathbf{x} = [x_1, x_2, \ldots, x_n] \in V_n$ can be written

$$\mathbf{x} = \sum_{m=1}^{n} x_m \mathbf{e}_m \tag{1.14}$$

so that

$$\mathbf{x} \cdot \mathbf{e}_k = \sum_{m=1}^{n} x_m \mathbf{e}_m \cdot \mathbf{e}_k = x_k \tag{1.15}$$

Example 1

Let $n = 4$, $\mathbf{x} = [1, 2, 3, 4]$, $\mathbf{y} = [2i, 1 - i, 1 + i, 0]$. Then

(i) $\mathbf{x} + \mathbf{y} = [1 + 2i, 3 - i, 4 + i, 4]$

(ii) $i\mathbf{x} + 2\mathbf{y} = [i, 2i, 3i, 4i] + [4i, 2 - 2i, 2 + 2i, 0]$
$= [5i, 2, 2 + 5i, 4i]$

(iii) $\mathbf{x} \cdot \mathbf{y} = -2i + 2 + 2i + 3 - 3i + 0 = 5 - 3i$

(iv) $\|\mathbf{x}\| = (1 + 4 + 9 + 16)^{1/2} = \sqrt{30}$

(v) $\|\mathbf{y}\| = (4 + 2 + 2)^{1/2} = \sqrt{8} = 2\sqrt{2}$

Example 2

Let $n = 4$, and, $\mathbf{x}_1 = [1, 0, 0, 1]$, $\mathbf{x}_2 = [0, 1, 1, 0]$, $\mathbf{x}_3 = [1, 1, 1, 0]$ and $\mathbf{x}_4 = [0, 0, 1, 1]$. Construct, using the Schmidt method, a system of four orthogonal vectors.

$$\mathbf{y}_1 = \mathbf{x}_1 = [1, 0, 0, 1]$$

$$\mathbf{y}_2 = \mathbf{x}_2 + a_{21}\mathbf{y}_1 = [0, 1, 1, 0] + [a_{21}, 0, 0, a_{21}] = [a_{21}, 1, 1, a_{21}]$$

Since $\mathbf{y}_2 \cdot \mathbf{y}_1 - 0$, it follows that

$$\mathbf{y}_2 \cdot \mathbf{y}_1 = 2a_{21} = 0, \quad a_{21} = 0$$

$\mathbf{y}_2 = \mathbf{x}_2 = [0, 1, 1, 0]$ This result could have been anticipated.

$\mathbf{y}_3 = \mathbf{x}_3 + a_{32}\mathbf{y}_2 + a_{31}\mathbf{y}_1 = [1, 1, 1, 0] + [0, a_{32}, a_{32}, 0]$
$+ [a_{31}, 0, 0, a_{31}]$

$\mathbf{y}_3 = [1 + a_{31}, 1 + a_{32}, 1 + a_{32}, a_{31}]$

$\mathbf{y}_3 \cdot \mathbf{y}_2 = 2 + 2a_{32} = 0 \qquad a_{32} = -1$

$\mathbf{y}_3 \cdot \mathbf{y}_1 = 1 + 2a_{31} - 0 \qquad a_{31} = -\frac{1}{2}$

$\mathbf{y}_3 = [\frac{1}{2}, 0, 0, -\frac{1}{2}]$

$\mathbf{y}_4 = \mathbf{x}_4 + a_{43}\mathbf{y}_3 + a_{42}\mathbf{y}_2 + a_{41}\mathbf{y}_1$

$\mathbf{y}_4 = [0, 0, 1, 1] + [\frac{1}{2}a_{43}, 0, 0, -\frac{1}{2}a_{43}] + [0, a_{42}, a_{42}, 0]$
$+ [a_{41}, 0, 0, a_{41}]$

$\mathbf{y}_4 = [\frac{1}{2}a_{43}, +a_{41}, a_{42}, 1 + a_{42}, 1 - \frac{1}{2}a_{43} + a_{41}]$

$\mathbf{y}_4 \cdot \mathbf{y}_3 = \frac{1}{2}a_{43} - \frac{1}{2} - 0 \qquad a_{43} = 1$

$\mathbf{y}_4 \cdot \mathbf{y}_2 = 2a_{42} + 1 = 0 \qquad a_{42} = -\frac{1}{2}$

$\mathbf{y}_4 \cdot \mathbf{y}_1 = 1 + 2a_{41} = 0 \qquad a_{41} = -\frac{1}{2}$

$\mathbf{y}_4 = [0, -\frac{1}{2}, \frac{1}{2}, 0]$

The vectors are

$$\begin{array}{ccc}
\mathbf{y}_1 = [1, 0, 0, 1] & & \mathbf{y}_1 = [1, 0, 0, 1] \\
\mathbf{y}_2 = [0, 1, 1, 0] & \text{or} & \mathbf{y}_2 = [0, 1, 1, 0] \\
\mathbf{y}_3 = [\frac{1}{2}, 0, 0, -\frac{1}{2}] & & \mathbf{y}_3 = [1, 0, 0, -1] \\
\mathbf{y}_4 = [0, \frac{1}{2}, -\frac{1}{2}, 0] & & \mathbf{y}_4 = [0, 1, -1, 0]
\end{array}$$

1.4 LINEAR TRANSFORMATIONS

Definition 1. *A linear transformation or a linear operator A on a vector space V is a correspondence that assigns to every vector* $\mathbf{x} \in V$ *a vector* $\mathbf{z} = A\mathbf{x} \in V$ *in such a way that for* α *and* β *being scalars,*

$$A(\alpha \mathbf{x} + \beta \mathbf{y}) = \alpha A\mathbf{x} + \beta A\mathbf{y} \tag{1.16}$$

Two transformations are specially to be considered:

(1) the transformation $\bigcirc$, such that $\bigcirc \mathbf{x} = 0$
(2) the transformation I, such that $I\mathbf{x} = \mathbf{x}$

for all $\mathbf{x} \in V$. $\bigcirc$ is called the zero operator and I the identity operator.

If A and B are linear operators such that $A\mathbf{x}_1 = \mathbf{y}_1$, $B\mathbf{x}_1 = \mathbf{y}_2$, then

$$A\mathbf{x}_1 + B\mathbf{x}_1 = (A + B)\mathbf{x}_1 = C\mathbf{x}_1 = \mathbf{y}_1 + \mathbf{y}_2 = \mathbf{y} \tag{1.17}$$

defines the sum of two operators, i.e., $A + B = C$. The difference of two operators is defined similarly.

If $\mathbf{x}$, $\mathbf{y}$, $\mathbf{z} \in V$, and A and B are operators such that $A\mathbf{x} = \mathbf{y}$ and $B\mathbf{y} = \mathbf{z}$, then $\mathbf{z} = B\mathbf{y} = B(A\mathbf{x}) = BA\mathbf{x} = P\mathbf{x}$ defines the product P of B and A. In general $P = BA \neq AB$; however, the following always hold:

$$\begin{cases} \bigcirc A = \bigcirc A = \bigcirc \\ IA = IA = A \end{cases} \tag{1.18}$$

It follows that for three operators A, B, C,

$$\begin{cases} A(B + C) = AB + AC \\ (A + B)C = AC + BC \\ A(BC) = (AB)C = ABC \end{cases} \tag{1.19}$$

which shows the distributivity of multiplication with respect to addition and the associativity of multiplication.

If $\mathbf{x}$, $\mathbf{y} \in V$, $A\mathbf{x} = \mathbf{y}$, and to every $\mathbf{y} \in V$ corresponds at least one $\mathbf{x}$, then the linear operation C such that $C\mathbf{y} = \mathbf{x}$ is said to be the inverse of A which is written

$$C = A^{-1} \tag{1.20}$$

which is equivalent to

$$A^{-1}A = AA^{-1} = I \tag{1.21}$$

1.5 MATRICES

A matrix is a rectangular array of numbers enclosed in a pair of brackets. Each number is called an element of the matrix. The position of the general element a_{ij} of the matrix A is indicated by the subscripts i

and j, referring to the ith row and the jth column, respectively. Thus we write

$$A = \begin{bmatrix} a_{11} & a_{12} & \cdots & a_{1j} & \cdots & a_{1n} \\ a_{21} & a_{22} & \cdots & a_{2j} & \cdots & a_{2n} \\ \cdot & \cdot & \cdots & \cdot & \cdots & \cdot \\ a_{i1} & a_{i2} & \cdots & a_{ij} & \cdots & a_{in} \\ \cdot & \cdot & \cdots & \cdot & \cdots & \cdot \\ a_{m1} & a_{m2} & \cdots & a_{mj} & \cdots & a_{mn} \end{bmatrix} \qquad (1.22)$$

In case $n = m$, the matrix is square and n is called the order of the matrix. The elements a_{kk} are said to form the diagonal of the matrix.

The following rules apply to the calculation of matrices:

R-1: Two matrices are equal if and only if all their elements are equal. Thus, if $A = B$, then $a_{ij} = b_{ij}$ for all i and j. This implies that both matrices have the same number of rows and columns.

R-2: The sum (or the difference) of two matrices A and B is a matrix C whose elements are the sums (or the differences) of the corresponding elements of A and B. Thus

$$A + B = C \Rightarrow a_{ij} + b_{ij} = c_{ij}$$
$$A - B = D \Rightarrow a_{ij} - b_{ij} = d_{ij} \qquad (1.23)$$
$$\text{for } 1 \leq i \leq m \quad \text{and} \quad 1 \leq j \leq n$$

R-3: The product of a matrix A by a scalar k is a matrix whose elements are obtained by multiplying all elements of A by the scalar k. Thus

$$kA = E \qquad ka_{ij} = e_{ij} \qquad (1.24)$$

R-4: The product of two matrices A and B is a matrix P whose elements are obtained by the following expressions

$$AB = P \qquad p_{ij} = \sum_{k=1}^{n} a_{ik} b_{kj} \qquad (1.25)$$

$$1 \leq k \leq m \qquad 1 \leq j \leq n$$

This implies that if the matrix A has m rows and n columns, then B must have n rows. If B has q columns, then the product P will have m rows and q columns.

Example I

Let

$$A = \begin{bmatrix} 1 & 2 & 3 \\ 4 & 3 & 2 \end{bmatrix} \qquad B = \begin{bmatrix} 0 & 1 & 2 \\ 3 & 5 & 4 \end{bmatrix} \qquad C = \begin{bmatrix} 1 & 2 & 0 \\ 3 & 4 & 1 \\ 5 & 6 & 2 \end{bmatrix}$$

Then

$$A + B = \begin{bmatrix} 1 & 3 & 5 \\ 7 & 8 & 6 \end{bmatrix} \qquad 2A - 3B = \begin{bmatrix} 2 & 1 & 0 \\ -1 & -9 & -8 \end{bmatrix}$$

AB does not exist, but

$$AC = \begin{bmatrix} 1+6+15 & 2+8+18 & 0+2+6 \\ 4+9+10 & 8+12+12 & 0+3+4 \end{bmatrix} = \begin{bmatrix} 22 & 28 & 8 \\ 23 & 32 & 7 \end{bmatrix}$$

If both A and B are $n \times n$ matrices, then clearly both AB and BA are defined, but *in general*

$$AB \neq BA \tag{1.26}$$

$AB = 0$ does not necessarily imply that $A = 0$, or $B = 0$ (1.27)

$AB = AC$ does not necessarily imply that $B = C$ (1.28)

On the other hand, both the left and the right distributive laws are verified, and so is the associative law, i.e.,

$$A(B + C) = AB + AC \tag{1.29}$$

$$(A + B)C = AC + BC \tag{1.30}$$

$$A(BC) = (AB)C = ABC \tag{1.30a}$$

Example 2

In the following, $AB = 0$, but $A \neq 0$, $B \neq 0$.

$$AB = \begin{bmatrix} 0 & 1 & 1 \\ 1 & 0 & 1 \\ 1 & 0 & 1 \end{bmatrix} \begin{bmatrix} 1 & 1 & 1 \\ 1 & 1 & 1 \\ -1 & -1 & -1 \end{bmatrix} = \begin{bmatrix} 0 & 0 & 0 \\ 0 & 0 & 0 \\ 0 & 0 & 0 \end{bmatrix} = 0$$

On the other hand,

$$BA = \begin{bmatrix} 1 & 1 & 1 \\ 1 & 1 & 1 \\ -1 & -1 & -1 \end{bmatrix} \begin{bmatrix} 0 & 1 & 1 \\ 1 & 0 & 1 \\ 1 & 0 & 1 \end{bmatrix} = \begin{bmatrix} 2 & 1 & 3 \\ 2 & 1 & 3 \\ -2 & -1 & -3 \end{bmatrix} \neq 0$$

For the following $CD = CE$, but $D \neq E$, thus

$$CD = \begin{bmatrix} 1 & 2 \\ 2 & 4 \end{bmatrix} \begin{bmatrix} 0 & 0 \\ 0 & 1 \end{bmatrix} = \begin{bmatrix} 0 & 2 \\ 0 & 4 \end{bmatrix}$$

$$CE = \begin{bmatrix} 1 & 2 \\ 2 & 4 \end{bmatrix} \begin{bmatrix} 2 & 0 \\ -1 & 1 \end{bmatrix} = \begin{bmatrix} 0 & 2 \\ 0 & 4 \end{bmatrix} = CD$$

Definition 1.   *The transpose of an n × m matrix A is the m × n matrix A′ obtained by interchanging the rows and columns of A.*

Thus if

$$A = \begin{bmatrix} 1 & 2 & 3 \\ 4 & 5 & 6 \end{bmatrix} \quad \text{then} \quad A' = \begin{bmatrix} 1 & 4 \\ 2 & 5 \\ 3 & 6 \end{bmatrix}$$

The main properties of transposed matrices are

P-1: $(A + B)' = A' + B'$
P-2: $(AB)' = B'A'$
                             $(ABC)' = C'B'A'$ etc.

Definition 2.   *A matrix A for which A′ = A is said to be symmetric. This clearly implies that the matrix is square.*

Definition 3.   *A matrix A for which A′ = −A is called skew-symmetric or anti-symmetric. Clearly a skew-symmetric is square and has zeros in the main diagonal.*

P-3: For any matrix A, $B = \frac{1}{2}(A + A')$ is symmetric and $C = \frac{1}{2}(A - A')$ is skew-symmetric, but since, $A = B + C$, every matrix can be considered as the sum of a symmetric matrix and a skew-symmetric matrix.

Definition 4.   *The (complex) conjugate of a matrix A is the matrix $\bar{A}$ where the elements of $\bar{A}$ are the complex conjugates of A.*

P-4: For any (complex) matrix $(\overline{A + B}) = \bar{A} + \bar{B}$

$$(\overline{AB}) = \bar{A}\bar{B} \qquad (\bar{A})' = \bar{A}'$$

Definition 5.   *If for a (complex) matrix A, $\bar{A}' = A$, the matrix is said to be Hermitian. It follows that a Hermitian matrix is square and has real diagonal elements.*

Definition 6.   *If for a complex matrix A, $\bar{A}' = -A$, the matrix is said to be skew-Hermitian or anti-Hermitian. A skew-Hermitian matrix is square and its diagonal elements are either zeros or pure imaginaries.*

P-5: For any matrix A, $D = \frac{1}{2}(A + \bar{A}')$ is Hermitian and $E = \frac{1}{2}(A - \bar{A}')$ is skew-Hermitian, but $A = E + D$, so that any matrix A is the sum of a Hermitian matrix and a skew-Hermitian matrix.

Definition 7.   *If A and B are square matrices such that AB = BA = I, then A and B are said to be inverses of each other and we write*

$$B = A^{-1} \qquad A = B^{-1}$$

P-6: If $P = AB$, then $P^{-1} = (AB)^{-1} = B^{-1}A^{-1}$.

Definition 8. *A square $n \times n$ matrix A is said to be singular if the determinant of its elements is equal to zero, i.e., if*

$$\det [A] = 0$$

The notation $\det [A]$ for the determinant of A will be consistently used throughout this book.

Definition 9. *Let A be a square matrix, $\det [A]$ its determinant, and let A_{ij} be the cofactor of the element a_{ij} of $\det [A]$. Then we call adjoint matrix of A, written adj A, the transpose matrix of the matrix of cofactors of the elements of $\det [A]$. Thus*

$$\text{adj } A = \begin{bmatrix} A_{11} & A_{21} & \cdots & A_{n1} \\ A_{12} & A_{22} & \cdots & A_{n2} \\ \cdot & & & \\ \cdot & & & \\ \cdot & & & \\ A_{1n} & A_{2n} & \cdots & A_{nn} \end{bmatrix}$$

P-7: Only nonsingular matrices have inverses.
P-8: For a nonsingular matrix

$$A^{-1} = \frac{1}{\det [A]} \text{ adj } A \tag{1.31}$$

Definition 10. *The trace of a square matrix is the sum of its diagonal elements, thus*

$$\text{tr } A = \sum_{m=1}^{n} a_{mm}$$

P-9: $$\text{tr } (A + B) = \text{tr } A + \text{tr } B$$

P-10: $$\text{tr } (AB) = \sum_{i=1}^{n} \sum_{j=1}^{n} a_{ij} b_{ji}$$

P-11: $$\text{tr } (BA) = \sum_{j=1}^{n} \sum_{i=1}^{n} b_{ji} a_{ij}$$

$$\text{tr } (AB) = \text{tr } (BA)$$

Definition 11. *A square matrix T is said to be triangular if $a_{ij} = 0$ for all i, j such that $j < i$ (in which case all elements below the diagonal are zero), or if $a_{ij} = 0$ for all i, j such that $j > i$ (in which case all elements above the diagonal are zero).*

Example 3

If

$$A = \begin{bmatrix} 1 & 2 & 0 \\ 2 & 3 & 1 \\ 1 & 0 & 3 \end{bmatrix} \qquad A' = \begin{bmatrix} 1 & 2 & 1 \\ 2 & 3 & 0 \\ 0 & 1 & 3 \end{bmatrix}$$

then

$$\det [A] = \begin{vmatrix} 1 & 2 & 0 \\ 2 & 3 & 1 \\ 1 & 0 & 3 \end{vmatrix} = -1$$

$$\text{adj } A = \begin{bmatrix} 9 & -6 & 2 \\ -5 & 3 & -1 \\ -3 & 2 & -1 \end{bmatrix}$$

$$A^{-1} = \frac{\text{adj } A}{\det [A]} = \begin{bmatrix} -9 & 6 & -2 \\ 5 & -3 & 1 \\ 3 & -2 & 1 \end{bmatrix}$$

$$AA^{-1} = \begin{bmatrix} 1 & 2 & 0 \\ 2 & 3 & 1 \\ 1 & 0 & 3 \end{bmatrix} \begin{bmatrix} -9 & 6 & -2 \\ 5 & -3 & 1 \\ 3 & -2 & 1 \end{bmatrix} = \begin{bmatrix} 1 & 0 & 0 \\ 0 & 1 & 0 \\ 0 & 0 & 1 \end{bmatrix} = I$$

$$\text{tr } A = 1 + 3 + 3 = 7$$
$$\text{tr adj } A = 9 + 3 - 1 = 11$$

Definition 12. *A matrix A is said to be orthogonal if*

$$AA' = I; \qquad \textit{i.e., if } A' = A^{-1}$$

P-12: The row vectors or column vectors of an orthogonal matrix are orthogonal.

P-13: The inverse and the transpose of an orthogonal matrix are orthogonal.

P-14: The product of two or more orthogonal matrices is orthogonal.

P-15: An orthogonal matrix applied to a vector preserves its norm, and conversely, if a matrix applied to a nonzero vector preserves its norm, then the matrix is orthogonal.

Definition 13. *The rank r of a matrix is the highest order minor that is not zero. In the case of a $n \times n$ square matrix we call its nullity the difference $n - r$.*

Example 4. Rank of a Matrix

(i)
$$A = \begin{bmatrix} 1 & 2 & 3 \\ 4 & 5 & 6 \\ 4 & 3 & 2 \end{bmatrix}$$

is singular since det $[A] = 0$. However, any one of its minors of order 2,

e.g., $\begin{vmatrix} 1 & 2 \\ 4 & 5 \end{vmatrix} = -3$ is different from zero; thus the rank of A is 2; its

nullity is 1.

(ii)
$$B = \begin{bmatrix} 1 & 0 & 0 & 0 \\ 1 & 0 & 0 & 0 \\ 0 & 1 & 0 & 0 \\ 1 & 0 & 0 & 0 \end{bmatrix}$$ is singular, since

det $[B] = 0$; it can be checked that all its minors of order 3 are zero, but the minor of order two

$$\begin{vmatrix} 0 & 1 \\ 1 & 0 \end{vmatrix} = -1$$

so that B is of rank 2, and of nullity $4 - 2 = 2$.

The problem of finding the rank of a matrix by direct calculation of the minors leads in general to impractically large computations. For example, in the case of a 10×10 matrix of rank 8, it is necessary to calculate not only the determinant of the matrix, but also at least 10^2 determinants of order nine. These inordinate calculations can be avoided by transforming the matrix into *normal form* using equivalence operations, as will be shown.

Definition 14

The normal form of a square matrix A of order n and of rank r is

$$N(A) = \begin{bmatrix} I_r & O_1 \\ O_2 & O_3 \end{bmatrix}$$

where I_r is the identity matrix of order r, O_1 the $(n - r) \times r$ zero matrix, O_2 is O_1', and O_3 is the $(n - r) \times (n - r)$ zero matrix.

Definition 15. *Elementary transformations on a matrix are those which change neither its rank nor its order.*

The following are elementary transformations:

(1) interchange of two rows
(2) interchange of two columns

(3) multiplication of the elements of a row by a scalar $\alpha \neq 0$

(4) multiplication of the elements of a column by a scalar $\alpha \neq 0$

(5) addition of a multiple of one row to another row

(6) addition of a multiple of one column to another column

(7) addition of the sum of the multiples of two or more rows to another row

(8) addition of the sum of the multiples of two or more columns to a column.

Definition 16. *Two matrices are* equivalent *if they can be obtained from each other by a sequence of elementary transformations. It is evident that equivalent matrices have the same rank and order. We write symbolically A equiv B.*

Example 5. Reduce the matrix A to normal form by a sequence of elementary transformations and thereby determine its rank.

$$A = \begin{bmatrix} 1 & 2 & 3 \\ 4 & 5 & 6 \\ 4 & 3 & 2 \end{bmatrix}$$

By applying a series of elementary transformations on A we obtain a second matrix B. Let R_1, R_2, R_3, be the rows and C_1, C_2, C_3 the columns of A, and perform the following elementary transformations:

(1) Add $-(C_1 + C_2)$ to C_3 $\qquad\qquad$ A equiv $\begin{bmatrix} 1 & 2 & 0 \\ 4 & 5 & -3 \\ 4 & 3 & -5 \end{bmatrix}$

(2) Add $-2C_1$ to C_2 $\qquad\qquad$ A equiv $\begin{bmatrix} 1 & 0 & 0 \\ 4 & -3 & -3 \\ 4 & -5 & -5 \end{bmatrix}$

(3) Add $-C_2$ to C_3 $\qquad\qquad$ A equiv $\begin{bmatrix} 1 & 0 & 0 \\ 4 & -3 & 0 \\ 4 & -5 & 0 \end{bmatrix}$

(4) Add $-R_2$ to R_3 $\qquad\qquad$ A equiv $\begin{bmatrix} 1 & 0 & 0 \\ 4 & -3 & 0 \\ 0 & -2 & 0 \end{bmatrix}$

(5) Add $-4R_1$ to R_2 $\qquad$ A equiv $\begin{bmatrix} 1 & 0 & 0 \\ 0 & -3 & 0 \\ 0 & -2 & 0 \end{bmatrix}$

(6) Multiply R_2 by $-\frac{1}{3}$ and R_3 by $-\frac{1}{2}$ $\quad$ A equiv $\begin{bmatrix} 1 & 0 & 0 \\ 0 & 1 & 0 \\ 0 & 1 & 0 \end{bmatrix}$

(7) Add $-R_2$ to R_3 $\qquad$ A equiv $\begin{bmatrix} 1 & 0 & 1 \\ 0 & 1 & 0 \\ 0 & 0 & 0 \end{bmatrix}$

We have reduced the given matrix to the normal form, from which it is apparent that $r = 2$.

Example 6. Reduce the matrix A to normal form by a sequence of elementary transformations.

$$A = \begin{bmatrix} 1 & 2 & 3 & 2 \\ 3 & 1 & 2 & 1 \\ 2 & 1 & 0 & 1 \\ 0 & 0 & 1 & 0 \end{bmatrix}$$

Using the same notation as in Ex. 5 we have:

add $-C_2$ to C_4 $\qquad$ A equiv $\begin{bmatrix} 1 & 2 & 3 & 0 \\ 3 & 1 & 2 & 0 \\ 2 & 1 & 0 & 0 \\ 0 & 0 & 1 & 0 \end{bmatrix}$

add $-3R_1$ to R_2 $\qquad$ A equiv $\begin{bmatrix} 1 & 2 & 3 & 0 \\ 0 & -5 & -7 & 0 \\ 2 & 1 & 0 & 0 \\ 0 & 0 & 1 & 0 \end{bmatrix}$

add $-2R_1$ to R_3 $\qquad$ A equiv $\begin{bmatrix} 1 & 2 & 3 & 0 \\ 0 & -5 & -7 & 0 \\ 0 & -3 & -6 & 0 \\ 0 & 0 & 1 & 0 \end{bmatrix}$

multiply R_2 by $-\frac{1}{5}$, and R_3 by $-\frac{1}{3}$ A equiv
$$\begin{bmatrix} 1 & 2 & 3 & 0 \\ 0 & 1 & \frac{7}{5} & 0 \\ 0 & 1 & 2 & 0 \\ 0 & 0 & 1 & 0 \end{bmatrix}$$

add $-R_2$ to R_3 A equiv
$$\begin{bmatrix} 1 & 2 & 3 & 0 \\ 0 & 1 & \frac{7}{5} & 0 \\ 0 & 0 & \frac{3}{5} & 0 \\ 0 & 0 & 1 & 0 \end{bmatrix}$$

multiply R_3 by $+\frac{5}{3}$ A equiv
$$\begin{bmatrix} 1 & 2 & 3 & 0 \\ 0 & 1 & \frac{7}{5} & 0 \\ 0 & 0 & 1 & 0 \\ 0 & 0 & 1 & 0 \end{bmatrix}$$

add $-R_3$ to R_4 A equiv
$$\begin{bmatrix} 1 & 2 & 3 & 0 \\ 0 & 1 & \frac{7}{5} & 0 \\ 0 & 0 & 1 & 0 \\ 0 & 0 & 0 & 0 \end{bmatrix}$$

add $-2R_2$ to R_1 A equiv
$$\begin{bmatrix} 1 & 0 & \frac{1}{5} & 0 \\ 0 & 1 & \frac{7}{5} & 0 \\ 0 & 0 & 1 & 0 \\ 0 & 0 & 0 & 0 \end{bmatrix}$$

add $-\frac{7}{5}R_3$ to R_2 and $-\frac{1}{5}R_3$ to R_1 A equiv
$$\begin{bmatrix} 1 & 0 & 0 & 0 \\ 0 & 1 & 0 & 0 \\ 0 & 0 & 1 & 0 \\ 0 & 0 & 0 & 0 \end{bmatrix}$$

which is the *normal form*. It shows that $r = 3$ and $n - r = 1$.

The elementary transformations can be written in matrix form. The matrices corresponding to the elementary transformation are called elementary matrices. Some of the elementary matrices premultiply, others postmultiply the matrix to which they apply. All elementary matrices have inverses and therefore are nonsingular. It follows that matrix-wise the transformation of a matrix into normal form can be written

$$N(A) = P_1 P_2 \cdots P_k A Q_1 Q_2 \cdots Q_n$$
$$= PAQ$$

P-16: The rank of the product of two singular matrices cannot exceed the rank of either matrix.

1.6 VECTORS AND MATRICES

A vector as defined in Sect. 1.3, i.e.,

$$\mathbf{x} = [x_1, x_2, \ldots, x_n] \tag{1.32}$$

can be considered as an $n \times 1$ or one-row matrix. Under these conditions its transpose

$$\mathbf{x}' = \begin{bmatrix} x_1 \\ x_2 \\ \cdot \\ \cdot \\ \cdot \\ x_n \end{bmatrix} \tag{1.33}$$

is a $1 \times n$ matrix and can be considered as a column or "vertical" vector in opposition to the row or "horizontal" vector defined by (1.32).

Let A be an $m \times n$ matrix. Then the matrix product

$$[x_1, x_2, \ldots, x_n] \begin{bmatrix} a_{11} & a_{12} & \cdots & a_{1m} \\ a_{21} & a_{22} & \cdots & a_{2m} \\ a & & & \\ a_{n1} & a_{n2} & \cdots & a_{nm} \end{bmatrix} = [y_1, y_2, \ldots, y_m] \tag{1.34}$$

where

$$y_k = \sum_{j=1}^{n} x_j a_{jk}, \qquad k = 1, 2, \ldots, m \tag{1.35}$$

is a horizontal vector. This relation can be written more compactly in the form

$$\mathbf{x}A = \mathbf{y} \tag{1.36}$$

Thus the $m \times n$ matrix A transforms the n-vector $\mathbf{x}$ into the m-vector $\mathbf{y}$.

Similarly,

$$A'\mathbf{x}' = \mathbf{y}' \tag{1.37}$$

We can write (1.37) in the form

$$B\mathbf{X} = \mathbf{Y} \tag{1.38}$$

where $B = A'$, $\mathbf{x}' = \mathbf{X}$, $\mathbf{y}' = \mathbf{Y}$.

The transformations defined by (1.36), (1.37), and (1.38) are linear since (1.35) is a linear relation.

It will be observed that an $n \times m$ matrix can be considered as vector of vectors, in fact,

$$A = \begin{bmatrix} a_{11} & a_{12} & \cdots & a_{1n} \\ a_{21} & a_{22} & \cdots & a_{2n} \\ \cdot & \cdot & \cdots & \cdot \\ \cdot & \cdot & \cdots & \cdot \\ \cdot & \cdot & \cdots & \cdot \\ a_{m1} & a_{m2} & \cdots & a_{mn} \end{bmatrix} = \begin{bmatrix} \boldsymbol{\alpha}_1 \\ \boldsymbol{\alpha}_2 \\ \cdot \\ \cdot \\ \cdot \\ \boldsymbol{\alpha}_m \end{bmatrix} = [\boldsymbol{\beta}_1, \boldsymbol{\beta}_2, \ldots, \boldsymbol{\beta}_n]$$

$$\boldsymbol{\alpha}_k = [a_{k1}, a_{k2}, \ldots, a_{kn}] \qquad k = 1, 2, \ldots, m$$

$$\boldsymbol{\beta}_h = \begin{bmatrix} a_{1h} \\ a_{2h} \\ \cdot \\ \cdot \\ \cdot \\ a_{mh} \end{bmatrix} \qquad h = 1, 2, \ldots, n$$

We finally state two important theorems.

Theorem 1. *A square orthogonal matrix applied to a vector leaves its norm unchanged.*

Theorem 2. *If $A\mathbf{x} = B\mathbf{x}$ for all $\mathbf{x}$, then $A = B$.*

1.7 EIGENVALUES AND EIGENVECTORS

Definition 1. *An eigenvector of a square matrix A is a vector $\mathbf{x} \neq 0$ such that*

$$A\mathbf{x} = \lambda\mathbf{x} \tag{1.39}$$

where λ is a scalar.

Definition 2. *The scalar λ is called an eigenvalue of the matrix A.*

Definition 3. *If $\lambda = 1$, the eigenvector $\mathbf{x}$ is called an invariant vector of the matrix A.*

Theorem 1. *The eigenvalues of a matrix A are the roots of the characteristic equation of the matrix A, which is the n-th degree algebraic equation*

$$\det [A - \lambda I] = \varphi(\lambda) = 0 \tag{1.40}$$

It is clear that $\varphi(\lambda)$ is of the degree n.

It should be observed that the eigenvalues need not be distinct, and that for a given eigenvalue there may exist more than one eigenvector. If $\mathbf{x}$ is an eigenvector of A so is $k\mathbf{x}$ for any scalar k. We shall assume that the eigenvectors are normalized. Although this is not necessary, it is convenient to do so.

Example I. Find the eigenvalues and eigenvectors of the matrix.

$$A = \begin{bmatrix} 1 & 0 & 1 \\ 0 & 1 & 1 \\ 1 & 1 & 0 \end{bmatrix}$$

SOLUTION

$$\varphi(\lambda) = \det\,[A - \lambda I] = \begin{vmatrix} 1 - \lambda & 0 & 1 \\ 0 & 1 - \lambda & 1 \\ 1 & 1 & -\lambda \end{vmatrix}$$

$$= (1 - \lambda)(1 + \lambda)(\lambda - 2) = 0$$

Thus the eigenvalues are $\lambda_1 = 1$, $\lambda_2 = -1$, $\lambda_3 = 2$.

To find the eigenvector $\mathbf{x}_1$ we use $\lambda = \lambda_1 = 1$, which gives

$$\begin{bmatrix} 0 & 0 & 1 \\ 0 & 0 & 1 \\ 1 & 1 & -1 \end{bmatrix} \begin{bmatrix} x_1 \\ x_2 \\ x_3 \end{bmatrix} = 0$$

Hence

$$0 + 0 + x_3 = 0$$
$$x_1 + x_2 - x_3 = 0$$
$$x_1 = -x_2$$

Normalizing gives

$$x_1^2 + x_2^2 = 2x_1^2 = 1$$

so that

$$\mathbf{x}_1 = \left[\frac{\sqrt{2}}{2}, -\frac{\sqrt{2}}{2}, 0 \right]$$

Similarly, to find $\mathbf{x}_2$ we use $\lambda = \lambda_2 = -1$, so that

$$\begin{bmatrix} 2 & 0 & 1 \\ 0 & 2 & 1 \\ 1 & 1 & 1 \end{bmatrix} \begin{bmatrix} x_1 \\ x_2 \\ x_3 \end{bmatrix} = 0$$

Therefore,

$$2x_1 + 0 + x_3 = 0$$
$$0 + 2x_2 + x_3 = 0$$
$$x_1 + x_2 + x_3 = 0$$
$$x_1^2 + x_2^2 + x_3^2 = 1$$

and

$$x_1 = -\frac{\sqrt{6}}{6} = x_2 \qquad x_3 = \frac{\sqrt{6}}{3}$$

Thus

$$\mathbf{X_2} = \left[-\frac{\sqrt{6}}{6}, -\frac{\sqrt{6}}{6}, \frac{\sqrt{6}}{3} \right]$$

For $\lambda = \lambda_3 = 2$, we have

$$\begin{bmatrix} -1 & 0 & 1 \\ 0 & -1 & 1 \\ 1 & 1 & -2 \end{bmatrix} \begin{bmatrix} x_1 \\ x_2 \\ x_3 \end{bmatrix} = 0$$

Hence

$$-x_1 + 0 + x_3 = 0$$
$$0 - x_2 + x_3 = 0$$
$$x_1 + x_2 - 2x_3 = 0$$
$$x_1^2 + x_2^2 + x_3^2 = 1$$

so that $x_1 = x_2 = x_3 = \frac{\sqrt{3}}{3}$ and $\mathbf{x_3} = \left[\frac{\sqrt{3}}{3}, \frac{\sqrt{3}}{3}, \frac{\sqrt{3}}{3} \right]$. We have normalized the eigenvectors in all three cases.

Example 2

Find the eigenvalues and eigenvectors for the matrix

$$A = \begin{bmatrix} 2 & 2 & 1 \\ 1 & 3 & 1 \\ 1 & 2 & 2 \end{bmatrix}$$

Write the characteristic equation and solve for the roots.

$$\varphi(\lambda) = \det [A - \lambda I] = \begin{bmatrix} 2 - \lambda & 2 & 1 \\ 1 & 3 - \lambda & 1 \\ 1 & 2 & 2 - \lambda \end{bmatrix}$$

$$= -(\lambda^3 - 7\lambda^2 + 11\lambda - 5) = -(\lambda - 5)(\lambda - 1)^2 = 0$$

Hence the eigenvalues are $\lambda_1 = 5$, $\lambda_2 = 1$, $\lambda_3 = 1$.

To find the eigenvector $\mathbf{x}_1$ corresponding to λ_1, we note that solution of

$$(A - \lambda_1 I)\,\mathbf{x}_1 = \begin{bmatrix} 2-5 & 2 & 1 \\ 1 & 3-5 & 1 \\ 1 & 2 & 2-5 \end{bmatrix} \begin{bmatrix} x_1 \\ x_2 \\ x_3 \end{bmatrix}$$

$$= \begin{bmatrix} -3 & 2 & 1 \\ 1 & -2 & 1 \\ 1 & 2 & -3 \end{bmatrix} \begin{bmatrix} x_1 \\ x_2 \\ x_3 \end{bmatrix} = 0$$

is

$$\mathbf{x}_1 = \begin{bmatrix} 1 \\ 1 \\ 1 \end{bmatrix}$$

Similarly, for $\lambda_2 = \lambda_3 = 1$ we find

$$(A - \lambda_2 I)\,\mathbf{x}_2 = \begin{bmatrix} 1 & 2 & 1 \\ 1 & 2 & 1 \\ 1 & 2 & 1 \end{bmatrix} \begin{bmatrix} x_1 \\ x_2 \\ x_3 \end{bmatrix} = 0$$

or

$$x_1 + 2x_2 + x_3 = 0$$
$$x_1 + 2x_2 + x_3 = 0$$
$$x_1 + 2x_2 + x_3 = 0$$

which among others has the two linearly independent solutions

$$\mathbf{x}_2 = \begin{bmatrix} 2 \\ -1 \\ 0 \end{bmatrix} \qquad \mathbf{x}_3 = \begin{bmatrix} 1 \\ 0 \\ -1 \end{bmatrix}$$

Normalizing the vectors we obtain the following three eigenvectors, written in transposed form:

$$\mathbf{x}_1' = \begin{bmatrix} \dfrac{1}{\sqrt{3}}, & \dfrac{1}{\sqrt{3}}, & \dfrac{1}{\sqrt{3}} \end{bmatrix}$$

$$\mathbf{x}_2' = \begin{bmatrix} \dfrac{1}{\sqrt{5}}, & -\dfrac{1}{\sqrt{5}}, & 0 \end{bmatrix}$$

$$\mathbf{x}_3' = \begin{bmatrix} \dfrac{1}{\sqrt{2}}, & 0, & -\dfrac{1}{\sqrt{2}} \end{bmatrix}$$

Definition 4. *Two matrices A and B are said to be similar if there exists a nonsingular matrix C such that*

$$B = C^{-1} AC \qquad (1.41)$$

or

$$A = CB \, C^{-1} \qquad (1.42)$$

Both relations are clearly equivalent.

We shall state some of the basic properties of similar matrices:

Q-1: Similar matrices have the same eigenvalues.

Q-2: If A and B are similar as defined by (1.41) or (1.42), if $\mathbf{x}$ is an eigenvector of B, i.e., $B\mathbf{x} = \lambda\mathbf{x}$, and if $C\mathbf{x} = \mathbf{y}$, then $A\mathbf{y} = \lambda\mathbf{y}$.

Q-3: A square matrix of order n that is similar to a diagonal matrix has n linearly independent eigenvectors.

Q-4: If a square matrix of order n has n linearly independent eigenvectors, it is similar to a diagonal matrix.

Q-5: The eigenvalues of a real symmetric matrix are all real.

Q-6: The eigenvectors of a real symmetric matrix of order n having n distinct eigenvalues are orthogonal to each other.

Q-7: Any matrix A is similar to a triangular matrix whose diagonal elements are the eigenvalues of A.

Q-8: If in Q-7 A is real and has real eigenvalues, then there exists an orthogonal matrix P such that $P^{-1} AP = T$, where T is triangular.

Q-9: If A is a real symmetric matrix, then there exists an orthogonal matrix P such that $P^{-1} AP = \Lambda$, where Λ is the diagonal matrix whose elements are the eigenvalues (distinct or not) of A.

Remark. It follows from the preceding properties that if A has distinct eigenvalues, it is similar to a diagonal matrix whose elements are the eigenvalues of A.

Example 3

Given

$$A = \begin{bmatrix} 2 & 2 & 1 \\ 1 & 3 & 1 \\ 1 & 2 & 2 \end{bmatrix} \quad \text{and} \quad C = \begin{bmatrix} 1 & 3 & 3 \\ 1 & 4 & 3 \\ 1 & 3 & 4 \end{bmatrix}$$

Find the similar matrix $B = C^{-1} AC$.

Using the method of determinants we find the inverse matrix

$$C^{-1} = \begin{bmatrix} 7 & -3 & -3 \\ -1 & 1 & 0 \\ -1 & 0 & 1 \end{bmatrix}$$

Hence, C^{-1} exists and C is nonsingular.

Calculation gives

$$B = C^{-1} AC = \begin{bmatrix} 7 & -3 & -3 \\ -1 & 1 & 0 \\ -1 & 0 & 1 \end{bmatrix} \begin{bmatrix} 2 & 2 & 1 \\ 1 & 3 & 1 \\ 1 & 2 & 2 \end{bmatrix} \begin{bmatrix} 1 & 3 & 3 \\ 1 & 4 & 3 \\ 1 & 3 & 4 \end{bmatrix}$$

$$= \begin{bmatrix} 5 & 14 & 13 \\ 0 & 1 & 0 \\ 0 & 0 & 1 \end{bmatrix}$$

To compare the eigenvalues of A and B we note that

$$\det [A - \lambda I] = \begin{vmatrix} 2 - \lambda & 2 & 1 \\ 1 & 3 - \lambda & 1 \\ 1 & 2 & 2 - \lambda \end{vmatrix} = (5 - \lambda)(1 - \lambda)^2 = 0$$

$$\det [B - \lambda I] = \begin{vmatrix} 5 - \lambda & 14 & 13 \\ 0 & 1 - \lambda & 0 \\ 0 & 0 & 1 - \lambda \end{vmatrix} = (5 - \lambda)(1 - \lambda)^2 = 0$$

Hence, A and B have the same eigenvalues

$$\lambda_1 = 5 \qquad \lambda_2 = 1 \qquad \lambda_3 = 1$$

as stated in Q-1 for similar matrices.

Example 4. Reduction of a Matrix to Triangular Form

We consider the matrix of Example 1

$$A = \begin{bmatrix} 1 & 0 & 1 \\ 0 & 1 & 1 \\ 1 & 1 & 0 \end{bmatrix}$$

having the eigenvalues 1, -1, 2, and the normalized eigenvectors

$$\mathbf{x}_1 = [\sqrt{2}/2, -\sqrt{2}/2, 0] \qquad \mathbf{x}_2 = [-\sqrt{6}/6, -\sqrt{6}/6, \sqrt{6}/3]$$
$$\mathbf{x}_3 = [\sqrt{3}/3, \sqrt{3}/3, \sqrt{3}/3]$$

We take

$$Q_1 = \begin{bmatrix} \sqrt{2}/2 & 0 & 0 \\ -\sqrt{2}/2 & 1 & 0 \\ 0 & 0 & 1 \end{bmatrix}$$

so that, $\det [Q_1] = \sqrt{2}/2$, and,

$$Q_1^{-1} = \begin{bmatrix} \sqrt{2} & 0 & 0 \\ 1 & 1 & 0 \\ 0 & 0 & 1 \end{bmatrix}$$

$$Q_1^{-1} A\, Q_1 = \begin{bmatrix} \sqrt{2} & 0 & 0 \\ 1 & 1 & 0 \\ 0 & 0 & 1 \end{bmatrix} \begin{bmatrix} 1 & 0 & 1 \\ 0 & 1 & 1 \\ 1 & 1 & 0 \end{bmatrix} \begin{bmatrix} \sqrt{2}/2 & 0 & 0 \\ -\sqrt{2}/2 & 1 & 0 \\ 0 & 0 & 1 \end{bmatrix} = \begin{bmatrix} 1 & 0 & \sqrt{2} \\ 0 & 1 & 2 \\ 0 & 1 & 0 \end{bmatrix}$$

Thus let $A_1 = \begin{bmatrix} 1 & 2 \\ 1 & 0 \end{bmatrix}$. The eigenvector corresponding to the eigenvalue -1 is found by writing

$$\begin{bmatrix} 1 & 2 \\ 1 & 0 \end{bmatrix} \begin{bmatrix} y_1 \\ y_2 \end{bmatrix} = \begin{bmatrix} y_1 + 2y_2 \\ y_1 \end{bmatrix} = \begin{bmatrix} -y_1 \\ -y_2 \end{bmatrix}$$

thus,

$$\left\{ \begin{array}{c} 2y_1 + 2y_2 - 0 \\ y_1 + y_2 = 0 \end{array} \right\}$$

with $y_1^2 + y_2^2 = 1$, hence, $y_1 = \sqrt{2}/2$; $y_2 = -\sqrt{2}/2$. We take

$$Q_2 = \begin{bmatrix} \sqrt{2}/2 & 0 \\ -\sqrt{2}/2 & 1 \end{bmatrix} \qquad \det [Q_2] = \sqrt{2}/2$$

$$Q_2^{-1} = \sqrt{2} \begin{bmatrix} 1 & 0 \\ \sqrt{2}/2 & \sqrt{2}/2 \end{bmatrix} = \begin{bmatrix} \sqrt{2} & 0 \\ 1 & 1 \end{bmatrix}$$

$$\begin{bmatrix} 1 & 0 \\ 0 & Q_2^{-1} \end{bmatrix} Q_1^{-1} A\, Q_1 \begin{bmatrix} 1 & 0 \\ 0 & Q_2 \end{bmatrix}$$

$$= \begin{bmatrix} 1 & 0 & 0 \\ 0 & \sqrt{2} & 0 \\ 0 & 1 & 1 \end{bmatrix} \begin{bmatrix} 1 & 0 & \sqrt{2} \\ 0 & 1 & 2 \\ 0 & 1 & 0 \end{bmatrix} \begin{bmatrix} 1 & 0 & 0 \\ 0 & \sqrt{2}/2 & 0 \\ 0 & -\sqrt{2}/2 & 1 \end{bmatrix} = \begin{bmatrix} 1 & -1 & \sqrt{2} \\ 0 & -1 & 2\sqrt{2} \\ 0 & 0 & 2 \end{bmatrix} = T$$

The diagonal elements are the eigenvalues of A. It follows that

$$Q^{-1} = \begin{bmatrix} 1 & 0 & 0 \\ 0 & \sqrt{2} & 0 \\ 0 & 1 & 1 \end{bmatrix} \begin{bmatrix} \sqrt{2} & 0 & 0 \\ 1 & 1 & 0 \\ 0 & 0 & 1 \end{bmatrix} = \begin{bmatrix} \sqrt{2} & 0 & 0 \\ \sqrt{2} & \sqrt{2} & 0 \\ 1 & 1 & 1 \end{bmatrix}$$

$$Q = \begin{bmatrix} \sqrt{2}/2 & 0 & 0 \\ -\sqrt{2}/2 & 1 & 0 \\ 0 & 0 & 1 \end{bmatrix} \begin{bmatrix} 1 & 0 & 0 \\ 0 & \sqrt{2}/2 & 0 \\ 0 & -\sqrt{2}/2 & 1 \end{bmatrix} = \begin{bmatrix} \sqrt{2}/2 & 0 & 0 \\ -\sqrt{2}/2 & \sqrt{2}/2 & 0 \\ 0 & -\sqrt{2}/2 & 1 \end{bmatrix}$$

and

$$Q^{-1} A Q = T$$

Example 5. Rotation Matrix

Consider the matrix

$$R(\theta) = \begin{bmatrix} \cos\theta & -\sin\theta \\ \sin\theta & \cos\theta \end{bmatrix}$$

which rotates a vector $\mathbf{r}$ counter-clockwise through an angle θ, Figure 1.1, where

$$\mathbf{r} = x\mathbf{i} + y\mathbf{j} = \begin{bmatrix} x \\ y \end{bmatrix}$$

Denote the rotated vector by $\mathbf{r}'$. Then

$$\mathbf{r}' = R\mathbf{r} = \begin{bmatrix} \cos\theta & -\sin\theta \\ \sin\theta & \cos\theta \end{bmatrix}\begin{bmatrix} x \\ y \end{bmatrix} = \begin{bmatrix} x\cos\theta & -y\sin\theta \\ x\sin\theta & +y\cos\theta \end{bmatrix}$$

(a) Show that R is orthogonal
(b) Prove that $R(\theta_1)R(\theta_2) = R(\theta_1 + \theta_2)$ and hence that $R^n(\theta) = R(n\theta)$.
(a) If R is orthogonal it must satisfy the condition

$$R'R = I$$

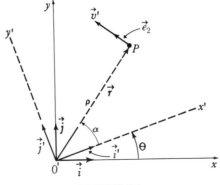

FIGURE 1.1

To test this calculate

$$R'R = \begin{bmatrix} \cos\theta & \sin\theta \\ -\sin\theta & \cos\theta \end{bmatrix} \begin{bmatrix} \cos\theta & -\sin\theta \\ \sin\theta & \cos\theta \end{bmatrix}$$

$$= \begin{bmatrix} 1 & 0 \\ 0 & 1 \end{bmatrix} = I$$

Hence, R is an orthogonal matrix.
(b) By direct calculation we find

$$R(\theta_1)R(\theta_2) = \begin{bmatrix} \cos\theta_1 & -\sin\theta_1 \\ \sin\theta_1 & \cos\theta_1 \end{bmatrix} \begin{bmatrix} \cos\theta_2 & -\sin\theta_2 \\ \sin\theta_2 & \cos\theta_2 \end{bmatrix}$$

$$= \begin{bmatrix} \cos\theta_1\cos\theta_2 - \sin\theta_1\sin\theta_2 & -\cos\theta_1\sin\theta_2 - \sin\theta_1\cos\theta_2 \\ \sin\theta_1\cos\theta_2 + \sin\theta_2\cos\theta_1 & -\sin\theta_1\sin\theta_2 + \cos\theta_1\cos\theta_2 \end{bmatrix}$$

$$= \begin{bmatrix} \cos(\theta_1 + \theta_2) & -\sin(\theta_1 + \theta_2) \\ \sin(\theta_1 + \theta_2) & \cos(\theta_1 + \theta_2) \end{bmatrix} = R(\theta_1 + \theta_2)$$

Letting $\theta_1 = \theta_2$ gives

$$R(\theta_1)R(\theta_1) = R^2(\theta_1) = R(2\theta_1)$$

Repeated application of this result leads to the general case

$$R^n(\theta) = R(n\theta)$$

which completes the proof.

Example 6. Reduction to Diagonal Form

Reduce the matrix

$$A = \begin{bmatrix} -\cos\theta & \sin\theta \\ \sin\theta & \cos\theta \end{bmatrix}$$

to diagonal form and check its invariant properties. Give a geometric interpretation of A.

Since the matrix is real and symmetric, it can be reduced to diagonal form by a nonsingular matrix and the diagonal terms are the eigenvalues of A. To find the eigenvalues we write

$$\det[A - \lambda I] = \begin{vmatrix} -\cos\theta - \lambda & \sin\theta \\ \sin\theta & \cos\theta - \lambda \end{vmatrix}$$

$$= -\cos^2\theta + \lambda^2 - \sin^2\theta = \lambda^2 - 1 = 0$$

$$\therefore \quad \lambda_1 = -1, \lambda_2 = 1$$

and the diagonal matrix D is

$$D = \begin{bmatrix} 1 & 0 \\ 0 & -1 \end{bmatrix}$$

We see that the invariant properties hold, i.e.,

$$\begin{cases} \text{tr } A = -\cos\theta + \cos\theta = 0 \\ \text{tr } D = -1 + 1 = 0 \end{cases}$$

$$\det [A] = \begin{vmatrix} -\cos\theta & \sin\theta \\ \sin\theta & \cos\theta \end{vmatrix} = -\cos^2\theta - \sin^2\theta = -1$$

$$\det [D] = \begin{vmatrix} -1 & 0 \\ 0 & 1 \end{vmatrix} = -1$$

GEOMETRIC INTERPRETATION OF A. Comparison with matrix R of Ex. 5 shows that A rotates a vector $\mathbf{r} = x\mathbf{i} + y\mathbf{j}$ counter clockwise through an angle θ and then multiplies the x-component by -1, i.e., reflects the rotated $\mathbf{r}$ in the y-axis.

Example 7. Matrix Which Cannot Be Reduced to Diagonal Form

Consider the rotational matrix

$$R = \begin{bmatrix} \cos\theta & -\sin\theta \\ \sin\theta & \cos\theta \end{bmatrix}$$

which rotates a vector $\mathbf{y} = R\mathbf{x}$ around the origin through an angle θ and thereby changes the directions of all vectors in the plane. It is evident from geometrical considerations that $R\mathbf{x} = \lambda\mathbf{x}$ cannot be satisfied and therefore that R has no eigenvectors.

This result follows analytically from the characteristic equation

$$\det [R - \lambda I] = \begin{vmatrix} \cos\theta - \lambda & -\sin\theta \\ \sin\theta & \cos - \lambda \end{vmatrix}$$

$$= (\cos\theta - \lambda)^2 + \sin^2\theta = 0$$

which has the complex roots

$$\lambda_1, \lambda_2 = \cos\theta \pm i\sin\theta = e^{\pm i\theta}$$

Since there are no real roots, there are no eigenvalues and therefore no eigenvectors, except for the trivial case $0 = 0, \pi, 2\pi, \ldots$

Example 8. Motion of a Rigid Body. In this example we apply the concepts and methods of matrices to rigid body mechanics.

The motion of a rigid body may be described in terms of the motions of the particles which comprise that body. To discuss the motions of these particles it is useful to consider a coordinate system fixed in the rigid body and moving with it. Then, since a rigid body by definition is invariable in shape and size, the position of any particle in the body is given by a position vector which is constant with respect to the coordinate system attached to the body. The position of the particle relative to an inertial or fixed frame of reference is then uniquely specified by the position vector of the origin and by the orientation of the moving coordinate system with respect to the inertial system.

To put these relations in quantitative form, let $Oxyz$ be a Cartesian coordinate system fixed in inertial space and $O'x'y'z'$ be a second Cartesian set attached to the body. Before the body is displaced, the two frames coincide and the position of a point P in the body is the same in both coordinate systems, i.e.,

$$\mathbf{r} = \mathbf{r}'$$

where $\mathbf{r}$ is the position vector of P in the inertial frame and $\mathbf{r}'$ the position in the body frame.

The displacement of the body consists of two parts, one a translation of the origin O' with respect to O represented by the vector $\mathbf{R}$, the second a rotation of $O'x'y'z'$ about the point O' represented by a matrix S. Then the position $\mathbf{r}$ of P in the inertial frame is given by the sum of the two motions

$$\mathbf{r} = \mathbf{R} + S\mathbf{r}' \tag{1}$$

as depicted in Figure 1.2. Since lengths in the two coordinate systems are preserved, we have also

$$\|\mathbf{r} - \mathbf{R}\| = \|S\mathbf{r}'\| = \|\mathbf{r}'\| \tag{2}$$

Hence, from property P-15 (p. 25) matrix S is orthogonal and satisfies the relation

$$S'S = I \tag{3}$$

Although S has nine components, Eq. (3) provides six conditions or constraints, which leaves only three free parameters. In rigid body mechanics these are usually chosen to be the three angles (φ, θ, ψ) shown in Figure 1.3 and known as Euler's angles. The physical significance of Euler's angles in a modern context is brought out by their application, for example, to the motion of an airplane where φ is called the yaw, θ the pitch, and ψ the roll of the plane.

The Eulerian angles represent three successive rotations of the $O'x'y'z'$ frame. Each rotation may be expressed analytically in matrix form, so that the final position is given by the product of the three matrices. These rotations must be made in a specified sequence since the matrix

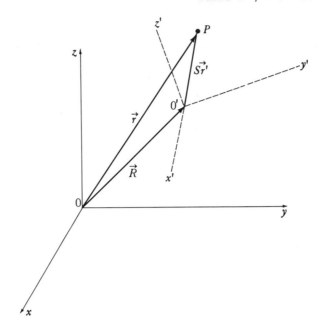

FIGURE 1.2

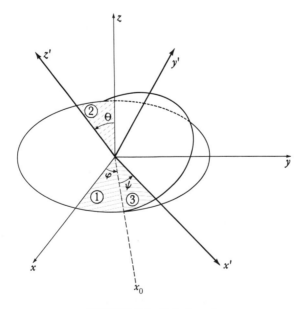

FIGURE 1.3 Euler's angles.

multiplications are not commutative, and different sequences will result in different final positions of the body. We use the sequence indicated in Figure 1.3, corresponding to the matrix product

$$S = S_\psi S_\theta S_\varphi \tag{4}$$

where

$$S_\varphi = \begin{bmatrix} \cos \varphi & \sin \varphi & 0 \\ -\sin \varphi & \cos \varphi & 0 \\ 0 & 0 & 1 \end{bmatrix}$$

$$S_\theta = \begin{bmatrix} 1 & 0 & 0 \\ 0 & \cos \theta & \sin \theta \\ 0 & -\sin \theta & \cos \theta \end{bmatrix} \tag{5}$$

$$S_\psi = \begin{bmatrix} \cos \psi & \sin \psi & 0 \\ -\sin \psi & \cos \psi & 0 \\ 0 & 0 & 1 \end{bmatrix}$$

Matrix S_φ represents a rotation through an angle φ about the z-axis, S_θ a rotation θ about the intermediate axis x_0, and S_ψ a rotation ψ about the z'-axis.
Calculation of the matrix products gives

$$S = \begin{bmatrix} \cos \psi \cos \varphi - \cos \theta \sin \varphi \sin \psi & \cos \psi \sin \varphi + \cos \theta \cos \varphi \sin \psi & \sin \varphi \sin \theta \\ -\sin \psi \cos \varphi - \cos \theta \sin \varphi \cos \psi & -\sin \psi \sin \varphi + \cos \theta \cos \varphi \cos \psi & \cos \psi \sin \theta \\ \sin \theta \sin \varphi & -\sin \theta \cos \varphi & \cos \theta \end{bmatrix}$$

$$\tag{6}$$

It can be shown that S can be expressed as a single rotation about a fixed axis of rotation. To prove this result, it is necessary to find an eigenvector **a** of S with eigenvalue $+1$, i.e.,

$$S\mathbf{a} = \lambda \mathbf{a} = \mathbf{a} \tag{7}$$

The eigenvalues are given by the characteristic equation

$$\det [S - \lambda I] = 0 \quad \text{(a cubic in } \lambda\text{)}$$

which has three roots λ_1, λ_2, λ_3. Since S is orthogonal,

$$\det [S] = \det [\lambda I] = \lambda_1 \lambda_2 \lambda_3 = 1 \tag{8}$$

and we conclude that at least one of the roots must have the value $+1$. Hence (7) is always satisfied and a vector **a** always exists which has the same components in both frames $Oxyz$ and $O'x'y'z'$.

The angle of rotation β about **a** is readily obtained from the invariance relation

$$\text{tr } S = \text{tr } T \qquad (9)$$

where

$$T = \begin{bmatrix} \cos \beta & \sin \beta & 0 \\ -\sin \beta & \cos \beta & 0 \\ 0 & 0 & 1 \end{bmatrix}$$

represents the rotation matrix about **a**. From (6) and (9), we find

$$\text{tr } S = (1 + \cos \theta) \cos (\varphi + \psi) + \cos \theta = \text{tr } T = 2 \cos \beta + 1$$

from which

$$\cos \beta = \frac{(1 + \cos \theta) \cos (\varphi + \psi) + \cos \theta - 1}{2} \qquad (10)$$

Next, assume that the rigid body is moving continuously with time. Then

$$\mathbf{r}(t) = \mathbf{R}(t) + S(t)\mathbf{r}'(t)$$

and the velocity of the point P is found by differentiation as follows

$$S\mathbf{r}' = \mathbf{r} - \mathbf{R}$$
$$S'S\mathbf{r}' = \mathbf{r}' = S'(\mathbf{r} - \mathbf{R})$$
$$\frac{d\mathbf{r}'}{dt} = \frac{d}{dt}[S'(\mathbf{r} - \mathbf{R})] = \dot{S}'(\mathbf{r} - \mathbf{R}) + S'(\dot{\mathbf{r}} - \dot{\mathbf{R}}) = 0$$

Multiplying by S and solving for $(\dot{\mathbf{r}} - \dot{\mathbf{R}})$ gives

$$SS'(\dot{\mathbf{r}} - \dot{\mathbf{R}}) = \dot{\mathbf{r}} - \dot{\mathbf{R}} = S\dot{S}'(\mathbf{r} - \mathbf{R}) \qquad (11)$$
$$= S\dot{S}'\mathbf{r}' = -\Omega\mathbf{r}'$$

where Ω is defined by

$$-\Omega = S\dot{S}'$$

But since

$$\frac{d}{dt}(SS') = \dot{S}S' + S\dot{S}' = 0$$

we see from P-2 of Sect. 1.5 that

$$S\dot{S}' = -\dot{S}S' = (S\dot{S}')'$$

or

$$\Omega = -\Omega'$$

Hence, Ω is anti-symmetric (skew-symmetric) and has the form

$$\Omega = \begin{bmatrix} 0 & \Omega_{xy} & -\Omega_{zx} \\ -\Omega_{xy} & 0 & \Omega_{yz} \\ \Omega_{zy} & -\Omega_{yz} & 0 \end{bmatrix} \qquad (12)$$

Since $\Omega_{xy}(t)$ is the rate of rotation about the z-axis it is conventional to denote this by ω_z, and similarly for Ω_{yz}, Ω_{zx}. Thus we write

$$\Omega = \begin{bmatrix} 0 & \omega_z & -\omega_y \\ -\omega_z & 0 & \omega_x \\ \omega_y & -\omega_x & 0 \end{bmatrix} \tag{13}$$

Substituting (13) in (11) and carrying out the matrix multiplication yields

$$\dot{\mathbf{r}} - \dot{\mathbf{R}} = -\Omega\mathbf{r}' = -\begin{bmatrix} 0 & \omega_z & -\omega_y \\ -\omega_z & 0 & \omega_x \\ \omega_y & -\omega_x & 0 \end{bmatrix}\begin{bmatrix} x' \\ y' \\ z' \end{bmatrix} = \begin{bmatrix} \omega_y z' - \omega_z y' \\ \omega_z x' - \omega_x z' \\ \omega_x y' - \omega_y x' \end{bmatrix}$$

$$= \boldsymbol{\omega} \times \mathbf{r}' \tag{14}$$

where $\boldsymbol{\omega}$ is the vector

$$\boldsymbol{\omega} = \omega_x \mathbf{i} + \omega_y \mathbf{j} + \omega_z \mathbf{k}$$

Solving for $\dot{\mathbf{r}}$ gives the velocity of point P

$$\mathbf{v} = \dot{\mathbf{r}} = \mathbf{V} + \boldsymbol{\omega} \times \mathbf{r}' \tag{15}$$

where $\mathbf{V}$ is the velocity of O' relative to O.

Rotation about a Fixed Axis

Let z be the fixed axis and choose

$$\mathbf{R} = 0 \qquad S = S_\varphi \qquad \varphi = \omega_z t$$

Then,

$$-\Omega = S_\varphi \dot{S}'_\varphi = \begin{bmatrix} \cos\varphi & \sin\varphi & 0 \\ -\sin\varphi & \cos\psi & 0 \\ 0 & 0 & 1 \end{bmatrix}\begin{bmatrix} -\sin\varphi & \cos\varphi & 0 \\ -\cos\varphi & -\sin\varphi & 0 \\ 0 & 0 & 0 \end{bmatrix}' \dot{\varphi}$$

$$= \begin{bmatrix} 0 & -\omega_z & 0 \\ \omega_z & 0 & 0 \\ 0 & 0 & 0 \end{bmatrix}$$

$$\mathbf{v} = -\Omega\mathbf{r}' = \begin{bmatrix} -\omega_z y' \\ \omega_z x' \\ 0 \end{bmatrix} = \omega_z(-y'\mathbf{i} + x'\mathbf{j}')$$

Expressing x', y' in polar coordinates gives

$$x' = \rho' \cos\alpha \qquad y' = \rho' \sin\alpha$$

Hence,
$$\dot{\mathbf{v}} = \omega_z \rho'(-\mathbf{i}' \cos \alpha + \mathbf{j}' \sin \alpha) = \omega_z \rho' \mathbf{e}_2$$

where $\mathbf{e}_2$ is a unit vector perpendicular to the radius ρ' and in the counter-clockwise direction as shown in Figure 1.1. It is seen that point P moves in a circle of radius ρ' about the fixed axis z (perpendicular to page) at an angular velocity ω_z. The unit vector $\mathbf{e}_2$ has a constant direction in frame $O'x'y'z'$, but a constantly changing direction in frame $Oxyz$.

1.8   THE CAYLEY-HAMILTON THEOREM

Theorem. *A square matrix C of elements c_{ij} satisfies its own characteristic equation.*

Thus, if C is the matrix, its characteristic equation is given by

$$\det [C - \lambda I] = \varphi(\lambda) = 0 \tag{1.43}$$

where, as we have seen,

$$\varphi(\lambda) = \sum_{m=0}^{n} a_m \lambda^m = 0 \tag{1.44}$$

Then, according to the theorem,

$$\varphi(C) = \sum_{m=0}^{n} a_m C^m = 0$$

Example 1.   Verify the Cayley-Hamilton theorem for the matrix of Example 1 of Sec. 1.7.

$$A = \begin{bmatrix} 1 & 0 & 1 \\ 0 & 1 & 1 \\ 1 & 1 & 0 \end{bmatrix}$$

whose eigenvalues are 1, -1, 2, and whose characteristic equation is

$$(1 - \lambda)(1 + \lambda)(2 - \lambda) = 0$$

The corresponding matrix equation is

$$[I - A][I + A][2I - A]$$

$$= \begin{bmatrix} 0 & 1 & -1 \\ 0 & 0 & -1 \\ -1 & -1 & 1 \end{bmatrix} \begin{bmatrix} 2 & 0 & 1 \\ 0 & 2 & 1 \\ 1 & 1 & 1 \end{bmatrix} \begin{bmatrix} 1 & 0 & -1 \\ 0 & 1 & -1 \\ -1 & -1 & 2 \end{bmatrix}$$

$$= \begin{bmatrix} -1 & -1 & -1 \\ -1 & -1 & -1 \\ -1 & -1 & -1 \end{bmatrix} \begin{bmatrix} 1 & 0 & -1 \\ 0 & 1 & -1 \\ -1 & -1 & 2 \end{bmatrix} = \begin{bmatrix} 0 & 0 & 0 \\ 0 & 0 & 0 \\ 0 & 0 & 0 \end{bmatrix} = 0$$

which verifies the theorem for this example.

Example 2

Evaluate $P(A) = A^4 + 6A$; where $A = \begin{bmatrix} 1 & -3 \\ 3 & 1 \end{bmatrix}$

The characteristic equation of A is

$$\varphi(\lambda) = \begin{vmatrix} 1-\lambda & -3 \\ 3 & 1-\lambda \end{vmatrix} = (1-\lambda)^2 + 9 = \lambda^2 - 2\lambda + 10 = 0$$

Hence according to the Cayley-Hamilton theorem:

$$A^2 - 2A + 10I = 0$$
$$A^2 = 2A - 10I$$

$$P(A) = A^4 + 6A = (2A - 10I)^2 + 6A$$
$$= 4A^2 - 40A + 100I + 6A$$
$$= 4[2A - 10I] - 40A + 100I + 6A$$

$$P(A) = -26A + 60I = -26 \begin{bmatrix} 1 & -3 \\ 3 & 1 \end{bmatrix} + 60 \begin{bmatrix} 1 & 0 \\ 0 & 1 \end{bmatrix}$$

$$= \begin{bmatrix} 34 & 78 \\ -78 & 34 \end{bmatrix}$$

Example 3. Find the inverse of the matrix A using the Cayley-Hamilton theorem.

$$A = \begin{bmatrix} 1 & 0 & 1 \\ 0 & 1 & 1 \\ 1 & 1 & 0 \end{bmatrix} \qquad \varphi(\lambda) = (1-\lambda)(1+\lambda)(2-\lambda) = 0$$

or, $\lambda^3 - 2\lambda^2 - \lambda + 2 = 0$. According to the Cayley-Hamilton theorem

$$A^3 - 2A^2 - A + 2I = 0$$

so that, multiplying by A^{-1}

$$A^2 - 2A - I + 2A^{-1} = 0$$

Thus,

$$A^{-1} = \tfrac{1}{2}[-A^2 + 2A + 1]$$

$$A^2 = \begin{bmatrix} 1 & 0 & 1 \\ 0 & 1 & 1 \\ 1 & 1 & 0 \end{bmatrix} \begin{bmatrix} 1 & 0 & 1 \\ 0 & 1 & 1 \\ 1 & 1 & 0 \end{bmatrix} = \begin{bmatrix} 2 & 1 & 1 \\ 1 & 2 & 1 \\ 1 & 1 & 2 \end{bmatrix}$$

$$A^{-1} = \tfrac{1}{2} \left(- \begin{bmatrix} 2 & 1 & 1 \\ 1 & 2 & 1 \\ 1 & 1 & 2 \end{bmatrix} + 2 \begin{bmatrix} 1 & 0 & 1 \\ 0 & 1 & 1 \\ 1 & 1 & 0 \end{bmatrix} + \begin{bmatrix} 1 & 0 & 0 \\ 0 & 1 & 0 \\ 0 & 0 & 1 \end{bmatrix} \right)$$

$$A^{-1} = +\tfrac{1}{2} \begin{bmatrix} 1 & -1 & 1 \\ -1 & 1 & 1 \\ 1 & 1 & -1 \end{bmatrix}$$

$$AA^{-1} = \tfrac{1}{2} \begin{bmatrix} 1 & 0 & 1 \\ 0 & 1 & 1 \\ 1 & 1 & 0 \end{bmatrix} \begin{bmatrix} 1 & -1 & 1 \\ -1 & 1 & 1 \\ 1 & 1 & -1 \end{bmatrix} = \tfrac{1}{2} \begin{bmatrix} 2 & 0 & 0 \\ 0 & 2 & 0 \\ 0 & 0 & 2 \end{bmatrix} = I$$

1.9 QUADRATIC FORMS

Definition 1. *A quadratic form of the variables* $x_1, x_2, \ldots, x_n$ *is an expression of the form*

$$Q(x_1, x_2, \ldots, x_n) = Q(\mathbf{x}) = \sum_{m=1}^{n} \sum_{k=1}^{n} a_{mk} x_m x_k \qquad \textbf{(1.45)}$$

It can be assumed without loss of generality that $a_{mk} = a_{km}$ and $Q(\mathbf{x})$ can be written

$$Q(\mathbf{x}) = \mathbf{x}\, A\, \mathbf{x}' \qquad \textbf{(1.46)}$$

where A is a symmetric matrix, i.e., $a_{ij} = a_{ji}$. Let n be the number of variables which is also the dimension of $\mathbf{x}$ and the order of A. Let r be the rank of A. If $n = r$, both A and Q are nonsingular, while if $n > r$, both A and Q are singular. Let B be a square matrix of order n such that

$$\mathbf{x} = \mathbf{y}B$$

or $\qquad\qquad\qquad\qquad\qquad\qquad\qquad\qquad\qquad\qquad\qquad\quad$ **(1.47)**

$$\mathbf{x}' = B\mathbf{y}'$$

It follows that by substituting $\mathbf{x}$ into (1.46) we obtain

$$R(\mathbf{y}) = \mathbf{y}BAB'\mathbf{y}' = \mathbf{y}[BAB']\mathbf{y}' \qquad \textbf{(1.48)}$$

which transforms the quadratic form Q into the quadratic form R. Both A and BAB' are symmetric.

Example 1

Transform the quadratic form

$$Q = x_1^2 + 2x_2^2 - 3x_3^2 - 6x_1x_2 + 8x_1x_3 - 4x_2x_3$$

by the change of variables

$$x_1 = \quad y_1 + 2y_2 + 3y_3$$

$$x_2 = -2y_1 + 0y_2 + y_3 \qquad \mathbf{x}' = B\mathbf{y}' \qquad B = \begin{bmatrix} 1 & 2 & 3 \\ -2 & 0 & 1 \\ 0 & 3 & -1 \end{bmatrix}$$

$$x_3 = \quad 0y_1 + 3y_2 - y_3$$

Rewrite Q in the form

$$Q = x_1^2 - 3x_1x_2 + 4x_1x_3 - 3x_2x_1 + 2x_2^2 - 2x_2x_3$$
$$+ 4x_1x_3 - 2x_2x_3 - 3x_3^2$$

$$Q = [x_1, x_2, x_3] \begin{bmatrix} 1 & -3 & 4 \\ -3 & 2 & -2 \\ 4 & -2 & -3 \end{bmatrix} \begin{bmatrix} x_1 \\ x_2 \\ x_3 \end{bmatrix} = \mathbf{x}\, A\, \mathbf{x}'$$

$Q(\mathbf{x})$ becomes $R(\mathbf{y})$ as follows

$$R(\mathbf{y}) = \mathbf{y}(B'AB)\mathbf{y}'$$

$$B'AB = \begin{bmatrix} 1 & -2 & 0 \\ 2 & 0 & 3 \\ 3 & 1 & -1 \end{bmatrix} \begin{bmatrix} 1 & -3 & 4 \\ -3 & 2 & -2 \\ 4 & -2 & -3 \end{bmatrix} \begin{bmatrix} 1 & 2 & 3 \\ -2 & 0 & 1 \\ 0 & 3 & -1 \end{bmatrix}$$

$$= \begin{bmatrix} 7 & -7 & 8 \\ 14 & -12 & -1 \\ -4 & -5 & 13 \end{bmatrix} \begin{bmatrix} 1 & 2 & 3 \\ -2 & 0 & 1 \\ 0 & 3 & -1 \end{bmatrix} = \begin{bmatrix} 21 & 38 & 6 \\ 38 & 25 & 31 \\ 6 & 31 & -30 \end{bmatrix}$$

Thus the expression for $R(\mathbf{y})$ is

$$R(\mathbf{y}) = 21y_1^2 + 25y_2^2 - 30y_3^2 + 76y_1y_2 + 12y_3y_1 + 62y_2y_3$$

Reduction of a quadratic form to canonical form (sum of squares) by Lagrange's method

The problem is to find a particular matrix B such that

$$Q_1(\mathbf{y}) = \mathbf{y}[B'AB]\mathbf{y}' = \sum_{m=1}^{n} h_m y_m^2 \tag{1.49}$$

where n is the order of the matrix A. This is to say we look for a particular, nonsingular matrix B such that $H = B'AB$ is diagonal. One way to obtain B is using elementary algebra and the method called *completing the squares* due to Lagrange. Before discussing the general case, we give an example.

Example 2

Reduce the quadratic form of Ex. 1 to a sum of squares

$$Q = x_1^2 + 2x_2^2 - 3x_3^2 - 6x_1x_2 + 8x_1x_3 - 4x_2x_3$$

SOLUTION.   We can write Q in terms of squares by the following steps.

$$
\begin{aligned}
Q &= [x_1^2 + 2x_1(-3x_2 + 4x_3) + (-3x_2 + 4x_3)^2] \\
&\quad + 2x_2^2 - 3x_3^2 - 4x_2x_3 - (-3x_2 + 4x_3)^2 \\
&= (x_1 - 3x_2 + 4x_3)^2 - 7x_2^2 - 19x_3^2 + 20x_2x_3 \\
&= (x_1 - 3x_2 + 4x_3)^2 - 7(x_2 - \tfrac{10}{7}x_3)^2 + \tfrac{100}{7}x_3^2 - 19x_3^2 \\
&= (x_1 - 3x_2 + 4x_3)^2 - 7(x_2 - \tfrac{10}{7}x_3)^2 - \tfrac{33}{7}x_3^2
\end{aligned}
$$

Let $\begin{cases} y_1 = x_1 - 3x_2 + 4x_3 \\ y_2 = x_2 - \tfrac{10}{7}x_3 \\ y_3 = x_3 \end{cases}$ so that $\begin{cases} x_1 = y_1 + 3y_2 + \tfrac{2}{7}y_3 \\ x_2 = 0y_1 + y_2 + \tfrac{10}{7}y_3 \\ x_3 = 0y_1 + 0y_2 + y_3 \end{cases}$

Thus:

$$
A = \begin{bmatrix} 1 & -3 & 4 \\ -3 & 2 & -2 \\ 4 & -2 & -3 \end{bmatrix} \qquad B = \begin{bmatrix} 1 & 3 & \tfrac{2}{7} \\ 0 & 1 & \tfrac{10}{7} \\ 0 & 0 & 1 \end{bmatrix}
$$

The reader will easily verify that

$$
\begin{aligned}
B'AB &= \begin{bmatrix} 1 & 0 & 0 \\ 3 & 1 & 0 \\ \tfrac{2}{7} & \tfrac{10}{7} & 1 \end{bmatrix} \begin{bmatrix} 1 & -3 & 4 \\ -3 & 2 & -2 \\ 4 & -2 & -3 \end{bmatrix} \begin{bmatrix} 1 & 3 & \tfrac{2}{7} \\ 0 & 1 & \tfrac{10}{7} \\ 0 & 0 & 1 \end{bmatrix} \\
&= \begin{bmatrix} 1 & 0 & 0 \\ 0 & -7 & 0 \\ 0 & 0 & -\tfrac{33}{7} \end{bmatrix} = H
\end{aligned}
$$

Hence, $Q_1(\mathbf{y}) = \mathbf{y}H\mathbf{y}' = y_1^2 - 7y_2^2 - \tfrac{33}{7}y_3^2$

For the general case, we operate the same way as long as no zero coefficients appear for the diagonal terms. For each operation there will be one variable less so that finally we shall obtain for Q a sum of squares. It is always possible to change the order of variables, i.e., change x_j into x_i in order to obtain nonzero diagonal terms. If at certain points all diagonal terms disappear we use the change of variable

$$
\begin{aligned}
x_i &= y_i - y_j \\
x_j &= y_i + y_j
\end{aligned}
$$

Thus,   (1.50)

$$
x_ix_j = y_i^2 - y_j^2
$$

and changes the mixed term into a difference of squares.

We illustrate the latter case by the following example:

Example 3

Let $Q = 2ax_1x_2 + 2bx_2x_3 + 2cx_3x_1$

We clearly have

$$A = \begin{bmatrix} 0 & a & c \\ a & 0 & b \\ c & b & 0 \end{bmatrix}$$

Let

$$x_1 = y_1 - y_2$$
$$x_2 = y_1 + y_2$$
$$x_3 = y_3$$

and obtain

$$x_1x_2 = y_1^2 - y_2^2 \qquad x_2x_3 = y_1y_3 + y_2y_3 \qquad x_3x_1 = y_1y_3 - y_2y_3$$

Hence, the steps are as follows:

$$Q_1 = 2ay_1^2 - 2ay_2^2 + 2by_1y_3 + 2by_2y_3 + 2cy_1y_3 - 2cy_2y_3$$

$$= 2ay_1^2 - 2ay_2^2 + 2(b+c)y_1y_3 + 2(b-c)y_2y_0$$

$$- 2a\left[y_1 + \frac{b+c}{2a}y_3\right]^2 - 2a\left[y_2 - \frac{b-c}{2a}y_3\right]^2 + \frac{(b-c)^2}{2a}y_0^2$$

$$- \frac{(b+c)^2}{2a}y_3^2$$

$$- 2a\left(y_1 + \frac{b+c}{2a}y_3\right)^2 - 2a\left(y_2 - \frac{b-c}{2a}y_3\right)^2 - \frac{2bc}{a}y_3^2$$

Now let

$$y_1 + \frac{b+c}{2a}y_3 = z_1$$

$$y_2 - \frac{b-c}{2a}y_3 = z_2$$

$$y_3 = z_3$$

or

$$y_1 = z_1 - \frac{b+c}{2a}z_3$$

$$y_2 = z_2 + \frac{b-c}{2a}z_3$$

$$y_3 = z_3$$

The quadratic thus becomes

$$Q_2(\mathbf{z}) = 2az_1^2 - 2az_2^2 - \frac{2bc}{a}z_3^2$$

We have performed two linear transformations, that is, $\mathbf{x}' = B_1\mathbf{y}'$ and $\mathbf{y}' = B_2\mathbf{z}'$. Hence $\mathbf{x}' = B_1B_2\mathbf{z}'$, so that

$$B = B_1B_2 = \begin{bmatrix} 1 & -1 & 0 \\ 1 & 1 & 0 \\ 0 & 0 & 1 \end{bmatrix}\begin{bmatrix} 1 & 0 & -\dfrac{b+c}{2a} \\ 0 & 1 & \dfrac{b-c}{2a} \\ 0 & 0 & 1 \end{bmatrix} = \begin{bmatrix} 1 & -1 & -\dfrac{b}{a} \\ 1 & 1 & -\dfrac{c}{a} \\ 0 & 0 & 1 \end{bmatrix}$$

We verify that

$$B'AB = \begin{bmatrix} 1 & 1 & 0 \\ -1 & 1 & 0 \\ -\dfrac{b}{a} & -\dfrac{c}{a} & 1 \end{bmatrix}\begin{bmatrix} 0 & a & c \\ a & 0 & b \\ c & b & 0 \end{bmatrix}\begin{bmatrix} 1 & -1 & -\dfrac{b}{a} \\ 1 & 1 & -\dfrac{c}{a} \\ 0 & 0 & 1 \end{bmatrix}$$

$$B'AB = \begin{bmatrix} a & a & c+b \\ a & -a & b-c \\ 0 & 0 & -\dfrac{2bc}{a} \end{bmatrix}\begin{bmatrix} 1 & -1 & -\dfrac{b}{a} \\ 1 & 1 & -\dfrac{c}{a} \\ 0 & 0 & 1 \end{bmatrix}$$

$$= \begin{bmatrix} 2a & 0 & 0 \\ 0 & -2a & 0 \\ 0 & 0 & -\dfrac{2bc}{a} \end{bmatrix} = H$$

Thus,

$$Q_1 = \mathbf{x}H\mathbf{x}' = 2ax_1^2 - 2ax_2^2 - \frac{2bc}{a}x_3^2 \qquad (1.51)$$

The form Q_1 is called the *canonical* form of Q. In its canonical form Q_1 is

$$Q_1 = \sum_{m=1}^{r} c_m x_m^2 \qquad (1.52)$$

If of the r terms c_m, p are positive, we say that p is the index of the given quadratic form Q. Given a nonsingular quadratic form, i.e., det $A \neq 0$. We say that the quadratic form is *positive* definite if $p = r = n$, i.e., it has the same rank, index, and order. It is said to be negative definite if $p = 0$. Then clearly $-Q$ is positive definite. More directly:

Definition 2. *A real quadratic form is said to be positive definite if it can be reduced to the form*

$$\| \mathbf{y} \|^2 = \sum_{k=1}^{n} y_k^2 > 0$$

it is negative definite if $-Q(\mathbf{x})$ *is positive definite.*

Theorem I. *The index of the canonical form of a quadratic form is an invariant, i.e., independent of the way in which it is obtained. This theorem is usually known as Sylvester's law of inertia.*

Reduction of a Quadratic Form to Canonical Form by Using Eigenvalues. This method uses property Q-8, Sect. 1.7. Since a real quadratic form can be written

$$Q = \mathbf{y}A\mathbf{y}'$$

where A is symmetric and real, we know according to Q 8 of Sect. 1.7 that there exists an orthogonal matrix P such that $P^{-1} AP$ is diagonal. We know also that for an orthogonal matrix, $P^{-1} = P'$, and $P'AP$ is diagonal, and its elements are the eigenvalues of A. Thus, in order to obtain the canonical form of Q we find the eigenvalues of A, and write

$$Q = \mathbf{y}A\mathbf{y}' = \mathbf{y}[P'AP]\mathbf{y}' = \mathbf{y} \begin{bmatrix} \lambda_1 & & & & \\ & \lambda_2 & & \bigcirc & \\ & & \cdot & & \\ & & & \cdot & \\ & \bigcirc & & & \cdot \\ & & & & \lambda_n \end{bmatrix} \mathbf{y}'$$

Example 4. Reduce the quadratic form of Ex. 2 to canonical form using the method of eigenvalues.

PROCEDURE
We have

$$Q = x_1^2 + 2x_2^2 - 3x_3^2 - 6x_1x_2 + 8x_1x_3 - 4x_2x_3$$

with

$$A = \begin{bmatrix} 1 & -3 & 4 \\ -3 & 2 & -2 \\ 4 & -2 & -3 \end{bmatrix}$$

To find the eigenvalues of A we write

$$\det [A - \lambda I] = \begin{bmatrix} (1 - \lambda) & -3 & 4 \\ -3 & (2 - \lambda) & -2 \\ 4 & -2 & (-3 - \lambda) \end{bmatrix} = \lambda^3 + 36\lambda + 33 = 0$$

which has the roots

$$\lambda_1 = 6.44 \qquad \lambda_2 = -0.94 \qquad \lambda_3 = -5.50$$

Thus a canonical form of Q is

$$Q_2(y) = 6.44y_1^2 - 0.94y_2^2 - 5.50y_3^2$$

It is observed that Q_2 is different from the Q_1 of Example 1, although both are canonical forms of the same quadratic form. The signs of the terms involved are however the same, as will be seen in Sylvester's law of inertia (Theorem 1 of this Section).

Example 5

A certain rigid body has a moment of inertia J_e with respect to a set of Cartesian axes $(Oxyz)$, given by the expression

$$J_e = 29\alpha_1^2 + 32\alpha_2^2 + 29\alpha_3^2 - 8\alpha_1\alpha_2 - 8\alpha_2\alpha_3 - 2\alpha_3\alpha_1$$

where the unit vector $\mathbf{e}$ defines the axis about which J_e is calculated, and $\mathbf{e}$ is given by

$$\mathbf{e} = \alpha_1\mathbf{i} + \alpha_2\mathbf{j} + \alpha_3\mathbf{k}$$

Find the principal axes and the principal moments of inertia for the body.

SOLUTION
The inertia matrix J is

$$J = \begin{bmatrix} 29 & -4 & -1 \\ -4 & 32 & -4 \\ -1 & -4 & 29 \end{bmatrix}$$

from which we find the characteristic equation

$$\det [J - \lambda I] = \begin{vmatrix} 29 - \lambda & -4 & -1 \\ -4 & 32 - \lambda & -4 \\ -1 & -4 & 29 - \lambda \end{vmatrix}$$

$$= 25920 - 2664\lambda + 90\lambda^2 - \lambda^3 = 0$$

The eigenvalues are the roots

$$\lambda_1 = 24 \qquad \lambda_2 = 36 \qquad \lambda_3 = 30$$

To find the eigenvectors, we use the conditions

$$J\mathbf{e}_1 = \lambda_1\mathbf{e}_1 \qquad J\mathbf{e}_2 = \lambda_2\mathbf{e}_2 \qquad J\mathbf{e}_3 = \lambda_3\mathbf{e}_3$$
$$\|\mathbf{e}_1\| = \|\mathbf{e}_2\| = \|\mathbf{e}_3\| = 1$$

For λ_1, we have

$$(J_{11} - \lambda_1)\alpha_1^1 + J_{12}\alpha_2^1 + J_{13}\alpha_3^1 = 0$$
$$J_{21}\alpha_1^1 + (J_{22} - \lambda_1)\alpha_2^1 + J_{23}\alpha_3^1 = 0$$
$$J_{31}\alpha_1^1 + J_{32}\alpha_2^1 + (J_{33} - \lambda_1)\alpha_3^1 = 0$$

plus six more corresponding to the eigenvalues λ_2 and λ_3. Calculation gives for the principal axes

$$\mathbf{e}_1 = \sqrt{\tfrac{1}{3}}\,(\mathbf{i} + \mathbf{j} + \mathbf{k})$$
$$\mathbf{e}_2 = \sqrt{\tfrac{1}{6}}\,(\mathbf{i} - 2\mathbf{j} + \mathbf{k})$$
$$\mathbf{e}_3 = \sqrt{\tfrac{1}{2}}\,(\mathbf{i} + 0 - \mathbf{k})$$

The inertial matrix in terms of these axes is

$$J = \begin{bmatrix} 24 & 0 & 0 \\ 0 & 36 & 0 \\ 0 & 0 & 30 \end{bmatrix}$$

Definition 3. *For the symmetric matrix A, the* principal minors *are the minors obtained by deleting the m-th row and the m-th column, i.e., by deleting a column and a row of the same number. We recall that a minor is always a determinant and not a matrix.*

Definition 4

The leading principal minors *of a matrix*

$$A = \begin{bmatrix} a_{11} & a_{12} & \cdots & a_{1n} \\ a_{21} & a_{22} & \cdots & a_{2n} \\ \cdot & \cdot & \cdots & \cdot \\ a_{n1} & a_{n2} & \cdots & a_{nn} \end{bmatrix}$$

are defined as follows

$$M_0 = 1, \qquad M_1 = a_{11}, \qquad M_2 = \begin{vmatrix} a_{11} & a_{12} \\ a_{21} & a_{22} \end{vmatrix},$$

$$M_3 = \begin{vmatrix} a_{11} & a_{12} & a_{13} \\ a_{21} & a_{22} & a_{23} \\ a_{31} & a_{32} & a_{33} \end{vmatrix} \cdots, \qquad M_n = \det[A]$$

Conditions for Positive Definite Quadratic Form

Theorem 2. *A necessary and sufficient condition for Q to be positive definite is that the leading principal minors M_1, M_2, ..., M_n are all positive.*

Conditions for Negative Definite Quadratic Form

Theorem 3. *A necessary and sufficient condition for Q to be negative definite is that the leading principal minors $M_k = \det [A_k]$, $k = 1, 2, \ldots, n$ have the sign $(-1)^k$*

Example 6. Test the following quadratic form for positive definiteness.

$$Q = x_1^2 + 5x_2^2 + 3x_3^2 + 4x_1x_2 - 2x_2x_3 - 2x_1x_3$$

$$A = \begin{vmatrix} 1 & 2 & -1 \\ 2 & 5 & -1 \\ -1 & -1 & 3 \end{vmatrix}$$

Then

$$M_1 = a_{11} = 1 > 0$$

$$M_2 = \begin{vmatrix} 1 & 2 \\ 2 & 5 \end{vmatrix} = 5 - 4 = 1 > 0$$

$$M_3 = \det A = 14 - 10 - 3 = 1 > 0$$

Hence, Q is positive definite.

Simultaneous Reduction of Two Quadratic Forms

Theorem 4. *If $\mathbf{x}A\mathbf{x}'$ and $\mathbf{x}B\mathbf{x}'$ are real quadratic forms, and if $\mathbf{x}B\mathbf{x}'$ is positive definite, there exists a real nonsingular matrix C such that $\mathbf{x}' = C\mathbf{y}'$, which transforms $\mathbf{x}A\mathbf{x}'$ into $\sum_{m=1}^{n} \lambda_m y_m^2$ and $\mathbf{x}B\mathbf{x}'$ into $\sum_{m=1}^{n} y_m^2$, where λ_m are the roots of the equation*

$$\det [\lambda B - A] = 0$$

1.10 DIFFERENTIATION AND INTEGRATION OF MATRICES

We study matrices whose elements are functions of t. Let

$$A(t) = \begin{bmatrix} a_{11}(t) & a_{12}(t) & \cdots & a_{1n}(t) \\ a_{21}(t) & a_{22}(t) & \cdots & a_{2n}(t) \\ \cdot & \cdot & \cdots & \cdot \\ a_{n1}(t) & a_{n2}(t) & \cdots & a_{nn}(t) \end{bmatrix} \tag{1.53}$$

be a matrix function of t. We define the derivative of A with respect to t to be

$$
\frac{dA(t)}{dt} = \lim_{\Delta t \to 0} \frac{1}{\Delta t} \left\{ \begin{bmatrix} a_{11}(t + \Delta t) & \cdots & a_{1n}(t + \Delta t) \\ \cdot & \cdot & \cdot \\ a_{n1}(t + \Delta t) & \cdots & a_{nn}(t + \Delta t) \end{bmatrix} \right.
$$

$$
\left. - \begin{bmatrix} a_{11}(t) & \cdots & a_{1n}(t) \\ \cdot & \cdot & \cdot \\ a_{n1}(t) & \cdots & a_{nn}(t) \end{bmatrix} \right\}
$$

$$
= \begin{bmatrix} \dfrac{da_{11}(t)}{dt} & \dfrac{da_{12}(t)}{dt} & \cdots & \dfrac{da_{1n}(t)}{dt} \\ \cdot & \cdot & \cdot \\ \dfrac{da_{n1}(t)}{dt} & \dfrac{da_{n2}(t)}{dt} & \cdots & \dfrac{da_{nn}(t)}{dt} \end{bmatrix} \tag{1.54}
$$

Thus, the derivative of a matrix is the matrix of the derivatives of its elements, provided that these exist.

Conversely, provided that the function $a_{ij}(t)$ can be integrated,

$$
\int_{\alpha}^{\beta} A(t) \, dt = \begin{bmatrix} \displaystyle\int_{\alpha}^{\beta} a_{11}(t) \, dt & \displaystyle\int_{\alpha}^{\beta} a_{12}(t) \, dt \, dt & \cdots & \displaystyle\int_{\alpha}^{\beta} a_{1n}(t) \, dt \\ \cdot & \cdot & \cdot \\ \displaystyle\int_{\alpha}^{\beta} a_{n1}(t) \, dt & \displaystyle\int_{\alpha}^{\beta} a_{n2}(t) \, dt & \cdots & \displaystyle\int_{\alpha}^{\beta} a_{nn}(t) \, dt \end{bmatrix}
$$

and if

$$
\frac{db_{ij}(t)}{dt} = a_{ij}(t)
$$

$$
\int A(t) \, dt = \begin{bmatrix} b_{11}(t) & b_{12}(t) & \cdots & b_{1n}(t) \\ b_{21}(t) & b_{22}(t) & \cdots & b_{2n}(t) \\ \cdot & \cdot & \cdot \\ b_{n1}(t) & b_{n2}(t) & \cdots & b_{nn}(t) \end{bmatrix} + C \tag{1.55}
$$

where C is a constant $n \times x$ matrix.

Example I

Find the differential and integral of the matrix

$$
A(t) = \begin{bmatrix} t & 2t^2 & +3t^3 \\ 1 & t & t + t^2 \\ \cos t & e^t & t \end{bmatrix}
$$

SOLUTION

$$\frac{dA}{dt} = \begin{bmatrix} 1 & 4t & +9t^2 \\ 0 & 1 & 1+2t \\ -\sin t & e^t & 1 \end{bmatrix}$$

and

$$\int A(t)\, dt = \begin{bmatrix} \dfrac{t^2}{2} & \dfrac{2}{3}t^3 & \dfrac{3}{4}t^4 \\ t & \dfrac{t^2}{2} & \dfrac{t^2}{2}+\dfrac{t^3}{3} \\ \sin t & e^t & \dfrac{t^2}{2} \end{bmatrix} + \begin{bmatrix} c_{11} & c_{12} & c_{13} \\ c_{21} & c_{22} & c_{23} \\ c_{31} & c_{32} & c_{33} \end{bmatrix}$$

where c_{ij} are arbitrary constants.

BIBLIOGRAPHY

1. A. C. Aitken, *Determinants and Matrices*. New York: Interscience Publishers, 1954.
2. F. Ayres, Jr., *Theory and Problems of Matrices*. New York: Schaum Publishing Co., 1962.
3. R. A. Beaumont and R. W. Ball, *Introduction to Modern Algebra and Matrix Theory*. New York: Holt, Rinehart and Winston, 1961.
4. G. Birkhoff and S. MacLane, *A Survey of Modern Algebra*. New York: Macmillan, 1960.
5. P. R. Halmos, *Finite Dimensional Vector Spaces*, Second Edition. New York: Van Nostrand, 1958.
6. F. B. Hildebrand, *Methods of Applied Mathematics*. Englewood Cliffs, N.J.: Prentice-Hall, 1958.
7. K. Hoffman and R. Kunze, *Linear Algebra*. Englewood Cliffs, N.J.: Prentice-Hall, 1961.
8. A. S. Householder, *The Theory of Matrices in Numerical Analysis*. New York: Blaisdell, 1964.
9. A. Lichnerowitz, *Algèbre et Analyse Linéaires*. Paris: Masson et Cie., 1947.
10. E. Madelung, *Die Mathematischen Hilfsmittel des Physikers*. New York: Dover, 1943.
11. N. H. McCoy, *Introduction to Modern Algebra*. Boston: Allyn and Bacon, 1960.
12. H. Weyl, *The Theory of Groups and Quantum Mechanics*. New York: Dover, 1931.
13. S. R. Searle, *Matrix Algebra for the Biological Sciences*. New York: Wiley, 1965.
14. Matrix and Tensor Quarterly, Published by The Tensor Society of Great Britain.
15. J. S. Frame, "Matrix Functions and Applications." Reprinted from *IEEE Spectrum*, March–July 1964 issues.
16. F. R. Gantmacher, *The Theory of Matrices*, 2 Volumes. New York: Chelsea, 1959. (Translated from the Russian. Comprehensive and contains an extensive bibliography.)
17. R. A. Fraser et al., *Elementary Matrices and Some Applications to Dynamics and Differential Equations*. Cambridge Press, 1938. (A pioneering work with many numerically solved illustrative problems.)
18. P. Horst, *Matrix Algebra for Social Scientists*. New York: Holt, Rinehart and Winston, 1963.
19. L. A. Pipes, *Matrix Methods for Engineering*. Englewood Cliffs, N.J.: Prentice-Hall, 1963.
20. R. K. Eisenschitz, *Matrix Algebra for Physicists*. New York: Plenum Press, 1966.

2

EXTREMA OF FUNCTIONS

2.1 EXTREMA PROBLEMS IN MATHEMATICS

Determination of the maxima and minima of functions was among the earliest problems studied in the calculus. Fermat (1629) devised a method applicable to such problems. The principles he used were in essence those of the modern calculus. Newton (1671) analytically conceived of the rate of change of a function and argued that zero rate of change corresponds to a maximum or a minimum. Leibniz (1684) thought of the problem geometrically and deduced the result that the tangent to a curve must be horizontal at a maximum or a minimum. He also gave the condition that the second derivative $f''(x)$ must be positive for a minimum, and negative for a maximum. Maclaurin (1698–1746) showed how a study of the higher derivatives is needed to distinguish between a maximum and a minimum when both $f'(x) = 0$ and $f''(x) = 0$.

Although the method of finding the maxima and minima of a function by setting its first derivative equal to zero is treated in elementary calculus, most of the significant and interesting problems in science and engineering require a more sophisticated approach. For example, even when dealing with functions of only one variable, the absolute maximum or absolute minimum may occur at a boundary point or at a singular point of the function. Since the absolute maximum or absolute minimum frequently is a consideration of prime interest in an engineering situation, methods for determination of these are essential. Further, many problems encountered in applications involve functions of many variables, and appropriate methods are necessary for dealing with these. Another common situation occurs when constraints or limiting conditions are

59

imposed on the functions under study. Accordingly, these problems, which require a more sophisticated analysis than is usually given in elementary calculus, are treated in this chapter.

The problem of finding the extrema of definite integrals arose almost simultaneously with the extrema problems of differential calculus. The fundamental problem in the case of definite integrals is to find the function which, integrated over some given domain, provides an extremum of the integral. Newton in his *Philosophiae Naturalis Principia Mathematica* (1686) stated without proof certain conditions for a solid of revolution to offer minimum resistance when moving through a fluid.

However, the real beginning of the *Calculus of Variations*, as this branch of mathematics is called, is due to Johann Bernoulli (1667–1748), who in 1696 proposed the celebrated problem of the brachistochrone (Greek brachisto = shortest, chronos = time). In essence, the brachistochrone problem is to find the path between two points *a* and *b* at different elevations down which a particle falls under gravity in the shortest time (see Fig. 3.2). Bernoulli's paper on the solution of this problem of finding the curve of quickest descent showed that the path is identical with the path of a light ray in an optical medium of suitable variable index of refraction. Thus an interesting analogy between particle mechanics and optics was established as early as 1696.

The earliest problems tackled by the eighteenth century mathematicians in the calculus of variations were by no means the simplest. Undoubtedly the two simplest problems of the calculus of variations are those of determining the shortest arc between two points on a given surface and of finding the surface of revolution of minimum area between two parallel, co-axial circles (the soap film problem).

Many problems of physics and engineering may be treated by the calculus of variations. Some of the most general and powerful principles of theoretical physics are most effectively formulated in terms of variational principles. Among these are Hamilton's principle, the principle of least action in mechanics, and Fermat's principle of least time in optics. Application of the calculus of variations to engineering problems is fairly recent, but since about 1950 has been widely exploited. The power, generality, and effectiveness of variational methods in the solution of engineering problems are certain to lead to ever increasing applications in today's complex technology. The calculus of variations is treated separately in Chap. 3.

In addition to the classic extrema problems, there are several basic extrema problems of more recent origin. Chief among these are (1) problems of maximizing or minimizing probability functions such as arise in queueing theory, (2) problems of minimizing risks when the strategy of the opponents is a major factor, or when decisions must be made on the basis of incomplete knowledge, and (3) problems of optimizing objectives, such as costs, profits, or performance, when the choice of methods or means is restricted. Problems of the third type commonly arise in engineering, physical science, and economics.

Problems in the third category generally come under the heading of linear programming which originated with the work of Dantzig and his associates beginning in 1948. In these methods, the mathematical equations, which represent the objective to be optimized and which usually are nonlinear, are approximated by linear equations. Similarly, the restrictions are approximated by linear inequalities. The resulting simplified mathematical problems may then be solved by linear methods and lead to useful results which have been widely applied to industrial and military problems. The parallel development of fast electronic computers made practicable the treatment of large-scale linear programming problems involving hundreds of variables. The underlying mathematical theory of linear programming is in essence the mathematics of linear inequalities.

2.2 FUNCTIONS OF ONE VARIABLE

The fundamental idea in the determination of extrema of functions rests on the simple concept of comparison of the values assumed by a function at points for which it is defined in a given interval. Application of this concept, especially in problems of a practical or physical nature, involves some rather subtle points and precise definitions having to do with the broad meaning of the terms function, limits and continuity, intervals or domains, types of discontinuities, boundary points, and other considerations of function theory and analysis. For this reason, it is desirable to review some of the concepts, definitions, and results of analysis which are needed throughout the book.

Recall of Analysis. The most general definition of a function is that of the correspondence between elements of two sets. Restrictions, such as continuity or differentiability, may be imposed on functions, but the general form is required for the broad treatment of extrema. Finite sets, which are useful in systems analysis, are introduced first, and then infinite sets are treated.

Let us consider the simplest case of a set (S) whose elements form an increasing sequence of numbers x_i, $i = 1, 2, \ldots, m$. If to every $x \in (S)$ corresponds a $y \in (T)$, we say that y is a *function* of x and write $y = f(x)$. We shall limit ourselves to the case where the correspondence is such that to a given x corresponds only one y, and call this a *single-valued function*. If, in addition to the preceding correspondence, it is possible to establish a correspondence of x values to given y values, then we say that x is a function of y and write $x = g(y)$. The two functions f and g are said to be inverse to each other.

In the majority of cases we shall assume that x takes all values in a given *interval* (X) defined by one of the following possibilities:

(1) $\qquad x_0 < x < x_1$, open interval, denoted as (x_0, x_1)

(2) $\qquad x_0 \leq x \leq x_1$, closed interval, denoted as $[x_0, x_1]$

(3) $\qquad x_0 \leq x < x_1$, semiclosed interval, denoted as $[x_0, x_1)$

(4) $\qquad x_0 < x \leq x_1$, semiclosed interval, denoted as $(x_0, x_1]$

Correspondingly, y will take values over the interval defined by the two numbers y_2 and y_3, defining the interval (J) which may be *open*, *closed*, or *semiclosed*. If then $y = f(x)$, the interval (J) is said to represent the *range* of function f.

As said earlier we shall be concerned primarily with single-valued functions, i.e., functions that for a given value of the variable x have a single value $f(x)$. In case $f(x)$ is multivalued each set of values is treated separately as a single-valued *branch* of a multi-valued function.

Definitions. *We state here some of the basic definitions which are required later.*

(1) The function $f(x)$ is said to tend to the limit L as $x \to x_0$, if, given a positive number h, however small, there exists a positive number k such that

$$|f(x) - L| < h \quad \text{for} \quad |x - x_0| < k$$

and we write

$$\lim_{x \to x_0} f(x) = L$$

It may happen that the limit is not the same when we approach x_0 from the right as from the left. In this case

(a) If $x \to x_0$ from the right, i.e., from the positive side of the real axis, we write

$$\lim_{x \to x_0^+} f(x) = L_1$$

(b) Similarly, if $x \to x_0$ from the left, i.e., from the negative side of the real axis, we write

$$\lim_{x \to x_0^-} f(x) = L_2$$

(2) The function $f(x)$ is said to be continuous at $x = x_0$, if L_1 and L_2 both exist (are finite) and if $L_1 = L_2 = f(x_0)$.

(3) The function $f(x)$ is said to be continuous over the open interval (a, b) if it is continuous for every x such that $a < x < b$.

(4) The function $f(x)$ is said to be continuous over the closed interval $[a, b]$ if it is continuous for every x such that $a \leq x \leq b$.

(5) The function $f(x)$ is said to have a discontinuity of the first kind at $x = x_0$ in any of the following cases:

    (a) L_1 does not exist, e.g., if $x \to x_0^+$, $f(x) \to \pm\infty$; L_2 is finite.
    (b) L_2 does not exist, e.g., if $x \to x_0^-$, $f(x) \to \pm\infty$; L_1 is finite.
    (c) Neither L_1 nor L_2 exists.

(6) The function $f(x)$ is said to have a discontinuity of the second kind at $x = x_0$ if L_1 and L_2 both exist but $L_1 \neq L_2$. We say in this case that the function has a finite jump $j = |L_1 - L_2|$ at $x = x_0$.

(7) The function $f(x)$ is said to be piecewise continuous over the closed interval $[a, b]$ if there exist a finite number of values $x_0 = a$, x_1, $x_2, \ldots, x_{n-1}$, $x_n = b$, such that over each of the open intervals (x_k, x_{k+1}), $k = 0, 1, 2, \ldots, n - 1$, $f(x)$ is continuous, while at $x_0, x_1, \ldots, x_n$ it has discontinuities of the second kind (see Fig. 2.1). The reader will verify the following properties:

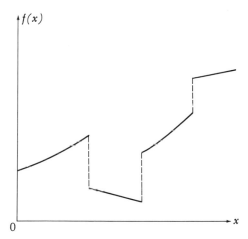

FIGURE 2.1   Piecewise continuous function.

    (a) If $f(x)$ is piecewise continuous it has no derivative at the points $x_0, x_1, \ldots, x_n$. However, it may have left-hand and right-hand derivatives at these points.

    (b) If $f(x)$ is piecewise continuous and $g(x)$ is continuous, then

$$\int_a^x f(t)g(t)\, dt = F(x) \qquad a \leq x \leq b$$

is a continuous function, over the closed interval $[a, b]$.

We give now several illustrative examples of these ideas, starting first with a polynomial, i.e., the simplest kind of function. However, instead of using a chain of theorems as is usually done, we treat the examples by direct methods.

Example 1. Study the continuity of the function $f(x) = 3x^2 + 2x + 1$. For some given value, $x = x_0$, we have $f(x_0) = 3x_0^2 + 2x_0 + 1 = L$. We shall prove that $\lim_{x \to x_0} f(x) = f(x_0)$. Let h be a positive number, then the condition

$$|f(x) - L| < h$$

is equivalent to

$$|3x^2 + 2x + 1 - 3x_0^2 - 2x_0 - 1| = |x - x_0| \, |3(x + x_0) + 2| < h$$

or alternatively to

$$|x - x_0| < \frac{h}{|3(x + x_0) + 2|}$$

Since the question of limit involves the neighborhood of $x = x_0$, we may assume that $x_0 - 1 < x < x_0 + 1$, hence, $2x_0 - 1 < x + x_0 < 2x_0 + 1$. But since $|2x_0 - 1| \leq 2 |x_0| + 1$, and $|2x_0 + 1| \leq 2 |x_0| + 1$, it follows that $|x + x_0| \leq 2 |x_0| + 1$, hence, that

$$|3(x + x_0) + 2| \leq 3 |x + x_0| + 2 < 3(2 |x_0| + 1) + 2 = 6 |x_0| + 5$$

Substituting we obtain

$$|x - x_0| < \frac{h}{6 |x_0| + 5} < \frac{h}{|3(x + x_0) + 2|}$$

By taking $k = \dfrac{h}{(6 |x_0| + 5)}$, we see that for $|x - x_0| < k$, $|f(x) - L| < h$. Since the sign of $x - x_0$ does not intervene, we can approach x_0 from the left or from the right. It follows that at $x = x_0$ the function $f(x)$ is continuous. Throughout this calculation we had no restriction on x_0. It follows that the function $f(x)$ is continuous for all values of x.

Example 2. Study the continuity of the function $f(x) = \dfrac{1}{x}$. At $x = x_0$, $f(x_0) = \dfrac{1}{x_0} = L$. $|f(x) - L| = \left| \dfrac{1}{x} - \dfrac{1}{x_0} \right| = \dfrac{|x - x_0|}{|x| \, |x_0|}$. Let us assume $x_0 \neq 0$. The condition $|f(x) - L| < h$, gives $|x - x_0| < h |x| \, |x_0|$. Following the procedure of Ex. 1, assume that

$$\frac{|x_0|}{2} < |x| < \frac{3 |x_0|}{2}$$

since we study the neighborhood of x_0. Multiplying by $h |x_0|$ gives

$$\frac{h |x_0|^2}{2} < h |x| \, |x_0| < \frac{3h |x_0|^2}{2}$$

Since we want

$$|x - x_0| < h |x| \, |x_0|$$

it is sufficient to assume that

$$|x - x_0| < \frac{h\,|x_0|^2}{2} < h\,|x|\,|x_0|$$

i.e., to take

$$k = \frac{h\,|x_0|^2}{2}$$

to insure

$$|f(x) - L| < h$$

provided that

$$x \neq 0, \qquad x_0 \neq 0$$

It therefore follows that $f(x)$ is continuous for all x except $x = 0$. At $x = 0$ there is a discontinuity of the first kind.

Example 3. Study the function $f(x) = e^{1/x}$. The reader will verify that the function is continuous for every $x_0 \neq 0$. However,

$$\lim_{x \to 0^+} e^{1/x} \text{ does not exist, for as } x \to 0^+, \, e^{1/x} \to \infty$$

$$\lim_{x \to 0^-} e^{1/x} = 0 \text{ for as } x \to 0^-, \, e^{1/x} \to 0$$

It follows that at $x = 0$, $e^{1/x}$ has a discontinuity of the first kind.

Example 4

The Heaviside step-function (cf. Fig. 2.2) is defined as

$$u(x - a) = \begin{cases} 0 \text{ for } x < a \\ 1 \text{ for } a < x \end{cases}$$

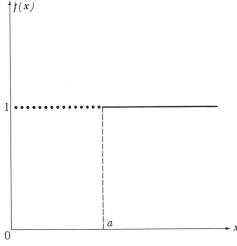

FIGURE 2.2 Heaviside step function, $u(x - a)$.

The reader will verify that the function is continuous everywhere except at $x = a$, where it has a discontinuity of the second kind. The jump is $j = 1$.

Example 5

Study the function

$$f(x) = \begin{cases} 0 \text{ for } x \text{ rational} \\ 1 \text{ for } x \text{ irrational} \end{cases}$$

It is intuitively clear that it is impossible to represent this function by a graph. The function is discontinuous for every value of x. To show this, it would be necessary to study the continuum of numbers. We will not undertake this study, but observe that by using the so-called Lebesgue integral it is possible to prove that

$$\int_0^1 f(x) \, dx = 1$$

Extrema of Discrete Functions. We consider first the case of a function defined over a set of values $x = x_1, x_2, \ldots, x_n$ in increasing order and such that $f(x_j) = y_j, j = 1, 2, \ldots, n$. Among these discrete values there may exist one value larger than all the others and one value smaller than all the others which we call respectively the maximum and the minimum values. In case there is only *one maximum* value and *one minimum* value, we write symbolically

$$y_{\text{Max}} = \text{Max}\,(y_1, y_2, \ldots, y_n) = f(x_k)$$
$$y_{\text{min}} = \min\,(y_1, y_2, \ldots, y_n) = f(x_h)$$

There may be, however, repeated maximum or repeated minimum values, and in the extreme case we may even have $y_1 = y_2 = \cdots = y_n = y_{\text{Max}} = y_{\text{min}}$, in which case the function is constant over the set of values $x_1, x_2, \ldots, x_n$.

It is clear that determination of the extrema of discrete functions is essentially a direct enumeration and comparison process which involves only elementary steps. The method may be extended to continuous functions by taking points at finite intervals and using iteration procedures to isolate and evaluate the actual extrema. This approach may be classified as a direct method and has the advantage of being adaptable to digital computer techniques as discussed in Sect. 2.5.

Extrema of Continuous Functions. In contrast to direct numerical methods, the classical approach is to explore the values of a function in an infinitesimal neighborhood and thereby to find necessary and sufficient conditions for the existence of extrema. From these conditions the locations and exact values of the extrema may be found.

Definitions (Continued). *To proceed with the analytical method it is first necessary to state additional basic definitions which pertain to extrema. These are covered in Definitions 8 through 15 which follow:*

(8) The function $f(x)$ is said to have a relative or local maximum at $x = x_0$ if there exists a positive number k such that

$$f(x) - f(x_0) \leq 0 \quad \text{for} \quad x_0 - k < x < x_0 + k$$

provided that $f(x_0)$ exists and is finite.

(9) The function $f(x)$ is said to have a relative or local minimum at $x = x_0$ if there exists a positive number k such that

$$f(x) - f(x_0) \geq 0 \quad \text{for} \quad x_0 - k < x < x_0 + k$$

provided that $f(x_0)$ exists and is finite.

(10) Maxima and minima are called extrema of functions. It will be observed that, according to our definitions, extrema can occur at points of discontinuity of the second kind.

(11) If $f(x)$ is continuous over the closed interval $[a, b]$, at $x = x_0$, $a < x_0 < b$, the equation of the tangent to the curve $y = f(x)$ is given by

$$g(x) = f(x_0) + (x - x_0)f'(x_0)$$

We say that $y = f(x)$ has an inflection point at $x = x_0$, if $f(x) - g(x)$ changes sign at $x = x_0$. The inflection point is horizontal if $f'(x_0) = 0$.

(12) A function $f(x)$ is said to have a stationary value at $x = x_0$, if $f'(x_0) = 0$ (see Fig. 2.3).

(13) A function $f(x)$ is said to belong to the class C^n of functions over the open interval (a, b) if the function as well as all its derivatives up to and including order n are continuous over (a, b). It is noted that a function

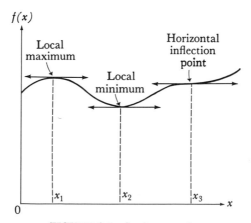

FIGURE 2.3 Stationary points.

of class C^n is, according to this definition, also of class C^{n-1}, where $n - 1 \geq 0$.

(14) A regular point of a function $f(x)$ is a value $x = x_0$ for which $f(x)$ and all its derivatives exist.

(15) All points of a function that are not regular are called singular.

(16) A function $f(x)$ is said to have an angular point at $x = x_0$, if for $x = x_0$, $f'(x)$ has a discontinuity of the second kind (cf. Def. 6) or of the first kind where either L_1 or L_2 exist (cf. Def. 5).

It is of interest to observe that the elementary transcendental functions sin x, cos x, sinh x, cosh x, e^x, as well as polynomials in x of any degree are continuous over the interval $(-\infty, +\infty)$ as well as all their derivatives. By way of contrast, the functions $f(x) = x^{-m}$, where m is a positive integer, are C^∞ over the intervals $(-\infty, 0)$ and $(0, +\infty)$, but not over the interval $(-\infty, +\infty)$, since these functions as well as their derivatives are discontinuous at $x = 0$.

For the basic theorems on limits and continuity of compound functions, that is, sums, products, quotients, and functions of functions, we refer the reader to any standard work on calculus, several of which are listed in the bibliography.

Extrema of Functions of Class C^n. Most of the problems of engineering and physics with which we are concerned may be formulated in terms of functions of class C^n, $n \geq 2$. In a number of cases it is necessary to consider functions that are either piecewise continuous (Def. 7, p. 63) or piecewise smooth, that is, functions that are continuous but whose first derivatives are piecewise continuous. The distinction between these cases can be eliminated by considering a finite interval not including points of discontinuity of the functions or of its derivative. For the study of extrema we do not need to consider any more than a finite interval, an interval sufficiently small to include only one extremum or stationary point. We write symbolically

$$f(x) \in C^n, \text{ for, } x \in (a, b)$$

Let us observe that physicists sometimes call functions of class C^∞ virtuous functions.

We treat the extrema of functions of class C^n with sufficient generality to include the bulk of problems of physical interest. We do not attempt complete generality because of the large number of special cases of limited interest which may be devised; however, some of these special cases are illustrated in the exercises to round out the treatment.

Let $f(x) \in C^n$ over the open interval (a, b). According to classical analysis $f(x)$ can be expanded in a Taylor's series with a remainder term as follows

$$f(x_0 + h) - f(x_0) = hf'(x_0) + \frac{h^2}{2!} f''(x_0)$$

$$+ \cdots + \frac{h^{n-1}}{(n-1)!} f^{(n-1)}(x_0) + R_{n-1}(x_0) \quad (2.1)$$

where

$$R_{n-1}(x_0) = \frac{h^n}{n!} f^{(n)}(x_0 + \theta h), \qquad 0 < \theta < 1$$

and h is sufficiently small so that $x_0 - h$ and $x_0 + h$ are also in the interval (a, b).

For $n = 1$, Eq. (2.1) reduces to the mean-value theorem,

$$f(x_0 + h) - f(x_0) = hf'(x_0 + \theta h) \qquad (2.2)$$

for $n = 2$, $\quad f(x_0 + h) = f(x_0) = hf'(x_0) + \frac{h^2}{2!} f''(x_0 + \theta h) \qquad (2.3)$

for $n = 3$, $\quad f(x_0 + h) - f(x_0) = hf'(x_0) + \frac{h^2}{2!} f''(x_0) + \frac{h^3}{3!} f''(x_0 + \theta h)$

and so on. $\qquad (2.4)$

The finite Taylor's series may be applied to the problem of determining necessary and sufficient conditions for extrema of functions of class C^n.

Theorem I. *A necessary condition for a local extremum of $f(x) \in C^n$, $n > 0$, at $x = x_0$, where $a < x_0 < b$, is that $f'(x_0) = 0$.*

Proof: Let us prove the theorem for a maximum at $x = x_0$. From Definition 8 and Eq. (2.2) we have the requirement

$$f(x_0 + h) - f(x_0) = hf'(x_0 + \theta h) \leq 0$$

To satisfy this condition it is necessary that $f'(x_0 + \theta h)$ and h have opposite signs. Since h may take both positive and negative values and since $f'(x_0 + \theta h)$ is continuous it follows that $f'(x_0) = 0$. A similar argument applies for a minimum at $x = x_0$.

We shall now show that $f'(x_0) = 0$ is not a sufficient condition for an extremum at $x = x_0$. Let us assume that $f'(x_0) = 0$ and that $f(x) \in C^2$. Then using (2.3) we may write

$$f(x_0 + h) - f(x_0) = \frac{h^2}{2!} f''(x_0 + \theta h)$$

If $f''(x_0 + \theta h)$ maintains a constant sign as h varies within its permissible limits, we clearly have an extremum. Since $f''(x)$ is continuous, it is always possible to choose an interval $x_0 - h < x < x_0 + h$ for which $f''(x_0 + \theta h)$ has a constant sign if $f''(x_0) \neq 0$. If on the contrary, $f''(x)$

changes sign at $x = x_0$ then $f''(x_0) = 0$ and there is no extremum at $x = x_0$. We can therefore state

Theorem 2. *A set of sufficient conditions for an extremum of* $f(x) \in C^n$, $n \geq 2$ *at* $x = x_0$ *is that* $f'(x_0) = 0$, *and* $f''(x_0) \neq 0$. *In particular if* $f''(x_0) < 0$ *there will be a maximum, while if* $f''(x_0) > 0$, *there will be a minimum.*

Let us now study the case $f'(x_0) = f''(x_0) = 0$. We assume that $f(x) \in C^n$, $n \geq 3$, and use (2.4) which reduces to

$$f(x_0 + h) - f(x_0) = \frac{h^3}{3!} \, f'''(x_0 + \theta h)$$

If $f'''(x)$ maintains a constant sign for $x - k < x < x + k$, i.e., as $-k < h < k$, then, since h^3 has the same sign as h, $f(x_0 + h) - f(x_0)$ will change sign with h, i.e., at $x = x_0$, there will be a sign change for $f(x) - f(x_0)$. But $y = f(x_0)$ is the equation of the tangent to the curve $y = f(x)$; it follows from Definition 11 that $y = f(x)$ has in this case an inflection point at $x = x_0$. By a discussion similar to the one leading to Theorem 2, it will be seen that if $f'''(x_0) \neq 0$ then for small enough k, $f'''(x)$ maintains a constant sign for $x_0 - k < x < x_0 + k$. We can state therefore

Theorem 3. *If the function* $f(x) \in C^n$, $n \geq 3$, *is such that* $f'(x_0) = f''(x_0) = 0$, *while* $f'''(x_0) \neq 0$, *the curve* $y = f(x)$ *has a horizontal inflection point at* $x = x_0$.

Considering now the case where $f'''(x_0) = 0$, we are in a situation similar to the one that led to Theorem 2. We can thus generalize the preceding results, leaving it to the reader to work out the details of the proofs.

Theorem 4. *Let* $f(x) \in C^n$, *over the interval* (a, b) *and* $a < x_0 < b$. *Then*
(i) *If* n *is even and if* $f'(x_0) = f''(x_0) = \cdots = f^{(n-1)}(x_0) = 0$, *while* $f^{(n)}(x_0) \neq 0$, *then* $f(x)$ *has an extremum at* $x = x_0$. *The extremum will be a maximum if* $f^{(n)}(x_0) < 0$ *and a minimum if* $f^{(n)}(x_0) > 0$.
(ii) *If* n *is odd and if* $f'(x_0) = f''(x_0) = \cdots = f^{(n-1)}(x_0) = 0$ *while* $f^{(n)}(x_0) \neq 0$ *then* $f(x)$ *has an inflection point at* $x = x_0$.

For geometrical reasons it is usual to say that under conditions (i) or (ii) the curve $y = f(x)$ has a contact of order n with its tangent and either an extremum or an inflection point of order n for $x = x_0$. In all cases, $f(x)$ has a stationary value at $x = x_0$.

Singularities, Absolute Extrema, Boundary Values. In the preceding section we have assumed that $f(x) \in C^n$ for $a < x < b$. Let us now assume that the function is to be studied over a closed interval $[c, d]$ which includes a and b such that $c < a < b < d$, and that at several points outside (a, b) the function has discontinuous derivatives of order $m \leq n$. We have seen

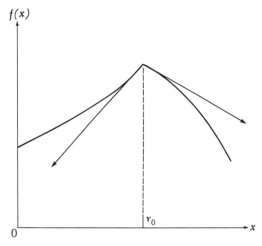

FIGURE 2.4 Angular point.

that such points are called singular points. We shall give first a few examples of singular points:

(i) *Angular points:* $f(x) \in C^0$ at $x = x_0$, but $f'(x)$ has a discontinuity of the second kind. It follows that we have a configuration as in Figure 2.4.

(ii) *Cusp point:* $f(x)$ is continuous, reaches an extremum at $x = x_0$, but $f'(x_0)$ does not exist (Fig. 2.5).

(iii) *Vertical inflection point:* $f(x)$ is continuous at $x = x_0$, but $f'(x_0)$ does not exist (Fig. 2.6).

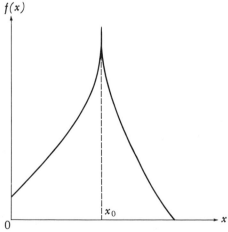

FIGURE 2.5 Cusp point.

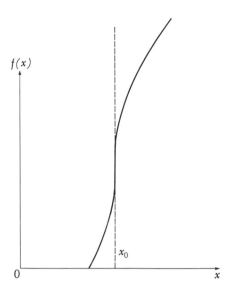

FIGURE 2.6   Vertical inflection point.

The detailed study of singular points requires a deeper analysis of neighborhoods, which is usually found in courses of differential geometry. As will be seen in the examples given, it is usually possible to study singular points without very powerful means.

In addition to singularities it is necessary to include the boundary values $f(c)$ and $f(d)$ of the function $f(x)$ which are frequently introduced by physical considerations.*

We now consider (1) the values of $f(x)$ at all local extrema, (2) the values of $f(x)$ at all singularities, and (3) the boundary values. The largest value obtained is called the absolute maximum and the smallest value the absolute minimum of the function $f(x)$ over the closed interval $[c, d]$ (cf. Fig. 2.7).

Let us terminate by stating the following theorem that can be found in Courant's *Differential and Integral Calculus*.

Theorem 5.   *Every function which is continuous in a closed interval* $a \le x \le b$ *assumes a greatest value at least once and a least value at least once, or, as we say, it possesses a greatest and a least value. This theorem can be generalized to the case of piecewise continuous functions, thus including the case of a finite number of finite discontinuities.*

---

* It is clear from the present discussion that there are two kinds of extrema—those due to the form of the function itself which we shall call *intrinsic* extrema, and those due to the boundaries which we shall call *boundary* extrema. This distinction is useful in certain applications.

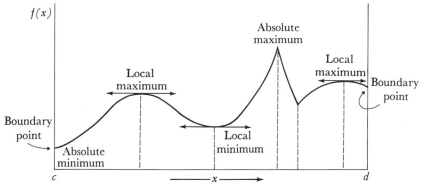

FIGURE 2.7 Extremum points.

Example 6

Find the stationary and extremum values of the function

$$f(x) = 3x^5 - 5x^3 + 3$$

ANALYTIC SOLUTION WITHOUT GRAPHICAL REPRESENTATION

The derivatives are

$$f'(x) = 15x^4 - 15x^2 = 15x^2(x - 1)(x + 1)$$
$$f''(x) = 60x^3 - 30x = 30x(2x^2 - 1)$$

Stationary points occur at $x = 0, 1, \quad 1$

$$f''(0) = 0$$
$f'''(0) = -30$, hence $x = 0$ is a horizontal inflection point
$f''(1) = 30 > 0$, hence $x = 1$ is a minimum
$f''(-1) = -30 < 0$, hence $x = -1$ is a maximum

Example 7

Examine the function $f(x) = x^5$ for stationary and extremum points.

SOLUTION. Again, we solve analytically without a graph. The first derivative is

$$f'(x) = 5x^4, \text{ hence } x = 0 \text{ is the only stationary point.}$$
$$f''(x) = 20x^3, \text{ hence } f''(0) = 0 \text{ and it is necessary}$$
$$\text{to look at higher derivatives.}$$

$$f'''(x) = 60x^2, f^{(4)}(x) = 120x, f^{(5)}(x) = 120$$

Hence, $f^{(4)}(0) = 0$, but $f^{(5)}(0) = 120$. Thus, $n = 5$ (odd) so that $x = 0$ is a horizontal inflection point of $f(x)$.

Example 8

Study the variation of the function $y = (x^3 - 3x + 2)^{1/3}$. We first observe that y is a continuous function for all values of x. It can be written

$$y = (x - 1)^{2/3}(x + 2)^{1/3}$$

which shows that it has two zeros: $x = 1$, which is a double zero, and $x = -2$. The derivative exists and is

$$y' = (x^2 - 1)(x^3 - 3x + 2)^{-2/3} = (x + 1)(x - 1)^{-1/3}(x + 2)^{-2/3}$$

We see therefore that $y' = 0$ for $x = -1$, while for $x = -2$ and $x = 1$, the derivative becomes infinite. The point $x = -1, y = 2^{2/3}$ is therefore a simple extremum. For $x = -2$, y' is infinite, and $y = 0$. Therefore, for this value of x we have a vertical inflection point. For $x = 1$, y' is infinite, $y = 0$, but since y contains the factor $(x - 1)^{2/3}$, which does not change sign for $x = 1$ it follows that for $x = 1$ we have a cusp point and therefore an extremum. By writing

$$y = x(1 - 3x^{-2} + 2x^{-3})^{1/3}$$

we see that when $x \to \infty$, y behaves like $y = x$. It follows that the line $y = x$ is an oblique asymptote to the curve. The function y is of class C^0 but not of class C^1 over the interval $(-\infty, +\infty)$ since y' is not continuous over this interval.

The following table of variation of $y = f(x)$ shows the intervals where the function is increasing and decreasing, as well as the important special values.

x	$-\infty$		-2		-1		$+1$		$+\infty$
y'		$+$	$+\infty$	$+$	0	$-$	$-\infty \| +\infty$	$+$	
y	$-\infty$	$\nearrow$	-2	$\nearrow$	$2^{2/3}$	$\searrow$	0	$\nearrow$	$+\infty$

This leads to the graph, Figure 2.8.

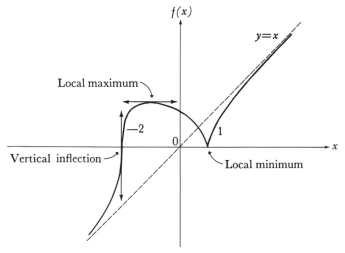

FIGURE 2.8 Graph of $= (x^3 - 3x + 2)^{1/3}$.

Example 9

Study the variation of the function $y = x^{1/x}$. The function is continuous for all positive values of x. For x negative it is defined only for x being a rational number of the form $\dfrac{p}{q}$, where both p and q are integers and p is odd; we limit therefore the study to the case $x > 0$. The function has a derivative given by

$$y' = x^{(1-2x)/x}(1 - \ln x)$$

The derivative has the same sign as $1 - \ln x$, which in turn has the same sign as $e - x$. For $x = 0$ we have an indeterminate form.

By taking the logarithm we have $\ln y = \dfrac{\ln x}{x}$ which for $x = 0$ yields $-\infty$, hence, as $x \to 0$, $x^{1/x} \to 0$. Similarly, it is seen that as $x \to +\infty$, $y \to 1$. We can now give the table of variation of y and obtain its graph, Figure 2.9.

x	0^+		e		$+\infty$
y'		$+$	0	$-$	
y	0	$\nearrow$	$e^{1/e}$	$\searrow$	1

It will be observed that $x = e$, $y = e^{1/e}$, is an absolute maximum for the function.

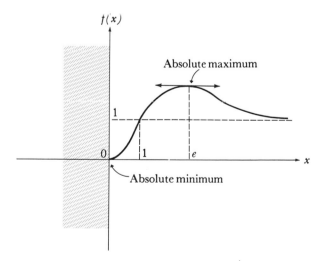

FIGURE 2.9 Graph of $y = x^{1/x}$.

Physical Boundaries. We have so far considered functions of the form $f(x)$ from a purely mathematical, i.e., analytical or geometrical point of view. We now introduce an additional element by assuming that the relationship, $y = f(x)$, represents a physical phenomenon. This often limits the set of values for x and y to bounded domains. Thus if for physical reasons $x_0 \leq x \leq x_1$, then $y_0 = f(x_0)$ and $y_1 = f(x_1)$ are boundary points of $y = f(x)$. They may be local extrema or absolute extrema according to the case. We illustrate by a few examples.

Example 10. A densimeter or hydrometer (cf. Fig. 2.10a) comprises a bulb B and a cylindrical stem C. Let M be the mass of the whole instrument, V_0 the volume of the bulb, and S the cross-section of the stem. Let x be the length of the stem that is submerged when the instrument floats in a liquid of density ρ and L the total length of the stem. Expressing the equilibrium of the floating densimeter we have

$$Mg = \rho(V_0 + xS)$$

Solving for x, gives

$$x = \frac{(Mg - V_0\rho)}{\rho S} = \frac{Mg}{\rho S} - \frac{V_0}{S}$$

The variation of ρ as a function of x is shown in Figure 2.10b. Considering the physical problem it is clear that $0 \leq x \leq L$, so that for the given

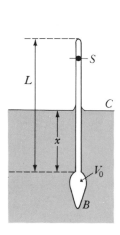

FIGURE 2.10a

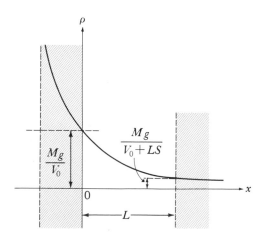

FIGURE 2.10b

densimeter:

$$\rho_{min} = \frac{Mg}{(V_0 + LS)}, \text{ limited by } x = L$$

$$\rho_{max} = \frac{Mg}{V_0}, \text{ limited by } x = 0$$

Example 11.   Given an electric generator of EMF $= V$ and internal resistance R connected to a load resistance R_L as depicted in Figure 2.11, find (a) the value of R_L which gives maximum power to the load, (b) the values of R_L for minimum and maximum current output, and (c) the value of R_L for maximum efficiency.

SOLUTION

(a) By definition, power output $P = I^2 R_L = \dfrac{V^2 R_L}{(R + R_L)^2}$

A necessary condition for an extremum is

$$P'(R_L) = 0$$

or

$$\frac{d}{dR_L}\left[\frac{V^2 R_L}{(R + R_L)^2}\right] = \left[\frac{V^2}{(R + R_L)^3}\right](R - R_L) = 0$$

from which $R_L = R$. To determine whether this condition yields a maximum or a minimum, apply the second test

$$P''(R_L) = \left[\frac{2V^2}{(R + R_L)^4}\right](R_L - 2R) < 0 \quad \text{for} \quad R = R_L$$

Hence, the power is a maximum,

(b) $$I = \text{current output} = \frac{V}{(R + R_L)}$$

$$I'(R_L) = -\frac{V}{(R + R_L)^2}$$

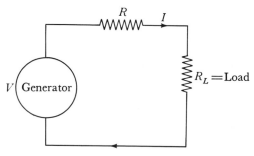

FIGURE 2.11   Generator with resistance load.

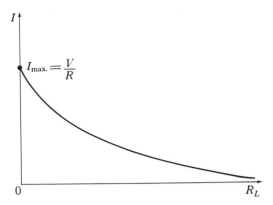

FIGURE 2.12 Current to load.

Since $I'(R_L)$ is not zero for any finite value of R_L, no stationary point exists, and therefore no local maximum or minimum. However, an absolute maximum of I exists corresponding to $R_L = 0$. There is no absolute minimum, although a lower limit exists.

Absolute maximum $I = \dfrac{V}{R}$, $\displaystyle\lim_{R_L \to \infty} I = 0$. See Figure 2.12.

(c) By definition, efficiency $= \dfrac{\text{power output}}{\text{power input}}$

$$= \eta = \frac{I^2 R_L}{I^2(R + R_L)}$$

Absolute minimum $\eta = 0$ for $R_L = 0$.

There is no maximum, but η has a limiting value as follows (Fig. 2.13)

$$\lim_{R_L \to \infty} \eta = 1$$

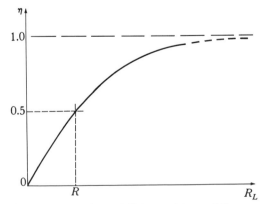

FIGURE 2.13 Efficiency characteristic.

Example 12. Find the maximum and minimum distances between a point (x_1, y_1) in the x, y plane and the circle $x^2 + y^2 = a^2$. See Figure 2.14.

SOLUTION 1. Let s = distance between (x_1, y_1) and a point (x, y) on the circle. Then

$$s^2 = (x_1 - x)^2 + (y_1 - y)^2 = (x_1 - x)^2 + (y_1 - \sqrt{a^2 - x^2})^2$$

$$s'(x) = \frac{1}{s}\left[\frac{xy_1}{\sqrt{a^2 - x^2}} - x_1\right] = 0$$

which has the roots $x = \pm\dfrac{ax_1}{s_1}$, where $s_1^2 = x_1^2 + y_1^2$.

To determine which of these gives a maximum and which a minimum, we could proceed to find $s''(x)$, but this is unnecessary for the present problem as the maximum and minimum solutions are evident from inspection of the diagram.

$$s_{\text{Max}} = \left[\left(x_1 + \frac{ax_1}{s_1}\right)^2 + \left(y_1 + \frac{ay_1}{s_1}\right)^2\right]^{1/2} = s_1 + a$$

$$s_{\text{min}} = \left[\left(x_1 - \frac{ax_1}{s_1}\right)^2 + \left(y_1 - \frac{ay_1}{s_1}\right)^2\right]^{1/2} = s_1 - a$$

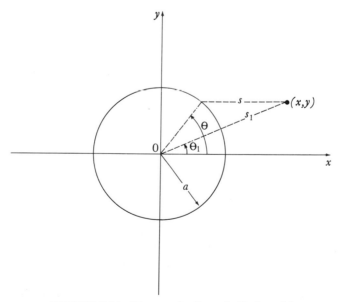

FIGURE 2.14 Diagram for Example 12, Sect. 2.2.

SOLUTION 2

A simpler solution is obtained by introducing a new variable as follows:

$$\theta = \tan^{-1} \frac{y}{x}$$

Then

$$x = a \cos \theta \qquad y = a \sin \theta$$

and

$$x_1 = s_1 \cos \theta_1 \qquad y_1 = s_1 \sin \theta_1$$

The distance s is given by

$$s^2 = s_1^2 - 2as_1 \cos (\theta - \theta_1) + a^2$$

For an extremum, we must have

$$s'(\theta) = \frac{as_1}{s} \sin (\theta - \theta_1) = 0$$

which has the solutions

$$\theta = \theta_1 \qquad \theta_1 + \pi$$

By inspection $\theta = \theta_1$ gives a minimum of s, and $\theta = \theta_1 + \pi$ a maximum, as follows:

$$s_{\text{Max}} = s_1 + a \qquad s_{\text{min}} = s_1 - a$$

SOLUTION 3. This problem is illustrative of certain cases for which the extrema may be found by algebraic methods. To see this we write

$$s^2 = s_1^2 - 2as_1 \cos (\theta - \theta_1) + a^2$$

If we choose $\theta = \theta_1$, the expression becomes

$$s^2 = s_1^2 - 2as_1 + a^2 = (s_1 - a)^2$$

an obvious minimum of s.

On the other hand, if $\theta = \theta_1 + \pi$, we have

$$s^2 = s_1^2 + 2as_1 + a^2 = (s_1 + a)^2$$

an obvious maximum of s.

2.3 FUNCTIONS OF SEVERAL VARIABLES

We consider now functions of an arbitrary number of variables. The stationary points and extrema of these functions may be found by generalization of the reasoning used in the treatment of functions of one variable. For this purpose it is advantageous to use vector spaces.

Let $f = f(x_1, x_2, \ldots, x_n)$ be a function of the n variables $(x_1, x_2, \ldots, x_n)$. Using the results of Chap. 1 we consider the n variables as components of a vector **x** and write

$$\mathbf{x} = [x_1, x_2, \ldots, x_n] \tag{2.5}$$

Thus $f = f(\mathbf{x})$ is a scalar function of the vector variable **x**.

From the set-theoretical point of view we can define a function as a correspondence of a set of vectors (V) and a set of numbers (S). Thus if $\mathbf{x} \in (V)$ and $y \in (S)$ we write, $y = f(\mathbf{x})$. The set (V) may be discrete, containing only a finite number of elements $\mathbf{x}_1, \mathbf{x}_2, \ldots, \mathbf{x}_n$ corresponding to the elements $y_1, y_2, \ldots, y_n$ of (S), or (V) and (S) can be infinite sets.

Extrema of Discrete Functions.   Consider a function defined over a discrete set of vectors (V), $\mathbf{x}_1, \mathbf{x}_2, \ldots, \mathbf{x}_n$. If $f(\mathbf{x}_k) = y_k, k = 1, 2, \ldots,$ n, we can easily define

$$y_{\mathrm{Max}} = \mathrm{Max} \, (y_1, y_2, \ldots, y_n)$$

and $\tag{2.6}$

$$y_{\mathrm{min}} = \mathrm{min} \, (y_1, y_2, \ldots, y_n)$$

as the largest and the smallest of the set of values of $y = f(\mathbf{x})$, whether these values are repeated or not.

Extrema of Continuous Functions.   We begin by giving definitions for functions of n variables which are required for the discussion of extrema.

Definitions

(1) The function $f(\mathbf{x})$ is said to tend to a limit L for $\mathbf{x} \to \mathbf{a}$ if, given a positive number ε no matter how small, there exists a positive number δ such that

$$|f(\mathbf{x}) - L| < \varepsilon \quad \text{for} \quad \|\mathbf{x} - \mathbf{a}\| < \delta$$

where

$$\mathbf{a} = [a_1, a_2, \ldots, a_n]$$

and

$$\|\mathbf{x} - \mathbf{a}\| = \left[\sum_{m=1}^{n} (x_m - a_m)^2 \right]^{1/2}$$

Let $\mathbf{h} = \mathbf{x} - \mathbf{a}$. As $\mathbf{x} \to \mathbf{a}$, it is evident that there are an infinite number of ways in which $\mathbf{h}$ can approach zero. In order to separate the directional element from the magnitude element of $\mathbf{h}$ we write $\|\mathbf{h}\| = t$. Thus,

$$\mathbf{x} - \mathbf{a} = \mathbf{h} = t\mathbf{e} \tag{2.7}$$

where **e** is the unit vector in the direction of **h**; in component form

$$\mathbf{e} = [e_1, e_2, \ldots, e_n]$$

Therefore

$$\mathbf{x} = \mathbf{a} + t\mathbf{e}$$

or

$$x_m = a_m + t e_m, \; m = 1, 2, \ldots, n \}$$

$$(2.8)$$

The condition $\|\mathbf{x} - \mathbf{a}\| < \delta$ becomes

$$0 < t < \delta$$

In the neighborhood of $\mathbf{a}$,

$$f(\mathbf{x}) = f(\mathbf{a} + t\mathbf{e}) = F(t) \tag{2.9}$$

for a given $\mathbf{a}$ and a given direction defined by $\mathbf{e}$. For this $\mathbf{e}$

$$\lim_{\mathbf{x} \to \mathbf{a}} f(\mathbf{x}) = \lim_{t \to 0} (\mathbf{a} + t\mathbf{e}) = \lim_{t \to 0} F(t) = L(\mathbf{a}, \mathbf{e})$$

which clearly indicates that the limit will, in general, depend on the direction $\mathbf{e}$ in which $\mathbf{h}$ approaches zero. The preceding condition can be expressed equivalently as follows:

$$|f(\mathbf{x}) - L(\mathbf{a}, \mathbf{e})| < \varepsilon \quad \text{for} \quad \|\mathbf{x} - \mathbf{a}\| = t < \delta$$

(2) The function $f(\mathbf{x})$ is said to be *continuous* at $\mathbf{x} = \mathbf{a}$, if $L = f(\mathbf{a})$; that is, if L exists, and is equal to the value of the function at $\mathbf{x} = \mathbf{a}$.*

(3) A function $f(\mathbf{x})$ is continuous over the domain $\|\mathbf{x} - \mathbf{a}\| < b$ if it is continuous at every point of the domain.

(4) A function $f(\mathbf{x})$ is said to have a relative maximum at the point $\mathbf{x} = \mathbf{a}$ if there exists a neighborhood about $\mathbf{a}$ defined by $0 < \|\mathbf{x} - \mathbf{a}\| < C$, where C is a given positive number for which $f(\mathbf{x}) \le f(\mathbf{a})$.

(5) A function $f(\mathbf{x})$ is said to have a relative minimum at the point $\mathbf{x} = \mathbf{a}$ if there exists a neighborhood about $\mathbf{a}$ defined by $0 < \|\mathbf{x} - \mathbf{a}\| < C$, where C is a given positive number for which $f(\mathbf{x}) \ge f(\mathbf{a})$.

(6) A function $f(\mathbf{x})$ belongs to the class C^p of functions over a domain (D) if the function and all its partial derivatives up to an including order p exist and are continuous over the given domain (D).

(7) A function $f(\mathbf{x})$ is said to have a stationary value at $\mathbf{x} = \mathbf{a}$ if $\nabla f(\mathbf{a}) = 0$, where ∇f is defined by Eq. (2.12).

Extrema of Functions of Class C^p.   To avoid possible confusion with the number of variables we use p instead of n to denote the class of functions for functions of several variables, where $p = 0, 1, 2, 3, \ldots$ corresponds to classes zero, one, two, three, and so on.

It is to be noted that determination of the stationary points and extrema of a function, whether of one or several variables, requires exploration of the infinitesimal neighborhood of a point. In the case of one

* This definition implies that L is independent $\vec{e}$.

variable, the exploration was conducted by means of Taylor's series expansions. The same technique is readily extended to functions of several variables.

Let $f(\mathbf{x})$ be a function of class C^p, where $p \geq 1$, and let $\mathbf{x} = \mathbf{a}$ be a point in the open region for which $f(\mathbf{x})$ is defined. Write

$$\mathbf{x} = \mathbf{a} + \mathbf{h} = \mathbf{a} + t\mathbf{e}$$

where $\mathbf{h}$ is defined as in Eq. (2.7). Then since $\mathbf{a}$ and $\mathbf{e}$ do not depend on t we can express $f(\mathbf{x})$ as a scalar function of t

$$f(\mathbf{x}) = f(\mathbf{a} + t\mathbf{e}) = F(t) \tag{2.9}$$

The problem is thus reduced to investigating the behavior of a scalar function $F(t)$ and may be treated by methods similar to those used for functions of one variable. Expansion of $F(t)$ in a Taylor's series about $t = 0$ gives

$$F(t) - F(0) = tF'(0) + \frac{t^2}{2!} F''(0) + \cdots + \frac{t^{p-1}}{(p-1)!} F^{(p-1)}(0) + R_{p-1} \tag{2.10}$$

where

$$R_{p-1} = \frac{t^p}{p!} F^{(p)}(\theta t) \qquad 0 < \theta < 1$$

and t is sufficiently small so that $\mathbf{a} + t\mathbf{e}$ is in the open domain for which $f(\mathbf{x})$ is defined.

The derivatives $F'(0), F''(0), \ldots$ may be evaluated by differentiation of (2.9). We find, with $\mathbf{e} = [e_1, e_2, \ldots, e_n]$, that

$$\frac{dF}{dt} = \frac{\partial f}{\partial x_1} \frac{dx_1}{dt} + \frac{\partial f}{\partial x_2} \frac{dx_2}{dt} + \cdots + \frac{\partial f}{\partial x_n} \frac{dx_n}{dt} = \frac{\partial f}{\partial x_1} e_1 + \frac{\partial f}{\partial x_2} e_2$$

$$+ \cdots + \frac{\partial f}{\partial x_n} e_n \tag{2.11}$$

$$\left(\frac{dF}{dt} \right)_{t=0} = F'(0) = e_1 \frac{\partial f}{\partial x_1} + e_2 \frac{\partial f}{\partial x_2} + \cdots + e_n \frac{\partial f}{\partial x_n} = \mathbf{e} \cdot \nabla f(\mathbf{a})$$

where

$$\nabla f = \left[\frac{\partial f}{\partial x_1}, \frac{\partial f}{\partial x_2}, \ldots, \frac{\partial f}{\partial x_n} \right] \tag{2.12}$$

is the generalized del or nabla operator in n dimensions acting on $f(\mathbf{x})$. The function ∇f may also be interpreted as the generalized gradient function and is an n-dimensional vector.

Similarly, we find

$$F''(t) = \frac{d^2 f}{dt^2} = \frac{d}{dt} \sum_{j=1}^{n} \frac{\partial f}{\partial x_j} e_j = \sum_{j=1}^{n} \frac{d}{dt} \left(\frac{\partial f}{\partial x_j} \right) e_j$$

$$= \sum_{j=1}^{n} \left(\frac{\partial^2 f}{\partial x_1 \, \partial x_j} e_1 e_j + \frac{\partial^2 f}{\partial x_2 \, \partial x_j} e_2 e_j + \cdots + \frac{\partial^2 f}{\partial x_n \, \partial x_j} e_n e_j \right)$$

$$= \sum_{i=1}^{n} \sum_{j=1}^{n} \frac{\partial^2 f}{\partial x_i \, \partial x_j} e_i e_j = (\mathbf{e} \cdot \mathbf{\nabla})^2 f(\mathbf{x})$$

Thus

$$F''(0) = (\mathbf{e} \cdot \mathbf{\nabla})^2 f(\mathbf{a}) \qquad (2.13)$$

More generally

$$F^{(q)}(0) = (\mathbf{e} \cdot \mathbf{\nabla})^q f(\mathbf{a}) \qquad (2.14)$$

Determination of the extrema of $f(\mathbf{x})$ thus reduces to the case of one variable, and by analogy with Theorem 4 of Sect. 2.2 we can write a set of sufficient conditions for an extremum as follows.

Theorem 6. *Let $f(\mathbf{x})$ be of class C^p, $p \geq 2$, over an open domain D and let $\mathbf{a}$ be a point in D. If $F'(0) = F''(0) = \cdots = F^{(q-1)}(0) = 0$ while $F^{(q)}(0) \neq 0$, where $q \leq p$ is an even number, then for*

(i) $F^{(q)}(0) < 0$, for every $\mathbf{e}$, $f(\mathbf{a})$ is a relative maximum
(ii) $F^{(q)}(0) > 0$, for every $\mathbf{e}$, $f(\mathbf{a})$ is a relative minimum.

The first condition $F'(0) = 0$ may, from Eq. (2.11), be written as

$$F'(0) = (\mathbf{e} \cdot \mathbf{\nabla}) f = 0$$

Since the unit vector $\mathbf{e}$ is arbitrary in direction, the condition can be satisfied only if the gradient $\mathbf{\nabla} f$ vanishes. We are thus led to the necessary condition

$$\mathbf{\nabla} f = \left[\frac{\partial f}{\partial x_1}, \frac{\partial f}{\partial x_2}, \ldots, \frac{\partial f}{\partial x_n} \right] = 0 \qquad (2.15)$$

for an extremum value of $f(\mathbf{x})$.

To study the sign of $F''(0)$, we use Eq. (2.13) and observe that this defines a quadratic form in the components e_i of $\mathbf{e}$ as follows

$$F''(0) = \sum_{i=1}^{n} \sum_{j=1}^{n} e_i e_j \frac{\partial^2 f}{\partial x_i \, \partial x_j} \qquad (2.16)$$

The second order partial derivatives evaluated at $\mathbf{x} = \mathbf{a}$ are constants which we may for convenience denote by a_{ij}

$$a_{ij} = \frac{\partial^2 f}{\partial x_i \, \partial x_j} \bigg|_{\mathbf{x}=\mathbf{a}}$$

Using this notation, (2.16) may be expressed as

$$F''(0) = \sum_{i=1}^{n} \sum_{j=1}^{n} a_{ij} e_i e_j = \mathbf{e} A \mathbf{e}' = Q \tag{2.17}$$

where A is the matrix of the quadratic form Q. By Theorem 6 the function $f(\mathbf{x})$ has a minimum at $\mathbf{x} = \mathbf{a}$ if the quadratic form is positive definite and maximum if the quadratic form is negative definite. Reference is made to Sect. 1.9 for the necessary and sufficient conditions for both cases.

Let us illustrate by several examples.

Example 1

Find the extrema of the function

$$f = x^2 - 2y + y^2$$

SOLUTION

$$\nabla f = \left[\frac{\partial f}{\partial x}, \frac{\partial f}{\partial y} \right] = [2x, -2 + 2y] = \mathbf{0}$$

Hence, $x - 0, y = 1$ is a stationary point.

$$A - \begin{bmatrix} \dfrac{\partial^2 f}{\partial x^2} & \dfrac{\partial^2 f}{\partial x \, \partial y} \\[2mm] \dfrac{\partial^2 f}{\partial y \, \partial x} & \dfrac{\partial^2 f}{\partial y^2} \end{bmatrix}$$

$$= \begin{bmatrix} 2 & 0 \\ 0 & 2 \end{bmatrix} = 4 > 0$$

Since the minors are $M_1 = \dfrac{\partial^2 f}{\partial x^2} = 2 > 0$ and $M_2 = \det[A] = 4 > 0$, the quadratic form is positive definite and $f(x, y)$ has a minimum at $(0, 1)$ of value $f(0, 1) = -1$.

Example 2

Find the extrema of the function

$$f(x, y, z) = 3(x^3 + y^3 - z^3) + (z - x - y)$$

SOLUTION

$$\frac{\partial f}{\partial x} = 9x^2 - 1 \qquad \frac{\partial f}{\partial y} = 9y^2 - 1 \qquad \frac{\partial f}{\partial z} = -9z^2 + 1$$

Taking $\dfrac{\partial f}{\partial x} = 0 \qquad \dfrac{\partial f}{\partial y} = 0 \qquad \dfrac{\partial f}{\partial z} = 0$ we obtain

$$x = \pm\tfrac{1}{3} \qquad y = \pm\tfrac{1}{3} \qquad z = \pm\tfrac{1}{3}$$

Thus there are eight stationary points

$$P_1: \quad (\tfrac{1}{3}, \tfrac{1}{3}, \tfrac{1}{3})$$
$$P_2: \quad (\tfrac{1}{3}, \tfrac{1}{3}, -\tfrac{1}{3})$$
$$P_3: \quad (\tfrac{1}{3}, -\tfrac{1}{3}, \tfrac{1}{3})$$
$$P_4: \quad (\tfrac{1}{3}, -\tfrac{1}{3}, -\tfrac{1}{3})$$
$$P_5: \quad (-\tfrac{1}{3}, \tfrac{1}{3}, \tfrac{1}{3})$$
$$P_6: \quad (-\tfrac{1}{3}, \tfrac{1}{3}, -\tfrac{1}{3})$$
$$P_7: \quad (-\tfrac{1}{3}, -\tfrac{1}{3}, \tfrac{1}{3})$$
$$P_8: \quad (-\tfrac{1}{3}, -\tfrac{1}{3}, -\tfrac{1}{3})$$

The character of the stationary points may be determined from the matrix A of the quadratic form (2.17) and the leading principal minors of det $[A]$.

$$A = \begin{bmatrix} 18x & 0 & 0 \\ 0 & 18y & 0 \\ 0 & 0 & -18z \end{bmatrix}$$

$$M_1 = a_{11} = 18x$$

$$M_2 = \begin{vmatrix} a_{11} & a_{12} \\ a_{21} & a_{22} \end{vmatrix} = 324xy$$

$$M_3 = \det [A] = -5832xyz$$

Evaluate these for the stationary points as follows:

Point	M_1	M_2	M_3	
P_1	$6 > 0$	$36 > 0$	$-216 < 0$	
P_2	$6 > 0$	$36 > 0$	$216 > 0$	Minimum
P_3	$6 > 0$	$-36 < 0$	$216 > 0$	
P_4	$6 > 0$	$-36 < 0$	$-216 < 0$	
P_5	$-6 < 0$	$-36 < 0$	$216 > 0$	
P_6	$-6 < 0$	$-36 < 0$	$-216 < 0$	
P_7	$-6 < 0$	$36 > 0$	$-216 < 0$	Maximum
P_8	$-6 < 0$	$36 > 0$	$216 > 0$	

There are eight stationary points, of which only two are extrema.

Physical Limitations. Here again we have to consider physical limitations that will introduce natural boundaries to the problem of the form, $a_k \leq x_k \leq b_k$; this in turn may produce local extrema at the boundary points which may require a special study. We illustrate by an example.

Example 3

Find the extrema of the function

$$f(x_1, x_2) = (x_1 - 3)^2 + (x_2 - \tfrac{1}{2})^2$$

in the region defined by

$$0 \leq x_1 \leq 2 \qquad 0 \leq x_2 \leq 2$$

SOLUTION

$$\nabla f = [2(x_1 - 3), 2(x_2 - \tfrac{1}{2})] = 0$$

Hence, f has only one stationary point, $(3, \tfrac{1}{2})$, which is outside the bounded region and therefore cannot be an extremum.

To examine the boundaries for extrema, it is noted that the circles represented by $f = a^2$ contact (but do not intersect) the line $x_1 = 2$ at $(2, \tfrac{1}{2})$. At the point $(2, \tfrac{1}{2})$, f has a local minimum:

$$f_{min} = (3 - 2)^2 = 1$$

The largest circle passes through $(0, 2)$ and hence f has a maximum:

$$f_{max} = (0 - 3)^2 + (2 - \tfrac{1}{2})^2 = \tfrac{45}{4}$$

f also has a local maximum at $(0, 0)$, $f(0, 0) = \tfrac{37}{4}$.

Summary   A stationary value of a function $f(\mathbf{x})$ of class C^p, $p \geq 2$, requires only that the gradient $\nabla f = 0$. An extremum exists if the quadratic form $Q \neq 0$; cf. Eq. (2.17). For a minimum $Q > 0$ and for a maximum $Q < 0$. Examination of all the maxima and minima, including those which may exist at the boundaries and singularities of a function, is necessary to determine its absolute maximum and its absolute minimum.

2.4 CONSTRAINED SYSTEMS: METHOD OF LAGRANGIAN MULTIPLIERS

Extrema problems treated in Sect. 2.3 involve functions, in the general case, of n independent variables. In many cases, the problem of maximizing or minimizing a function may include restrictions or auxiliary conditions which limit the freedom of the variables. The problem considered here is more general than the case studied earlier concerning physical limitations. If restrictions exist in the form of auxiliary equations, it should be possible to eliminate m of the variables from the function f, thereby reducing the problem to $n - m$ variables. Unfortunately, the elimination process frequently is formidable or even impracticable when dealing with implicit functions. Moreover, there are cases where the elimination is impossible because the auxiliary equations are in the form of nonintegrable differentials.*

A mathematically elegant method for coping with auxiliary conditions was devised by Lagrange, the method of the undetermined or Lagrangian multipliers. In order to justify the procedure, we consider first the case of a function of two variables

$$u = f(x, y) = f(\mathbf{x})$$

which we assume to be of class C^1 and whose stationary values are to be determined with the constraint $g(x, y) = 0$.

Let us consider the family of curves $u = C$, for different values of C. If at a certain point $P(x_0, y_0)$, $g(x, y)$ is tangent to a curve $u = C_0 = f(x_0, y_0)$, clearly at this point f is stationary. This means that f and g have the same tangent at P_0, so that

$$\begin{cases} (x - x_0) \dfrac{\partial f}{\partial x} + (y - y_0) \dfrac{\partial f}{\partial y} = 0 \\[3mm] (x - x_0) \dfrac{\partial g}{\partial x} + (y - y_0) \dfrac{\partial g}{\partial y} = 0 \end{cases}$$

should represent the same line, which we can express by writing

$$\frac{\partial f}{\partial x} + \lambda \frac{\partial g}{\partial x} = 0 \qquad \frac{\partial f}{\partial y} + \lambda \frac{\partial g}{\partial y} = 0$$

or, using vector notation,

$$\mathbf{\nabla} f + \lambda \mathbf{\nabla} g = \left[\frac{\partial f}{\partial x}, \frac{\partial f}{\partial y} \right] + \lambda \left[\frac{\partial g}{\partial x}, \frac{\partial g}{\partial y} \right] = 0 \qquad \textbf{(2.18)}$$

The coefficient λ is called the *Lagrangian multiplier* and is actually an additional unknown. The unknowns of the problem x_0, y_0, λ are determined by the three equations

$$\frac{\partial f}{\partial x} + \lambda \frac{\partial g}{\partial x} = 0 \qquad \frac{\partial f}{\partial y} + \lambda \frac{\partial g}{\partial y} = 0 \qquad g = 0 \qquad \textbf{(2.19)}$$

More generally, let us consider the function

$$u = f(\mathbf{x}) = f[x_1, x_2, \ldots, x_n] \qquad f \in C^1$$

---

* Such cases occur in mechanics and are called nonholonomic (cf. Chap. 6).

whose stationary points are to be found with the following m constraints:

$$g_k(\mathbf{x}) = 0 \qquad k = 1, 2, \ldots, m$$

Generalizing the preceding result, we consider the function

$$H(\mathbf{x}) = f(\mathbf{x}) + \sum_{k=1}^{m} \lambda_k g_k(\mathbf{x})$$

and its stationary points defined by

$$\nabla H = \nabla f + \sum_{k=1}^{m} \lambda_k \nabla g_k(\mathbf{x}) = 0 \qquad\qquad (2.20)$$

To determine the $n + m$ unknowns $\mathbf{x}$ and $\lambda_1, \lambda_2, \ldots, \lambda_m$, we add to the n equations of (2.20) the m conditions

$$g_k(\mathbf{x}) = 0 \qquad k = 1, 2, \ldots, m \qquad\qquad (2.21)$$

It is clear that $m < n$; otherwise the system would have as many con-straints as variables, which is logically impossible.

Example 1. Given n point masses $m_1, m_2, \ldots, m_n$ situated at points $(x_1, y_1), (x_2, y_2), \ldots, (x_n, y_n)$ in the x, y plane, prove that the axis about which the moment of inertia J is a minimum passes through the center of mass (C.M.) of the particles.

SOLUTION

The moment of inertia about an axis perpendicular to the (x, y) plane is

$$J = f(x, y) = \sum_{s=1}^{n} m_s \left[(x_s - x)^2 + (y_s - y)^2 \right]$$

A necessary condition for a minimum is $\nabla f' = 0$, or

$$\sum_{s=1}^{n} m_s(x_s - x) = 0 \qquad \sum_{s=1}^{n} m_s(y_s - y) = 0$$

This pair of equations may be written in vector form by defining the vectors

$$\mathbf{r}_s = x_s \mathbf{i} + y_s \mathbf{j} \qquad \mathbf{r} = x\mathbf{i} + y\mathbf{j}$$

Then the necessary condition for a minimum of J may be expressed as

$$\sum_{s=1}^{n} m_s \mathbf{r}_s = \sum_{s=1}^{n} m_s \mathbf{r} = \mathbf{r} \sum_{s=1}^{n} m_s = M\mathbf{r}$$

where M is the total mass of the particles. Solving for $\mathbf{r}$ gives

$$\mathbf{r} = \frac{1}{M} \sum_{s=1}^{n} m_s \mathbf{r}_s = \mathbf{r}_0$$

which by definition is the C.M. of the system of point masses.

To show that this value of **r** gives a minimum of J, we calculate the principal minors of $|A|$.

$$a_{11} = \left.\frac{\partial^2 f}{\partial x^2}\right|_{\mathbf{r}_0} = \sum_{s=1}^{n} 2m_s = 2M > 0$$

$$\det [A] = \begin{vmatrix} 2M & 0 \\ 0 & 2M \end{vmatrix} = 4M^2 > 0$$

Since both principal minors are positive, the solution gives the minimum value for the moment of inertia J.

Example 2

Determine the stationary points of the function $f(x_1, x_2)$ with the restriction $F(x_1, x_2) = 0$, where

$$f = x_1^2 + x_1 x_2 + 4x_2^2$$

$$F = x_1 + x_2 - 1 = 0$$

SOLUTION

$$\frac{\partial(f + \lambda F)}{\partial x_1} = 2x_1 + x_2 + \lambda = 0$$

$$\frac{\partial(f + \lambda F)}{\partial x_2} = x_1 + 8x_2 + \lambda = 0$$

$$F = x_1 + x_2 - 1 = 0$$

These comprise a set of three equations which may be solved for the three unknowns x_1, x_2, λ, giving

$$x_1 = \tfrac{7}{8} \qquad x_2 = \tfrac{1}{8} \qquad \lambda = -\tfrac{15}{8}$$

To investigate the nature of the stationary point, find the coefficients a_{11}, a_{12}, a_{22} of Q_2. These are $a_{11} = 2$, $a_{12} = 1$, $a_{22} = 8$, from which we further find

$$a_{11} > 0 \qquad \begin{vmatrix} a_{11} & a_{12} \\ a_{12} & a_{22} \end{vmatrix} = \begin{vmatrix} 2 & 1 \\ 1 & 8 \end{vmatrix} = 15 > 0$$

Therefore from Eq. (2.17), Q_2 is positive definite and the stationary point is a minimum.

Example 3.

Given an electric circuit comprised of two resistors R_1 and R_2 in parallel through which a total current i_0 flows, if the currents i_1 and i_2 are such that the total heat generated in R_1 and R_2 is a minimum, find the value of i_1 and i_2.

SOLUTION.   There are two unknowns, i_1 and i_2, and one equation of constraint, $i_1 + i_2 = i_0$. Consequently we can write

$$f(i_1, i_2) = i_1^2 R_1 + i_2^2 R_2 = \text{heat generated}$$

$$F(i_1, i_2) = i_1 + i_2 - i_0 = 0 \text{ (constraint on currents)}$$

$$f + \lambda F = i_1^2 R_1 + i_2^2 R_2 + \lambda(i_1 + i_2 - i_0)$$

$$\frac{\partial(f + \lambda F)}{\partial i_1} = 2i_1 R_1 + \lambda = 0$$

$$\frac{\partial(f + \lambda F)}{\partial i_2} = 2i_2 R_2 + \lambda = 0$$

Solving for i_1, i_2, λ gives

$$i_1 = \frac{R_2 i_0}{(R_1 + R_2)} \qquad i_2 = \frac{R_1 i_0}{(R_1 + R_2)} \qquad \lambda = \frac{-2 R_1 R_2 i_0}{(R_1 + R_2)}$$

Note that the current values found agree with Kirchhoff's laws of electric network theory. In this problem it is not necessary to find λ, but in many instances we shall see that λ has physical significance. The assumption of minimum heat generation made for this example is a special case of a more general minimal principle applicable to electric networks.

Example 4.   Find the maximum volume of a rectangular parallelepiped which fits into a sphere of radius a with center at the origin.

SOLUTION

Use coordinates x_1, x_2, x_3. Then

$$f = \text{volume} = 8x_1 x_2 x_3$$

$$x_1^2 + x_2^2 + x_3^2 = a^2 \text{ (equation of sphere)}$$

$$f + \lambda F = 8x_1 x_2 x_3 + \lambda(x_1^2 + x_2^2 + x_3^2 - a^2)$$

$$\left.\begin{array}{l}
\dfrac{\partial(f + \lambda F)}{\partial x_1} = 8x_2 x_3 + 2\lambda x_1 = 0 \\[2mm]
\dfrac{\partial(f + \lambda F)}{\partial x_2} = 8x_1 x_3 + 2\lambda x_2 = 0 \\[2mm]
\dfrac{\partial(f + \lambda F)}{\partial x_3} = 8x_1 x_2 + 2\lambda x_3 = 0 \\[2mm]
F = x_1^2 + x_2^2 + x_3^2 - a^2 = 0
\end{array}\right\}$$

Since these equations are symmetrical in (x_1, x_2, x_3), one has immediately the relations

$$x_1^2 = x_2^2 = x_3^2$$

from which

$$x_1 = \pm a/\sqrt{3} \qquad x_2 = \pm a/\sqrt{3} \qquad x_3 = \pm a/\sqrt{3}$$

Hence, the solution gives a cube of side $2a/\sqrt{3}$ inscribed in the sphere. The volume is

$$f = \text{volume} = (2a/\sqrt{3})^3 = 8a^3/3\sqrt{3}$$

2.5 OPTIMIZATION TECHNIQUES

During the past 10 or 15 years there has developed a rising interest in mathematical optimization techniques motivated by a host of new problems in technology, science, and economics. Many of these problems involve large numbers of variables, complex and expensive systems, and the necessity for rapid solutions. The advent of high speed electronic computers has greatly broadened the scope and capability of optimization methods both in classical areas and in the development of new areas. Most prominent of the newer methods are the various programming techniques—linear programming, nonlinear programming, dynamic programming, and the like. Direct enumeration and search methods for solution of large problems have also become practicable through the agency of the fast computer.

In this section, we sketch some of the newer techniques and relate them to the classic analytical methods. Many of the difficulties encountered in the practical application of optimization techniques occur not because of gaps or defects in the theory but rather because of the complications of dealing with large numbers of variables. For instance, in Ex. 2, Sect. 2.3, which involves finding the extrema of a function of three variables, solution of $\nabla f = 0$ yields eight stationary points corresponding to two roots for each of the three scalar equations $\dfrac{\partial f}{\partial x_i} = 0$, $i = 1, 2, 3$. A similar problem in, say, 10 variables would yield $2^{10} = 1024$ stationary points; more generally, for n variables there would be 2^n stationary points. Since each of these points must be further examined to discover actual extrema, the detailed work may easily become prohibitive in time and cost even on a fast computer.

Further difficulties arise in the calculation of extrema in the bounded domains characteristic of physical and economic processes. For example, in a three-dimensional extremum problem in a bounded region, $c_i \leq x_i \leq d_i$, $i = 1, 2, 3$, it is necessary to examine separately the six boundary surfaces, the twelve edges, and the eight corners or vertices of the bounding parallelepiped, a total of 26 separate investigations. Of course, by no means do all multivariable problems involve these difficulties to an extreme or even burdensome degree, but the incidence is sufficient to warrant or even require alternate approaches.

A continuous variational problem may be approximated by a finite difference problem for which extrema may be found by brute force enumeration of all cases or points. However, consider 3 variables, each ranging over 20 different values; the number of cases is $20^3 = 8000$. More generally, for n variables ranging over m different values, the number of points to be compared is m^n. In actual practice, the number of cases may be substantially reduced because of limitations imposed on choices. Nevertheless, it is apparent that the brute force approach is intrinsically inefficient and fails for all except relatively small problems.

A far more efficient direct process is the method of gradients or method of steepest descent in which the search for an extremum point is guided by choosing a series of finite increments, each in the direction of maximum gradient. The process ends either at a boundary or at a point of zero gradient which by definition is a stationary point of the function. Further exploration will determine whether or not the stationary point is also an extremum. The method of steepest descent (or ascent) can be programmed on a digital computer, thus permitting the handling of almost any input function. In one of the variations of the method of gradients a time parameter is introduced and continuous motion along the gradient is considered. The analytical expressions describing the motion become particularly simple if the magnitude of the velocity of the point is taken proportional to the magnitude of the gradient. The process then stops automatically at a stationary point (or a boundary). Details of various refinements and formulations of the method of gradients, on which there is a large literature, are described in the references listed in the bibliography.

A very interesting and important class of extrema problems is encompassed under the category of linear programming. These problems are concerned with finding extrema of the linear form

$$f(x) = \sum_{i=1}^{n} b_i x_i \qquad (2.22)$$

in a region bounded by the inequality constraints

$$\sum_{j=1}^{n} a_{ij} x_i \le c_j \qquad j = 1, 2, \ldots, m \qquad (2.23)$$

Since the gradient ∇f is constant throughout the bounded region, the extrema of $f(x)$ must occur at the boundaries, and, in general, at the vertices of the bounded region. This is illustrated for an elementary two-dimensional case in Figure 2.15. The lines $f = 1, f = 2, \ldots$ clearly show that f has a maximum at vertex 3 and a minimum at vertex 0.

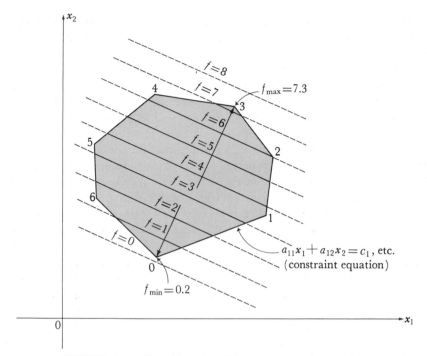

FIGURE 2.15   Two-dimensional linear programming problem.

Numerous methods have been devised for solving linear programming problems. In terms of the theory of extrema presented in this chapter, it is possible to prove that the extrema of a linear function with consistent linear inequality constraints always occur at the vertices of the bounded region. Since the vertices for any given problem are finite in number, the extrema may be determined by simple enumeration of the values of the linear function at the vertex points. This method may become quite cumbersome for large problems, and other more efficient techniques are desirable.

One of the earliest techniques (ca. 1950), the Simplex Method of Dantzig (Ref. 5), is still among the most popular. In the Simplex method, the variables are taken to be non-negative (corresponding to the essentially positive character of many economic and physical quantities such as cost, distance, and time of transport), and the inequalities are replaced by equalities through the medium of so-called slack variables. The resulting equations may be then dealt with by iterative techniques to yield extrema in a finite number of steps.

Nonlinear programming may be thought of as an extension of linear programming in the sense that the functions and inequality constraints are nonlinear. Thus the problem is to find the extrema of a nonlinear function $f(\mathbf{x})$ subject to the nonlinear constraints

$$F_j(\mathbf{x}) \leq 0 \qquad j = 1, 2, \ldots, m \tag{2.24}$$

Note that the problem is also a generalization of the extrema problems considered in Sects. 2.3 and 2.4 in that the bounded domain is defined by nonlinear equations instead of by $c_i \leq x_i \leq d_i$.

Solution of nonlinear programming problems can be obtained by classical methods. The interior of the bounded region is examined for extrema by the methods of Sect. 2.3. Extrema on the boundaries may be found by (1) expressing the equations for the bounding surfaces in parametric form which then reduces the problem to that of finding extrema in a $(n-1)$-dimensional space, or (2) using the Lagrange multiplier technique.

An example of a two-dimensional nonlinear programming problem is illustrated in Figure 2.16. The bounded region is defined by four nonlinear constraints. From the contours of the function, it is apparent that there is an absolute maximum at the interior point P, an absolute minimum at vertex 0, and a relative minimum at vertex 2.

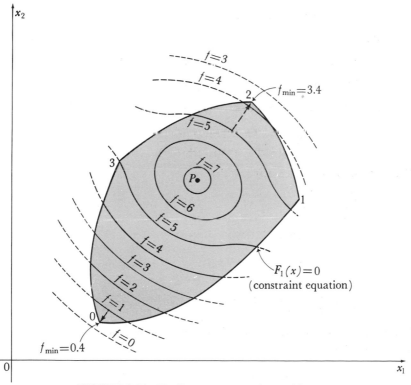

FIGURE 2.16 Nonlinear programming problem.

As stated in the Introduction, we do not treat numerical methods in this book. Optimization techniques so far have been numerical. This is why our treatment of these methods is on a purely descriptive level. On the other hand, it is necessary at least to mention these methods in a chapter on extrema. We refer the interested reader to the specialized treatments given in the bibliography.

2.6 ADDITIONAL ILLUSTRATIVE EXAMPLES

The purpose of this section is to present several additional examples, both to illustrate the mathematical methods of optimization and to show applications of interest to industrial and economic problems. Other illustrative problems are included in the applications chapters.

Example I

Find the extrema of the function

$$f(\mathbf{x}) = x_1 x_2 x_3 \cdots x_n$$

subject to the constraint

$$F(\mathbf{x}) = x_1 + x_2 + x_3 + \cdots + x_n - 1 = 0$$

SOLUTION

Eliminate x_n from $f(\mathbf{x})$ by means of $F(\mathbf{x}) = 0$. Then

$$f(\mathbf{x}) = x_1 x_2 x_3 \cdots x_{n-1}(1 - x_1 - x_2 - x_3 - \cdots - x_{n-1})$$

A necessary condition for an extremum is

$$\nabla f = \left[\frac{\partial f}{\partial x_1}, \frac{\partial f}{\partial x_2}, \ldots, \frac{\partial f}{\partial x_{n-1}} \right] = 0$$

Taking the partial derivatives gives

$$\frac{\partial f}{\partial x_1} = (x_2 x_3 \cdots x_{n-1})(1 - x_1 - x_2 - x_3 - \cdots - x_{n-1})$$

$$- x_1 x_2 x_3 \cdots x_{n-1}$$

$$= \frac{f}{x_1} - \frac{f}{x_n} = 0$$

$$\frac{\partial f}{\partial x_2} = \frac{f}{x_2} - \frac{f}{x_n} = 0 \quad \text{etc.}$$

From these $n - 1$ equations, it follows that the necessary condition for an extremum is

$$x_1 = x_2 = x_3 = \cdots = x_n = \frac{1}{n}$$

To determine the nature of the extremum, calculate the matrix

$$A = \left[\frac{\partial^2 f}{\partial x_i \, \partial x_j}\right] \quad \text{at} \quad x_i = \frac{1}{n}, x_j = \frac{1}{n}$$

$$\frac{\partial^2 f}{\partial x_1^2}\bigg]_{1/n} = -x_2 x_3 \cdots x_{n-1} - x_2 x_3 \cdots x_{n-1} = -2\left(\frac{1}{n}\right)^{n-2}$$

$$\frac{\partial^2 f}{\partial x_1 \, \partial x_2}\bigg]_{1/n} = x_3 x_4 \cdots x_n - x_1 x_3 x_4 \cdots x_{n-1} = 0$$

etc.   Thus A has the form

$$A - \begin{bmatrix} \dfrac{-2}{n^{n-2}} & & & & \\ & \dfrac{-2}{n^{n-2}} & & & 0 \\ & & \cdot & & \\ & & & \cdot & \\ 0 & & & & \cdot \\ & & & & \dfrac{-2}{n^{n-2}} \end{bmatrix}$$

which (cf. Sect. 1.9) is negative definite. The extremum is therefore a maximum given by

$$f_{\text{max}} = \left(\frac{1}{n}\right)^n$$

*Example 2.** A projectile starting from rest, subject to a monotonically decreasing acceleration, traverses a distance s_0, attaining a final velocity v_0. Determine the maximum time of traverse.

SOLUTION

Since

$$v = \int_0^{t_0} a \, dt$$

and acceleration a is monotonically decreasing, it follows that v is a convex function of time t. Moreover, since s_0 is a fixed quantity and is represented by the area under the velocity-time curve, the time will be a maximum when v is a minimum subject to the constraint of being convex. Since the "least convex" function is $v = kt$, where k is a constant, it is evident that

$$v_0 = kt_0 \qquad s_0 = \tfrac{1}{2}kt_0^2 = \tfrac{1}{2}v_0 t_0$$

---

* From Siam Review *4* 259, 1962.

and

$$t_0 = \frac{2s_0}{v_0} = \text{maximum value of time for } v_0 \text{ convex}$$

*Example 3.** In an electrostatic precipitator for removing fine particles from gases, the particle-laden gases are passed through an electric corona field maintained by grids of corona wires centered between pairs of plates. Each pair of plates and its associated corona electrodes comprise a duct. Large electrical precipitators are built up by using many such ducts in parallel.

The problem arises in the practice of determining the optimum width ducts which can be fitted into a precipitator of given width. Increasing the number of ducts raises cleaning efficiency because more collection surface is provided. On the other hand, narrow ducts are detrimental because, in operation, dust particles build up in thick layers on the plate surfaces and limit operating voltages and field strengths. The question thus arises as to the best or optimum duct width.

TREATMENT OF PROBLEM. Let there be a total of n ducts, each of width $2s$. Then the width W of the precipitator is given by

$$W = 2sn$$

The collection efficiency η of the precipitator is known from theory and practice to be represented by the equation

$$\eta = 1 - e^{-Aw/V}$$

where A is the total collection surface of the plates, V is the volume rate of gas flow through the precipitator, and w is the drift velocity of the dust particles through the electric field due to electrostatic force on the charged particles.

The optimum value of s is determined by the condition that η be a maximum with respect to s. Hence

$$\frac{d\eta}{ds} = -e^{-Aw/V}\left[\frac{d}{ds}\left(-\frac{Aw}{V}\right)\right] = 0$$

which is equivalent to the condition

$$\frac{d}{ds}(Aw) = 0$$

The total plate area A may be expressed in terms of the dimensions

$$A = 2Lhn = \frac{LhW}{s}$$

---

* From H. J. White, *Industrial Electrostatic Precipitation*. Addison-Wesley, 1963, p. 177.

where L is the width of the plates and h the height. With this substitution,

$$\frac{d}{ds}(Aw) = \frac{d}{ds}\left(\frac{LhWw}{s}\right)$$

$$= LhW\left(\frac{1}{s}\frac{dw}{ds} - \frac{w}{s^2}\right) = 0$$

or

$$\frac{dw}{ds} = \frac{w}{s} \text{ (condition for an extremum)}$$

It is now necessary to relate w to the electric field E and the operating voltage V_p. This relationship is known from semiempirical observations to be of the form

$$w = kE^m = k\left(\frac{V_p}{s}\right)^m$$

where k is a constant and m an exponent of value between 4 and 5.

In practice V_p is always limited by dust build-up on the plates. Denote this build-up or thickness of the accumulated dust layers on the plates by Δs. Then V_p can be expressed by

$$V_p = V_0\left(1 - \frac{\Delta s}{s}\right)$$

where V_0 is the maximum operating voltage (as limited by electric spark over) for clean plates.

It is now possible to write w as a function of s and Δs as follows

$$w = kV_0^m\left(1 - \frac{\Delta s}{s}\right)^m$$

from which

$$\frac{dw}{ds} = mkV_0^m\left(1 - \frac{\Delta s}{s}\right)^{m-1}\left(\frac{\Delta s}{s^2}\right)$$

$$= mw\frac{\Delta s}{s}\left(\frac{1}{1 - \frac{\Delta s}{s}}\right)$$

For an extremum

$$\frac{dw}{ds} = \frac{w}{s} = mw\frac{\Delta s}{s^2}\left(\frac{1}{1 - \frac{\Delta s}{s}}\right)$$

Solving for s yields the relation

$$s = (m + 1)\,\Delta s$$

Thus the optimum spacing depends on the dust build-up Δs. For clean conditions, it is apparent that s can be quite small. However, for heavy duty industrial precipitators, such as would be used in steam electric power plants or steel mills, Δs will be of the order of 1 inch. Hence with $m = 4$

$$s = (4 + 1)(1) = 5 \text{ inches}$$

and $2s = 10$ inches $=$ duct width, a value which conforms to practice.

COMMENT. The solution of this problem, although not complicated mathematically, does require a rather intimate knowledge of the overall system, both with respect to theory and to practice. Such problems are often met in industrial systems, and the solution is critically dependent on establishing realistic and workable mathematical models.

Example 4

Find the extrema of the function

$$f(x_1, x_2) = (x_1 - 3)^2 + (x_2 - \tfrac{1}{2})^2$$

in the region bounded by

$$0 \leq x_1 \leq 2 \qquad x_2 \geq 0$$
$$x_1^2 - 2x_1 - 3 + 4x_2 \leq 0$$

SOLUTION. The function $f(x_1, x_2)$ has only one extremum, i.e., the minimum at $(3, \tfrac{1}{2})$ which is outside the bounded region (see Fig. 2.17). Therefore, if there are any extrema, they must occur on the boundary. The smallest circle which contacts the boundary has a radius of unity for which

$$f_{\min} = (3 - 2)^2 + (\tfrac{1}{2} - \tfrac{1}{2})^2 = 1$$

The largest circle contacts the boundary at the origin, and for this point

$$f_{\max} = (0 - 3)^2 + (0 - \tfrac{1}{2})^2 = \tfrac{37}{4}$$

There is also a secondary maximum at $(0, \tfrac{3}{4})$, for which $f = \tfrac{147}{16}$ (cf. Fig. 2.17).

Example 5.* Optimum Allocation of Thermal Plants in an Electric Utility System. Given n thermal electric generating plants interconnected by a high-voltage transmission network serving various load areas, determine the optimum generating plant loading pattern necessary to minimize the total fuel cost at each instant. In this problem, thermodynamic efficiencies, fuel costs, and transmission line losses differ for each

---

* From L. E. Kirchmeyer, "System Optimization Techniques." *Electrical Engineering,* *81* 618, 1962.

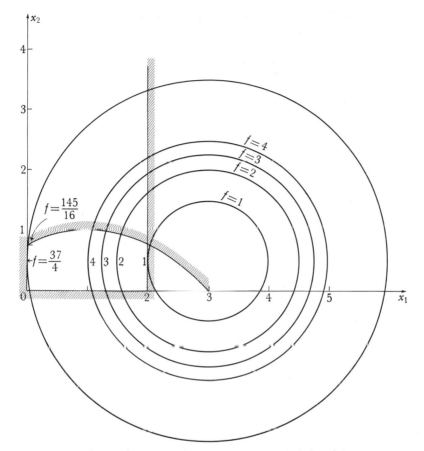

FIGURE 2.17 Bounded region for Example 4, Sec. 2.6.

plant and also may change with time. Furthermore, each plant has certain maximum and minimum ratings which cannot be exceeded.

The mathematical problem is to minimize the expression

$$f_t = \sum_{i=1}^{n} f_i \tag{1}$$

where f_t is the total fuel cost per hour and f_i is the fuel cost per hour for plant i. The generation system output is subject to the constraint equation

$$\sum_{i=1}^{n} P_i = P_L + P_R \tag{2}$$

where P_i is the electric power output of plant i, P_L is the total transmission line loss and P_R the total customer load.

Solution of the problem may be carried out by means of the Lagrange multiplier method. Form the function

$$g = \sum_{i=1}^{n} f_i + \lambda \left\{ P_L + P_R - \sum_{i=1}^{n} P_i \right\} \tag{3}$$

and write the necessary condition for an extremum, i.e.,

$$\nabla g = \left[\frac{\partial g}{\partial P_1}, \frac{\partial g}{\partial P_2}, \dots, \frac{\partial g}{\partial P_n} \right] = 0$$

or

$$\frac{\partial g}{\partial P_i} = \frac{\partial f_i}{\partial P_i} + \lambda \left\{ \frac{\partial P_L}{\partial P_i} - 1 \right\} = 0 \qquad i = 1, 2, \dots n$$

Rearranging gives

$$\frac{1}{1 - \dfrac{\partial P_L}{\partial P_i}} \cdot \frac{\partial f_i}{\partial P_i} = \lambda \qquad i = 1, 2, \dots, n \tag{4}$$

In these equations, the terms $\dfrac{\partial P_L}{\partial P_i}$ represent the rate of change of transmission line loss due to a loading change in plant i. In fact, $\dfrac{\partial P_L}{\partial P_i} \Delta P_i$ may be identified as the incremental transmission loss caused by an incremental change in power ΔP_i. Similarly, $\dfrac{\partial f_i}{\partial P_i} \Delta P_i$ is the incremental change in production cost of plant i, and $\lambda \Delta P_i$ is the incremental change in the cost of received power.

Incremental power generation costs for the various plants are readily determined in practice from known input-output characteristics. But the corresponding costs for incremental changes in transmission network losses have proved to be much more difficult to evaluate, and for many years impeded effective optimization of power generation scheduling. Successful solution of the transmission loss problem was effected by means of the Kron tensor analysis method which leads to the expression

$$P_L = \frac{1}{2} \sum_{i=1}^{n} \sum_{j=1}^{n} B_{ij} P_i P_j \tag{5}$$

where the B_{ij} are power loss coefficients. For the optimization problem, it is assumed that these coefficients are known for a given system. With this understanding

$$\frac{\partial P_L}{\partial P_i} = \sum_{j=1}^{n} B_{ij} P_j$$

and (4) becomes

$$\frac{\dfrac{\partial f_i}{\partial P_i}}{1 - \sum_{j=1}^{n} B_{ij} P_j} = \lambda \qquad i = 1, 2, \ldots, n \qquad (6)$$

Expressions (6) comprise a set of n simultaneous nonlinear algebraic equations which may be solved and implemented by various computer-control arrangements now available and in operation. These control systems have the capability of maintaining optimum-generation patterns on an essentially continuous basis.

BIBLIOGRAPHY

1. E. F. Beckenbach, ed., *Modern Mathematics for the Engineer*. New York: McGraw-Hill, 1956.
2. R. Bellman, ed., *Mathematical Optimization Techniques*. Berkeley: University of California Press, 1963.
3. R. C. Buck, *Advanced Calculus*. New York: McGraw-Hill, 1956.
4. R. Courant, *Differential and Integral Calculus*. London: Blackie, 1944.
5. G. B. Dantzig, *Linear Programming and Extensions*. Princeton: Princeton University Press, 1963.
6. W. Fulks, *Advanced Calculus*. New York: Wiley, 1961.
7. E. Goursat, *Cours d'Analyse Mathematique*, Vol. I. Paris: Gauthier-Villars, 1942.
8. H. Hancock, *Theory of Maxima and Minima*. New York: Dover, 1960.
9. G. Leitmann, ed., *Optimization Techniques with Applications to Aerospace Systems*. New York: Academic Press, 1962.
10. J. Pierpont, *Lectures on the Theory of Functions of Real Variables*, Vols. I and II. New York: Dover, 1959.
11. L. S. Pontryagin et al., *The Mathematical Theory of Optimal Processes*. New York: Interscience Publishers, 1962.
12. T. L. Saaty and J. Bram, *Nonlinear Mathematics*. New York: McGraw-Hill, 1964.
13. B. V. Shah et al., "Some Algorithms for Minimizing a Function of Several Variables." *Journal Society of Industrial and Applied Mathematics, 12* 74, 1964.
14. J. L. Walsh, *A Rigorous Treatment of Maximum Minimum Problems in the Calculus*. Boston: Heath, 1962.
15. D. J. Wilde, *Optimum Seeking Methods*. Englewood Cliffs, N.J.: Prentice-Hall, 1964.
16. *Journal of Optimization Theory and Applications*. Published bimonthly by Plenum Publishing Corporation, New York.
17. A. Lavi and T. P. Vogl, eds., *Recent Advances in Optimization Techniques*. New York: Wiley, 1966.
18. G. P. McCormick, "Second Order Conditions for Constrained Minima." *SIAM Journal on Applied Mathematics, 15* 641, 1967.
19. R. Frisch, *Maxima and Minima*. Chicago: Rand McNally and Co., Distributors for U.S.A., 1966. (Translation of French Edition, published by Dunod, Paris, 1960.)
20. D. J. Wilde and C. S. Beightler, *Foundations of Optimization*. Englewood Cliffs, N.J.: Prentice-Hall, 1967.

3

CALCULUS OF VARIATIONS

The calculus of variations, although closely related to the extrema of functions, deals with the far more extensive problem of finding stationary values and extrema of definite integrals. As shown in the preceding chapter, calculus and certain aspects of linear algebra are adequate to deal with the extrema problems of functions. Extension of the problem to definite integrals requires not just the determination of a finite number of points for which a function has extrema, but rather the entire course of one or more functions for which given integrals assume maximum or minimum values. In geometric language, the calculus of variations deals with the problem of finding paths or regions of integration which yield extrema values for definite integrals.

The quantity to be extremized is a functional form or *functional*. Functionals, or functions of functions, are very important in both pure and applied mathematics, and occur throughout analysis and its applications. In dealing with a finite number of variables, as in the preceding chapter, it is necessary to specify a region of definition. Similarly, in dealing with functionals, it is necessary to define the domain of admissible functions from which the argument functions can be chosen. This domain or space of admissible functions may, for instance, be the set of all functions of class C^n, $n \geq 0$. A functional may be regarded as a function of infinitely many variables. This may be seen by considering argument functions which, for example, may be expanded in an infinite power series or an infinite Fourier series. The functional will depend on the series coefficients which then comprise the infinity of variables. This concept proves useful in finding direct solutions to calculus of variations problems.

Most of the classical problems of the calculus of variations, such as the brachistochrone and minimum surfaces of revolution, are of the general type

$$I = \int_a^b f(x, y, y') \, dx$$

where f is a given functional form. The integration is along the arc of a curve $y = y(x)$ between points A and B, and y is a single-valued function of x in the interval $[a, b]$. The problem is to determine the arc of the curve $y = y(x)$ from a domain of admissible functions which makes I an extremum.

This basic problem may be generalized in many ways, a few of which follow.

(1) Argument functions of several dependent variables, that is,

$$f - f(x, \mathbf{y}, \mathbf{y}')$$

where

$$\mathbf{y} = [y_1(x), y_2(x), \ldots, y_n(x)]$$

(2) Isoperimetric problems. The integral I is to be extremized subject to constraints or conditions in the form of integrals. The classical problem of the geodesic curve, where it is required to find the arc of a curve of minimum length between two given points on a given surface, is of this form.

(3) Variable limits. The points A and B, instead of being held fixed as in integral I, are allowed to vary along two fixed curves L_1 and L_2. The problem of finding the shortest distance between any two curves L_1 and L_2 is of this type.

(4) Argument functions of several independent variables, that is, the problem of determining extrema of multiple integrals. This problem leads to one or more partial differential equations in place of the ordinary differential equations associated with argument functions of one independent variable. Many of the partial differential equations of mathematical physics and engineering are readily formulated using this approach.

(5) Higher derivatives. The integral I is generalized to the form

$$I = \int_a^b f(x, y, y', y'', \ldots, y^{(m)}) \, dx$$

where the domain of admissible functions is the set of all functions of class C^{2m}.

An intrinsic difficulty encountered in the calculus of variations is that there is no general existence theorem for solutions of problems, such as is the case for extrema of functions where a fundamental theorem of Weierstrass (Theorem 5, Chap. 2) guarantees the existence of solutions.

In the absence of a general existence theorem, it cannot be assumed that an extremum actually exists for any given problem. Instead it is necessary to establish existence proofs for each problem or class of problems. Because of this essential difficulty, it is usual in the calculus of variations to formulate necessary conditions for the existence of extrema and to leave the question of sufficiency open. In problems of geometric or physical origin it is often possible to assert the existence of solutions on intuitive grounds or to give special proofs. Moreover, for many physical applications it is necessary only to find stationary solutions and the question of an actual extremum does not arise.

The calculus of variations is a large and extensive field of mathematics and only the foundations needed for systems analysis can be covered in this book. We therefore treat the case of argument functions of several dependent variables and also isoperimetric problems, both of which are fundamental to the analysis of discrete systems of many variables. For the remaining topics we refer the reader to specialized treatises listed in the bibliography.

3.1 NOTATION

We introduce here certain classical notation used in the calculus of variations.

Let $y = y(x)$ and $\eta = \eta(x)$, $a \leq x \leq b$, be two functions of class C^n (cf. Chap. 2), $n \geq 2$, and let ε be a small positive number. Consider the function $\varphi = \varphi(x, y, y', \ldots, y^{(n)})$, $\varphi \in C^q$ with respect to $x, y', y'', \ldots, y^{(n)}$.

Consider the difference

$$\Delta\varphi = \varphi(x, y + \varepsilon\eta, y' + \varepsilon\eta', \ldots, y^{(n)} + \varepsilon\eta^{(n)})$$
$$- \varphi(x, y, y', \ldots, y^{(n)}) \tag{3.1}$$

The quantity $\Delta\varphi$ is called the *total variation* of φ. If we perform an expansion of $\Delta\varphi$ in powers of ε we obtain

$$\Delta\varphi = \frac{\varepsilon}{1}\,\varphi_1 + \frac{\varepsilon^2}{2!}\,\varphi_2 + \cdots + \frac{\varepsilon^q}{q!}\,R_q \tag{3.2}$$

where

$$\varphi_1 = \varphi_y\eta + \varphi_{y'}\eta' + \cdots + \varphi_{y^{(n)}}\eta^{(n)}$$
$$\varphi_2 = \varphi_{yy}\eta^2 + \varphi_{y'y'}\eta'^2 + \cdots + 2\varphi_{yy'}\eta\eta' + \cdots \tag{3.3}$$

and R_q is a remainder term.

The terms of the expansion (3.2) are called first, second, $\ldots$, n-th variation. Lagrange introduced the following operational notation for the

variations:
$$\varepsilon\varphi_1 = \delta\varphi \qquad \varepsilon^2\varphi_2 = \delta^2\varphi, \ldots \qquad \varepsilon^{q-1}\varphi_{q-1} = \delta^{q-1}\varphi \qquad (3.4)$$

In the special case where $\varphi = \varphi(x, y, y')$ we clearly have for $q = 2$

$$\Delta\varphi = \varepsilon\left[\eta\,\frac{\partial}{\partial y} + \eta'\,\frac{\partial}{\partial y'}\right]\varphi(x, y, y')$$

$$+ \frac{\varepsilon^2}{2!}\left[\eta^2\,\frac{\partial^2}{\partial y^2} + 2\eta\eta'\,\frac{\partial^2}{\partial y\,\partial y'} + \eta'^2\,\frac{\partial^2}{\partial y'^2}\right]\varphi(x, y + \theta_1\varepsilon\eta, y' + \theta_2\varepsilon\eta')$$

$$0 < \theta_1 < 1 \qquad 0 < \theta_2 < 1 \qquad (3.5)$$

and for $q = 3$

$$\Delta\varphi = \varepsilon\left[\eta\,\frac{\partial}{\partial y} + \eta'\,\frac{\partial}{\partial y'}\right]\varphi(x, y, y')$$

$$+ \frac{\varepsilon^2}{2!}\left[\eta^2\,\frac{\partial^2}{\partial y^2} + 2\eta\eta'\,\frac{\partial^2}{\partial y\,\partial y'} + \eta'^2\,\frac{\partial^2}{\partial y'^2}\right]\varphi(x, y, y')$$

$$+ \frac{\varepsilon^3}{3!}\left[\eta^3\,\frac{\partial^3}{\partial y^3} + 3\eta^2\eta'\,\frac{\partial^3}{\partial y^2\,\partial y'} + 3\eta\eta'^2\,\frac{\partial^3}{\partial y\,\partial y'^2} + \eta'^3\,\frac{\partial^3}{\partial y'^3}\right]$$

$$\times\,\varphi(x, y + \theta_1\varepsilon\eta, y' + \theta_2\varepsilon\eta')$$

$$0 < \theta_1 < 1 \qquad 0 < \theta_2 < 1 \qquad (3.6)$$

The same nomenclature applies to integrals. Thus if

$$I = \int_a^b f(x, y, y')\,dx \qquad (3.7)$$

its total variation is

$$\Delta I = \int_a^b [f(x, y + \varepsilon\eta, y' + \varepsilon\eta') - f(x, y, y')]\,dx \qquad (3.8)$$

and its k-th variation is given by

$$\delta^k I = \int_a^b \delta^k f(x, y, y')\,dx \qquad (3.9)$$

Theorem I. *If φ is the function defined at the beginning of this section then (with the notation of Eq. 3.4)*

$$\delta^k\varphi = \left[\frac{\partial^k}{\partial\varepsilon^k}\,(\varphi + \Delta\varphi)\right]_{\varepsilon=0}\varepsilon^k \qquad (3.10)$$

Proof

In Eq. (3.2) let $q \geq k$, then

$$\frac{\partial^k}{\partial\varepsilon^k}\,(\varphi + \Delta\varphi) = \varphi_k + \frac{\varepsilon}{1!}\,\varphi_{k+1} + \frac{\varepsilon^2}{2!}\,\varphi_{k+2} + \cdots + \frac{\varepsilon^{q-k}}{(q-k)!}\,\varphi_k$$

Thus

$$\left[\frac{\partial^k}{\partial \varepsilon_k}(\varphi + \Delta\varphi)\right]_{\varepsilon=0} = \varphi_k$$

and

$$\left[\frac{\partial^k}{\partial \varepsilon_k}(\varphi + \Delta\varphi)\right]_{\varepsilon=0}\varepsilon^k = \varepsilon^k \varphi_k = \delta^k \varphi$$

In Eq. (3.8) clearly $(\Delta I + I)$ is a function of ε only so that (3.10) can be adapted to the integral I by writing

$$\delta^k I = \left[\frac{\partial}{\partial \varepsilon^k}(I + \Delta I)\right]_{\varepsilon=0}\varepsilon^k$$

We terminate with the following definitions.

Definition 1.   *The integral of Eq. (3.7) is said to be* stationary *if its first variation is identically zero, i.e.,*

$$\delta I = 0$$

Definition 2.   *The total variation* $\Delta\varphi = \omega(x, \varepsilon)$, *of the function* φ, *as defined in Sect. 3.1 is said to be*

(1) weak for $a \leq x \leq b$, if both

$$\lim_{\varepsilon \to 0} \omega(x, \varepsilon) = 0 \qquad \lim_{\varepsilon \to 0} \frac{\partial\omega(x, \varepsilon)}{\partial x} = 0$$

(2) strong for $a \leq x \leq b$, if only

$$\lim_{\varepsilon \to 0} \omega(x, \varepsilon) = 0$$

3.2   THE EULER-LAGRANGE EQUATION

We study first the simplest problem of the calculus of variations: Find the extrema of the definite integral

$$I = \int_a^b f(x, y, y')\, dx \tag{3.7}$$

where f is a given functional, whose properties will be defined later, $y = y(x)$ and $y'(x)$ are single-valued functions defined over $[a, b]$ and such that $y(a) = c, y(b) = d$.

As in the case of finding the extrema of functions, we first find the stationary values of I and then consider further conditions which determine the character of the stationary values.

Definition. $y = y(x)$ *is said to be an* admissible arc *for* $a \leq x \leq b$, *if:*

(i) $y(a) = c$, $y(b) = d$, i.e., the curve $y = f(x)$ passes through the points $A(a, c)$ and $B(b, d)$ of the (x, y)-plane;

(ii) $y(x)$ is a single valued function;

(iii) $y(x) \in C^n$, $n \geq 2$.

It will be assumed that for y being an admissible arc and $a \leq x \leq b$, $f \in C^p$, $p \geq 3$.

The variational problem is to find the admissible arc which makes I stationary.

Let $y = \varphi(x)$ be the equation of the admissible curve that makes I stationary and let

$$Y = \varphi(x) + \varepsilon\eta(x) \tag{3.11}$$

be the equation of a varied admissible curve, (cf. Fig. 3.1), where ε is an arbitrary number independent of x, y and η, $\eta(x)$ is a function of class C^n, $n \geq 1$, over $[a, b]$ such that $\eta(a) = \eta(b) = 0$, and $\eta(x)$ is independent of ε. The difference in the y-coordinate between $y = \varphi(x)$ and $Y = \varphi(x) + \varepsilon\eta(x)$ is given by

$$\Delta y = Y - y = \varepsilon\eta(x) \tag{3.12}$$

Since ε and $\eta(x)$ are independent of each other, and since $\eta(x) \in C^n$, $n \geq 1$,

$$\lim_{\varepsilon \to 0} Y' = \lim_{\varepsilon \to 0} [\varphi'(x) + \varepsilon\eta'(x)] = \varphi'(x) \tag{3.13}$$

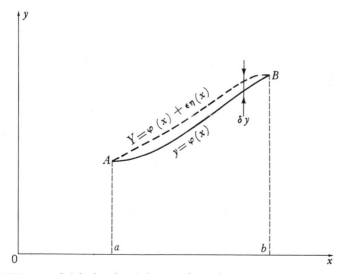

FIGURE 3.1 Original and varied curves for study of stationary value of integral I.

Using the definition of the preceding section, we have

$$\Delta y = \omega(x, \varepsilon) = \varepsilon\eta(x)$$

$$\frac{\partial\omega(x, \varepsilon)}{\partial x} = \varepsilon\eta'(x)$$

so that

$$\lim_{\varepsilon \to 0} \omega(x, \varepsilon) = 0 \qquad \lim_{\varepsilon \to 0} \frac{\partial\omega(x, \varepsilon)}{\partial x} = 0$$

which shows that the problem is one of weak variation. According to the definition given in Sect. 3.1, the total variation of I is given by

$$\Delta I = \int_a^b [f(x, Y, Y') - f(x, y, y')]\, dx$$

$$= \int_a^b [f(x, y + \varepsilon\eta, y' + \varepsilon\eta') - f(x, y, y')]\, dx \qquad (3.14)$$

Similarly, the total variation of f is given by

$$\Delta f = f(x, y + \varepsilon\eta, y' + \varepsilon\eta') - f(x, y, y')$$

Expanding $\Delta f'$ in a Taylor series with a remainder term, we obtain in agreement with Eq. (3.5)

$$\Delta f = \varepsilon\left[\eta\frac{\partial}{\partial y} + \eta'\frac{\partial}{\partial y'}\right]f(x, y, y') + R_1 = \delta f + R_1 \qquad (3.15)$$

$$R_1 = \frac{\varepsilon^2}{2}\left[\eta^2\frac{\partial^2}{\partial y^2} + 2\eta\eta'\frac{\partial^2}{\partial y\,\partial y'} + \eta'^2\frac{\partial^2}{\partial y'^2}\right]f(x, y + \theta_1\varepsilon\eta, y' + \theta_2\varepsilon\eta')$$

where

$$0 < \theta_1 < 1 \qquad 0 < \theta_2 < 1$$

According to the definition of the preceding section, I will be stationary if $\delta I = 0$, i.e.,

$$\delta I = \varepsilon\int_a^b\left[\eta\frac{\partial}{\partial y} + \eta'\frac{\partial}{\partial y'}\right]f(x, y, y')\, dx = 0 \qquad (3.16)$$

The second part of this integral can be transformed by integration by parts:

$$\int_a^b \eta'\frac{\partial f}{\partial y'}\, dx = \left[\eta\frac{\partial f}{\partial y'}\right]_a^b - \int_a^b \eta\frac{d}{dx}\left[\frac{\partial f}{\partial y'}\right]\, dx$$

where $\eta(a) = \eta(b) = 0$, and $\partial f/\partial y'$ is continuous. Thus the quantity in brackets is equal to zero, so that

$$\int_a^b \eta'\frac{\partial f}{\partial y'}\, dx = -\int_a^b \eta\frac{d}{dx}\left[\frac{\partial f}{\partial y'}\right]\, dx$$

Substituting this expression in (3.16) we obtain

$$\delta I = \varepsilon \int_a^b \eta \left[\frac{\partial f}{\partial y} - \frac{d}{dx}\left(\frac{\partial f}{\partial y'}\right) \right] dx = 0 \tag{3.17}$$

for any ε and η. This is possible if and only if

$$\frac{\partial f}{\partial y} - \frac{d}{dx}\left(\frac{\partial f}{\partial y'}\right) = 0 \tag{3.18}$$

Let

$$I + \Delta I = \psi(\varepsilon)$$

Then if I has an extremal value for $y = \varphi$, the extremum will occur for $\varepsilon = 0$, i.e., $\psi(\varepsilon)$ will have an extremum for $\varepsilon = 0$, which is to say that

$$\left[\frac{d\psi}{d\varepsilon}\right]_{\varepsilon=0} = 0$$

Since I does not contain ε, $\dfrac{d\psi}{d\varepsilon} = \dfrac{\partial}{\partial \varepsilon}[\Delta I]$. From (3.14),

$$\Delta I = \int_a^b \Delta f \, dx = \int_a^b \delta f \, dx + \int_a^b R_1 \, dx$$

Thus, according to (3.15),

$$\left[\frac{d\psi}{d\varepsilon}\right]_{\varepsilon=0} - \left[\frac{d}{d\varepsilon}\Delta I\right]_{\varepsilon=0} = \int_a^b \left[\eta \frac{\partial}{\partial y} - \eta' \frac{\partial}{\partial y'}\right] f(x,y,y') \, dx$$

which shows that (3.18) is a necessary condition for an extremum of I.

Equation (3.18) leads to a second order differential equation in y whose general solution contains two constants. The constants are determined by the boundary conditions expressing that $y(a) = c, y(b) = d$.

We can state the following theorem:

Theorem 2. *The integral $\int_a^b f(x,y,y') \, dx$, whose end points are fixed, is stationary for weak variations if y satisfies the differential equation (3.18) with the boundary conditions $y(a) = c, y(b) = d$.*

Theorem 2 gives a necessary condition for an extremum of I. The differential equation (3.18) was first discovered by Euler in 1744, and was later (1762–1770) derived in a more rigorous fashion by Lagrange, whose method is essentially that presented here. Accordingly, it is appropriate to designate this basic result of the calculus of variations as the Euler-Lagrange equation. In the classical treatises on the calculus of variations this equation is variously referred to as the Euler equation, the Lagrange equation, or the Euler-Lagrange equation. In the field of dynamics, equations of this form are universally known as Lagrange's equations.

Example 1.   Find the stationary values of the integral

$$I = \int_0^1 (x^2 + y^2 + y'^2)\, dx,$$

where $y(0) = 0, y(1) = 1$.

SOLUTION.   Apply the Euler-Lagrange equation, then

$$f = x^2 + y^2 + y'^2$$
$$\partial f/\partial y - (d/dx)(df/dy') = 2y - (d/dx)(2y') = 0$$

Hence, the differential equation is

$$y'' - y = 0$$

which has the solution

$$y = C_1 \cosh x + C_2 \sinh x$$

We must determine the constants of integration, C_1 and C_2, so that the curve $y = y(x)$ passes through the end points $(0, 0)$ and $(1, 1)$. Thus

$$y(0) = C_1 = 0 \qquad y(1) = C_2 \sinh 1 = 1$$

and the solution, therefore, is

$$y = \frac{\sinh x}{\sinh 1}$$

The Euler-Lagrange equation simplifies for those cases where either x or y does not appear explicitly in the function f. In the first circumstance, $f = f(y, y')$ and the Euler-Lagrange equation integrates to

$$f - y'(\partial f/\partial y') = C \tag{3.19}$$

where C is an arbitrary constant. To prove this result, we note that

$$(d/dx)[f - y'(\partial f/\partial y')] = 0$$

Carrying out the indicated differentiation gives

$$(\partial f/\partial y)y' + \cancel{(\partial f/\partial y')y''} - \cancel{y''(\partial f/\partial y')} - y'(d/dx)(\partial f/\partial y')$$
$$= y'[\partial f/\partial y - (d/dx)(\partial f/\partial y')] = 0$$

Therefore, since $y' \neq 0$

$$\partial f/\partial y - (d/dx)(\partial f/\partial y') = 0$$

which shows that (3.19) in this case is equivalent to the Euler-Lagrange equation.

In the second special case, $f = f(x, y')$ and the Euler-Lagrange equation reduces to

$$(d/dx)(\partial f/\partial y') = 0$$

which has the integral

$$\partial f / \partial y' = C, \quad \text{where} \quad C \text{ is a constant} \tag{3.20}$$

These results may be stated as corollaries to Theorem 2.

Corollary 1. The integral $\int_a^b f(y, y')\, dx$, whose end points are fixed, is stationary for weak variations if y satisfies the differential equation

$$f - y'(\partial f / \partial y') = C, \quad \text{where} \quad C \text{ is a constant}$$

Corollary 2. The integral $\int_a^b f(x, y')\, dx$, whose end points are fixed, is stationary for weak variations if y satisfies the differential equation

$$\frac{\partial f}{\partial y'} = C, \quad \text{where} \quad C \text{ is a constant} \tag{3.21}$$

Example 2. Problem of the Brachistochrone. This problem was first solved by Johann Bernoulli in 1691. Its name comes from brachus ($=$ short) and chronon ($=$ time), or curve of quickest descent of a particle moving without friction in a vertical plane between two given points under the effect of gravity. It is thus desired to find the plane curve along which a particle falls in shortest time from point A to point B (cf. Fig. 3.2)

Let $y = \psi(x)$ be the unknown arc of the curve, s the distance measured along the curve, and v the speed of the particle. Then $v = \dfrac{ds}{dt}$ and

$$\int_a^b dt = t_b - t_a = \int_a^b ds/v = \int_a^b \sqrt{\frac{dx^2 + dy^2}{2gy}} = \frac{1}{2\sqrt{g}} \int_a^b \sqrt{\frac{1 + y'^2}{y}}\, dx$$

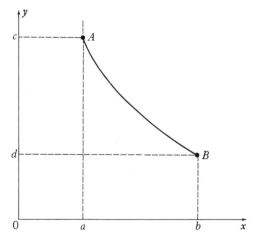

FIGURE 3.2 Curve between two points (a, c) and (b, d) in a vertical plane.

Thus the integral to be minimized is

$$I = \int_a^b \sqrt{\frac{1 + y'^2}{y}}\, dx$$

Since $f = f(y, y')$ we can use Corollary 1, i.e., we can write

$$f - y' \frac{\partial f}{\partial y'} = C$$

with

$$f = \sqrt{\frac{1 + y'^2}{y}}$$

Hence

$$\sqrt{\frac{1 + y'^2}{y}} - y' \frac{\partial}{\partial y'} \sqrt{\frac{1 + y'^2}{y}} = C$$

which gives

$$\sqrt{\frac{1 + y'^2}{y}} - \frac{y'^2}{\sqrt{y(1 + y'^2)}} = C$$

or

$$y(1 + y'^2) = \frac{1}{C^2} = C_1$$

To complete the integration it is convenient to make the change of variable $y' = \tan \theta$. We find

$$y = \frac{C_1}{1 + y'^2} = \frac{C_1}{1 + \tan^2 \theta} = C_1 \cos^2 \theta = \frac{C_1}{2}[1 + \cos 2\theta]$$

$$dx = \frac{dy}{\tan \theta} = -C_1 \frac{\sin 2\theta\, d\theta}{\tan \theta} = -2C_1 \cos^2 \theta\, d\theta = -C_1(1 + \cos 2\theta)\, d\theta$$

$$x = -\frac{C_1}{2}[2\theta + \sin 2\theta] + C_2$$

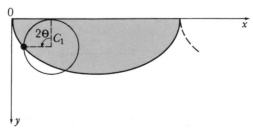

FIGURE 3.3   Cycloid generated by rolling circle.

where C_2 is another constant of integration. The equations for x and y are those for a cycloid in parametric form with 2θ as angular parameter and C_1 the radius of the generating circle (cf. Fig. 3.3). Note the cusps at the points where the marked point on the generating circle touches the x-axis.

Example 3.   Find the stationary value of the integral I which has the end points $(1, 0)$ and $(2, 2)$, where

$$I = \int_1^2 (x^2 + y'^2) \, dx$$

SOLUTION.   In this example, the integrand does not contain y explicitly. Hence, we can apply Corollary 2 which gives the solution

$$\frac{\partial f}{\partial y'} = 2y' = C$$

where C is a constant. This is readily integrated to yield

$$y' = \frac{dy}{dx} = \frac{C}{2} = C_1$$

$$y = C_1 x + C_2$$

where C_1 and C_2 are constants of integration. To evaluate C_1 and C_2 we have the end point conditions

$$y(1) = C_1 + C_2 = 0$$
$$y(2) = 2C_1 + C_2 = 2$$

Hence

$$C_1 = 2 \qquad C_2 = -2$$

and the solution is

$$y = 2(x - 1)$$

Example 4.   Soap Film Problem.   Find the surface of revolution S between two circles of radii R_1 and R_2 (Figure 3.4) about the axis Ox such that its surface area is minimum.

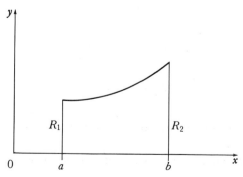

FIGURE 3.4   Minimal surface.

SOLUTION.   The area of a surface of revolution about Ox passing through the two given circles is

$$S = 2\pi \int_a^b y \, ds = 2\pi \int_a^b y(1 + y'^2)^{1/2} \, dx$$

Since $f = f(y, y')$, we can apply Corollary 1 and write

$$y\sqrt{1 + y'^2} - y' \frac{\partial}{\partial y'} (y\sqrt{1 + y'^2}) = C_1$$

from which

$$y\sqrt{1 + y'^2} - \frac{yy'^2}{\sqrt{1 + y'^2}} = C_1$$

or

$$y = C_1\sqrt{1 + y'^2}$$

Solving for y'

$$y' = \frac{dy}{dx} = \sqrt{\frac{y^2}{C_1^2} - 1}$$

or

$$\frac{C_1 \, dy}{\sqrt{y^2 - C_1^2}} = dx$$

Integration gives

$$C_1 \operatorname{arg cosh} \frac{y}{C_1} = x + K$$

or

$$y = C_1 \cosh \left(\frac{x}{C_1} + C_2 \right)$$

To find C_1 and C_2 we use the boundary conditions $y(a) = R_1, y(b) = R_2$, i.e.,

$$R_1 = C_1 \cosh \left(\frac{a}{C_1} + C_2 \right)$$

$$R_2 = C_1 \cosh \left(\frac{b}{C_1} + C_2 \right)$$

Thus

$$\frac{a}{C_1} + C_2 = \operatorname{arg cosh} \frac{R_1}{C_1}$$

$$\frac{b}{C_1} + C_2 = \operatorname{arg cosh} \frac{R_2}{C_1}$$

$$\frac{(a - b)}{C_1} = \operatorname{arg cosh} \frac{R_1}{C_1} - \operatorname{arg cosh} \frac{R_2}{C_1}$$

which gives C_1. The constant C_2 is found similarly from the relation

$$C_2 = \operatorname{arg cosh} \frac{R_1}{C_1} - \frac{a}{C_1}$$

3.3 LEGENDRE CONDITIONS

Let us now consider the expression of Δf as given by (3.15) and its integral. Thus, using the notation of Eq. (3.14) we have

$$\Delta I(\varepsilon) = \int_a^b \Delta f\, dx = \varepsilon \int_a^b \left[\eta\,\frac{\partial}{\partial y} + \eta'\,\frac{\partial}{\partial y'}\right] f(x,y,y')$$

$$+ \frac{\varepsilon^2}{2!}\int_a^b\left[\eta^2\,\frac{\partial^2}{\partial y^2} + 2\eta\eta'\,\frac{\partial^2}{\partial y\,\partial y'} + \eta'^2\,\frac{\partial^2}{\partial y'^2}\right] f(x,y,y')\,dx$$

$$+ \frac{\varepsilon^3}{3!}\int_a^b\left[\eta'\,\frac{\partial^3}{\partial y^3} + 3\eta^2\eta'\,\frac{\partial^3}{\partial y^2\,\partial y'} + 3\eta\eta'^2\,\frac{\partial^3}{\partial y\,\partial y'^2} + \eta'^3\,\frac{\partial^3}{\partial y'^3}\right]$$

$$\times\, f(x, y + \theta_1\varepsilon\eta,\, y' + \theta_2\varepsilon\eta')\,dx$$

$$0 < \theta_1 < 1 \qquad 0 < \theta_2 < 1$$

$$\Delta I(\varepsilon) = \int_a^b \delta f(x)\, dx + \int_a^b \delta^2 f(x)\, dx + \int_a^b R_2(x)\, dx \qquad \textbf{(3.22)}$$

We have seen that a necessary condition for an extremum is that the first integral vanishes identically, and that this condition gives the Euler-Lagrange equation. In order to investigate the nature of the extremum, we study

$$A = \left[\frac{\partial^2}{\partial\varepsilon^2}\,\Delta I(\varepsilon)\right]_{\varepsilon=0} = \int_a^b\left[\eta^2\,\frac{\partial^2}{\partial y^2} + 2\eta\eta'\,\frac{\partial^2}{\partial y\,\partial y'} + \eta'^2\,\frac{\partial^2}{\partial y'^2}\right] f(x,y,y')\,dx$$

$$\textbf{(3.23)}$$

It is shown in Chapter 2, that the extremum occurring for $\varepsilon = 0$ will be a maximum if $A < 0$ and a minimum if $A > 0$. If $A = 0$ the investigation has to be deepened by an expansion using an additional term in Eq. (3.22).

In order to study the sign of A, we transform it somewhat. We first observe that B as defined below equals A, for,

$$B = \int_a^b\left[\eta^2\,\frac{\partial^2}{\partial y^2} - \eta^2\,\frac{d}{dx}\left(\frac{\partial^2}{\partial y\,\partial y'}\right) - \eta\,\frac{d}{dx}\left(\eta'\cdot\frac{\partial^2}{\partial y'^2}\right)\right] f(x,y,y')\,dx$$

$$= \int_a^b\left[\eta^2\,\frac{\partial^2}{\partial y^2} + \eta\eta'\,\frac{\partial^2}{\partial y\,\partial y'} - \eta\,\frac{d}{dx}\left(\eta\,\frac{\partial^2}{\partial y\,\partial y'} + \eta'\,\frac{\partial^2}{\partial y'^2}\right)\right] f(x,y,y')\,dx$$

$$= \int_a^b\left[\eta^2\,\frac{\partial^2}{\partial y^2} + \eta\eta'\,\frac{\partial^2}{\partial y\,\partial y'}\right] f(x,y,y')\,dx$$

$$- \eta\left[\eta\,\frac{\partial^2}{\partial y\,\partial y'} + \eta'\,\frac{\partial^2}{\partial y'^2}\right] f(x,y,y')\Big|_a^b + \int_a^b \eta'\left[\eta\,\frac{\partial^2}{\partial y\,\partial y'} + \eta'\,\frac{\partial^2}{\partial y'^2}\right] f(x,y,y')\,dx$$

where the negative terms have been integrated by parts. Since $\eta(b) = \eta(a) = 0$, the second term above cancels out, so that

$$B = \int_a^b \left[\eta^2 \frac{\partial^2}{\partial y^2} + 2\eta\eta' \frac{\partial^2}{\partial y\,\partial y'} + \eta'^2 \frac{\partial^2}{\partial y'^2} \right] f(x, y, y')\, dx = A$$

Let $u(x)$ be a function of class C^2 which is a solution of the differential equation

$$\left[\frac{\partial^2 f}{\partial y^2} - \frac{d}{dx} \frac{\partial^2 f}{\partial y\,\partial y'} \right] u - \frac{d}{dx}\left(\frac{\partial^2 f}{\partial y'^2} \frac{du}{dx} \right) = 0 \qquad \textbf{(3.24)}$$

This is said to be a Jacobi differential equation. According to (3.24),

$$\eta^2 \left[\frac{\partial^2 f}{\partial y^2} - \frac{d}{dx} \frac{\partial^2 f}{\partial y\,\partial y'} \right] = \frac{\eta^2}{u} \frac{d}{dx}\left(\frac{\partial^2 f}{\partial y'^2} \frac{du}{dx} \right)$$

so that substituting into the original expression for B we obtain

$$A = B = \int_a^b \frac{\eta}{u} \left[\eta \frac{d}{dx}\left(\frac{\partial^2 f}{dy'^2} \frac{du}{dx} \right) - u \frac{d}{dx}\left(\eta' \frac{\partial^2 f}{\partial y'^2} \right) \right] dx$$

But

$$\frac{d}{dx}\left[\frac{\partial^2 f}{\partial y'^2}\left(\eta \frac{du}{dx} - u \frac{d\eta}{dx} \right) \right] = \eta \frac{d}{dx}\left(\frac{\partial^2 f}{dy'^2} \frac{du}{dx} \right) + \frac{\partial^2 f}{\partial y'^2} \frac{du}{dx} \frac{d\eta}{dx}$$

$$- \frac{\partial^2 f}{\partial y'^2} \frac{du}{dx} \frac{d\eta}{dx} - u \frac{d}{dx}\left(\frac{\partial^2 f}{\partial y'^2} \frac{d\eta}{dx} \right)$$

$$= \eta \frac{d}{dx}\left(\frac{\partial^2 f}{\partial y'^2} \frac{du}{dx} \right) - u \frac{d}{dx}\left(\frac{\partial^2 f}{\partial y'^2} \frac{d\eta}{dx} \right)$$

and hence

$$A = \int_a^b \frac{\eta}{u} \frac{d}{dx}\left[\frac{\partial^2 f}{\partial y'^2} (\eta u' - \eta' u) \right] dx$$

$$= \frac{\eta}{u} \frac{\partial^2 f}{\partial y'^2} (\eta u' - \eta' u) \bigg]_a^b - \int_a^b \frac{\partial^2 f}{\partial y'^2} (\eta u' - \eta' u) \frac{d}{dx}\left(\frac{\eta}{u} \right) dx \qquad \textbf{(3.25)}$$

Since $\eta(a) = \eta(b) = 0$ the first quantity vanishes, so that, provided $u(a) \neq 0$ and $u(b) \neq 0$ (although we shall prove later that this is not necessary)

$$A = -\int_a^b \frac{\partial^2 f}{\partial y'^2} (\eta u' - \eta' u) \frac{d}{dx}\left(\frac{\eta}{u} \right) dx$$

$$= -\int_a^b \frac{\partial^2 f}{\partial y'^2} \frac{(\eta u' - \eta' u)(\eta' u - \eta u')}{u^2} dx$$

$$A = \int_a^b \frac{\partial^2 f}{\partial y'^2}\left[\eta' - \eta \frac{u'}{u} \right]^2 dx \qquad \textbf{(3.26)}$$

Let us now discuss the new form of A, as given by (3.26), taking into account all secondary conditions assumed during the deduction. We first observe that the Jacobi equation (3.24) is actually identical to the condition obtained by writing the Euler-Lagrange equation for A as given by (3.23), where η is the unknown function. For if A is considered to be

$$A = \int_a^b \varphi(x, \eta, \eta')\, dx$$

then A will be an extremum provided

$$\frac{\partial \varphi}{\partial \eta} - \frac{d}{dx}\left(\frac{\partial \varphi}{\partial \eta'}\right) = 0 \qquad \varphi = \eta^2 \frac{\partial^2 f}{\partial y^2} + 2\eta\eta' \frac{\partial^2 f}{\partial y\, \partial y'} + \eta'^2 \frac{\partial^2 f}{\partial y'^2}$$

i.e.,

$$2\eta' \frac{\partial^2 f}{\partial y\, \partial y'} + 2\eta \frac{\partial^2 f}{\partial y^2} - \frac{d}{dx}\left[2\eta \frac{\partial^2 f}{\partial y\, \partial y'} + 2\eta' \frac{\partial^2 f}{\partial y'^2}\right] = 0$$

This expression can be simplified as follows

$$\eta' \frac{\partial^2 f}{\partial y\, \partial y'} + \eta \frac{\partial^2 f}{\partial y^2} - \eta' \frac{\partial^2 f}{\partial y\, \partial y'} - \eta \frac{d}{dx}\frac{\partial^2 f}{\partial y\, \partial y'} - \frac{d}{dx}\left[\eta' \frac{\partial^2 f}{\partial y'^2}\right] = 0$$

or

$$\left[\frac{\partial^2 f}{\partial y^2} - \frac{d}{dx}\frac{\partial^2 f}{\partial y\, \partial y'}\right]\eta - \frac{d}{dx}\left[\frac{\partial^2 f}{\partial y'^2}\frac{d\eta}{dx}\right] = 0$$

which shows that η is actually a solution of (3.24).

Next we consider the Jacobi equation (3.24) which is of the form

$$\frac{d}{dx}\left[P(x)\frac{du}{dx}\right] - Q(x)u = 0 \qquad\qquad \textbf{(3.27)}$$

where

$$P(x) = \frac{\partial^2 f}{\partial y'^2} \quad \text{and} \quad Q(x) = \frac{\partial^2 f}{\partial y^2} - \frac{d}{dx}\frac{\partial^2 f}{\partial y\, \partial y'}$$

This is a second order linear differential equation, since it can be written

$$Pu'' + P'u - Qu = 0 \qquad\qquad \textbf{(3.28)}$$

and is of the Sturm-Liouville type. Let u_1 and u_2 be two linearly independent solutions of this equation. Then clearly

$$\frac{d}{dx}[Pu_1'] - Qu_1 = 0$$

$$\frac{d}{dx}[Pu_2'] - Qu_2 = 0$$

Multiplying the first of these equations by u_2, the second by u_1, and subtracting, we obtain

$$u_2 \frac{d}{dx}[Pu_1'] - u_1 \frac{d}{dx}[Pu_2'] = 0 \qquad (3.29)$$

But since

$$\frac{d}{dx}[P(u_2 u_1' - u_1 u_2')] = u_2 \frac{d}{dx}[Pu_1'] + \cancel{Pu_1' u_2'} - \cancel{Pu_1' u_2'} - u_1 \frac{d}{dx}[Pu_2']$$

it follows that (3.29) can be written

$$\frac{d}{dx}[P(u_2 u_1' - u_1 u_2')] = 0$$

Thus

$$u_2 u_1' - u_1 u_2' = \frac{C}{P} \qquad (3.30)$$

where C is an arbitrary constant. Since f is of class C^p, $p \geq 3$, P is finite for $a \leq x \leq b$ so that $C/P \neq 0$. This means that neither u_1 and u_1' nor u_2 and u_2' can cancel together for $a \leq x \leq b$, so that u_1 and u_2 cannot have double roots for $a \leq x \leq b$. In the deduction of A as given by (3.26) we have assumed that $u(a)$ and $u(b)$ are both different from zero. Actually, even if $u(a)$ or $u(b)$ or both are zero in the integrated part of (3.25), the first term $\dfrac{\eta^2}{u} u' \dfrac{\partial^2 f}{\partial y'^2}$ will still vanish at a and b since u cannot have a double root for $a \leq x \leq b$.

We now consider A as given by (3.26). Assuming first that $\eta' - \eta u'/u$ is not identically zero, then the integrand will have a constant sign if $\dfrac{\partial^2 f}{\partial y'^2}$ has a constant sign throughout the interval $[a, b]$.

If now $\dfrac{\partial^2 f}{\partial y'^2}$ has not a constant sign throughout $[a, b]$, then it will change sign at least once. Suppose it is positive for $a \leq x < p$, negative for $p < x \leq b$, and zero for $x = p$. Since the choice of $\eta(x)$ is arbitrary (except for the fact that it is of class C^n, $n \geq 1$, and $\eta(a) = \eta(b) = 0$), we take first

$$\eta(x) = \eta_1(x) = \begin{cases} = 0 & \text{for} \quad a \leq x \leq p \\ \neq 0 & \text{for} \quad p < x \leq b \end{cases}$$

This will make A, as given by Eq. (3.26), negative. We then take

$$\eta(x) = \eta_2(x) = \begin{cases} \neq 0 & a \leq x < p \\ = 0 & p \leq x \leq b \end{cases}$$

which will make A positive. Thus by changing $\eta(x)$ we can have a different sign for A, which shows that I will have neither a maximum nor a minimum.

Now consider the quantity $\left(\eta' - \eta \dfrac{u'}{u}\right)^2$. We must make sure that this quantity does not vanish, for if it vanishes $\delta^2 I$ would be identically zero and we would have to consider the expression of Eq. (3.6) in the case $q = 4$. Actually it would be necessary to have $\delta^3 I = 0$ and to study the sign of $\delta^4 I$. In order to avoid this we establish the following test due to Jacobi.

The Jacobi equation, as seen earlier, has a general solution of the form

$$u = C_1 u_1 + C_2 u_2$$

where C_1 and C_2 are arbitrary constants, and u_1 and u_2 are two linearly independent solutions of Eq. (3.24). If $\eta' - \eta u/u' = 0$, then clearly $\eta = Ku$, where K is an arbitrary constant. This is what we actually want to avoid. We know that $\eta(a) = \eta(b) = 0$. Let us assume that $u(a) = 0$. Then clearly

$$0 = C_1 u_1(a) + C_2 u_2(a)$$

so that

$$u_1(a)/u_2(a) = -C_2/C_1$$

Let r be the first root of u that is larger than a, i.e., $r > a$ and $u(r) = 0$; r is called the conjugate point of a. Clearly

$$u(r) = 0 = C_1 u_1(r) + C_2 u_2(r)$$

and therefore

$$\frac{u_1(r)}{u_2(r)} = -\frac{C_1}{C_2} = \frac{u_1(a)}{u_2(a)}$$

If $r > b$, then u, not having a root at b, cannot be proportional to η, i.e., $\left(\eta' - \eta \dfrac{u'}{u}\right)^2 > 0$, except at *points* where η and η' would vanish simultaneously. Thus if $r > b$, $\left(\eta' - \eta \dfrac{u'}{u}\right)^2 \geq 0$ so that $\delta^2 I$ will have the sign of $\dfrac{\partial^2 f}{\partial y'^2}$. We can thus state:

Theorem 3. *If $y = y(x)$ is the equation of an extremal through (a, c) and (b, d) for which the integral $I = \int_a^b f(x, y, y')\, dx$ is stationary, and if r, the first root of the solution of the Jacobi equation to the right of a, is larger than b, and if $\partial^2 f/\partial y'^2$ has a constant sign throughout the interval $[a, b]$, then I is a maximum if $\partial^2 f/\partial y'^2 < 0$, and a minimum if $\partial^2 f/\partial y'^2 > 0$.*

If $r < b$, it can be shown that I has neither a maximum nor a minimum. We refer for this proof to more specialized treatises.

Theorem 3 states what is called the Legendre test, although it is often stated in the form that the "interval is not too large," instead of explicitly $r > b$.

In a number of cases, especially in problems with a physical background, the nature of the extremum can be deduced from the problem itself.

Example I

Study the extrema of $I = \int_0^1 (x + y^2 + y'^2)\, dx$, with $y(0) = 0$, $y'(0) = 1$.

We clearly have

$$f = x + y^2 + y'^2 \qquad \frac{\partial f}{\partial y} = 2y \qquad \frac{\partial^2 f}{\partial y^2} = 2$$

$$\frac{\partial f}{\partial y'} = 2y' \qquad \frac{\partial^2 f}{\partial y'^2} = 2$$

$$\frac{\partial^2 f}{\partial y\, \partial y'} = 0$$

The Euler equation leads to

$$\frac{d}{dx}\left[\frac{\partial f}{\partial y'}\right] - \frac{\partial f}{\partial y} = \frac{d}{dx}\,[2y'] - 2y = 0$$

or to

$$y'' - y = 0$$

Thus

$$y = A \cosh x + B \sinh x$$

With the given boundary conditions $y(0) = 0$, $y'(0) = 1$, we find

$$y = \sinh x$$

Let us now find the Jacobi equation

$$\left(\frac{\partial^2 f}{\partial y} - \frac{d}{dx}\,\frac{\partial^2 f}{\partial y\, \partial y'}\right) u - \frac{d}{dx}\left(\frac{\partial^2 f}{\partial y'^2}\,\frac{du}{dx}\right) = 0$$

which reduces to

$$u'' - u = 0$$

Its general solution is

$$u = C_1 \cosh x + C_2 \sinh x$$

The solution such that $u(0) = 0$ is clearly

$$u = C_2 \sinh x$$

This function has clearly no other root, so that r does not exist. Since on the other hand $\partial^2 f/\partial y'^2 = 2$, it follows that the solution

$$y = \sinh x$$

makes I a minimum.

Example 2. Apply the Legendre test to the problem of the brachistrochrone.

SOLUTION. We are to determine the second order partial derivative of the integrand of

$$I = \int_a^b \sqrt{(1 + y'^2)/y}\, dx$$

which gives

$$\frac{\partial^2}{\partial y'^2} \sqrt{(1 + y'^2)/y} = y^{-1/2} (1 + y'^2)^{-3/2}$$

If the positive values of the square roots are chosen and $y > 0$, it follows that $\partial^2 f/\partial y'^2 > 0$ throughout the range of integration. Therefore the integrals obtained in Ex. 2 of Sect. 3.2 admit a minimum for a sufficiently small range of integration. The actual range of integration permitted can be determined by the method explained, but would lead to a very elaborate calculation. This can be avoided by observing that it is possible to choose a path making I infinite; thus I cannot have a maximum.

3.4 LAGRANGE MULTIPLIER

There are many problems in the calculus of variations where the variables are subject to constraints or condition equations in the form of integrals. One of the oldest of these problems is to find the shape of a closed plane curve of given length which encloses maximum area. That is to say, to find the maximum value of $\int dA$ subject to the condition $\int ds = L$, where L is the given length of the curve. A second example is to find the arc of a curve of minimum length between two given points, where the curve and the given points lie on a given surface. In mathematical terms, it is required to find the minimum value of $\int ds$ subject to the condition $S(x, y, z) = 0$, where ds is the element of arc and the equation is that of the surface S. This is the classical problem of the geodesic curve.

We consider the case of a single dependent variable where it is required to make

$$I = \int_a^b f(x, y, y')\, dx$$

a maximum or a minimum, subject to a constraint of the form

$$F(x, y, y') = 0, \quad \text{or}$$

$$I_1 = \int_a^b F(x, y, y')\, dx = \text{constant} = C$$

This problem is conveniently handled by the method of the Lagrangian multiplier. Define an integral

$$J = I + \lambda I_1 \tag{3.31}$$

where λ is an unknown constant called the Lagrange multiplier. A necessary condition for an extremum is

$$\delta J = \delta(I + \lambda I_1) = 0 \tag{3.32}$$

Suppose that curves

$$g(x, y, C_1, C_2, \lambda) = 0 \tag{3.33}$$

are found which satisfy condition (3.32). Two of the arbitrary constants may be used in general to pass the curve through the two fixed points A and B, and the third may be determined so that the integral I_1 has its prescribed value. The stationary value of I is then assured because of (3.32).

We recognize (3.32) as the first variation of an integral with the integrand $f + \lambda F$. Hence we may apply Theorem 1 directly and obtain the Euler-Lagrange equation

$$\partial(f + \lambda F)/\partial y - (d/dx)[\partial(f + \lambda F)/\partial y'] = 0 \tag{3.34}$$

Equation (3.34) is a second order ordinary differential equation whose solution in general contains two arbitrary constants C_1 and C_2. Hence, the solution is of the form of (3.33) which satisfies the conditions that the arc of the curve represented by the equation $y = y(x)$ passes through the end points A and B (assuming this to be possible), and that the condition integral I_1 equals the prescribed constant.

Example 1. In the x, y plane find the curve of constant length L passing through the points $A(0, 1)$, $B(1, 0)$ (Fig. 3.5) such that the area

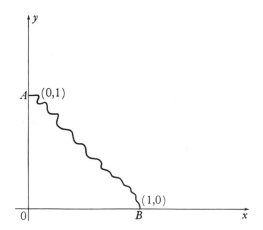

FIGURE 3.5   Illustration for Example 1, Sect. 3.4.

enclosed by the curve, Ox, and Oy is an extremum. Consider the special case where $L = \dfrac{\pi}{2}$.

SOLUTION. Let $y = y(x)$ be the solution. Then the area to be extremized is

$$S = \int_0^1 y\, dx$$

and the length of the curve is

$$L = \int_0^1 \sqrt{1 + y'^2}\, dx$$

Using the Lagrange multiplier λ, we have to extremize

$$J = S + \lambda L = \int_0^1 [y + \lambda\sqrt{1 + y'^2}]\, dx$$

Since x does not appear explicitly in $f = f(y, y')$ we can write

$$y + \lambda\sqrt{1 + y'^2} - \lambda\frac{y'^2}{\sqrt{1 + y'^2}} = C_1$$

or

$$y' = \frac{dy}{dx} = \sqrt{\frac{\lambda^2 - (y - C_1)^2}{(y - C_1)^2}}$$

By separating variables, we obtain

$$\frac{(y - C_1)\, dy}{\sqrt{\lambda^2 - (y - C_1)^2}} = dx$$

which can be integrated to give

$$\lambda^2 = (x - C_2)^2 + (y - C_1)^2$$

By introducing the boundary conditions $y(0) = 1$, $y(1) = 0$, we obtain

$$\lambda^2 = C_2^2 + (C_1 - 1)^2 = C_1^2 + (C_2 - 1)^2$$

from which it is seen that

$$C_1 = C_2$$

We take $C_1 = C_2 = C$, and obtain $\lambda^2 = C^2 + (C - 1)^2$. This is the relation between λ and C. It follows that

$$\lambda^2 = (x - C)^2 + (y - C)^2$$

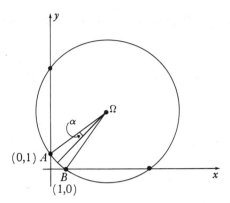

FIGURE 3.6 Illustration for Example 1, Sect. 3.4.

The solutions are circles centered on the line $y = x$. To determine λ, we find the length of the curve

$$L = \int_0^1 \sqrt{1 + y'^2}\, dx = \lambda \int_0^1 \frac{dx}{\sqrt{\lambda^2 - (x - C)^2}} = \lambda \arcsin \frac{x - C}{\lambda}\Big|_0^1$$

or

$$\lambda\left[\arcsin \frac{1 - C}{\lambda} + \arcsin \frac{C}{\lambda}\right]$$

$$= \lambda \arcsin \frac{(1 - C)\sqrt{\lambda^2 - C^2} + C\sqrt{\lambda^2 - (1 - C)^2}}{\lambda^2} = L$$

This transcendental equation connects λ and C for a given L. A simpler expression can be obtained by observing (cf. Fig. 3.6) that since the radius of the circle is λ,

$$\sin \alpha = \frac{\sqrt{2}}{2\lambda}$$

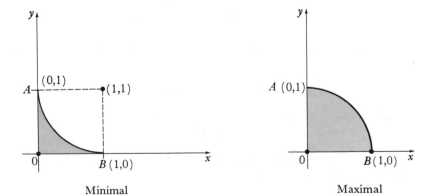

Minimal   Maximal

FIGURE 3.7 Maximum and minimum areas.

and $L = 2\alpha\lambda = 2\lambda \arcsin\left(\dfrac{\sqrt{2}}{2\lambda}\right)$.   It follows that $\sin(L/2\lambda) = \dfrac{\sqrt{2}}{2\lambda}$,

and

$$\lambda^2 = (C-1)^2 + C^2$$

which is the relation between L and λ.

The special case $L = \pi/2$ is shown in Figure 3.7. If $C = \lambda = 1$, the surface is minimum with $L = \pi/2$. If $C = 0$, $\lambda = 1$, the surface is maximum with $L = \pi/2$.

3.5   SEVERAL DEPENDENT VARIABLES

We generalize the Euler-Lagrange equation to the case of n functions and their first derivatives, i.e., we find the extrema of the integral

$$I = \int_a^b f(x, y_1, y_2, \ldots, y_n, y_1', y_2', \ldots, y_n')\, dx \tag{3.35}$$

The notation is simplified by using the concept of vector spaces. Let

$$\mathbf{y} = [y_1, y_2, \ldots, y_k, \ldots, y_n]$$

and

$$\mathbf{y}' = \frac{d\mathbf{y}}{dx} = \left[\frac{dy_1}{dx}, \frac{dy_2}{dx}, \ldots, \frac{dy_k}{dx}, \ldots, \frac{dy_n}{dx}\right]$$

where

$$y_k = y_k(x) \qquad k = 1, 2, \ldots, n,$$

Then I of (3.35) may be written

$$I = \int_a^b f(x, \mathbf{y}, \mathbf{y}')\, dx \tag{3.36}$$

The integral is taken between the two points A and B of the $(n + 1)$-dimensional space $(x, \mathbf{y})$ defined as follows:

(i) at A: $(x, \mathbf{y}) = (a, \mathbf{a})$ $x = a$ $y_k = a_k$ $k = 1, 2, \ldots, n$

(ii) at B: $(x, \mathbf{y}) = (b, \mathbf{b})$ $x = b$ $y_k = b_k$ $k = 1, 2, \ldots, n$

along a curve in the $(n + 1)$-dimensional vector space defined by $(x, \mathbf{y})$. The problem is to find $\mathbf{y}(x)$ so that I is stationary. As in the case of extrema of functions, the solution procedure is, first, to find the stationary values of I, and, second, to seek further conditions which determine the character of the stationary values.

The arc of the curve $\mathbf{y} = \mathbf{y}(x)$ is called an *admissible* arc for $a \leq x \leq b$, if:

(i) $\mathbf{y}(a) = \mathbf{a}$, $\mathbf{y}(b) = \mathbf{b}$, i.e., the curve $\mathbf{y} = \mathbf{y}(x)$ passes through $A(a, \mathbf{a})$ and $B(b, \mathbf{b})$;

(ii) $\mathbf{y}(x)$ is a single-valued vector function, i.e., $y_k = y_k(x)$ are single valued functions; and

(iii) $\mathbf{y}(x) \in C^m$, $m \geq 2$, i.e., $y_k \in C^m$, $m \geq 2$, for $k = 1, 2, \ldots, n$.

It will be assumed that for $\mathbf{y}$ being an admissible arc that $f(x, \mathbf{y}, \mathbf{y}') \in C^p$, $p \geq 3$, $a \leq x \leq b$. The problem then reduces to finding the admissible arc or arcs joining A and B in the $(n + 1)$-dimensional space for which I is stationary.

Let $\mathbf{y} = \boldsymbol{\varphi}(x)$ be the equation of the admissible arc which makes I an extremum, and let $\mathbf{Y} = \boldsymbol{\varphi}(x) + \varepsilon\boldsymbol{\eta}(x)$ be the equation of a varied admissible curve, as illustrated in Figure 3.1 (for $n = 1$). The quantity ε is an arbitrary parameter independent of $\mathbf{x}, \mathbf{y}$, and $\boldsymbol{\eta}$. The vector function $\boldsymbol{\eta}(x) = [\eta_1(x), \eta_2(x), \ldots, \eta_k(x), \ldots, \eta_n(x)]$ is of class C^m, $m \geq 1$, is independent of ε, and such that $\boldsymbol{\eta}(a) = \boldsymbol{\eta}(b) = 0$.

The difference $\Delta\mathbf{y}$ between the original curve $\mathbf{y} = \boldsymbol{\varphi}(x)$ and the varied curve $\mathbf{Y} = \boldsymbol{\varphi}(x) + \varepsilon\boldsymbol{\eta}(x)$ is called the variation of $\mathbf{y}$, and is written as follows

$$\Delta\mathbf{y} = \mathbf{Y} - \mathbf{y} = \varepsilon\boldsymbol{\eta}(x) \tag{3.37}$$

Since ε and $\boldsymbol{\eta}(x)$ are independent of each other, and since $\eta(x) \in C^m$, $m \geq 1$, it follows that

$$\lim_{\varepsilon \to 0} \mathbf{Y}' = \lim_{\varepsilon \to 0} [\boldsymbol{\varphi}'(\mathbf{x}) + \boldsymbol{\varepsilon}\boldsymbol{\eta}'(x)] = \boldsymbol{\varphi}'(\mathbf{x}) \tag{3.38}$$

Generalizing the notion of weak variation (as given in Sect. 3.1) to the case of a vector function, we have

$$\Delta\mathbf{y} = \boldsymbol{\omega}(x, \varepsilon) = \varepsilon\boldsymbol{\eta}(x)$$

$$\frac{\partial\boldsymbol{\omega}(x, \varepsilon)}{\partial x} = \varepsilon\eta'(x)$$

so that

$$\lim_{\varepsilon \to 0} \boldsymbol{\omega}(x, \varepsilon) = 0 \qquad \lim_{\varepsilon \to 0} \frac{\partial\boldsymbol{\omega}(x, \varepsilon)}{\partial x} = 0$$

This shows that the problem is one of weak variation. The total variation of I is then given by

$$\Delta I = \int_a^b [f(x, \mathbf{y} + \varepsilon\boldsymbol{\eta}, \mathbf{y}' + \varepsilon\boldsymbol{\eta}') - f(x, \mathbf{y}, \mathbf{y}')]\, dx \tag{3.39}$$

In order to use the results of Sect. 2.3, we homogenize the notation by introducing the $2n$-dimensional vectors

$$\mathbf{Z} = [\mathbf{y}, \mathbf{y}'] \qquad \mathbf{H} = [\boldsymbol{\eta}, \boldsymbol{\eta}']$$

Then

$$\mathbf{Z} + \varepsilon\mathbf{H} = [\mathbf{y} + \varepsilon\boldsymbol{\eta}, \mathbf{y}' + \varepsilon\boldsymbol{\eta}']$$
$$= [y_1 + \varepsilon\eta_1, y_2 + \varepsilon\eta_2, \ldots, y_n + \varepsilon\eta_n, y_1'$$
$$+ \varepsilon\eta_1', y_2' + \varepsilon\eta_2', \ldots, y_n' + \varepsilon\eta_n']$$

Thus, using the same procedure as in Sect. 3.2, we consider the total variation of I when $\mathbf{y}$ changes to $\mathbf{Y}$, i.e.,

$$\Delta I = \int_a^b [f(x, \mathbf{Y}, \mathbf{Y}') \quad f(x, \mathbf{y}, \mathbf{y}')] \, dx$$
$$= \int_a^b [f(x, \mathbf{Z} + \varepsilon\mathbf{H}) - f(x, \mathbf{Z})] \, dx$$

Following the method of Sect. 2.3, we can write

$$\Delta f = f(x, \mathbf{Z} + \varepsilon\mathbf{H}) - f(x, \mathbf{Z}) = F(\varepsilon) - F(0)$$
$$= \varepsilon F'(0) + \frac{\varepsilon^2}{2!} F''(\theta\varepsilon) \qquad \textbf{(3.40)}$$

where $0 < \theta < 1$.

In Eq. (3.40),

$$F'(\varepsilon) = \mathbf{H} \cdot \nabla f(x, \mathbf{Z} + \varepsilon\mathbf{H})$$

and

$$F''(\varepsilon) = (\mathbf{H} \cdot \nabla)^2 f(x, \mathbf{Z} + \varepsilon\mathbf{H})$$

Thus

$$\Delta I = \varepsilon \int_a^b \mathbf{H} \cdot \nabla f(x, \mathbf{Z}) \, dx + \frac{\varepsilon^2}{2!} \int_a^b (\mathbf{H} \cdot \nabla)^2 f(x, \mathbf{Z} + \theta\varepsilon\mathbf{H}) \, dx \quad \textbf{(3.41)}$$

with

$$\frac{\varepsilon^2}{2!} (\mathbf{H} \cdot \nabla)^2 f(x, \mathbf{Z} + \theta\varepsilon\mathbf{H}) = R_1$$

According to the definition given in Sect. 3.1, the integral I will be stationary if

$$\delta I = \varepsilon \int_a^b \mathbf{H} \cdot \nabla f(x, \mathbf{Z}) \, dx = \varepsilon \int_a^b \delta f \, dx = 0 \qquad \textbf{(3.42)}$$

Since $\varepsilon \neq 0$, Eq. (3.42) can be written

$$\int_a^b \mathbf{H} \cdot \nabla f(x, \mathbf{Z}) \, dx = \int_a^b \sum_{k=1}^n \left(\eta_k \frac{\partial f}{\partial y_k} + \eta_k' \frac{\partial f}{\partial y_k'} \right) dx$$

Integration by parts gives

$$\int_a^b \eta_k' \frac{\partial f}{\partial y_k'} \, dx = \left[\eta_k \frac{\partial f}{\partial y_k'} \right]_a^b - \int_a^b \eta_k \frac{d}{dx} \left(\frac{\partial f}{\partial y_k'} \right) dx$$

But $\eta_k(a) = \eta_k(b) = 0$ and $\dfrac{\partial f}{\partial y_k'}$ is continuous, so that the quantity in brackets is equal to zero. Hence,

$$\int_a^b \eta_k' \frac{\partial f}{\partial y_k'} \, dx = - \int_a^b \eta_k \frac{d}{dx} \left(\frac{\partial f}{\partial y_k'} \right) dx \qquad k = 1, 2, \ldots, n$$

and it follows that

$$\delta I = \varepsilon \int_a^b \sum_{k=1}^n \left[\frac{\partial f}{\partial y_k} - \frac{d}{dx} \left(\frac{\partial f}{\partial y_k'} \right) \right] \eta_k \, dx = 0 \qquad \textbf{(3.43)}$$

Now let

$$A_k = \frac{\partial f}{\partial y_k} - \frac{d}{dx} \left(\frac{\partial f}{\partial y_k'} \right)$$

and define

$$\mathbf{A} = [A_1, A_2, \ldots, A_n] = \boldsymbol{\nabla}_{y_k} f - \frac{d}{dx} \boldsymbol{\nabla}_{y_k'} f$$

where

$$\boldsymbol{\nabla}_{y_k} = \left[\frac{\partial}{\partial y_1}, \frac{\partial}{\partial y_2}, \ldots, \frac{\partial}{\partial y_n} \right]$$

and

$$\boldsymbol{\nabla}_{y_k'} = \left[\frac{\partial}{\partial y_1'}, \frac{\partial}{\partial y_2'}, \ldots, \frac{\partial}{\partial y_n'} \right]$$

Hence

$$\delta I = \varepsilon \int_a^b \mathbf{A} \cdot \boldsymbol{\eta} \, dx = 0$$

for any ε and $\boldsymbol{\eta}$. This is possible if and only if

$$A_k = \frac{\partial f}{\partial y_k} - \frac{d}{dx} \left(\frac{\partial f}{\partial y_k'} \right) = 0 \qquad k = 1, 2, \ldots, n \qquad \textbf{(3.44)}$$

Let again $I + \Delta I = \psi(\varepsilon)$, where I has been defined by Eq. (3.8). Since $\mathbf{y} = \boldsymbol{\varphi}$ gives an extremal value to I, the extremum will occur for $\varepsilon = 0$, i.e., $\psi(\varepsilon)$ will have an extremum for $\varepsilon = 0$. This is to say that

$$\left[\frac{d\psi}{d\varepsilon} \right]_{\varepsilon=0} = 0$$

But since I does not contain ε, $\dfrac{d\psi}{d\varepsilon} = \dfrac{\partial}{\partial \varepsilon} [\Delta I]$. Using Eqs. (3.41) and (3.42)

$$\Delta I = \varepsilon \int_a^b \delta f \, dx + \frac{\varepsilon^2}{2} \int_a^b R_1 \, dx$$

which shows that for an extremum of I

$$\left[\frac{d\psi}{d\varepsilon}\right]_{\varepsilon=0} = \left[\frac{\partial}{\partial\varepsilon}\Delta I\right]_{\varepsilon=0} = \int_a^b \delta f \, dx = 0$$

that is,

$$\delta I = 0$$

It follows that Eq. (3.44) gives the necessary conditions for an extremum of I.

We can state:

Theorem 4. *The integral $\int_a^b f(x, \mathbf{y}, \mathbf{y}') \, dx$, whose end points are fixed, is stationary for weak variations if the components of $\mathbf{y}$ satisfy the differential equation (3.44). These conditions are necessary for an extremum of the integral.*

The general solution of the system of simultaneous differential equations contains in general two arbitrary constants for each y_k which must be chosen so that the solution curve passes through $A(a, \mathbf{a})$ and $B(b, \mathbf{b})$.

Corollaries 1 and 2 of Sect. 3.2 can be applied individually to all y_k, as we shall show. In the first case

$$\frac{d}{dx}\left[f(\mathbf{y}, \mathbf{y}') - \sum_{k=1}^n y_k' \frac{\partial f}{\partial y_k'}\right]$$

$$= \sum_{k=1}^n \left[\frac{\partial f}{\partial y_k} y_k' + \frac{\partial f}{\partial y_k'} y_k'' - y_k'' \frac{\partial f}{\partial y_k'} - y_k' \frac{d}{dx}\frac{\partial f}{\partial y_k'}\right] = \sum_{k=1}^n y_k'\left[\frac{\partial f}{\partial y_k} - \frac{d}{dx}\left(\frac{\partial f}{\partial y_k'}\right)\right]$$

and therefore

$$\frac{d}{dx}\left[f(\mathbf{y}, \mathbf{y}') - \sum_{k=1}^n y_k' \frac{\partial f}{\partial y_k'}\right] = \sum_{k=1}^n y_k'\left[\frac{\partial f}{\partial y_k} - \frac{d}{dx}\left(\frac{\partial f}{\partial y_k'}\right)\right]$$

Since the last expression contains in brackets the Lagrange equation which is identically zero for all k, we can write

$$\frac{d}{dx}\left[f(\mathbf{y}, \mathbf{y}') - \sum_{k=1}^n y_k' \frac{\partial f}{\partial y_k'}\right] = \sum_{k=1}^n y_k'\left[\frac{\partial f}{\partial y_k} - \frac{d}{dx}\frac{\partial f}{\partial y_k'}\right] = 0$$

and the expression can be integrated to give

$$f(\mathbf{y}, \mathbf{y}') - \sum_{k=1}^n y_k' \frac{\partial f}{\partial y_k'} = \text{const.} \qquad (3.45)$$

In the second case, assume that f does not contain a variable y_k explicitly. Then the corresponding Euler-Lagrange equation reduces to the form

$$\frac{d}{dx}\left(\frac{\partial f}{\partial y_k'}\right) = 0$$

which has the integral

$$\frac{\partial f}{\partial y'_k} = C_k = \text{const.} \tag{3.46}$$

It is clear that there is an integral of the form of Eq. (3.21) for each missing variable y_k in f.

These results can be stated as corollaries to Theorem 4.

Corollary 3. The integral $\int_a^b f(\mathbf{y}, \mathbf{y}') \, dx$, whose end points are fixed, is stationary for weak variations if y satisfies the differential equation

$$f = \sum_{k=1}^{n} y'_k \frac{\partial f}{\partial y'_k} = C = \text{const.}$$

Corollary 4. If the integral $\int_a^b f(x, \mathbf{y}, \mathbf{y}') \, dx$, whose end points are fixed, does not contain explicitly certain y'_ks, then it will be stationary for weak variations if for the y'_ks that *do not* appear explicitly in f the differential equation (3.46) is satisfied while for those which appear explicitly in f the differential equation (3.44) is satisfied.

3.6 LEGENDRE CONDITIONS FOR SEVERAL DEPENDENT VARIABLES

The analysis of Sect. 3.3 can be repeated in the case of several variables. The calculations are rather long and the discussion complicated. In many applications, one can do without the Legendre condition by using physical conditions. The complete discussion of the Legendre condition is outside the framework of this book and we refer to more specialized treatises for the proofs and discussions (see Ref. 12). We limit ourselves here to the statement of the Legendre conditions in their simplest form:

Theorem 5. *The integral*

$$\int_a^b f(x, \mathbf{y}, \mathbf{y}') \, dx$$

whose end points are fixed has an extremum for weak variations of $\mathbf{y}$ *if:*

(i) $\dfrac{\partial^2 f}{\partial y_k'^2}$ $\quad k = 1, 2, \ldots, n$, all have constant signs throughout the interval $[a, b]$ for x

(ii) $\left(\dfrac{\partial^2 f}{\partial y_k'^2}\right)\left(\dfrac{\partial^2 f}{\partial y_s'^2}\right) - \left(\dfrac{\partial^2 f}{\partial y_k' \, \partial y_s'}\right)^2 > 0, \; k, s = 1, 2, \ldots, n$

throughout the interval $[a, b]$ for x

(iii) the range of integration is sufficiently small; and
(iv) the extremum is

$$\left.\begin{array}{l} \text{a minimum if } \dfrac{\partial^2 f}{\partial y_k'^2} > 0 \\[3em] \text{a maximum if } \dfrac{\partial^2 f}{\partial y_k'^2} < 0 \end{array}\right\} \quad \text{for} \quad k = 1, 2, \ldots, n$$

Condition (iii) corresponds to a generalization of the Jacobi differential equation which relates to the position of the conjugate point a.

3.7   CONSTRAINED   VARIATIONS

Let us consider the problem in the general case. Given

$$I = \int_a^b f(x, \mathbf{y}, \mathbf{y}') \, dx$$

where $\mathbf{y} = [y_1, y_2, \ldots, y_n]$, which is to be extremized with the conditions

$$F_k(x, \mathbf{y}, \mathbf{y}') = 0 \qquad k = 1, 2, \ldots, m < n$$

Consider the integral

$$J = I + \sum_{k=1}^m \lambda_k I_k$$

where

$$I_k = \int_a^b F_k(x, \vec{y}, \vec{y}') \, dx = 0 \qquad k = 1, 2, \ldots, m < n$$

and the λ_k are called the Lagrange multipliers (cf. Chap. 2). A necessary condition for an extremum is that $\delta J = 0$. This condition gives n differential equations of the second order containing $2n$ constants that will be determined by the conditions $\mathbf{y}(a) = \mathbf{a}$ and $\mathbf{y}(b) = \mathbf{b}$, and $(a, \mathbf{a})$ and $(b, \mathbf{b})$ correspond to the points A and B. We obtain, therefore, n relations for $\mathbf{y}$ and $\lambda_1, \lambda_2, \ldots, \lambda_m$. The relations $F_k = 0, k = 1, 2, \ldots, m$ give the m additional equations necessary to determine $\mathbf{y}$ and the m Lagrange multipliers.

Example I

Let

$$I = \int_0^1 (x + y + z + y'^2 + z'^2) \, dx$$

where $y = y(x)$, $z = z(x)$, and I is to be extremized with the conditions

$$
\begin{aligned}
y(0) &= 0 & z(0) &= 1 \\
y(1) &= 1 & z(1) &= 0 \\
y + z &= 1
\end{aligned}
$$

SOLUTION

Here

$$
\begin{aligned}
J &= \int_0^1 [x + y + z + y'^2 + z'^2 + \lambda(y + z - 1)] \, dx \\
&= \int_0^1 [x - \lambda + y(1 + \lambda) + z(1 + \lambda) + y'^2 + z'^2] \, dx
\end{aligned}
$$

so that

$$
\frac{\partial f}{\partial y} - \frac{d}{dx}\left(\frac{\partial f}{\partial y'}\right) = 1 + \lambda - \frac{d}{dx}(2y') = 1 + \lambda - 2y'' = 0
$$

and

$$
\frac{\partial f}{\partial z} - \frac{d}{dx}\left(\frac{\partial f}{\partial z'}\right) = 1 + \lambda - \frac{d}{dx}(2z') = 1 + \lambda - 2z'' = 0
$$

Next solve the differential equations, obtaining

$$
y = \frac{1 + \lambda}{4} x^2 + C_1 x + C_2 \qquad z = \frac{1 + \lambda}{4} x^2 + C_3 x + C_4
$$

To evaluate the constants of integration, substitute the boundary conditions in the equations for y and z as follows:

$$
\begin{aligned}
y(0) &= C_2 = 0 & z(0) &= C_4 = 1 \\
y(1) &= \frac{1 + \lambda}{4} + C_1 = 1 & z(1) &= \frac{1 + \lambda}{4} + C_3 + 1 = 0
\end{aligned}
$$

The latter two conditions give for C_1 and C_3 the values

$$
C_1 = \frac{3 - \lambda}{4} \qquad C_3 = -\frac{5 + \lambda}{4}
$$

Hence

$$
y = \frac{1 + \lambda}{4} x^2 + \frac{3 - \lambda}{4} x \qquad z = \frac{1 + \lambda}{4} x^2 - \frac{5 + \lambda}{4} x + 1
$$

Using the condition $y + z = 1$ gives

$$
y + z = 1 = \frac{1 + \lambda}{2} x^2 - \frac{1 + \lambda}{2} x + 1 = 1
$$

which is possible only for $1 + \lambda = 0$; hence, the solutions for y and z are

$$y = x$$

$$z = 1 - x$$

Example 2.   Optimum Gas Flow Problem*

In the removal of smoke and dust particles from gases by electrostatic precipitation, the gases are passed through a high voltage corona field maintained between grids of fine wires and parallel flat plates. The particles become electrically charged by the corona ions and are carried by electrical forces to the plate surfaces. The theory of the electrostatic precipitation process shows that the chance of a given particle passing through the precipitator and escaping collection is given by

$$Q = e^{-Aw/A_c v}$$

where Q is the chance of escaping capture, A is the collection surface area of the plates, w is the velocity of the particle due to electric force, A_c is the cross-sectional area of the precipitator for gas flow, and v is the gas velocity through the precipitator.

The above equation holds for uniform gas velocity distribution through the precipitator. An important basic question arises as to the effect of nonuniform flow through the precipitator. If the gas velocity is nonuniform and is distributed according to some statistical distribution function $\gamma(v)$, the fraction of particles escaping collection is then given by

$$Q = \int_0^{v_{max}} e^{-Aw/A_c v} \gamma(v) \, dv$$

where now Q is the fraction of particles lost and v_{max} is the maximum gas velocity at any point over the face of the precipitator.

The problem then is to determine $\gamma(v)$ such that Q is a minimum, i.e., that the collection efficiency is a maximum.

SOLUTION

(1) Since the total gas flux is fixed and $\gamma(v)$ is normalized, the integral Q is to be minimized subject to the constraints

$$\int_0^{v_{max}} \gamma(v) \, dv = 1 \tag{1}$$

---

* From H. J. White, *Industrial Electrostatic Precipitation*. Reading, Mass.: Addison-Wesley, 1963, page 258.

and

$$V = \int v \, dA_c = A_c \int_0^{v_{max}} v\gamma(v) \, dv = V_0 \tag{2}$$

where A_c is the cross-sectional area of the precipitator and V_0 is a constant.

(2) Using Lagrangian multipliers, we minimize the integral

$$J = \int_0^{v_{max}} [e^{-Aw/A_c v} + \lambda_1 A_c v + \lambda_2] \gamma(v) \, dv$$

The Euler-Lagrange equation gives

$$\frac{\partial f}{\partial \gamma} = e^{-Aw/A_c v} + \lambda_1 A_c v + \lambda_2 = 0$$

which has the solution

$$v = \text{const.} = v_0$$

There is only one solution possible since for physical reasons $v > 0$.

From the relation

$$\int v \, dA_c = v_0 A_c = V_0$$

it follows that

$$v_0 = \frac{V_0}{A_c}$$

We deduce that this value of v_0 minimizes Q, since from the form of the integral for Q there is no maximum value, although the upper limit of Q is unity.

BIBLIOGRAPHY

1. E. Goursat, *Cours d'Analyse Mathematique*, Vol. III, Fifth Edition. Paris: Gauthier-Villars, 1942.
2. R. Courant, *Differential and Integral Calculus*, Vol. II. London: Blackie and Son, 1944.
3. R. Courant and D. Hilbert, *Methods of Mathematical Physics*, Vol. I. New York: Interscience Publishers, 1953.
4. H. Margenau and G. M. Murphy, *The Mathematics of Physics and Chemistry*, Second Edition. Princeton: Van Nostrand, 1956.
5. F. B. Hildebrand, *Methods of Applied Mathematics*. Englewood Cliffs, New Jersey: Prentice-Hall, 1958.
6. P. Frank and R. V. Mises, *Die Differential- und Integralgleichungen der Mechanik und Physik*. New York: Dover, 1961.
7. E. F. Beckenbach, *Modern Mathematics for the Engineer*. New York: McGraw-Hill, 1956.

8. O. Bolza, *Lectures on the Calculus of Variations*, Second Edition. New York: Chelsea, reprint 1904.
9. N. I. Akhiezer, *The Calculus of Variations*. New York: Blaisdell, 1962.
10. O. Bolza, *Vorlesungen über Variationsrechnung*. Leipzig: B. G. Teubner, 1909.
11. I. M. Gelfand and S. V. Fomin, *Calculus of Variations*. Englewood Cliffs, New Jersey: Prentice-Hall, 1963. (Translated from the Russian by R. A. Silverman.)
12. A. R. Forsyth, *Calculus of Variations*. New York: Dover Publications, 1960.

4

SYSTEMS OF DIFFERENTIAL EQUATIONS

Major objectives of this chapter are to familiarize the reader with the modern approach to differential equations based on the unifying methods of linear algebra, and to introduce the concept of state space. Vector and matrix formulation of systems of differential equations has in recent years emerged as a basic formalism highly useful in both the theory and applications of differential equations. This formalism leads naturally to the special case of autonomous systems. Hamilton's canonical equations of dynamics and the associated ideas of phase space provide the basis for the modern concept of state space covered in the latter part of this chapter.

4.1 VECTOR FORMULATION OF DIFFERENTIAL EQUATIONS

We first consider a system of n first-order ordinary differential equations in the normal form

$$\left.\begin{aligned} Dx_1 &= f_1(x_1, x_2, \ldots, x_n, t) \\ Dx_2 &= f_2(x_1, x_2, \ldots, x_n, t) \\ \cdot \quad &\cdot \quad \cdot \quad \cdot \quad \cdot \quad \cdot \quad \cdot \quad \cdot \\ Dx_n &= f_n(x_1, x_2, \ldots, x_n, t) \end{aligned}\right\} \qquad \textbf{(4.1)}$$

where the f_i are given functions of the $n + 1$ real variables $x_1, x_2, \ldots,$ x_n, t. It is assumed that the functions f_i are continuous and real-valued in a given region R of the $(n + 1)$-dimensional space of the variables

138

$x_1, x_2, \ldots, x_n, t$. A solution of this system is a set of functions $x_1(t)$, $x_2(t), \ldots, x_n(t)$ of class C^1 which satisfies (4.1).

We can consider the n functions x_m, $m = 1, 2, \ldots, n$, as components of a vector $\mathbf{x}$ in an n-dimensional vector space and write

$$\mathbf{x} = [x_1, x_2, \ldots, x_n] \tag{4.2}$$

The derivative of $\mathbf{x}(t)$ is a vector whose components are the derivatives of the components, i.e., $Dx_m(t)$, $m = 1, 2, \ldots, n$. Hence

$$D\mathbf{x} = [Dx_1, Dx_2, \ldots, Dx_n] \tag{4.3}$$

The n functions $f_1, f_2, \ldots, f_n$ can also be considered as the components of a vector $\mathbf{f}(\mathbf{x}, t) = [f_1, f_2, \ldots, f_n]$. Hence, in vector notation the system (4.1) may be displayed in the very concise form

$$D\mathbf{x} = \mathbf{f}(\mathbf{x}, t) \tag{4.4}$$

We shall show that all ordinary differential equations and systems of differential equations can be reduced to the normal form (4.1) or its equivalent (4.4). First consider the n-th order differential equation $f(x, Dx, D^2x, \ldots, D^nx, t) = 0$ and assume that D^nx can be expressed in explicit form

$$D^nx = f(x, Dx, D^2x, D^3x, \ldots, D^{n-1}x, t) \tag{4.5}$$

Let $x(t) = x_1$ be any solution of (4.5) and define the functions

$$\left.\begin{aligned}
Dx_1 &= x_2 = f_1(x_1, x_2, \ldots, x_n, t) = Dx \\
Dx_2 &= x_3 = f_2(x_1, x_2, \ldots, x_n, t) = D^2x \\
&\cdot \quad \cdot \quad \cdot \quad \cdot \quad \cdot \quad \cdot \quad \cdot \quad \cdot \quad \cdot \quad \cdot \quad \cdot \quad \cdot \\
Dx_{n-1} &= x_n = f_{n-1}(x_1, x_2, \ldots, x_n, t) = D^{n-1}x
\end{aligned}\right\} \tag{4.6}$$

so that (4.5) can be written as

$$Dx_n = f(x_1, x_2, \ldots, x_{n-1}, t) = D^nx \tag{4.7}$$

It follows that (4.6) and (4.7) can be written in the same form as (4.4). Next consider the system of simultaneous differential equations

$$\left.\begin{aligned}
f_1(x, Dx, D^2x, \ldots, D^ax, y, Dy, D^2y, \ldots, D^by, z, Dz, \ldots, D^cz, t) &= 0 \\
f_2(x, Dx, D^2x, \ldots, D^hx, y, Dy, D^2y, \ldots, D^jy, z, Dz, \ldots, D^kz, t) &= 0 \\
f_3(x, Dx, D^2x, \ldots, D^px, y, Dy, D^2y, \ldots, D^qy, z, Dz, \ldots, D^rz, t) &= 0
\end{aligned}\right\} \tag{4.8}$$

where x, y, z are functions of t.

We limit ourselves to the case of three equations with three unknown functions for notational reasons only. The proof is identical for n functions and n equations.

We assume that this system can be solved with respect to the highest order derivative of the functions x, y, z. Thus, if max $(a, h, p) = \alpha$, max $(b, j, q) = \beta$, max $(c, k, r) = \gamma$, (4.8) is equivalent to

$$
\left.
\begin{aligned}
D^\alpha x &= g_1[x, Dx, \ldots, D^{\alpha-1}x, y, Dy, \ldots, D^{\beta-1}y, z, Dz, \ldots, D^{\gamma-1}z, t] \\
D^\beta y &= g_2[x, Dx, \ldots, D^{\alpha-1}x, y, Dy, \ldots, D^{\beta-1}y, z, Dz, \ldots, D^{\gamma-1}z, t] \\
D^\gamma z &= g_3[x, Dx, \ldots, D^{\alpha-1}x, y, Dy, \ldots, D^{\beta-1}y, z, Dz, \ldots, D^{\gamma-1}z, t]
\end{aligned}
\right\}
$$

$$(4.9)$$

We now introduce the following notation:

$$
\left.
\begin{aligned}
x &= x_1, \; Dx = Dx_1 = x_2, \; D^2x = Dx_2 = x_3, \ldots, \; D^{\alpha-1}x = Dx_{\alpha-1} = x_\alpha, \\
y &= x_{\alpha+1}, \; Dy = Dx_{\alpha+1} = x_{\alpha+2}, \ldots, \; D^{\beta-1}y = Dx_{\alpha+\beta-1} = x_{\alpha+\beta} \\
z &= x_{\alpha+\beta+1}, \; Dx = Dx_{\alpha+\beta+1} = x_{\alpha+\beta+2}, \ldots, \; D^{\gamma-1}z = Dz_{\alpha+\beta+\gamma-1} = x_{\alpha+\beta+\gamma},
\end{aligned}
\right\}
$$

$$(4.10)$$

and

$$\mathbf{x} = [x_1, x_2, \ldots, x_\alpha, x_{\alpha+1}, \ldots, x_{\alpha+\beta}, x_{\alpha+\beta+1}, \ldots, x_{\alpha+\beta+\gamma}]$$

We can rewrite (4.9)

$$
\begin{aligned}
D^\alpha x &= Dx_\alpha \\
&= g_1[x_1, x_2, \ldots, x_\alpha, x_{\alpha+1}, \ldots, x_{\alpha+\beta}, x_{\alpha+\beta+1}, \ldots, x_{\alpha+\beta+\gamma}, t] = g_1 \\
D^\beta y &= Dx_{\alpha+\beta} \\
&= g_2[x_1, x_2, \ldots, x_\alpha, x_{\alpha+1}, \ldots, x_{\alpha+\beta}, x_{\alpha+\beta+1}, \ldots, x_{\alpha+\beta+\gamma}, t] = g_2 \\
D^\gamma z &= Dx_{\alpha+\beta+\gamma} \\
&= g_3[x_1, x_2, \ldots, x_\alpha, x_{\alpha+1}, \ldots, x_{\alpha+\beta}, x_{\alpha+\beta+1}, \ldots, x_{\alpha+\beta+\gamma}, t] = g_3
\end{aligned}
$$

It follows that

$$
\begin{aligned}
D\mathbf{x} &= [x_2, x_3, \ldots, x_\alpha, g_1, x_{\alpha+2}, \ldots, x_{\alpha+\beta}, g_2, x_{\alpha+\beta+2}, \ldots, x_{\alpha+\beta+\gamma}, g_3] \\
&= \mathbf{f}(\mathbf{x}, t)
\end{aligned}
$$

$$(4.11)$$

where both $\mathbf{x}$ and $\mathbf{f}$ are vectors in $\alpha + \beta + \gamma = n$ dimensions. Thus, we obtain again the normal form

$$D\mathbf{x} = \mathbf{f}(\mathbf{x}, t)$$

The reader will observe that the reduction of differential equations and systems of simultaneous differential equations to normal form is purely formal. The normal form (4.4) has, however, the advantage of containing all possible cases of differential equations and systems of simultaneous differential equations of any order in one single vector differential equation of first order.

We have assumed that all equations or systems of equations can be written in the form (4.5) or (4.9). If this is not the case, then there exist several expressions for the quantity to be solved for explicitly. In this case each expression represents a *different branch* of the equation and each branch is to be treated separately. Thus, even if the explicit form does not exist there are several branches, and each one leads to an equation of the form (4.4).

We illustrate by the following examples:

Example 1

Given the differential equation

$$D^3x + 3(D^2x)(Dx) - (Dx)^2 + xDx - 2t(t-1)x + 1 = 0 \qquad \textbf{(1)}$$

find its vector form.

Let

$$x = x_1 \qquad Dx = Dx_1 = x_2$$

$$D^2x = Dx_2 = x_3$$

and

$$\mathbf{x} = [x_1, x_2, x_3]$$

(1) can be written

$$D^3x + 3x_3x_2 - x_2^2 + x_1x_2 - 2t(t-1)x_1 + 1 = 0$$

or

$$D^3x = Dx_3 = -3x_2x_3 + x_2^2 - x_1x_2 + 2t(t-1)x_1 - 1 = f(x_1, x_2, x_3, t)$$

so that

$$D\mathbf{x} = [x_2, x_3, f(x_1, x_2, x_3, t)] = \mathbf{f}(\mathbf{x}, t)$$

Example 2

Given the system of simultaneous differential equations

$$\begin{aligned} -D^2x + Dx\,Dy + 2Dx\,D^2z + xyz &= 0 & \text{(a)} \\ x + D^2y + z + 2t\,Dx &= 0 & \text{(b)} \\ -x - y + D^3z - Dz &= 0 & \text{(c)} \end{aligned} \right\} \qquad \textbf{(1)}$$

find its vector form.

We observe that (a) contains D^2x, the highest order of derivative in x, (b) contains D^2y, the highest order derivative in y, and (c) contains D^3z, the highest order derivative in z. Thus, this system can be written as follows:

$$\begin{aligned} D^2x &= Dx\,Dy + 2Dx\,D^2z + xyz \\ D^2y &= -x - z - 2t\,Dx \\ D^3z &= x + y + Dx \end{aligned} \right\} \qquad \textbf{(2)}$$

Let

$$x = x_1 \qquad Dx = Dx_1 = x_2 \qquad y = x_3 \qquad Dy = Dx_3 = x_4$$
$$z = x_5 \qquad Dz = Dx_5 = x_6 \qquad D^2z = Dx_6 = x_7$$

so that (2) can be written in terms of $x_1, x_2, x_3, x_4, x_5, x_6, x_7$

$$D^2x = Dx_2 = x_2x_4 + 2x_2x_7 + x_1x_3x_5 = \varphi_1(x_1, x_2, x_3, x_4, x_5, x_6, x_7, t)$$
$$D^2y = Dx_4 = -x_1 - x_5 - 2tx_2 = \varphi_2(x_1, x_2, x_3, x_4, x_5, x_6, x_7, t)$$
$$D^3z = x_1 + x_3 + x_6 = \varphi_3(x_1, x_2, x_3, x_4, x_5, x_6, x_7, t)$$

Thus, if

$$\mathbf{x} = [x_1, x_2, x_3, x_4, x_5, x_6, x_7]$$

the system (1) may be exhibited in the form

$$Dx = [Dx_1, Dx_2, Dx_3, Dx_4, Dx_5, Dx_6, Dx_7]$$
$$= [x_2, \varphi_1, x_4, \varphi_2, x_6, x_7, \varphi_3]$$
$$= \mathbf{f}(\mathbf{x}, t)$$

Example 3

Write in vector form the differential equation

$$(D^2x)^2 - t^2x^2 = 0$$

This is equivalent to the two equations

$$D^2x - tx = 0 \qquad\qquad\qquad\qquad (1)$$
$$D^2x + tx = 0 \qquad\qquad\qquad\qquad (2)$$

For both we take $x = x_1$, $Dx_1 = x_2$, $\mathbf{x} = [x_1, x_2]$; then for (1) we obtain

$$D^2x = Dx_2 = tx_1$$
$$Dx = [Dx_1, Dx_2] = [x_2, tx_1] = \mathbf{f}_1[\mathbf{x}, t] \qquad\qquad (3)$$

for (2) we obtain

$$D^2x = Dx_2 = -tx_1$$
$$Dx = [Dx_1, Dx_2] = [x_2, -tx_1] = \mathbf{f}_2[\mathbf{x}, t] \qquad\qquad (4)$$

Thus, two distinct vector equations are obtained, i.e., two different branches of the initial equation, which we can write as

$$[Dx - \mathbf{f}_1][Dx - \mathbf{f}_2] = 0$$

4.2   MATRIX FORMULATION OF LINEAR SYSTEMS OF SIMULTANEOUS DIFFERENTIAL EQUATIONS

We consider the system of simultaneous linear differential equations

$$Dx_h = \sum_{m=1}^{n} a_{hm}(t)x_m(t) + b_h(t) \qquad h = 1, 2, \ldots, n \qquad \textbf{(4.12)}$$

Using vector notation as before, we can write $\mathbf{x} = [x_1, x_2, \ldots, x_n]$, and $D\mathbf{x} = [Dx_1, Dx_2, \ldots, Dx_n]$, so that (4.12) may be expressed in the form

$$D\mathbf{x} = A\mathbf{x} + \mathbf{b} \qquad (4.13)$$

and the corresponding homogeneous system

$$D\mathbf{x} = A\mathbf{x} \qquad (4.13a)$$

where A is the square matrix of the functions $a_{hm}(t)$ and $\mathbf{b}$ the vector of the functions $b_h(t)$.

We shall now give a vectorial interpretation of some results concerning linear differential equations, and, at the same time, generalize these results to systems of simultaneous linear homogeneous differential equations.

(i) If $\mathbf{x}$ is a solution of (4.13a), so is $a\mathbf{x}$. If $\mathbf{x}$ and $\mathbf{y}$ are solutions, so is $a\mathbf{x} + b\mathbf{y}$. It follows that the solutions are vectors in an n-dimensional vector space.

(ii) If $\mathbf{x}_1, \mathbf{x}_2, \ldots, \mathbf{x}_n$ is a set of n linearly independent solutions, then any solution $\mathbf{x}$ can be expressed as a linear combination of $\mathbf{x}_1, \mathbf{x}_2, \ldots, \mathbf{x}_n$, for were it not so, there would be $n + 1$ linearly independent vectors in an n-dimensional vector space, which is impossible. Thus

$$\mathbf{x} = \sum_{m=1}^{n} b_m \mathbf{x}_m$$

where the b_m's are numbers that are not all zero.

(iii) The vectors $\mathbf{x}_1, \mathbf{x}_2, \ldots, \mathbf{x}_n$ being linearly independent form a basis. Thus, a system of n linearly independent solutions of (4.13a), a fundamental system of solutions, forms a basis in the n-dimensional vector space.

Since every solution is a vector, let us write a solution $\mathbf{x}_h$ in the form $\mathbf{x}_h = (x_{1h}, x_{2h}, \ldots, x_{nh})$, $h = 1, 2, \ldots, n$. We can therefore write (4.13) in component form as follows:

$$Dx_{m,h} = \sum_{k=1}^{n} a_{m,k}(t) x_{k,h} \qquad (4.14)$$

(4.14) shows that we can write the system in matrix form, i.e.,

$$DX = AX \qquad (4.15)$$

where X is the square matrix of general component $x_{k,h}$. Thus, a fundamental system of solutions of (4.13) is a solution of the matrix differential equation (4.15).

The following theorem is a generalization of a basic theorem on linear differential equations.

Theorem. *If det $[X(t)]$ is the determinant of the square matrix X, then*

$$\det [X] = \det [X(t_0)] \cdot \exp \int_{t_0}^{t} (trA) \, dt \qquad (4.16)$$

We leave the proof of this theorem as an exercise to the reader. Let us consider now the case of the linear differential equation of order n, both homogeneous and nonhomogeneous,

$$\sum_{m=0}^{n} a_m(t) \, D^m x = b(t) \qquad (4.17)$$

$$\sum_{m=0}^{n} a_m(t) \, D^m x = 0 \qquad (4.18)$$

Since the equation is of order n, $a_n(t) \neq 0$, we divide by $a_n(t)$ and take $a_m(t)/a_n(t) = c_m(t)$. Then let $x = x_1$, $Dx = Dx_1 = x_2$, $Dx_2 = x_3, \ldots$, $Dx_{n-1} = x_n$. Finally, we can write

$$D^n x = Dx_n = -\sum_{m=0}^{n-1} c_m(t) \, D^m x = -\sum_{m=0}^{n-1} c_m(t) x_{m+1} \qquad (4.19)$$

It follows that

$$A = \begin{bmatrix} 0 & 1 & 0 & 0 & \cdot & \cdot & \cdot & 0 \\ 0 & 0 & 1 & 0 & \cdot & \cdot & \cdot & 0 \\ 0 & 0 & 0 & 1 & \cdot & \cdot & \cdot & 0 \\ \cdot & \cdot & \cdot & \cdot & \cdot & \cdot & \cdot & \cdot \\ 0 & 0 & 0 & 0 & \cdot & \cdot & \cdot & 1 \\ -c_0 & -c_1 & -c_2 & -c_3 & \cdot & \cdot & \cdot & -c_{n-1} \end{bmatrix} \qquad (4.20)$$

The matrix solution of the differential equation $DX = AX$ will be of the form

$$X = \begin{bmatrix} u_1 & u_2 & \cdot & \cdot & \cdot & u_n \\ Du_1 & Du_2 & \cdot & \cdot & \cdot & Du_n \\ D^2 u_1 & D^2 u_2 & \cdot & \cdot & \cdot & D^2 u_n \\ \cdot & \cdot & \cdot & \cdot & \cdot & \cdot \\ D^{n-1} u_1 & D^{n-1} u_2 & \cdot & \cdot & \cdot & D^{n-1} u_n \end{bmatrix} \qquad (4.21)$$

where u_m, $m = 1, 2, \ldots, n$, are linearly independent solutions of (4.18).

It follows that

$$P = AX = \begin{bmatrix} 0 & 1 & 0 & 0 & \cdot & \cdot & \cdot & \cdot & 0 \\ 0 & 0 & 1 & 0 & \cdot & \cdot & \cdot & \cdot & 0 \\ \cdot & \cdot & \cdot & \cdot & \cdot & \cdot & \cdot & \cdot & \cdot \\ 0 & 0 & 0 & 0 & \cdot & \cdot & \cdot & \cdot & 1 \\ -c_0 & -c_1 & -c_2 & -c_3 & \cdot & \cdot & \cdot & \cdot & -c_{n-1} \end{bmatrix}$$

$$\begin{bmatrix} u_1 & \cdot & \cdot & \cdot & \cdot & u_n \\ Du_1 & \cdot & \cdot & \cdot & \cdot & Du_n \\ \cdot & \cdot & \cdot & \cdot & \cdot & \cdot \\ \cdot & \cdot & \cdot & \cdot & \cdot & \cdot \\ D^{n-1}u_1 & \cdot & \cdot & & & D^{n-1}u_n \end{bmatrix}$$

The general element of P is

$$p_{hk} = \sum_{m=1}^{n} c_{hm} x_{mk} = \sum_{m-1}^{n} c_{hm} D^{m-1} u_k$$

Since for $h \neq m - 1$, $c_{nm} = 0$, we have in this case for $h \neq n$,

$$p_{hk} = u_{m-1,m} D^{m-1} u_k$$

On the other hand, tr A reduces to $-c_{n-1}$, so that according to the theorem just stated

$$\det [X(t)] = [\det X(t_0)] \exp \left[- \int_{t_0}^{t} c_{n-1}(t)\, dt \right]$$

a result which is known in the theory of linear differential equations, in which case det $[X(t)]$ is called the Wronskian of the differential equation (4.18).

4.3 SPECIAL CASE OF LINEAR SYSTEMS OF SIMULTANEOUS DIFFERENTIAL EQUATIONS WITH CONSTANT COEFFICIENTS

In this case, A in (4.13) is a constant square matrix of order n. If the equation is homogeneous we have

$$D\mathbf{x} = A\mathbf{x}$$

We look for a solution of the form

$$\mathbf{x} = \mathbf{c}e^{\lambda t} \tag{4.22}$$

where $\mathbf{c}$ is a constant vector. In order to substitute into (4.13a) we find

$Dx = c\lambda e^{\lambda t} = \lambda x = \lambda I x$, where I is the identity matrix, so that

$$Dx = \lambda I x = Ax \tag{4.23}$$

or

$$(A - \lambda I)x = 0 \tag{4.24}$$

This means that x is an eigenvector of A and λ the corresponding eigenvalue (cf. Chap. 1). The values of λ are found by solving the characteristic equation

$$\det [A - \lambda I] = f(\lambda) = 0 \tag{4.25}$$

We consider two cases:

(i) $f(\lambda)$ has n distinct roots $\lambda_1, \lambda_2, \ldots, \lambda_n$ in which case

$$x = \sum_{m=1}^{n} c_m e^{\lambda_m t} \tag{4.26}$$

is the general solution of (4.13a).

(ii) $f(\lambda)$ has repeated roots. Let us assume that λ_1 is a $(p-1)$-times repeated root, i.e., the roots of (4.25) are

$$\underbrace{\lambda_1, \lambda_1, \lambda_1, \ldots, \lambda_1,}_{p\text{-times}} \lambda_{p+1}, \lambda_{p+2}, \ldots, \lambda_n$$

In this case we look for a solution of (4.13a) of the form

$$x = e^{\lambda_1 t}[c_1 + c_2 t + \cdots + c_p t^{p-1}] + \sum_{s=p+1}^{n} c_s e^{\lambda_s t}$$

$$x = P(t)e^{\lambda_1 t} + \sum_{s=p+1}^{n} c_s e^{\lambda_s t} \tag{4.27}$$

To obtain this result we operate by induction. Let k_1 be an (constant) eigenvector of A with eigenvalue λ_1, i.e., $Ak_1 = \lambda_1 k_1$.

$$x_1 = e^{\lambda_1 t} k_1$$

is a solution of (4.13a), for

$$Dx_1 = \lambda_1 e^{\lambda_1 t} k_1 = Ak_1 e^{\lambda_1 t} = Ax_1$$

Let k_2 then be a constant vector defined by

$$Ak_2 = \lambda_1 k_2 + k_1$$

then

$$x_2 = e^{\lambda_1 t}[k_2 + tk_1]$$

is a solution of (4.13a), for

$$Dx_2 = \lambda_1 e^{\lambda_1 t}[k_2 + tk_1] + e^{\lambda_1 t} k_1$$
$$= e^{\lambda_1 t}[(\lambda_1 k_2 + k_1) + \lambda_1 t k_1]$$
$$= e^{\lambda_1 t}[Ak_2 + Atk_1] = Ae^{\lambda_1 t}[k_2 + tk_1]$$

i.e.,

$$Dx_2 = Ax_2$$

Let k_3 be a constant vector defined by

$$Ak_3 = \lambda_1 k_3 + 2k_2$$

then

$$x_3 = e^{\lambda_1 t}k_3 + 2te^{\lambda_1 t}k_2 + t^2 e^{\lambda_1 t}k_1$$

is a solution, for

$$Dx_3 = \lambda_1 e^{\lambda_1 t}k_3 + 2e^{\lambda_1 t}k_2 + 2t\lambda_1 e^{\lambda_1 t}k_2 + 2te^{\lambda_1 t}k_1 + t^2\lambda_1 e^{\lambda_1 t}k_1$$
$$= (\lambda_1 k_3 + 2k_2)e^{\lambda_1 t} + 2(\lambda_1 k_2 + k_1)te^{\lambda_1 t} + t^2\lambda_1 k_1 e^{\lambda_1 t}$$
$$= e^{\lambda_1 t}(Ak_3 + 2Atk_2 + At^2 k_1) = A[k_3 + 2tk_2 + t^2 k_1]e^{\lambda_1 t}$$
$$Dx_3 = Ax_3$$

Continuing the same way it is easily checked that with

$$Ak_p = \lambda_1 k_p + (p-1)k_{p-1}$$

$$x_p = \sum_{m=0}^{p-1} \binom{p-1}{m} t^m k_{p-m} e^{\lambda_1 t}$$

where $\binom{p-1}{m} = \dfrac{(p-1)!}{m!\,(p-m-1)!}$ is the binominal coefficient, x_p is a solution of (4.13a). Let $a_1, a_2, \ldots, a_p$ be p arbitrary constants; then clearly

$$y = a_1 x_1 + a_2 x_2 + \cdots + a_p x_p \qquad (4.28)$$

is a solution of (4.13a) which can be written

$$y = a_1 k_1 e^{\lambda_1 t}$$
$$+ a_2[k_2 + tk_1]e^{\lambda_1 t}$$
$$+ a_3[k_3 + 2tk_2 + t^2 k_1]e^{\lambda_1 t}$$
$$+ \cdots$$
$$+ a_p\left[k_p + \binom{p-1}{1}tk_{p-1} + \binom{p-1}{2}t^2 k_{p-2} + \cdots\right]e^{\lambda_1 t}$$

Let

$$a_1 k_1 + a_2 k_2 + \cdots + a_p k_p = c_1$$

$$a_2 k_1 + 2a_3 k_2 + \cdots + \binom{p-1}{1}a_p k_{p-1} = c_2$$

$$a_3 k_1 + 3a_4 k_2 + \cdots + \binom{p-1}{2}a_p k_{p-2} = c_3$$

$$\cdot \quad \cdot \quad \cdot \quad \cdot \quad \cdot \quad \cdot \quad \cdot \quad \cdot \quad \cdot$$

$$a_j k_1 + ja_{j+1}k_2 + \cdots + \binom{p-1}{j-1}a_p k_{p-j+1} = c_j$$

$$\cdot \quad \cdot \quad \cdot \quad \cdot \quad \cdot \quad \cdot \quad \cdot \quad \cdot \quad \cdot$$

$$a_p k_1 = c_p$$

then
$$\mathbf{y} = (\mathbf{c}_1 + \mathbf{c}_2 t + \mathbf{c}_3 t^2 + \cdots + \mathbf{c}_p t^{p-1}) e^{\lambda_1 t}$$

which proves (4.27) for the first part. The second part is obtained as in the case of distinct roots of the characteristic equation.

4.4   AUTONOMOUS SYSTEMS: PHASE PLANE

Consider the vector equation

$$\frac{d\mathbf{x}}{dt} = \mathbf{F}(\mathbf{x}, t) \tag{4.29}$$

where $\mathbf{x} = [x_1, x_2, \ldots, x_n]$, $x_k = x_k(t)$, $k = 1, 2, \ldots, n$, and $\mathbf{F} = [F_1, F_2, \ldots, F_n]$, $F_k = F_k(x_1, x_2, \ldots, x_n, t) = F_k(\mathbf{x}, t)$. This vector differential equation is obviously equivalent to a system of n simultaneous differential equations. We shall say that (4.29) or the corresponding system of simultaneous differential equations is *autonomous* if $\mathbf{F} = \mathbf{F}(\mathbf{x})$, i.e., t does not appear explicitly in (4.29),

$$\frac{d\mathbf{x}}{dt} = \mathbf{F}(\mathbf{x}) \tag{4.30}$$

The general solution of (4.29) will be of the form

$$\mathbf{x} = \boldsymbol{\varphi}(\mathbf{C}, t) \tag{4.31}$$

The constant vector $\mathbf{C}$ will be determined by the boundary condition $t = t_0$, $\mathbf{x} = \mathbf{x}_0$.

In the case $n = 3$ we have

$$x_1 = x = x(t) \qquad x_2 = y = y(t) \qquad x_3 = z = z(t) \tag{4.32}$$

$$\frac{dx}{dt} = F_1(x, y, z) \qquad \frac{dy}{dt} = F_2(x, y, z) \qquad \frac{dz}{dt} = F_3(x, y, z) \tag{4.33}$$

(4.33) can be considered to represent a vector field. If t is the time then $\dot{\mathbf{x}} = \left[\dfrac{dx}{dt}, \dfrac{dy}{dt}, \dfrac{dz}{dt}\right]$ is a velocity field. It can be thought to represent the velocity of a flowing fluid at the point (x, y, z). (4.32) would then represent the trajectory of a fluid particle. The general solution (4.31) would be

$$x = x(C_1, C_2, C_3, t) \qquad y = y(C_1, C_2, C_3, t) \qquad z = z(C_1, C_2, C_3, t) \tag{4.34}$$

In order to find $[C_1, C_2, C_3] = \mathbf{C}$ we express the fact that for $t = t_0$, $\mathbf{x} = [x_0, y_0, z_0]$, then obtain a solution of the form (4.32). This would represent the trajectory of a particle of fluid through the point (x_0, y_0, z_0).

This trajectory is clearly time independent, so that the solution can be written either $\mathbf{x} = \boldsymbol{\varphi}(\mathbf{C}, t)$, or $\mathbf{x} = \boldsymbol{\varphi}(\mathbf{C}, t - t_0)$. Thus the change of t into $t - t_0$ would alter the time origin without changing the trajectory. This is an important property of autonomous systems.

This property is clearly not limited to the case $n = 3$, although in this case the geometric image is clearer.

The autonomous system of Eq. (4.30), i.e.,

$$\frac{d\mathbf{x}}{dt} = \mathbf{F}(\mathbf{x})$$

is said to be linear if it can be written

$$D\mathbf{x} = \frac{d\mathbf{x}}{dt} = A\mathbf{x} \tag{4.35}$$

where A is a matrix of general constant elements a_{ij}. Eq. (4.35) can be written

$$\frac{dx_k}{dt} = \sum_{m=1}^{n} a_{km}x_m \qquad k = 1, 2, \ldots, n \tag{4.36}$$

A critical point of the autonomous system is a point for which

$$D\mathbf{x} = 0$$

This will clearly be the case for $\mathbf{x} = 0$, i.e., it follows that the origin is always a critical point. If in addition det $[A] \neq 0$, i.e., A is nonsingular, then there is no other critical point. In hydrodynamics the origin or any other point for which $D\mathbf{x} = 0$, i.e., a critical point, corresponds to a point of zero velocity, called a *point of stagnation.*

Two-dimensional Linear Systems. In this section we shall study the behavior of linear systems in the neighborhood of a critical point. We consider systems of simultaneous linear differential equations

$$\frac{dx}{dt} = ax + by \qquad \frac{dy}{dt} = px + qy \tag{4.37}$$

when a, b, p, q are constants. The system (4.37) has a critical point at $(0, 0)$. Any other critical point will be such that $ax + by = 0, px + qy = 0$, which *will have* a solution besides $(0, 0)$ provided

$$\Delta = \det [A] = \begin{vmatrix} a & b \\ p & q \end{vmatrix} = 0 \tag{4.38}$$

i.e., A is singular. Let us assume that there is no critical point besides $(0, 0)$, which is to say $\Delta \neq 0$. The characteristic equation of the matrix

A is given by

$$\begin{vmatrix} a - \lambda & q \\ p & b - \lambda \end{vmatrix} = \lambda^2 - (a + q)\lambda + (aq - bp) = 0 \qquad \textbf{(4.39)}$$

We observe that the last term is Δ, which we have assumed to be different from zero, so that none of the roots of Eq. (4.39) can be zero. If we introduce new variables in (4.37) as follows

$$\left. \begin{array}{l} X = \alpha x + \beta y \\ Y = \gamma x + \delta y \end{array} \right\} \qquad \begin{vmatrix} \alpha & \beta \\ \gamma & \delta \end{vmatrix} \neq 0 \qquad \textbf{(4.40)}$$

where α, β, γ, δ are constants

or

$$\mathbf{X} = \begin{bmatrix} X \\ Y \end{bmatrix} = \begin{bmatrix} \alpha & \beta \\ \gamma & \delta \end{bmatrix} \begin{bmatrix} x \\ y \end{bmatrix} = M\mathbf{x} \qquad \det [M] \neq 0 \qquad \textbf{(4.41)}$$

then

$$D\mathbf{X} = MD\mathbf{x} = MA\mathbf{x} = MAM^{-1}\mathbf{X}$$

$$D\mathbf{X} = B\mathbf{X}$$

where $B = MAM'$ is nonsingular. We have seen in Chap. 1 that B has the same characteristic values as A.

Let us solve Eq. (4.37). We write it

$$\left. \begin{array}{l} (D - aI)x - bIy = 0 \\ -pIx + (D - qI)y = 0 \end{array} \right\} \qquad \textbf{(4.42)}$$

$$\begin{array}{l} [(D - aI)(D - qI) - bpI]x = 0 \\ [(D - aI)(D - qI) - bpI]y = 0 \end{array} \qquad \textbf{(4.43)}$$

or

$$[D^2 - (a + q)D + (aq - bp)I]\begin{bmatrix} x \\ y \end{bmatrix} = 0$$

Thus

$$x = C_1 e^{\lambda_1 t} + C_2 e^{\lambda_2 t}$$

$$y = C_3 e^{\lambda_1 t} + C_4 e^{\lambda_3 t}$$

where λ_1 and λ_2 are the roots of (4.39), and thus the roots of the characteristic equation. By resubstituting into the first of Eq. (4.42) we have

$$[C_1(\lambda_1 - a) - bC_3]e^{\lambda_1 t} + [C_2(\lambda_2 - a) - bC_4]e^{\lambda_2 t} = 0$$

so that

$$C_3 = C_1 \frac{\lambda_1 - a}{b} \qquad C_4 = C_2 \frac{\lambda_2 - a}{b}$$

Similarly, by resubstituting into the second of (4.42) we have

$$[C_1 p - (\lambda_1 - q)C_3]e^{\lambda_1 t} + [C_2 p - (\lambda_2 - q)C_4]e^{\lambda_2 t} = 0$$

so that

$$C_3 = C_1 \frac{p}{\lambda_1 - q} \qquad C_4 = C_2 \frac{p}{\lambda_2 - q} \tag{4.44}$$

It is clear that the values obtained for C_3 and C_4 as functions of C_1 and C_2 are identical, for

$$\frac{a - \lambda_1}{b} = \frac{p}{q - \lambda_1}$$

yields $(a - \lambda_1)(q - \lambda_1) - bp = 0$, or $\lambda_1^2 - (a + q)\lambda_1 + (aq - bp) = 0$, which shows that λ_1 is a root of (4.39), the same being true for λ_2. The general solution of (4.37) is thus

$$x = C_1 e^{\lambda_1 t} + C_2 e^{\lambda_2 t}$$

$$y = C_1 \frac{p}{\lambda_1 - q} e^{\lambda_1 t} + C_2 \frac{p}{\lambda_2 - q} e^{\lambda_2 t}$$

Let us assume that $\lambda_1 < \lambda_2 < 0$ and find

$$\lim_{t \to \infty} \frac{y}{x} = \lim_{t \to \infty} \frac{C_1 \dfrac{p}{\lambda_1 - q} e^{\lambda_1 t} + C_2 \dfrac{p}{\lambda_2 - q} e^{\lambda_2 t}}{C_1 e^{\lambda_1 t} + C_1 e^{\lambda_2 t}}$$

$$= \lim_{t \to \infty} \frac{C_1 \dfrac{p}{\lambda_1 - q} e^{(\lambda_1 - \lambda_2)t} + C_2 \dfrac{p}{\lambda_2 - q}}{C_1 e^{(\lambda_1 - \lambda_2)t} + C_2} = \frac{p}{\lambda_1 - q} \tag{4.45}$$

On the other hand, $\lim_{t \to \infty} x = \lim_{t \to \infty} y = 0$, so that all trajectories pass through

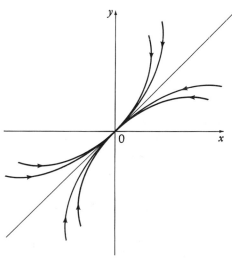

FIGURE 4.1

the origin and are tangent to a line of slope $p/(\lambda_1 - q)$, where λ_1 is the smaller root. We say in this case that the origin is a nodal point. If $0 < \lambda_2 < \lambda_1$ we change t into $-t$ and obtain the same result (Fig. 4.1).

Other cases to be considered are:

(ii) $\lambda_1 < 0 < \lambda_2$

(iii) $\lambda_1 = \lambda_2$, the real matrix $A - \lambda I$ is of rank 0,

(iv) $\lambda_1 = \lambda_2$, the real matrix $A - \lambda I$ is of rank 1,

(v) $\lambda_1 = \lambda_2$, the real part of λ_1 and $\lambda_2 \neq 0$,

(vi) λ_1 and λ_2 both imaginary.

We leave the discussion of these cases to the exercises.

Poincaré Phase-Plane. Consider the case of a single function $x = x(t)$ and a second order differential equation not containing t:

$$\ddot{x} = \varphi(x, \dot{x}) \tag{4.46}$$

Let $\dot{x} = v$ (since in general $\dfrac{dx}{dt}$ is a velocity). Then

$$\ddot{x} = \frac{dv}{dt} = \frac{dv}{dx}\frac{dx}{dt} = v\frac{dv}{dx} = \varphi(x, \dot{x}) = \varphi(x, v)$$

i.e., (4.46) becomes

$$\left.\begin{array}{c} \dfrac{dv}{dt} = \varphi(x, v) \\[3mm] \dfrac{dx}{dt} = v \end{array}\right\} \tag{4.47}$$

which we can write in vector form

$$\left.\begin{array}{l} \mathbf{x} = [x, v] \\[2mm] D\mathbf{x} = [Dx, Dv] = [v, \varphi(x, v)] \end{array}\right\} \tag{4.48}$$

Thus (4.48) is the autonomous system corresponding to (4.46).

The plane (x, v) is called the Poincaré phase-plane and permits a useful presentation of the solution. (See Ref. 14.)

Example I

Consider the simple pendulum which has the differential equation (cf. Chap. 6)

$$\frac{d^2\theta}{dt^2} = -k^2 \sin\theta \qquad k^2 = g/L$$

where L is the length of the simple pendulum and θ the angle of deflection. Although the pendulum equation will be studied several times throughout this book, this is the only place where the phase-plane is used. Let $y = d\theta/dt$, then

$$\left. \begin{array}{l} d\theta/dt = y \\ dy/dt = -k^2 \sin \theta \end{array} \right\}$$

It is clear that $y = 0$, $\theta = n\pi$, $n = 0$, ± 1, ± 2, ..., are *critical points* according to the definition given earlier.

Since

$$\frac{dy}{dt} = \frac{dy}{d\theta} \frac{d\theta}{dt}$$

and

$$\frac{dy}{d\theta} = \frac{dy}{dt} \bigg/ \frac{d\theta}{dt} = -k^2 \sin \theta / y$$

it follows that

$$y \, dy = -k^2 \sin \theta \, d\theta$$

or

$$y^2 = 2k^2 \cos \theta + C \tag{1}$$

where C is a constant that must be such that

$$2k^2 \cos \theta + C \geq 0 \tag{2}$$

Since $\cos \theta = \cos(-\theta)$, we may assume $\theta > 0$. The part of the graph in the (θ, y)-plane corresponding to $\theta < 0$ will be obtained from the other by symmetry with respect to the y axis. Considering the periodicity of $\cos \theta$ it is sufficient to study the interval $0 \leq \theta \leq \pi$. For every admissible value of C there will be a different curve in the (θ, y)-plane. Let us find the admissible values of C. From (2) we clearly have

$$-\frac{C}{2k^2} \leq \cos \theta \leq 1 \tag{3}$$

This is possible if and only if $-C/2k^2 \leq 1$, i.e., $-C \leq 2k^2$ or $C \geq -2k^2$.

On the other hand, if $-C/2k^2 \leq -1$, i.e., $C \leq 2k^2$, θ can take all the values in the interval, $0 \leq \theta \leq \pi$. For $-2k^2 < C < 2k^2$, the values of θ will be limited so that (cf. Fig. 4.2) $0 \leq \theta < \alpha$, where $\alpha = $ Arc cos $(-C/2k^2)$.

Let us now study the values of θ for $y = 0$. From (2) we see that $y = 0$, if $2k^2 \cos \theta + C = 0$, and since $0 < \theta < \pi$, we see that roots are given by $\theta = \alpha$, provided that $-1 \leq -C/2k^2 \leq +1$, or $-2k^2 \leq C \leq 2k^2$.

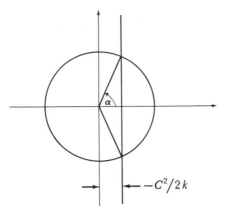

$$-C^2/2k$$

FIGURE 4.2 Circle diagram for Example 1.

We arrive, therefore, at the following conclusions:

C	Curve	Intersection with 0θ
$C < -2k^2$	no curve	none
$C = -2k^2$	reduces to $\theta = 0, y = 0$	$\theta = 0$
$-2k^2 < C < 2k^2$	right $\frac{1}{2}$ of closed curve	$\theta = \alpha$
$C = 2k^2$	$y = \pm 2k \cos \dfrac{\theta}{2}$	$\theta = \pi$
$2k^2 < C$	curve exists	none

Results are shown in Figure 4.3.

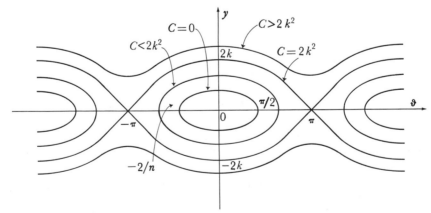

FIGURE 4.3 Poincaré phase-plane diagram for plane pendulum.

4.5 PHASE SPACE AND STATE VARIABLES

The concept of phase space is an outgrowth of Hamiltonian mechanics, and has proved useful in various theoretical and applied fields of science such as statistical mechanics, theory of vibrations, and, more recently, in feedback and other systems. Phase space, in the context of Hamiltonian mechanics, refers to an abstract $2f$-dimensional space associated with the $2f$ components $(q_1, q_2, \ldots, q_f; p_1, p_2, \ldots, p_f)$ used to define the state of a dynamical system having f degrees of freedom in the sense of Lagrange.

One of the advantages of the phase space representation is that the totality of solutions of a given problem for a range of initial conditions can be represented geometrically. The state of the entire system, at any time t, can be pictured by a single point in phase space. As time varies, this point traces out a curve which represents the motion of the system. To each distinct set of initial conditions there corresponds a distinct curve in phase space. The family of curves thus defined has the property that, in general, one and only one curve of the family passes through each point in phase space. These curves represent the totality of motions for the system; no two curves overlap or cross each other, and each curve has a tangent at every point. This family of curves in phase space may be thought of as streamlines along which the points representing the solutions for different initial conditions move.

Perhaps the greatest importance of the phase space concept of Hamiltonian mechanics is that it provides a means for representing all possible solutions of a set of first order differential equations (which can be linear or nonlinear). Poincaré capitalized on this advantage in his work on the phase-plane and limit cycles (see preceding section). More recently Bellman, Kalman (Ref. 13), and others have translated these ideas with great success to control theory, stability problems, network theory, space mechanics, guidance, and other fields.

State Variables and State Space. The extension of the mathematical ideas of Hamilton's canonical equations and Poincaré's phase-plane to systems problems in the general sense has become known as the *state-variable approach*. In this newer context, the *state* of a system is defined as a minimal set of variables $x_1(t), x_2(t), \ldots, x_n(t)$ that uniquely describes the behavior of the system for any time $t \geq t_0$, the initial state $x_1(t_0), x_2(t_0), \ldots, x_n(t_0)$ being known. The *state variables* $x_1(t), x_2(t), \ldots, x_n(t)$ can be considered as the components of a *state vector* $\mathbf{x}(t)$ in an n-dimensional vector or *state space*.

State variables and the concept of state can be applied to a wide variety of processes, including deterministic, probabalistic, discrete, continuous, and sequential binary processes. In keeping with the scope of this book, we consider primarily those processes which can be described by ordinary differential or algebraic equations, but note the broader applicability of the state method.

In the modern literature it is becoming customary to describe processes in terms of the generalized Hamiltonian concept of canonical equations and variables. This state-variable approach not only has the theoretical advantage of encompassing the totality of all possible solutions for a process or system, but also has the great practical advantage of being readily adapted to simulation on an analog computer or to numerical solution on a digital computer. Such adaptability to solution by analog or digital computer is decisive for most problems above the elementary level.

BIBLIOGRAPHY

1. R. Bellman, *Stability Theory of Differential Equations*. New York: McGraw-Hill, 1953.
2. E. A. Coddington and N. Levinson, *Theory of Ordinary Differential Equations*. New York: McGraw-Hill, 1955.
3. E. Goursat, *Cours d'Analyse*, Vol. II, Fifth Edition. Paris: Gauthier Villard, 1942.
4. F. B. Hildebrand, *Advanced Calculus for Applications*. Englewood Cliffs, New Jersey: Prentice-Hall, 1962.
5. E. L. Ince, *Ordinary Differential Equations*. New York: Dover, 1956.
6. E. Jahnke and F. Emde, *Tables of Functions*. New York: Dover, 1945.
7. E. Kamke, *Differentialgleichungen, Lösungsmethoden und Lösungen*. New York: Chelsea, 1959.
8. W. Kaplan, *Ordinary Differential Equations*. Reading, Mass.: Addison-Wesley, 1958.
9. H. Margenau and G. M. Murphy, *The Mathematics of Physics and Chemistry*, Second Edition. New York: Van Nostrand, 1955.
10. G. M. Murphy, *Ordinary Differential Equations*. New York: Van Nostrand, 1960.
11. E. T. Whittaker and G. N. Watson, *A Course of Modern Analysis*. Cambridge: Cambridge University Press, 1952.
12. E. D. Rainville, *Special Functions*. New York: Macmillan, 1960.
13. R. E. Kalman, "On the General Theory of Control Systems." *Proceedings of the International Federation of Automatic Control*, Moscow, 1960, Butterworths, pp. 481–492, 1961.
14. H. Poincaré, *Les Nouvelles Méthodes de la Mécanique Céleste*, Vol. 1–3. New York: Dover, 1957.

Part II

Physical Concepts

5

ORIGINS OF THE SYSTEM CONCEPT

"Who knows only his own generation remains always a child." *

A basic knowledge of the history and significance of a scientific or technical field is becoming more and more desirable for creative or even functional work in that field. This knowledge provides an effective medium for sifting out the new from the merely resurrected; it provides a matrix for placing new developments in perspective; and, above all, it provides many leads, insights, and simplifications to guide and accelerate the new.

The concept of a system in its broadest sense is embedded in the general culture, and was certainly known to the Greeks and Romans, as indicated by the classical origin of our English word *system*. Systems science as we know it today, however, has its roots in the physics, astronomy, and mathematics of the Renaissance. In a very broad sense systems methods in physical science and astronomy may be traced back at least to the French philosopher and mathematician René Descartes (1596–1650), who had the concept of describing the world order and system in broad mathematical terms (Ref. 1). Maupertuis (1698–1759) also contributed to the ideas of the passage from philosophy to scientific systems.

Copernicus (1473–1543) developed the idea of a world system in which the Earth moves around the Sun rather than vice versa. The Copernican or heliocentric theory was confirmed by the work of Kepler (1571–1630) in his famous three laws of planetary motion. These laws

---

* Inscription from Norlin Library, University of Colorado.

were based on very extensive analysis of the highly accurate astronomical observations of Tycho Brahe (1546–1601).

Galileo (1564–1642) generally is credited with establishing the science of mechanics in the modern sense. His experiments on pendulums, falling bodies, and the concept of force inspired Newton's formulation of the laws of mechanics, the foundation of all later work in the field. Galileo is also the author of the famous dictum that mathematics is the language of science. The following quotation from Lagrange epitomizes Galileo's work in mechanics:

> *"Dynamics is a science due entirely to the moderns. Galileo laid its foundations. Before him philosophers considered the forces which act on bodies in a state of equilibrium only. Although they attributed in a vague way the acceleration of falling bodies and the curvilinear movement of projectiles to the constant action of gravity, nobody had yet succeeded in determining the laws of these phenomena. Galileo made the first important steps, and thereby opened a way, new and immense, to the advancement of mechanics as a science."*

> (Lagrange: "Mécanique Analytique," 1788)

Newton (1642–1727) achieved the first notable success in applying this concept when, having formulated his laws of motion and gravitation and having perfected his new system of fluxions, he was able to accurately describe the intricate motions of the planetary system in highly precise mathematical terms. This analysis is published in Book Three of the *Principia Mathematica* (Ref. 2), and in it are exhibited all the features and sophistication of the archetype for subsequent study of systems.

The *Principia* is generally regarded as one of the greatest of all works of science and one of the most influential books any man has yet conceived. Newton's own thoughts in creating the *Principia* are revealed by the opening lines of the Preface, "Since the ancients esteemed the science of mechanics of greatest importance in the investigation of natural things, and the moderns, rejecting substantial forms and occult qualities, have endeavored to subject the phenomena of nature to the laws of mathematics, I have in this treatise cultivated mathematics as far as it relates to philosophy." Newton's achievement established the model for much of the scientific work in physics and astronomy throughout the eighteenth and nineteenth centuries.

The great philosopher and mathematician Leibniz (1646–1716) developed the idea of the vis viva (living force), which is equivalent to the quantity we know today as kinetic energy, and which has been generalized as one of the basic concepts in the treatment of physical systems. Also to be mentioned during this period are Johann Bernoulli

(1667–1748) and Euler (1707–1783), both of whom contributed prominently to the development of rational mechanics. The concept of a precisely predictable world order probably reached its peak, however, with the work of Laplace (1749–1827) on celestial mechanics, and, at the end of the eighteenth century, with the work of Lagrange (1736–1813) on generalized mechanics.

The second great area of advancement in the development of the physical-mathematical science which underlies our modern concept of a system is unquestionably the generalization of Newton's mechanics, achieved primarily by Lagrange, by Hamilton (1805–1865), and by Jacobi (1804–1851). There is ample reason to rank generalized mechanics, in terms of its concepts, methods, generality, and scope of application, as the most highly advanced and elegant model of systems analysis even to this day.

Lagrange's treatise, *Mécanique Analytique* (1788), which appeared 100 years after Newton's *Principia*, is one of the great landmarks of science (Ref. 3). In the "avertissement" to his work Lagrange proudly states, "On ne trouvera point de figures dans cet ouvrage" (One will find no figures in this work). Indeed, so successful was he in achieving his aim of reducing mechanics to a branch of mathematical analysis that classical mechanics to this day is given as a basic course in the mathematics departments of many universities and colleges.

Hamilton brought mechanics to a new pinnacle of perfection with the formulation of Hamilton's principle, the development of the canonical equations, and the identification of the dual particle-wave properties of matter and of light waves (Ref. 4). His extension of the formalism of generalized mechanics was the inspiration for further extension in the twentieth century to quantum mechanics. Jacobi extended Hamilton's theory in the form now known as the Hamilton-Jacobi equation. These and other contributions are covered in Jacobi's *Vorlesungen über Dynamik*, a series of famous lectures given in 1842–1843 and published in 1866.

The third phase in tracing the origins of the systems concept may be regarded as that of enlarging the scope of generalized mechanics to include certain electromagnetic and electromechanical phenomena. The pioneering work in this phase was done by Maxwell (1831–1879). Maxwell conceived electromagnetism to be a dynamical phenomenon, and developed a dynamical theory of electromagnetism based on this concept (Ref. 5). In this theory, a system of conductors carrying currents may be treated as a dynamical system in which the energy may be in part kinetic and in part potential. Electric currents play the role of generalized velocities, and electromotive forces correspond to generalized forces. Maxwell thus showed how it is possible to treat a system of electric currents, including their interactions with mechanical elements, by the abstract methods of generalized mechanics. A fascinating account of this historic work is given by Maxwell in Vol. II, Chap. VI of his *Treatise on Electricity and Magnetism* (1873).

With the unification of electrical and mechanical phenomena within the framework of generalized mechanics, the foundations were laid for later generalizations and methods, such as the generalization of Kirchhoff's laws and Ohm's law to produce a theory of electrical and mechanical networks or generalized circuits, and Kron's theory of generalized machines (Ref. 6).

Essentially, twentieth century development of systems science may be regarded as having its origin and impetus in the massive scientific efforts associated with World War II. Developments during this important period included operations research (originally meaning application of sophisticated mathematical methods to military systems and problems), complex radar systems, advanced types of servomechanisms, and various other large-scale operations. Immediately following World War II, systems concepts and methods were applied on a broad scale to cybernetics, management science, information and decision theory, process dynamics, chemical process control, linear programming, and a wide variety of engineering design problems. The development of analog and digital computers during this period provided the necessary tools for wide practical application of systems methods.

While it is true that the origins of these developments in most instances antedate the war period, as for example the work of Bush (1932) on the differential analyzer (a form of analog or simulation computer) (Ref. 7), and the early work of Maxwell (1868) on the theory of automatic feedback control systems (Ref. 8), the fact remains that the decisive developments and the widespread acceptance of the newer methods are clearly associated with the scientific and technical renaissance generated by the war effort.

More recently, a great use of systems methods and philosophy in other fields has occurred in the rush to exploit the proved successes of the systems approach. Especially notable are the applications of this approach in cases of power generation and transmission through the organization of super-scale power grids, of complex inertial control devices for guided missiles, Earth satellites, and space ships, and of hard-core problems in the life sciences and social sciences. The practicability of successfully coping with these problems rests in large measure on the availability of high-capacity, high-speed digital computers and on the related advances in mathematical methods of numerical analysis which are essential for the efficient use of these computers.

Mathematics has always been fundamental to the development of the system concept and systems methods. Newton created the calculus to deal with the time-flowing quantities which are central to mechanics. His contemporary Leibniz was the co-creater of the calculus (Ref. 9) and also the originator of energy methods in mechanics.

The extrema of functions and the calculus of variations were treated by Newton, but the major early advances in these fields are credited to Euler, Johann Bernoulli, and Lagrange (Ref. 10). Matrix algebra, which is fundamental to organizing and dealing with the multivariable problems characteristic of systems, was originated by Cayley (1821–1895) (Ref. 11), with related developments by Hamilton (quarternions and Cayley-Hamilton theorem), and Sylvester (1814–1897) (theory of algebraic invariants). Operational calculus, particularly as applied to electromagnetic theory and problems, was largely the creation of Heaviside (1850–1925), though later the mathematical basis of his work was identified with transform theory. The stability theory of differential equations is associated with Routh and with Hurwitz (Ref. 12) and, in more general form, with Liapunov (Ref. 13). In the twentieth century mathematical developments pertaining to systems center mainly on the theory of optimal processes and on optimization methods adapted for digital computers. References 14 through 17 give an indication of these trends.

This chapter represents, of necessity, only the slimmest outline of what might be presented on the origins of the system concept and the history (growth) of systems methods. In addition to the account given here, various aspects are amplified in subsequent chapters. The reader is also referred to the original literature and representative later works listed in the bibliographies of this and other chapters in the book.

BIBLIOGRAPHY

1. R. Descartes, *Discours de la méthode pour bien conduire sa raison et chercher la vérité dans les sciences.* Published at Leyden, 1637.
2. *Philosophiae Naturalis Principia Mathematica.* London: 1687. (The *Principia* was translated into English by Motte in 1729, and in modern times has been revised and rendered in modern phraseology by Cajori in the following work: F. Cajori, *Newton's Principia, A Revision of Motte's Translation.* Berkeley: University of California Press, 1947.)
3. J. L. Lagrange, *Mécanique Analytique.* Paris: 1788.
4. W. R. Hamilton, "General Method in Dynamics." *Philosophical Transactions*, p. 307, 1834.
5. J. C. Maxwell, "A Dynamical Theory of the Electromagnetic Field." *Transactions of the Royal Society*, *155* Dec. 1864.
6. G. Kron, "Non-Riemannian Dynamics of Rotating Electrical Machinery." *Journal of Mathematics and Physics*, *13* 103, 1934.
7. V. Bush and S. H. Caldwell, "A New Type of Differential Analyzer." *Journal of the Franklin Institute*, *240* 255, 1945.
8. J. C. Maxwell, "On Governors." *Proceedings of the Royal Society of London*, *16* 270, 1868.
9. C. B. Boyer, *The Concepts of the Calculus.* New York: Hafner, 1939. (This book gives an historical account of the origins and development of the calculus.)
10. The early history of the calculus of variations is contained in the following two authoritative works: (A) R. Woodhouse, *A Treatise on Isoperimetrical Problems and the Calculus of Variations.* Cambridge University Press, 1810. (B) I. Todhunter, *A History of the Calculus of Variations During the Nineteenth Century.* Cambridge University Press, 1861.

11. A Cayley, "A Memoir on the Theory of Matrices." *Philosophical Transactions*, London, *148* 17, 1857.

12. Hurwitz' original paper was published in the Mathematische Annalen, *46* 273, 1895. A translation is given in a recent book: *Selected Papers on Mathematical Trends in Control Theory*, pp. 70–82, edited by R. Bellman and R. Kalaba, and published by Dover, 1964.

13. J. La Salle and S. Lefshetz, *Stability by Liapunov's Direct Method with Applications.* New York: Academic Press, 1961.

14. L. S. Pontryagin et al., *The Mathematical Theory of Optimal Processes.* New York: Interscience, 1961. (Translated from the Russian.)

15. R. Bellman, ed., *Mathematical Optimization Techniques.* Berkeley: University of California Press, 1963.

16. G. B. Dantzig, *Linear Programming and Extensions.* Princeton, New Jersey: Princeton University Press, 1963.

17. R. E. Bellman and S. E. Dreyfus, *Applied Dynamic Programming.* Princeton, New Jersey: Princeton University Press, 1962.

18. G. E. Forsythe, "Today's Computational Methods of Linear Algebra." *Siam Review*, *9* 489, July 1967.

6

GENERALIZED MECHANICS

The purpose of this chapter is to formulate a generalized theory of mechanics which provides means for dealing effectively and efficiently with a broad range of physical systems. The methods of generalized mechanics cover not only mechanical systems, but also a vast range of nonmechanical and mixed systems—electromagnetic, feedback, space technology, and so forth.

The methods of generalized mechanics are probably derived most concisely by taking Hamilton's principle as a basic postulate; this is the procedure followed in this book. While Hamilton's principle is equivalent mathematically to Newton's laws for mechanics, it has the advantage of being applicable also to nonmechanical systems. Thus, the use of Hamilton's principle as a basic postulate is consistent with systems philosophy because it unifies physical principles and therefore avoids having to deal with them separately. By treating Hamilton's principle abstractly, and then identifying various applications, the concept of analogs (as between electrical and mechanical, for example) follows immediately. Thus, it is not necessary to introduce this useful concept as a separate idea. Generalized mechanics, therefore, provides a basis for analog simulation of systems. The abstract approach also provides powerful means for synthesis of systems through the process of constructing abstract systems with desirable properties and then devising real systems to match these properties.

The axioms of mechanics, which summarize in precise fashion the experimental or physical basis of the science, may be stated in several different forms. Historically, the axioms were first formulated by Newton in his famous laws of motion. These laws are framed in terms of the vector concepts of force, momentum, and acceleration. Later formulations, due principally to d'Alembert, Lagrange, and Hamilton, are based on the scalar concepts of energy and work and, in some respects, are more elegant. Many applications of mechanics may be based directly on Newton's laws, but experience has shown that the generalized methods based on energy concepts and variational principles are more general and more powerful.

6.1 GENERALIZED COORDINATES

The number of independent quantities necessary to specify uniquely the position or configuration of a system is known as the number of degrees of freedom of the system. In mechanics, for example, a single unconstrained particle has three degrees of freedom since three quantities are necessary to specify its position in space. If, however, the particle is constrained to move on a surface, the number of degrees of freedom is reduced to two. If the motion is limited to a curve, only one degree of freedom remains. More generally, the number of degrees of freedom f for a system of n particles subject to k independent constraints is given by Eq. (6.1)

$$f = 3n - k \tag{6.1}$$

A rigid body has in the most general case six degrees of freedom, three of translation and three of rotation. Restricting one point of the body to move on a surface, as for example a top spinning on a floor, reduces the degrees of freedom to five. Further restriction to motion about a fixed point leaves only three rotational degrees of freedom. Fixing two points in the body permits only one degree of freedom, that of rotation about the axis through the points.

For deformable bodies, which include elastic and plastic solids as well as fluids (gases and liquids), the number of degrees of freedom cannot be enumerated in the same way as for systems of discrete particles or for rigid bodies. Instead, deformable bodies are visualized phenomeno-logically as being infinitely sub-divisible. Therefore, the number of degrees of freedom theoretically becomes infinite, and the equations of motion become partial differential equations characteristic of continuous systems. By contrast, the motion of a system with a finite number of degrees of freedom f is expressed by a system of f second-order ordinary differential equations. Treatment of deformable bodies, frequently re-ferred to as continuum mechanics, requires a rather sophisticated foundation in partial differential equations and is, therefore, outside the scope of this book.

Any set of f quantities $(q_1, q_2, \ldots, q_f)$ which define uniquely the position of a system of f degrees of freedom are called generalized co-ordinates; the corresponding time derivatives $(\dot{q}_1, \dot{q}_2, \ldots, \dot{q}_f)$ are called generalized velocities for the system. Although the total number of degrees of freedom f is a property of the system, actual choice of the generalized coordinates usually can be made in a number of ways. For example, they may represent lengths, angles, or some other quantities associated with the system such as momenta or energies. Most problems are simplified by judicious choice of coordinates. For example, the Kepler problem of planetary motion is far simpler in polar coordinates than in Cartesian coordinates.

For constrained systems, the advantages of generalized coordinates are conspicuous. These advantages may be illustrated by considering the example of a particle system of $f = 3n - k$ degrees of freedom. The configuration of such a system may be specified by n position vectors

$$\mathbf{r}_i = \mathbf{r}_i(x_i, y_i, z_i) \qquad i = 1, 2, \dots, n \tag{6.2}$$

subject to k equations of constraint

$$f_j = f_j(\mathbf{r}_1, \mathbf{r}_2, \dots, \mathbf{r}_n, t) = 0 \qquad j = 1, 2, \dots, k \tag{6.3}$$

where, in general, the constraints may depend explicitly on time t. Physically, the presence of t signifies a moving constraint. The k equations of constraint may, at least in principle, be solved for k of the coordinates x_i, y_i, z_i, substituted in Eq. (6.2), and thus eliminated from the problem. However, in practice this elimination process is not only tedious but often impossible to carry out. Therefore, it is much better to eliminate these constraints by transformation to generalized coordinates for which the number of coordinates is just equal to the number of degrees of freedom for the sytem. The position vectors then are functions of the f generalized coordinates $(q_1, q_2, \dots, q_f)$ and also of the time if moving constraints or moving coordinates are involved. Hence, we have the transformation equations

$$\mathbf{r}_i = \mathbf{r}_i(q_1, q_2, \dots, q_f, t) \qquad i = 1, 2, \dots, n \tag{6.4}$$

which give the configuration of the system at any time t in terms of the generalized coordinates $(q_1, q_2, \dots, q_f)$; the constraints no longer appear.

Examples of Generalized Coordinates

(1) Generalized coordinates for a spherical pendulum (Fig. 6.1a). The pendulum bob is suspended on a string of length a, and has one constraint given by the equation,

$$x^2 + y^2 + z^2 = a^2$$

There are $f = 3 - 1 = 2$ degrees of freedom. Using spherical coordinates (r, φ, θ) allows the equation of constraint to be expressed in the form $r = a$. Hence, the number of independent coordinates is reduced to two (φ, θ) because of the identity $r = a$.

(2) Three particles connected by bars of negligible mass (Fig. 6.1b). There are 3 constraints; thus, $f = 3n - k = 9 - 3 = 6$ degrees of

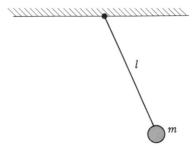

(a) Particle with one constraint (spherical pendulum)

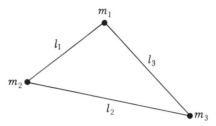

(b) Three particles with three constraints

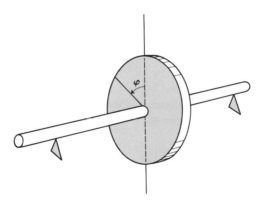

(c) Rigid body with two fixed points, i.e., five constraints (flywheel mounted on fixed bearings)

FIGURE 6.1 Examples of generalized coordinates.

freedom. The generalized coordinates may be chosen, for example, as the position (x, y, z) of the center of mass and the Eulerian angles (φ, θ, ψ).

(3) A flywheel rotating on fixed bearings (Fig. 6.1c). In this case there is only one degree of freedom; the generalized coordinate is conveniently chosen as the angle of rotation φ.

(4) Parallel plate capacitor connected through a resistor to a source of constant emf and with variable plate separation (Fig. 6.2a). There

are two degress of freedom and two generalized coordinates which may be chosen conveniently as the plate separation x and the capacitor charge q.

(5) Electric network with two meshes and variable magnetic coupling (Fig. 6.2b). The system has three degrees of freedom. For generalized coordinates, one may choose the electric charges q_1, q_2 and coil separation x.

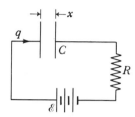

(a) Parallel plate capacitor with variable plate separation

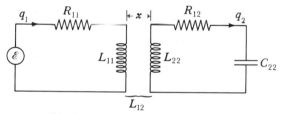

(b) Magnetically coupled network

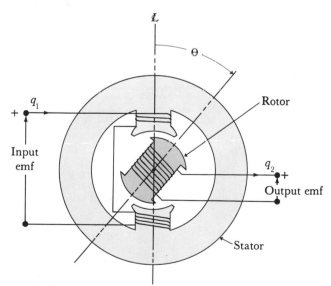

(c) Rotating electromechanical energy converter

FIGURE 6.2

(6) Rotating electromechanical energy converter (Fig. 6.2c). Machines of this type form the basis for a great variety of generators, motors, feedback control system components, electric power control units, and the like. The unit shown has three degrees of freedom represented by the generalized coordinates (q_1, q_2, θ).

Holonomic and Nonholonomic* Constraints. We have seen that constraints are of the form (6.3). According to Heinrich Hertz, the well-known nineteenth century physicist, there are two possibilities: (1) The constraints contain the quantities $\mathbf{r}_1, \mathbf{r}_2, \ldots, \mathbf{r}_n$ in the entire (non-infinitesimal) form only, in which case we speak of holonomic constraint (from Greek holos meaning integer or entire); (2) the constraints contain differentials of the quantities $\mathbf{r}_1, \mathbf{r}_2, \ldots, \mathbf{r}_n$ that cannot be integrated, in which case we speak of nonholonomic constraints. In this case, the conditions or constraints are on the velocities, and can be given in integrated form only after the problem is solved.

In a holonomic system the configuration is specified by f independent coordinates q_i which is just equal to the number of degrees of freedom. Every arbitrary infinitesimal change of the coordinates represents a possible displacement of the system. By contrast, a nonholonomic system has fewer degrees of freedom than the number of independent coordinates required to specify its position.

A classic example of a nonholonomic dynamical system is a vertical disk or wheel rolling on a perfectly rough plane; slippage cannot occur and the motion is that of pure rolling. Four independent coordinates, (x, y, θ, φ) are needed to specify the position of the disk. Coordinates x and y define the point of contact of the disk on the plane, θ is the angle between the plane of the disk and the x-axis, and φ is the angle of rotation of the disk, as shown in Figure 6.3. The nonholonomic constraint requires that the velocity of the center of the disk be

$$\mathbf{v} = a\dot{\varphi}\mathbf{e}_1$$

where $\mathbf{e}_1$ is a unit vector parallel to the $x - y$ plane and perpendicular to the radius of the disk. Expressing $\mathbf{v}$ and $\mathbf{e}_1$ in Cartesian coordinates, then, gives the required result

$$\mathbf{v} = \mathbf{i}\dot{x} + \mathbf{j}\dot{y} = a\dot{\varphi}(\mathbf{i}\sin\theta + \mathbf{j}\cos\theta)$$

which is equivalent to the two scalar equations of constraint

$$dx - a\sin\theta\, d\varphi = 0$$
$$dy - a\cos\theta\, d\varphi = 0 \tag{6.5}$$

The constraint equations (6.5) are of differential form and nonintegrable. They reduce the degrees of freedom from $f = 4$ to $f = 2$. In general, if a

---

* Also called anholonomic.

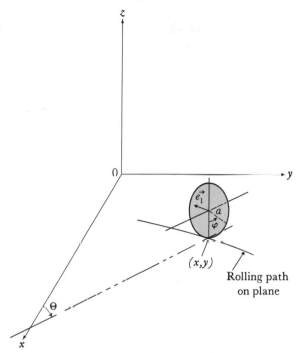

FIGURE 6.3 Example of nonholonomic constraint; vertical disc rolling on a rough plane.

system subject to r nonholonomic conditions has f degrees of freedom in finite motion, it has only $f - r$ degrees of freedom in infinitesimal motion.

Configuration Space. Treatment of generalized mechanics gains both clarity and conciseness by introducing the concept of configuration space. The sequence of f generalized coordinates may be written in the form of an f-component vector as follows:

$$\mathbf{q} = [q_1, q_2, \ldots, q_f] \tag{6.6}$$

Moreover, we may associate with the f components $q_1, q_2, \ldots, q_f$ an abstract f-dimensional space with a point P in this space. Thus, the position or configuration of the entire dynamical system may be pictured by a single point in the f-dimensional space, which is called configuration space.

Since the solution of a dynamical problem is represented by the equations

$$q_i = q_i(t) \qquad i = 1, 2, \ldots, f \tag{6.7}$$

the point P moves along some trajectory in the f-dimensional configuration space, with time t as a parameter. A descriptive terminology is to refer to the point P as a C-point and the curve traced out by P as a C-curve, both in C-space. The motions of complicated systems may be represented by the curve traced by a single point in C-space.

It is evident that, in general, the C-space for a given system is not unique; the generalized position vector $\mathbf{q}$ is not determined beyond the requirement that it completely specifies the state of the system. Let us, therefore, choose another set of generalized coordinates, denoted by $\mathbf{q}'$, and examine the relations between the two sets, $\mathbf{q}$ and $\mathbf{q}'$. Since the two sets represent the same system, there must be a functional relationship between them of the form

$$\mathbf{q}' = \mathbf{f}(\mathbf{q}) \tag{6.8}$$

or in terms of the components

$$\left.\begin{aligned}
q_1' &= q_1'(q_1, q_2, \ldots, q_f) \\
q_2' &= q_2'(q_1, q_2, \ldots, q_f) \\
&\;\cdots \cdots \cdots \cdots \\
q_f' &= q_f'(q_1, q_2, \ldots, q_f)
\end{aligned}\right\} \tag{6.9}$$

To investigate the behavior of the transformation on an infinitesimal scale, consider an infinitesimal displacement $d\mathbf{q}'$

$$\left.\begin{aligned}
dq_1' &= \frac{\partial q_1'}{\partial q_1}\,dq_1 + \frac{\partial q_1'}{\partial q_2}\,dq_2 + \cdots + \frac{\partial q_1'}{\partial q_f}\,dq_f \\[1ex]
dq_2' &= \frac{\partial q_2'}{\partial q_1}\,dq_1 + \frac{\partial q_2'}{\partial q_2}\,dq_2 + \cdots + \frac{\partial q_2'}{\partial q_f}\,dq_f \\[1ex]
&\;\cdots \cdots \cdots \cdots \cdots \cdots \cdots \cdots \\[1ex]
dq_f' &= \frac{\partial q_f'}{\partial q_1}\,dq_1 + \frac{\partial q_f'}{\partial q_2}\,dq_2 + \cdots + \frac{\partial q_f'}{\partial q_f}\,dq_f
\end{aligned}\right\} \tag{6.10}$$

which may be expressed in vector-matrix form

$$d\mathbf{q}' = A\,d\mathbf{q} \tag{6.11}$$

where A is the matrix with elements a_{ij} defined by

$$a_{ij} = \frac{\partial q_i'}{\partial q_j} \tag{6.12}$$

Since the elements a_{ij} are constants for a given point $\mathbf{q}_0$, we see that transformation (6.11) is linear, and the infinitesimal displacements $d\mathbf{q}$ and $d\mathbf{q}'$ are linearly dependent.

The quantity $\det A$ is called the Jacobian of the transformation and is expressed by the symbol

$$J = \det [A] = \frac{\partial(q_1', q_2', \ldots, q_f')}{\partial(q_1, q_2, \ldots, q_f)}$$

Its importance derives from the fact that Eqs. (6.9) cannot be solved for $(q_1, q_2, \ldots, q_f)$ in terms of $(q_1', q_2', \ldots, q_f')$ if the Jacobian vanishes; in this case the transformation is not unique. The result follows directly from (6.11), for to express $d\mathbf{q}$ in terms of $d\mathbf{q}'$ requires

$$d\mathbf{q} = A^{-1}\, d\mathbf{q}'$$

which is possible only if A^{-1} exists, that is, if det $[A] \neq 0$. If the Jacobian of a transformation vanishes identically, then it may be shown that a functional relationship exists among the quantities $q_1, q_2, \ldots, q_f$. On the other hand, if the Jacobian vanishes only at certain points, these points are singular points for the transformation. (See Ex. 2 of this section.)

Example I. Consider an Earth satellite in a circular orbit with constant angular speed ω and radius r_0 measured from the center of the Earth. We may closely approximate an inertial frame by choosing a set of Cartesian coordinates with origin at the center of the Earth and with directions fixed with respect to the stars. We can orient this reference system so that the orbit equations are

$$x = r_0 \cos \omega t \qquad y = r_0 \sin \omega t$$

Transformation to polar coordinates $(x, y) \rightarrow (\rho, \varphi)$ yields

$$\rho = (x^2 + y^2)^{1/2} = r_0$$

$$\varphi - \tan^{-1}\frac{x}{y} = \tan^{-1}\left(\frac{\sin \omega t}{\cos \omega t}\right) = \omega t$$

In this example, it is evident that the number of degrees of freedom is zero since both ρ and φ are specified. Nevertheless, the problem has more than trivial interest as may be seen by calculation of the acceleration. In polar coordinates the acceleration $\mathbf{a}$, which may be found by direct differentiation $\mathbf{a} = \ddot{\mathbf{r}} = \dfrac{d^2}{dt^2}(\rho\mathbf{e}_1)$ where $\mathbf{e}_1$ is the unit vector along ρ, is

$$\mathbf{a} = \begin{bmatrix} a_\rho \\ a_\varphi \end{bmatrix} = \begin{bmatrix} \ddot{\rho} - \rho\dot{\varphi}^2 \\ 2\,\dot{\rho}\dot{\varphi} + \rho\ddot{\varphi} \end{bmatrix} = \begin{bmatrix} -\rho\dot{\varphi}^2 \\ 0 \end{bmatrix} = \begin{bmatrix} -r_0\omega^2 \\ 0 \end{bmatrix}$$

A body of mass m in the space ship is subject to a force

$$\mathbf{F} = \begin{bmatrix} F_\rho \\ F_\varphi \end{bmatrix} = \begin{bmatrix} -mg \\ 0 \end{bmatrix}$$

and since $\mathbf{F} - m\mathbf{a} = 0$, we have the result

$$\mathbf{F} - m\mathbf{a} = \begin{bmatrix} -mg + mr_0\omega^2 \\ 0 \end{bmatrix} = 0$$

It is thus clear that a stable circular orbit of radius r_0 is possible only if the period (time for one rotation) has the value

$$\tau = \frac{2\pi}{\omega} = 2\pi \sqrt{\frac{r_0}{g}}$$

which, curiously, is also the expression for the period of vibration of a simple pendulum of length r_0. A second result of considerable practical interest is that an astronaut riding in such a space vehicle will experience no external force, that is, he is "weightless." The acceleration of gravity is just balanced by the acceleration of rotation.

Example 2. Transformation from Cartesian to Polar Coordinates in a Plane

Let

$$\mathbf{q}' = \begin{bmatrix} x \\ y \end{bmatrix} = \begin{bmatrix} \rho \cos \varphi \\ \rho \sin \varphi \end{bmatrix} = \mathbf{f}(\mathbf{q})$$

where

$$\mathbf{q} = \begin{bmatrix} \rho \\ \varphi \end{bmatrix}$$

Calculation of the differentials gives

$$d\mathbf{q}' = \begin{bmatrix} dx \\ dy \end{bmatrix} = \begin{bmatrix} \dfrac{\partial x}{\partial \rho} & \dfrac{\partial x}{\partial \varphi} \\ \dfrac{\partial y}{\partial \rho} & \dfrac{\partial y}{\partial \varphi} \end{bmatrix} \begin{bmatrix} d\rho \\ d\varphi \end{bmatrix}$$

$$= \begin{bmatrix} \cos \varphi & -\rho \sin \varphi \\ \sin \varphi & \rho \cos \varphi \end{bmatrix} \begin{bmatrix} d\rho \\ d\varphi \end{bmatrix} = A \, d\mathbf{q}$$

The Jacobian of the transformation is

$$J = \det [A] = \rho$$

Hence, the transformation exists and is unique everywhere except at $\rho = 0$ which is a singular point. It is observed that a particle trajectory which is a circle about the origin of radius $(x_0^2 + y_0^2)^{1/2}$ in $[x, y]$ space transforms to the straight line $\rho = \text{const.} = (x_0^2 + y_0^2)^{1/2}$ in $[\rho, \varphi]$ space.

Example 3. Curvilinear Coordinates. A more general application, essential for determining kinetic energy and potential functions of mechanical systems, is the transformation in Euclidean space (E_3) from Cartesian to curvilinear coordinates.

Let (u, v, w) be a set of coordinates defined by the transformation

$$u = u(x, y, z)$$
$$v = v(x, y, z)$$
$$w = w(x, y, z)$$

The position of a point $P(x, y, z)$ may be specified by the intersection of the three surfaces $u = \text{const.} = C_1$, $v = \text{const.} = C_2$, $w = \text{const.} = C_3$. These surfaces are called coordinate surfaces, while the space curves defined by their intersections are correspondingly designated coordinate lines. The tangents to the coordinate lines at any point of intersection determine three directions in space which we denote by a set of unit vectors $(\mathbf{e_1}, \mathbf{e_2}, \mathbf{e_3})$. It is noted that these directions are not in general fixed in space as is true for Cartesian coordinates but, rather, change orientation as the point P moves.

To investigate the properties of curvilinear coordinates, write the inverse transformation

$$x = x(u, v, w)$$
$$y = y(u, v, w)$$
$$z = z(u, v, w)$$

and denote

$$\mathbf{q'} = \mathbf{r} = \begin{vmatrix} x \\ y \\ z \end{vmatrix} \qquad \mathbf{q} = \begin{vmatrix} u \\ v \\ w \end{vmatrix}$$

The Jacobian matrix A in the relation $d\mathbf{r} = A\, d\mathbf{q}$ is, from Eq. (6.10), given by

$$A = \begin{bmatrix} \dfrac{\partial x}{\partial u} & \dfrac{\partial x}{\partial v} & \dfrac{\partial x}{\partial w} \\[2mm] \dfrac{\partial y}{\partial u} & \dfrac{\partial y}{\partial v} & \dfrac{\partial y}{\partial w} \\[2mm] \dfrac{\partial z}{\partial u} & \dfrac{\partial z}{\partial v} & \dfrac{\partial z}{\partial w} \end{bmatrix}$$

The line elements $d\mathbf{r}$ and ds may now be expressed in terms of the matrix A. First, write $d\mathbf{r}$ in Cartesian components and transform to curvilinear components.

$$d\mathbf{r} = \mathbf{i}\, dx + \mathbf{j}\, dy + \mathbf{k}\, dz = \left(\mathbf{i}\frac{\partial x}{\partial u} + \mathbf{j}\frac{\partial y}{\partial u} + \mathbf{k}\frac{\partial z}{\partial u} \right) du$$

$$+ \left(\mathbf{i}\frac{\partial x}{\partial v} + \mathbf{j}\frac{\partial y}{\partial v} + \mathbf{k}\frac{\partial z}{\partial v} \right) dv + \left(\mathbf{i}\frac{\partial x}{\partial w} + \mathbf{j}\frac{\partial y}{\partial w} + \mathbf{k}\frac{\partial z}{\partial w} \right) dw$$

$$= \mathbf{h_1}\, du + \mathbf{h_2}\, dv + \mathbf{h_3}\, dw$$

where

$$\mathbf{h}_1 = \mathbf{i}\,\frac{\partial x}{\partial u} + \mathbf{j}\,\frac{\partial y}{\partial u} + \mathbf{k}\,\frac{\partial z}{\partial u}$$

$$\mathbf{h}_2 = \mathbf{i}\,\frac{\partial x}{\partial v} + \mathbf{j}\,\frac{\partial y}{\partial v} + \mathbf{k}\,\frac{\partial z}{\partial v}$$

$$\mathbf{h}_3 = \mathbf{i}\,\frac{\partial x}{\partial w} + \mathbf{j}\,\frac{\partial y}{\partial w} + \mathbf{k}\,\frac{\partial z}{\partial w}$$

These relations may be written in matrix form as follows

$$
\begin{bmatrix} \mathbf{h}_1 \\ \mathbf{h}_2 \\ \mathbf{h}_3 \end{bmatrix}
=
\begin{bmatrix}
\dfrac{\partial x}{\partial u} & \dfrac{\partial y}{\partial u} & \dfrac{\partial z}{\partial u} \\[2mm]
\dfrac{\partial x}{\partial v} & \dfrac{\partial y}{\partial v} & \dfrac{\partial z}{\partial v} \\[2mm]
\dfrac{\partial x}{\partial w} & \dfrac{\partial y}{\partial w} & \dfrac{\partial z}{\partial w}
\end{bmatrix}
\begin{bmatrix} \mathbf{i} \\ \mathbf{j} \\ \mathbf{k} \end{bmatrix}
= A'
\begin{bmatrix} \mathbf{j} \\ \mathbf{j} \\ \mathbf{k} \end{bmatrix}
$$

where A' is the transpose of the transformation matrix A.

  The line element ds may be calculated from the equations

$$ds^2 = dx^2 + dy^2 + dz^2 = d\mathbf{r}' \cdot d\mathbf{r}$$
$$= (A\,d\mathbf{q})' \cdot (A\,d\mathbf{q}) = d\mathbf{q}'A'A\,d\mathbf{q}$$

The term $d\mathbf{q}'A'A\,d\mathbf{q}$ represents a quadratic form in (du, dv, dw), and $A'A$ is the matrix of the quadratic form.

  Calculation of $A'A$ in terms of its elements yields

$$
A'A =
\begin{bmatrix}
\dfrac{\partial x}{\partial u} & \dfrac{\partial y}{\partial u} & \dfrac{\partial z}{\partial u} \\[2mm]
\dfrac{\partial x}{\partial v} & \dfrac{\partial y}{\partial v} & \dfrac{\partial z}{\partial v} \\[2mm]
\dfrac{\partial x}{\partial w} & \dfrac{\partial y}{\partial w} & \dfrac{\partial z}{\partial w}
\end{bmatrix}
\begin{bmatrix}
\dfrac{\partial x}{\partial u} & \dfrac{\partial x}{\partial v} & \dfrac{\partial x}{\partial w} \\[2mm]
\dfrac{\partial y}{\partial u} & \dfrac{\partial y}{\partial v} & \dfrac{\partial y}{\partial w} \\[2mm]
\dfrac{\partial z}{\partial u} & \dfrac{\partial z}{\partial v} & \dfrac{\partial z}{\partial w}
\end{bmatrix}
= G
$$

where $G = A'A$ is introduced to simplify notation. By direct calculation it is found that

$$g_{11} = \left(\frac{\partial x}{\partial u}\right)^2 + \left(\frac{\partial y}{\partial u}\right)^2 + \left(\frac{\partial z}{\partial u}\right)^2 = \mathbf{h}_1 \cdot \mathbf{h}_1 = h_1^2$$

$$g_{12} = g_{21} = \frac{\partial x}{\partial u}\frac{\partial x}{\partial v} + \frac{\partial y}{\partial u}\frac{\partial y}{\partial v} + \frac{\partial z}{\partial u}\frac{\partial z}{\partial v} = \mathbf{h}_1 \cdot \mathbf{h}_2$$

with similar equations for the remaining elements.

For orthogonal coordinate systems, the dot products $\mathbf{h}_1 \cdot \mathbf{h}_2$, $\mathbf{h}_2 \cdot \mathbf{h}_3$, and $\mathbf{h}_3 \cdot \mathbf{h}_1$ vanish and G reduces to diagonal form. For this important case

$$ds^2 = h_1^2 \, du^2 + h_2^2 \, dv^2 + h_3^2 \, dw^2$$

and the Jacobian reduces to

$$J = \frac{\partial(x, y, z)}{\partial(u, v, w)} = \det A = \det A' = (\det G)^{1/2}$$

$$= h_1 h_2 h_3$$

6.2 HAMILTON'S PRINCIPLE

In this book Hamilton's principle is taken as an independent basic postulate, applicable to nonmechanical as well as to mechanical systems. The advantages of this generalization depend, of course, on the range of systems and problems to which the principle can be applied and on the corpus of physical laws which can be derived from the principle. It will be shown that the generalization leads to unified treatment of many electrodynamic, electromechanical, electric network, and thermal systems and is, therefore, in harmony with the basic aims of systems science.

Hamilton's principle is stated abstractly, although because of the origins of the principle it will be convenient to retain the language of mechanics. Consider a system of f degrees of freedom characterized by a set of time-dependent generalized coordinates

$$\mathbf{q} = (q_1, q_2, \ldots, q_f) \qquad q_k = q_k(t)$$

where $q_k \in C^2$, $k = 1, 2, \ldots, f$. Let the configuration of the system at time t_0 and time t_1 be given by the two sets of values of the coordinates $\mathbf{q}(t_0)$ and $\mathbf{q}(t_1)$. Then, the motion of the system between these two configurations occurs in such a way that the integral

$$\int_{t_0}^{t_1} (\delta T + \delta W) \, dt = 0 \tag{6.13}$$

where T is the kinetic energy and W the work function, and the symbol δ has the same meaning as in Chap. 3. Expression (6.13) is Hamilton's principle, valid for nonconservative as well as conservative systems.

The quantity δW, called virtual work, is defined by the work done by a system during a virtual displacement. In generalized coordinates,

$$\delta W = \sum_{i=1}^{f} Q_i \delta q_i = \mathbf{Q} \cdot \delta \mathbf{q} \tag{6.14}$$

where the Q_i's are known as the generalized forces and $\mathbf{Q}$ is the f-component vector

$$\mathbf{Q} = [Q_1, Q_2, \ldots, Q_f] \tag{6.15}$$

The generalized forces may or may not have the physical dimensions of force, but the product $Q_i \, \delta q_i$ always has the dimensions of work or energy. If, for example, δq_i is a length, then Q_i is a force, while if δq_i is an angle, Q_i is a torque.

The work done by the forces acting in a dynamical system will in general depend on the path or trajectory along which the motion occurs. However, there are certain classes of forces for which the work done depends only on the end points of the motion and not on the path. Such forces are called conservative, and systems which have only conservative forces are correspondingly called conservative systems. Examples of conservative forces are gravitational force, electrostatic force, and elastic force. Nonconservative forces are those which depend on velocity, time, or factors other than position.

Mathematically, the condition for conservative forces is that $\mathbf{Q} \cdot d\mathbf{q}$ be a perfect differential or, equivalently, that $\mathbf{Q}$ be derivable as the gradient of a scalar function V; that is,

$$\mathbf{Q} = -\boldsymbol{\nabla} V \tag{6.16}$$

The function $V = V(\mathbf{q})$ is called the potential, a term used by Lagrange. The negative sign is conventional; it is useful in practice since the positive direction of the force is then always in the direction of decreasing potential energy. It follows from Eq. (6.16) that the work W done by conservative forces is

$$W = \int \mathbf{Q} \cdot d\mathbf{q} = -\int (\boldsymbol{\nabla} V) \cdot d\mathbf{q} = -V + \text{const.} \tag{6.17}$$

where the additive constant depends on the datum or initial point from which the work is calculated. It is evident from (6.16) that adding an arbitrary constant to V has no effect on $\mathbf{Q}$.

Hamilton's principle for conservative systems may be written in simplified form as follows:

$$\int_{t_0}^{t_1} (\delta T + \delta W) \, dt = \delta \int_{t_0}^{t_1} (T - V) \, dt = 0 \tag{6.18}$$

where $W = -V + \text{const.}$ from Eq. (6.17), and it is permissible to move the variation operator δ outside the integral sign because both T and V are entire functions. In the more general case of a nonconservative system, the virtual work δW is not integrable and δ cannot be moved outside the integral sign. In works on mechanics it is customary to introduce the Lagrangian function L defined by

$$L = T - V$$

and write Eq. (6.18) in the form

$$\delta \int_{t_0}^{t_1} L \, dt = 0 \qquad \textbf{(6.19)}$$

The Lagrangian is a function

$$L = L(\dot{\mathbf{q}}, \mathbf{q}, t)$$

where the generalized velocities $\dot{\mathbf{q}}$ enter into the expression for kinetic energy, and the generalized coordinates $\mathbf{q}$ enter into the potential energy and also frequently into the kinetic energy expression. The time t does not appear explicitly for conservative systems, but can be present for systems involving moving constraints, moving coordinate systems, or time-dependent potential functions; such systems are not conservative but can be shown to obey (6.19).

6.3 LAGRANGE'S EQUATIONS

Lagrange's equations are a direct consequence of Hamilton's principle, and assume several distinct forms corresponding to the different forms of Hamilton's principle. In the most general form, no restrictions are imposed on the system other than that the constraints be holonomic and that a kinetic energy function T and a virtual work function δW can be formulated in terms of a set of generalized coordinates. The first step in deriving Lagrange's equations is to write Hamilton's principle, Eq. (6.13), in the form

$$\int_{t_0}^{t_1} (\delta T + \delta W) \, dt = \delta \int_{t_0}^{t_1} T \, dt + \int_{t_0}^{t_1} \delta W \, dt$$

Carrying out the indicated variations in terms of generalized coordinates in accord with the calculus of variations (cf. Chap. 3), gives

$$\delta \int_{t_0}^{t_1} T \, dt = \int_{t_0}^{t_1} \sum_{i=1}^{f} \left[\frac{\partial T}{\partial q_i} - \frac{d}{dt} \left(\frac{\partial T}{\partial \dot{q}_i} \right) \right] \delta q_i \, dt$$

and

$$\int_{t_0}^{t_1} \delta W \, dt = \int_{t_0}^{t_1} \sum_{i=1}^{f} Q_i \delta q_i \, dt$$

Collecting terms gives

$$\int_{t_0}^{t_1} \left\{ \sum_{i=1}^{f} \left[\frac{\partial T}{\partial q_i} - \frac{d}{dt} \left(\frac{\partial T}{\partial \dot{q}_i} \right) + Q_i \right] \delta q_i \right\} dt = 0$$

Since the δq_i's are arbitrary, the integral can vanish only if each of the f bracket expressions under the integral sign vanishes identically, that is, if

$$\frac{\partial T}{\partial q_i} - \frac{d}{dt} \left(\frac{\partial T}{\partial \dot{q}_i} \right) + Q_i = 0 \qquad i = 1, 2, \ldots, f \qquad \textbf{(6.20)}$$

Equation (6.20) is the most general form of Lagrange's equations for systems with holonomic constraints. The generalized forces Q_i are found from Eq. (6.14) for the virtual work δW.

Hamilton's principle for conservative systems takes the simpler form

$$\delta \int_{t_0}^{t_1} L \, dt = 0$$

where $L = T - V$. The Euler-Lagrange condition for a stationary value of this integral (Chap. 3) leads directly to Lagrange's equations of motion

$$\frac{d}{dt}\left(\frac{\partial L}{\partial \dot{q}_i}\right) - \frac{\partial L}{\partial q_i} = 0 \qquad i = 1, 2, \ldots, f \qquad \textbf{(6.21)}$$

Lagrange's equations of motion comprise a set of f second-order ordinary differential equations which, together with the boundary conditions, provide the complete set of differential equations for a dynamical problem. The general solution contains $2f$ arbitrary constants. To specify uniquely the motion of the system, it is necessary to determine the values of these constants. This requires a knowledge of the state of the system, i.e., the values of all the coordinates and velocities at some given instant of time, which is usually taken as the time t_0 at the start of the motion.

Example I. Motion of a Free Particle. The particle of mass m has three degrees of freedom and is subject to no forces. Choosing Cartesian coordinates, the Lagrangian function is

$$L = m(\dot{x}^2 + \dot{y}^2 + \dot{z}^2)$$

Substitution in Lagrange's equations (6.21) gives the equations of motion

$$\frac{d}{dt}\left(\frac{\partial L}{\partial \dot{x}}\right) = \frac{d}{dt}(m\dot{x}) = 0$$

$$\frac{d}{dt}\left(\frac{\partial L}{\partial \dot{y}}\right) = \frac{d}{dt}(m\dot{y}) = 0$$

$$\frac{d}{dt}\left(\frac{\partial L}{\partial \dot{z}}\right) = \frac{d}{dt}(m\dot{z}) = 0$$

These have the obvious solutions

$$m\dot{x} = \text{const.} = m\dot{x}_0$$
$$m\dot{y} = \text{const.} = m\dot{y}_0$$
$$m\dot{z} = \text{const.} = m\dot{z}_0$$

where $\dot{x}_0, \dot{y}_0, \dot{z}_0$ are the velocities at time t_0. Thus, the particle moves with constant velocity

$$\mathbf{v}_0 = \dot{x}_0\mathbf{i} + \dot{y}_0\mathbf{j} + \dot{z}_0\mathbf{k}$$

a result in agreement with Newton's first law of motion.

Example 2. Motion of a Particle Under a Given Force. Since the force is not necessarily conservative, we use Lagrange's equations in the form (6.20) and write, again using Cartesian coordinates,

$$T = \tfrac{1}{2}m(\dot{x}^2 + \dot{y}^2 + \dot{z}^2)$$

$$\mathbf{F} = [F_x, F_y, F_z]$$

The equations of motion evidently are

$$\frac{d}{dt}(m\dot{x}) = \dot{p}_x = F_x$$

$$\frac{d}{dt}(m\dot{y}) = \dot{p}_y = F_y$$

$$\frac{d}{dt}(m\dot{z}) = \dot{p}_z = F_z$$

where

$$\mathbf{p} = [p_x, p_y, p_z]$$

defines the linear momentum of the particle. The result expresses Newton's second law of motion, i.e.,

$$\dot{\mathbf{p}} = \mathbf{F}$$

It is instructive to extend the example to the case where the frame of reference (the coordinate system) is moving at a constant velocity $\mathbf{v}_c$. Then

$$T = \tfrac{1}{2}m(\mathbf{v} + \mathbf{v}_c)^2$$

$$= \tfrac{1}{2}mv^2 + m\mathbf{v}\cdot\mathbf{v}_c + \tfrac{1}{2}mv_c^2$$

and, since the second and third terms for T in the preceding equation contribute nothing when substituted in Lagrange's equations, it follows that the equations of motion remain unchanged. This result is in essence Galileo's principle of relativity.*

Example 3. Lagrange's Equations for Spherical Coordinates. Set up Lagrange's equations of motion in spherical coordinates r, θ, φ, for a particle of mass m subject to a force whose components are F_r, F_θ, F_φ.

SOLUTION. There are three degrees of freedom, one for each coordinate. Use Lagrange's equations with generalized forces Q_r, Q_θ, Q_φ. The energy expressions are

$$T = \tfrac{1}{2}m(\dot{r}^2 + r^2\dot{\theta}^2 + r^2\dot{\varphi}^2 \sin^2\theta)$$

$$\delta W = F_r\delta r + F_\theta r\delta\theta + F_\varphi r \sin\theta\delta\varphi$$

_____

* Also called Newtonian relativity.

Substitution in Lagrange's equations gives

(1) $\dfrac{d}{dt}\left(\dfrac{\partial T}{\partial \dot r}\right) - \dfrac{\partial T}{\partial r} = m\ddot r - mr\dot\theta^2 - mr\dot\varphi^2 \sin^2\theta = F_r$

(2) $\begin{cases} \dfrac{d}{dt}\left(\dfrac{\partial T}{\partial \dot\theta}\right) - \dfrac{\partial T}{\partial \theta} = \dfrac{d}{dt}(mr^2\dot\theta) - mr^2\dot\varphi^2\cos\theta\sin\theta = F_\theta r \\[2mm] \text{which simplifies to} \\[1mm] mr\ddot\theta + 2m\dot r\dot\theta - mr\dot\varphi^2\sin\theta\cos\theta = F_\theta \end{cases}$

(3) $\begin{cases} \dfrac{d}{dt}\left(\dfrac{\partial T}{\partial \dot\varphi}\right) - \dfrac{\partial T}{\partial \varphi} = \dfrac{d}{dt}(mr^2\dot\varphi\sin^2\theta) = F_\varphi r\sin\theta \\[2mm] mr^2\sin^2\theta\,\ddot\varphi + 2mr\dot r\dot\varphi\sin^2\theta + 2mr^2\dot\varphi\sin\theta\cos\theta\,\dot\theta = F_\varphi r\sin\theta \\[1mm] \text{which simplifies to} \\[1mm] mr\sin\theta\,\ddot\varphi + 2m\dot r\dot\varphi\sin\theta + 2mr\dot\varphi\dot\theta\cos\theta = F_\varphi \end{cases}$

Example 4. Moving Coordinate System. Find the equations of motion for a particle of mass m on a horizontal platform rotating with constant angular velocity ω.

SOLUTION

Choose as generalized coordinates

$$q_1 = r \qquad q_2 = \theta + \omega t$$

Then we have for **v**, denoting unit vectors along q_1 and q_2 as $\mathbf{e}_1$ and $\mathbf{e}_2$ respectively,

$$\mathbf{v} = \dot r\mathbf{e}_1 + r(\dot\theta + \omega)\mathbf{e}_2$$

$$v^2 = \dot r^2 + r^2(\dot\theta + \omega)^2$$

$$T = \tfrac{1}{2}mv^2 = \tfrac{1}{2}m[\dot r^2 + r^2(\dot\theta + \omega)^2]$$

$$\frac{d}{dt}\left(\frac{\partial T}{\partial \dot r}\right) - \frac{\partial T}{\partial r} = m\ddot r - mr(\dot\theta + \omega)^2 = Q_r = F_r$$

$$\frac{d}{dt}\left(\frac{\partial T}{\partial \dot\theta}\right) - \frac{\partial T}{\partial \theta} = \frac{d}{dt}[mr^2(\dot\theta + \omega)] = Q_\theta = rF_\theta$$

where F_r and F_θ are the components of the applied force along $\mathbf{e}_1$ and $\mathbf{e}_2$ respectively.

Example 5. Simple Harmonic Oscillator

A system with one degree of freedom and Lagrangian

$$L = \tfrac{1}{2}m\dot q^2 - \tfrac{1}{2}kq^2$$

obviously has the equation of motion

$$\frac{d}{dt}\left(\frac{\partial L}{\partial \dot{q}}\right) - \frac{\partial L}{\partial q} = m\ddot{q} + kq = 0$$

Here m and k are constants, but are not identified with any specific system. The equation of motion has the general solution

$$q = q_0 \sin \sqrt{\frac{k}{m}}\,(t - t_0)$$

as may be shown by direct substitution in the differential equation.

The result thus obtained represents a simple harmonic (sine-wave) vibration of angular frequency $\sqrt{\dfrac{k}{m}}$ and amplitude q_0. Many physical embodiments exist, as for example, the vibrations of a mass m suspended from a spring of constant k, or an electric circuit of inductance $L = m$ and capacitance $C = \dfrac{1}{k}$. In the latter case, the angular frequency clearly is $\sqrt{\dfrac{1}{LC}}$, in agreement with the well-known result for an $L - C$ oscillator. These and other embodiments are studied further in Chap. 11.

Example 6. Damped Oscillator. Find the equation of motion for a plane pendulum subjected to viscous damping.

SOLUTION. Denote the length of the pendulum by l, its mass by m, and the damping coefficient by k. There is one degree of freedom and, therefore, one generalized coordinate which is chosen as θ, the angle of deflection from equilibrium. The kinetic energy T and virtual work δW are given by

$$T = \tfrac{1}{2}mv^2 = \tfrac{1}{2}ml^2\dot{\theta}^2$$
$$\delta W = (-mgl \sin \theta - kv)\,\delta\theta$$
$$= -(mgl \sin \theta + kl\dot{\theta})\,\delta\theta$$

Using Lagrange's equation

$$\frac{\partial T}{\partial \theta} - \frac{d}{dt}\left(\frac{\partial T}{\partial \dot{\theta}}\right) + Q_\theta = 0$$

there is obtained

$$-\frac{d}{dt}\,(ml^2\dot{\theta}) - mgl \sin \theta - kl\dot{\theta} = 0$$

or

$$ml^2\ddot{\theta} + kl\dot{\theta} + mgl \sin \theta = 0$$

Dividing by ml^2 gives

$$\ddot{\theta} + \frac{k}{lm}\,\dot{\theta} + \frac{g}{l} \sin \theta = 0$$

For small angular displacements, $\sin \theta \simeq \theta$, and the equation is approximated by

$$\ddot{\theta} + \frac{k}{lm}\dot{\theta} + \frac{g}{l}\theta = 0$$

which is the differential equation for the damped harmonic oscillator.

Example 7.   Find Lagrange's equations of motion for the coupled-mass system shown in Figure 6.4.

SOLUTION.   There are two degrees of freedom, y_1 and y_2. Write expressions for T and V.

$$T = \tfrac{1}{2}m_1\dot{y}_1^2 + \tfrac{1}{2}m_2(\dot{y}_1 + \dot{y}_2)^2$$
$$V = -m_1gy_1 - m_2g(y_1 + y_2) + \tfrac{1}{2}k_1y_1^2 + \tfrac{1}{2}k_2y_2^2$$

where k_1 and k_2 are the spring constants. The equations of motion are

$$\frac{d}{dt}\left(\frac{\partial L}{\partial \dot{y}_1}\right) - \frac{\partial L}{\partial y_1} = m_1\ddot{y}_1 + m_2(\ddot{y}_1 + \ddot{y}_2) - m_1g - m_2g + k_1y_1$$

$$= (m_1 + m_2)\ddot{y}_1 + m_2\ddot{y}_2 + k_1y_1 - (m_1 + m_2)g = 0$$

and

$$\frac{d}{dt}\left(\frac{\partial L}{\partial \dot{y}_2}\right) - \frac{\partial L}{\partial y_2} = m_2(\ddot{y}_1 + \ddot{y}_2) - m_2g + k_2y_2$$

$$= m_2\ddot{y}_1 + m_2\ddot{y}_2 + k_2y_2 - m_2g = 0$$

These may also be written as

$$\ddot{y}_1 + \frac{m_2}{m_1 + m_2}\ddot{y}_2 + \frac{k_1}{m_1 + m_2}y_1 = g$$

$$\ddot{y}_1 + \ddot{y}_2 + \frac{k_2}{m_2}y_2 = g$$

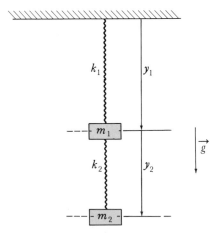

FIGURE 6.4   Problem of coupled masses.

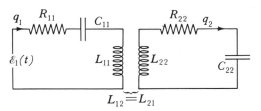

FIGURE 6.5 Inductively coupled electric network.

Example 8. Single-Mesh Electric Network*. Find the mesh equation for an electric circuit consisting of an inductance L, capacitance C, resistance R, and EMF source ε, all in series, using the Lagrangian procedure.

SOLUTION. Write the expressions for kinetic energy T and virtual work δW, and substitute in Lagrange's equation. Using q for the electric charge, we have

$$T = \tfrac{1}{2}LI^2 = \tfrac{1}{2}L\dot{q}^2$$

$$\delta W = \left(-R\dot{q} - \frac{q}{C} + \varepsilon\right)\delta q = Q\delta q$$

Then

$$\frac{\partial T}{\partial q} - \frac{d}{dt}\left(\frac{\partial T}{\partial \dot{q}}\right) + Q = -L\ddot{q} - R\dot{q} - \frac{q}{C} + \varepsilon = 0$$

or

$$L\ddot{q} + R\dot{q} + \frac{q}{C} = \varepsilon$$

Note that Hamilton's principle, from which this mesh equation is derived, leads to Kirchhoff's law for the sum of the voltage drops around a closed loop of an electrical network. It is clear that the result can be readily generalized to cover multiloop networks.

Example 9. Dual-Mesh Coupled Electric Network. Derive the network equations for the circuit shown in Figure 6.5.

SOLUTION. Set up expressions for the kinetic energy and virtual work.

$$T = \tfrac{1}{2}[L_{11}\dot{q}_1^2 - (L_{12} + L_{21})\dot{q}_1\dot{q}_2 + L_{22}\dot{q}_2^2] = \tfrac{1}{2}[L_{11}\dot{q}_1^2 - 2L_{12}\dot{q}_1\dot{q}_2 + L_{22}\dot{q}_2^2]$$

$$\delta W = -R_{11}\dot{q}_1\delta q_1 - \frac{q_1}{C_{11}}\delta q_1 + \varepsilon_1\delta q_1 - R_{22}\dot{q}_2 + \frac{q_2}{C_{22}}\delta q_2$$

$$= \left(-R_{11}\dot{q}_1 - \frac{q_1}{C_{11}} + \varepsilon_1\right)\delta q_1 + \left(-R_{22}\dot{q}_2 - \frac{q_2}{C_{22}}\right)\delta q_2 = Q_1\,\delta q_1 + Q_2\,\delta q_2$$

---

* The theory and methods for dealing with electric networks by generalized mechanics is covered more fully in Chap. 8. The purpose of this and the next example is to show the general procedure and to illustrate electrical applications.

From these, write Lagrange's equations as follows:

$$\frac{\partial T}{\partial q_1} - \frac{d}{dt}\left(\frac{\partial T}{\partial \dot{q}_1}\right) + Q_1 = -L_{11}\ddot{q}_1 + L_{12}\ddot{q}_2 - R_{11}\dot{q}_1 - \frac{q_1}{C_{11}} + \varepsilon_1 = 0$$

$$\frac{\partial T}{\partial q_2} - \frac{d}{dt}\left(\frac{\partial T}{\partial \dot{q}_2}\right) + Q_2 = L_{12}\ddot{q}_1 - L_{22}\ddot{q}_2 - R_{22}\dot{q}_2 - \frac{q_2}{C_{22}} = 0$$

Rearranging terms gives

$$L_{11}\ddot{q}_1 + R_{11}\dot{q}_1 + \frac{q_1}{C_{11}} - L_{12}\ddot{q}_2 = \varepsilon_1$$

$$-L_{21}\ddot{q}_1 + L_{22}\ddot{q}_2 + R_{22}\dot{q}_2 + \frac{q_2}{C_{22}} = 0$$

Note that, as in the preceding example, these mesh equations agree with Kirchhoff's law of voltage drops around a closed mesh.

It is sometimes convenient to express these equations in operator form. Let

$$a_{11} = L_{11}\frac{d^2}{dt_2} + R_{11}\frac{d}{dt} + \frac{1}{C_{11}}$$

$$a_{12} = a_{21} = -L_{12}\frac{d^2}{dt^2}$$

$$a_{22} = L_{22}\frac{d^2}{dt^2} + R_{22}\frac{d}{dt} + \frac{1}{C_{22}}$$

Then, we may write the mesh equations as

$$a_{11}q_1 + a_{12}q_2 = \varepsilon_1$$
$$a_{21}q_1 + a_{22}q_2 = 0$$

These may be further given in matrix form

$$\begin{bmatrix} a_{11} & a_{12} \\ a_{21} & a_{22} \end{bmatrix} \begin{bmatrix} q_1 \\ q_2 \end{bmatrix} = \begin{bmatrix} \varepsilon_1 \\ 0 \end{bmatrix}$$

or

$$A\mathbf{q} = \boldsymbol{\epsilon}$$

Constants or Invariants of Motion. The Lagrangian formalism provides a systematic plan for finding the equations of motion for any dynamical system, and thereby avoids the necessity of searching for ingenious special devices and methods for finding these equations for the endless variety of systems which occur in practice. In addition, the general methods are useful in the integration or solution of the equations of motion. Integrals for dynamics problems are often referred to as constants or invariants of motion.

There are two classes of problems for which invariants of motion are readily found for Hamilton's principle or Lagrange's equations. One of these is where the Lagrangian L does not contain a coordinate q_k explicitly, in which case the corresponding Lagrange equation simplifies to

$$\frac{d}{dt}\left(\frac{\partial L}{\partial \dot{q}_k}\right) = \frac{\partial L}{\partial q_k} = 0$$

This equation has the obvious integral

$$\frac{\partial L}{\partial \dot{q}_k} = \text{constant} = p_k \qquad (6.22)$$

where p_k is known as the generalized momentum. Such a coordinate q_k, which does not occur explicitly in L, is called an ignorable or cyclic coordinate. Each ignorable coordinate furnishes an integral of motion. Furthermore, Eq. (6.22) may be solved to eliminate the corresponding q_k, thereby reducing by one the number of unknowns, as illustrated by the elimination of φ in Ex. 10 for the motion of an electron about a fixed nucleus.

A second invariant of motion occurs whenever L does not contain t explicitly, for which case the Euler-Lagrange equations have the integral (cf. Chap. 3)

$$\sum_{i=1}^{f} \dot{q}_i \frac{\partial L}{\partial \dot{q}_i} - L = \text{constant} = E' \qquad (6.23)$$

In the special case of a stationary coordinate system, the kinetic energy is a quadratic form in the generalized velocities $\dot{q}_i$ and we can use Euler's result

$$\sum_{i=1}^{f} \dot{q}_i \frac{\partial T}{\partial \dot{q}_i} = 2T \qquad (6.24)$$

Then, if a potential energy V exists which is a function of position only, i.e., the system is conservative, we have the result

$$\sum_{i=1}^{f} \dot{q}_i \frac{\partial L}{\partial \dot{q}_i} - L = \sum_{i=1}^{f} \dot{q}_i \frac{\partial T}{\partial \dot{q}_i} - L$$
$$= 2T - (T - V) = T + V = E \qquad (6.25)$$

where E is the total energy of the system. Thus, the result expresses the law of conservation of energy.

Example 10. Motion of an Electron About a Fixed Nucleus (Fig. 6.6). Let the nucleus have a positive charge Ze. Use polar coordinates and write the Lagrangian $L = T - V$.

$$T = \tfrac{1}{2}mv^2 = \tfrac{1}{2}m(\dot{r}^2 + r^2\dot{\varphi}^2) \qquad V = -\frac{Ze^2}{r}$$

$$L = T - V = \tfrac{1}{2}m(\dot{r}^2 + r^2\dot{\varphi}^2) + \frac{Ze^2}{r}$$

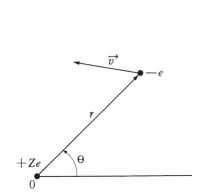

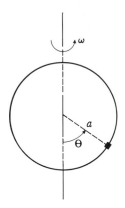

FIGURE 6.6 Motion of an electron about a fixed nucleus.

FIGURE 6.7 Moving constraint, bead sliding on smooth rotating circular wire loop.

There are two degrees of freedom, with generalized coordinates (r, φ). The corresponding Lagrange's equations are

$$\frac{d}{dt}\left(\frac{\partial L}{\partial \dot{\varphi}}\right) - \frac{\partial L}{\partial \varphi} = \frac{d}{dt}\,(mr^2\dot{\varphi}) = 0 \tag{1}$$

$$\frac{d}{dt}\left(\frac{\partial L}{\partial \dot{r}}\right) - \frac{\partial L}{\partial r} = m\ddot{r} - mr\dot{\varphi}^2 + \frac{Ze^2}{r^2} = 0 \tag{2}$$

Equation (1) shows that φ is an ignorable coordinate and, therefore,

$$mr^2\dot{\varphi} = \text{constant} = k \tag{3}$$

Eliminating $\dot{\varphi}$ from (2) by using Eq. (3) yields

$$m\ddot{r} - \frac{k^2 mr}{m^2 r^4} + \frac{Ze^2}{r^2} = 0 \tag{4}$$

or

$$\ddot{r} - \frac{k^2}{m^2 r^3} + \frac{Ze^2}{mr^2} = 0 \tag{5}$$

The first equation states that the angular momentum, $mr^2\dot{\varphi}$, is constant for the particle. It also is a statement of Kepler's law of equal areas swept by the radius in equal times.

Example 11. Moving Constraint. Consider a bead of mass m which slides without friction on a circular wire loop of radius a. (See Fig. 6.7.) The loop rotates about the vertical axis with constant angular velocity ω. There is only one degree of freedom since the angular motion ω is specified. The Lagrangian L is

$$L = T - V = \tfrac{1}{2}m(a^2\dot{\theta}^2 + a^2\omega^2 \sin^2 \theta) - mga(1 - \cos \theta)$$

This problem is illustrative of a large class of dynamical problems for which the labor of deriving and solving the equation of motion can be avoided by taking advantage of the invariants of motion. Therefore, applying Eq. (6.23), since L does not contain t explicitly, we obtain

$$\dot{\theta}\left(\frac{\partial L}{\partial \dot{\theta}}\right) - L = \text{constant} = E'$$

or

$$ma^2\dot{\theta}^2 - \tfrac{1}{2}ma^2\dot{\theta}^2 - \tfrac{1}{2}ma^2\omega^2 \sin^2\theta + mga(1 - \cos\theta)$$
$$= \tfrac{1}{2}ma^2\dot{\theta}^2 - \tfrac{1}{2}ma^2\omega^2 \sin^2\theta + mga(1 - \cos\theta) = E'$$

The constant E' is not the total energy, since the middle term of the equation for E' has the wrong sign; work is done by the force of constraint between bead and wire, so that total energy is not conserved unless we take account of the work done by the constraint.

Dynamical Equations for Systems with Dissipative and Exchange Forces. In this section we consider systems which have frictional or dissipative forces and also are acted on by external forces. The dynamical equations for such systems are obtained from Lagrange's equations (6.20) by putting

$$\mathbf{Q} = \mathbf{Q}_p + \mathbf{Q}_d + \mathbf{Q}_e \qquad (6.26)$$

where

$$\mathbf{Q}_p = -\nabla V = \text{forces derivable from a conservative}$$
$$\text{potential function}$$

$$\mathbf{Q}_d = \text{frictional or dissipative forces}$$

$$\mathbf{Q}_e = \text{exchange or external forces}$$

Using this notation, we write Lagrange's equations in the form

$$\frac{d}{dt}\left(\frac{\partial L}{\partial \dot{q}_i}\right) - \frac{\partial L}{\partial q_i} = Q_{di} + Q_{ei} \qquad i = 1, 2, \ldots, f \qquad (6.27)$$

where

$$L = T - V, \text{ with } V = V(\mathbf{q}).$$

In many problems the retarding or frictional forces are proportional to the velocities (viscous damping) or may be adequately represented by such an approximation. For these cases Q_{di} may be expressed in the form

$$-Q_{di} = \sum_{j=1}^{f} R_{ij}\dot{q}_j \qquad i = 1, 2, \ldots, f \qquad (6.28)$$

which may also be represented in matrix form,

$$-\mathbf{Q}_d = R\dot{\mathbf{q}}' \qquad (6.29)$$

where R is the matrix of the coefficients R_{ij} and is symmetric. Lord Rayleigh introduced a dissipation function F defined by

$$F = -\tfrac{1}{2}\mathbf{Q}_d \cdot \dot{\mathbf{q}} = \tfrac{1}{2}\dot{\mathbf{q}}R\dot{\mathbf{q}}' \tag{6.30}$$

Lagrange's equations for systems of this class assume the convenient form

$$\frac{d}{dt}\left(\frac{\partial L}{\partial \dot{q}_i}\right) - \frac{\partial L}{\partial q_i} + \frac{\partial F}{\partial \dot{q}_i} = Q_{ei} \qquad i = 1, 2, \ldots, f \tag{6.31}$$

Energy equations for systems with dissipative and exchange forces may be deduced from Hamilton's principle and Lagrange's equations as follows: Multiply Eq. (6.27) by $\dot{q}_i$ and sum over the coordinates

$$\sum_{i=1}^{f} \dot{q}_i \left[\frac{d}{dt}\left(\frac{\partial L}{\partial \dot{q}_i}\right) - \frac{\partial L}{\partial q_i}\right] = \sum_{i=1}^{f} (Q_{di} + Q_{ei})\dot{q}_i = (\mathbf{Q}_d + \mathbf{Q}_e) \cdot \dot{\mathbf{q}}$$

But from Chap. 3 and Eq. (6.25), we find that

$$\sum_{i=1}^{f} \dot{q}_i \left[\frac{d}{dt}\left(\frac{\partial L}{\partial \dot{q}_i}\right) - \frac{\partial L}{\partial q_i}\right] = \frac{d}{dt}\left[\sum_{i=1}^{f} \dot{q}_i \frac{\partial L}{\partial \dot{q}_i} - L\right] = \frac{d}{dt}(T + V)$$

This leads to the energy equation

$$\frac{d}{dt}(T + V) = \mathbf{Q}_d \cdot \dot{\mathbf{q}} + \mathbf{Q}_e \cdot \dot{\mathbf{q}} \tag{6.32}$$

where the first term is the rate at which energy is stored in the system, the second is the rate of energy dissipation, and the third is the rate at which work is done on the system by the external forces. If, further, a Rayleigh dissipation function exists, we can write (6.32) in the form

$$\frac{d}{dt}(T + V) + 2F = \mathbf{Q}_e \cdot \dot{\mathbf{q}} \tag{6.33}$$

In this expression, the first term represents the rate at which energy is stored in the system, $2F$ is the rate of energy dissipation, and the sum of the two is the rate at which work is done on the system by the forces $\mathbf{Q}_e$.

Example 12. System with Frictional Dissipation (Fig. 6.8). Given a particle moving through the atmosphere with a drag force proportional to the velocity, find the motion.

SOLUTION. The particle has two degrees of freedom which are most conveniently specified by the Cartesian coordinates x, z. The energy functions are as follows:

$$T = \tfrac{1}{2}mv^2 = \tfrac{1}{2}m(\dot{x}^2 + \dot{z}^2)$$
$$V = mgz$$
$$F = \tfrac{1}{2}Rv^2 = \tfrac{1}{2}R(\dot{x}^2 + \dot{z}^2)$$

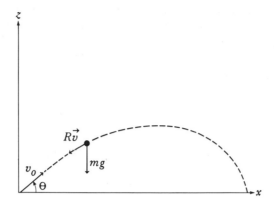

FIGURE 6.8 Particle moving through viscous atmosphere.

Lagrange's equations are

$$\frac{d}{dt}\left(\frac{\partial L}{\partial \dot{q}_i}\right) - \frac{\partial L}{\partial q_i} + \frac{\partial F}{\partial q_i} = 0 \qquad i = 1, 2$$

or in terms of the x, z coordinates

$$
\begin{cases}
\dfrac{d}{dt}\left(\dfrac{\partial L}{\partial \dot{x}}\right) - \dfrac{\partial L}{\partial x} + \dfrac{\partial F}{\partial \dot{x}} = 0 \\[2mm]
\dfrac{d}{dt}(m\dot{x}) + R\dot{x} = 0
\end{cases}
\quad \text{and} \quad
\begin{cases}
\dfrac{d}{dt}\left(\dfrac{\partial L}{\partial \dot{z}}\right) - \dfrac{\partial L}{\partial z} + \dfrac{\partial F}{\partial \dot{z}} = 0 \\[2mm]
\dfrac{d}{dt}(m\dot{z}) + mg + R\dot{z} = 0
\end{cases}
$$

Thus, we have the two differential equations of motion

$$m\ddot{x} + R\dot{x} = 0$$
$$m\ddot{z} + R\dot{z} + mg = 0$$

with the initial conditions

$$x(0) = 0 \qquad \dot{x}(0) = v_0 \cos\theta$$
$$z(0) = 0 \qquad \dot{z}(0) = v_0 \sin\theta$$

These equations may be integrated by elementary means to yield the solutions

$$x = \left[\frac{(mv_0 \cos\theta)}{R}\right](1 - e^{-Rt/m})$$

$$z = \frac{m^2 g}{R^2} + \frac{mv_0 \sin\theta}{R}(1 - e^{-Rt/m}) - \frac{mg}{R}t$$

Eliminating t between these two equations gives the trajectory

$$z = \frac{mg}{Rv_0}\cos\theta + (\tan\theta)x + \frac{m^2 g}{R^2}\ln\left[\frac{1 - Rx}{mv_0}\right]\cos\theta$$

The range is found by solving for x with $z = 0$. The resulting equation is transcendental and may be solved numerically for given values of the parameters. Note that for large values of t, the term $e^{-Rt/m}$ becomes very small, x approaches a constant, and z decreases linearly with time, i.e., the body falls at constant velocity.

6.4 EQUILIBRIUM

Lagrange's equations for systems in static equilibrium reduce to

$$\mathbf{Q} = [Q_1, Q_2, \ldots, Q_f] = 0 \qquad (6.34)$$

for the general case of systems with work functions, and to

$$\nabla V = \left[\frac{\partial V}{\partial q_1}, \frac{\partial V}{\partial q_2}, \ldots, \frac{\partial V}{\partial q_f}\right] = 0 \qquad (6.35)$$

for conservative systems. These results follow immediately from Eqs. (6.20) and (6.21) by setting the kinetic energy $T = 0$.

In theoretical mechanics, it is customary to discuss equilibrium under a very general principle known as the principle of virtual work. This principle follows from Hamilton's principle by setting the kinetic energy T equal to zero in Eq. (6.13). There results

$$\delta W = \sum_{i=1}^{f} Q_i \, \delta q_i = \mathbf{Q} \cdot \delta \mathbf{q} = 0 \qquad (6.36)$$

Since the work done by the forces of constraint is zero, constraint forces can be omitted and δW expressed in terms of the applied or external forces only. For example, the principle of virtual work for a system of n particles subject to a set of applied forces $\mathbf{F}_1, \mathbf{F}_2, \ldots, \mathbf{F}_n$, where $\mathbf{F}_1$ acts on particle 1, etc., may be written in the form

$$\delta W = \sum_{j=1}^{n} \mathbf{F}_j \cdot \delta \mathbf{r}_j = 0 \qquad (6.37)$$

The virtual displacement of the system is given by $\delta \mathbf{r}_1, \delta \mathbf{r}_2, \ldots, \delta \mathbf{r}_n$ and thus involves in the general case $3n$ coordinates, of which only $3n - k$ are independent, k being the number of constraints.

The equilibrium state of the system defined by Eq. (6.35) will be stable, unstable, or neutral in accord with the character of the stationary point of the potential function. Displacement from a point of stable equilibrium produces forces which tend to return the system to its original state, that is, to increase the potential energy. For unstable equilibrium the situation is reversed and the potential energy decreases. For neutral

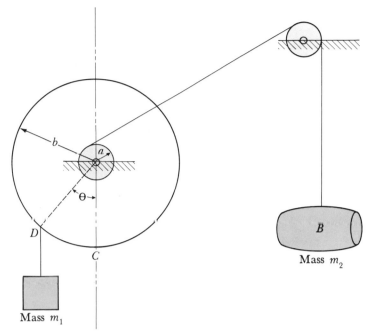

FIGURE 6.9 Diagram of treadmill crane, adapted from Stevin's *Practice of Weighing*, Leyden, 1681.

equilibrium the potential energy remains constant under small displacements. It follows, therefore, that stable equilibrium occurs when the potential energy V has a relative minimum, and unstable equilibrium when V has a relative maximum (cf. Chap. 2).

Example 1. A man at D on the treadmill of a crane (Fig. 6.9) counterbalances the weight of the barrel B. If he proceeds toward E the barrel will rise. Conversely, it will fall if he proceeds toward C. Find the angle θ for the equilibrium position.

SOLUTION

The principle of virtual work leads to a concise, simple solution:

$$\delta W = \sum_{j=1}^{2} \mathbf{F}_j \cdot \delta \mathbf{r}_j = \mathbf{F}_1 \cdot \delta \mathbf{r}_1 + \mathbf{F}_2 \cdot \delta \mathbf{r}_2$$

$$= m_1 g \, \delta(b \cos \theta) + m_2 g \, \delta(a\theta)$$

$$= (-m_1 g b \sin \theta + m_2 g a) \, \delta\theta = 0$$

from which we find

$$\sin \theta = \frac{m_2 a}{m_1 b}$$

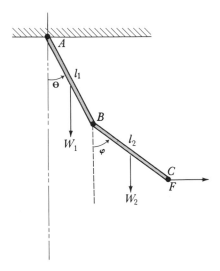

FIGURE 6.10 Suspended jointed bars with horizontal deflecting force.

where m_1 is the mass of the man and m_2 is the mass of the barrel. Hence, there are two values of θ as follows:

$$\theta_1 = \sin^{-1} \frac{m_2 a}{m_1 b} \text{ (stable equilibrium)}$$

$$\theta_2 = \pi - \theta_1 \text{ (unstable equilibrium)}$$

provided that $\frac{m_2 a}{m_1 b} < 1$

Example 2. Two uniform rods AB and BC are freely jointed at B, and hang vertically from A as shown in Figure 6.10. If a horizontal force F is applied at C, find the angles of inclination θ and φ.

SOLUTION. Let the lengths of the bars be l_1 and l_2, and the weights W_1 and W_2. Then, choosing θ and φ as generalized coordinates,

$$\delta W = F \, \delta x + W_1 \, \delta y_1 + W_2 \, \delta y_2$$
$$= \left(F \frac{\partial x}{\partial \theta} + W_1 \frac{\partial y_1}{\partial \theta} + W_2 \frac{\partial y_2}{\partial \theta} \right) \delta\theta$$
$$+ \left(F \frac{\partial x}{\partial \varphi} + W_1 \frac{\partial y_1}{\partial \varphi} + W_2 \frac{\partial y_2}{\partial \varphi} \right) \delta\varphi$$
$$= (F l_1 \cos\theta - \tfrac{1}{2} W_1 l_1 \sin\theta - \tfrac{1}{2} W_2 l_1 \sin\theta) \delta\theta$$
$$+ (F l_2 \cos\varphi - \tfrac{1}{2} W_2 l_2 \sin\varphi) \delta\varphi = 0$$

Since $\delta\theta$ and $\delta\varphi$ are independent variations, two conditions for equilibrium are found which yield the solution

$$\tan \theta = \frac{2F}{W_1 + 2W_2}$$

$$\tan \varphi = \frac{2F}{W_2}$$

Example 3. Indicating meters, used extensively in science and technology to measure physical quantities such as voltage, current, gas pressure, force, and the like, are excellent examples of equilibrium systems. These meters usually operate on the principle of a balance between a deflecting force or torque produced by the quantity to be measured and a restoring force or torque produced by a spring, torsion element, or weight.

A moving-coil galvanometer used to measure small electric currents illustrates these principles and is analyzed in this example. Construction of the galvanometer is depicted in Figure 6.11. Basically, it comprises a rectangular coil suspended in a uniform magnetic field. A current flowing through the coil causes it to rotate in the field to a position where the turning torque is just balanced by the restoring torque of the torsion fiber. Viscous damping due to the air, or sometimes imposed externally, suppresses vibrations, but has no effect on the balance point since the damping depends only on velocity, not on position of the coil.

The turning moment M on the coil is readily calculated from the Biot-Savart law and is given by

$$M = ANIH \cos \theta$$

where A is the area (height by width) of the coil, N is the number of turns on the coil, I is the current, H is the magnetic field, and θ is the

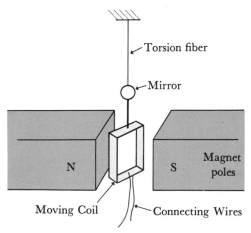

FIGURE 6.11 Galvanometer equilibrium.

angle between the coil and the field. In the usual case the restoring moment of the torsion fibre is proportional to θ. The principle of virtual work, then, gives the condition for equilibrium

$$\delta W = (ANIH \cos\theta - k\theta)\delta\theta = 0$$

or

$$I = \frac{k\theta}{NAH \cos\theta}$$

where k is the torque constant of the fiber. It is apparent that the reading is linear only for small values of θ. In practice, this deficiency may be overcome by resetting the top of the fiber to the point where $\theta = 0$; the angle of reset, then, is a direct measure of the current.

Example 4. There is given an electrometer consisting of a conducting fiber stretched midway between two parallel plates maintained at opposite potentials. A charge placed on the fiber will cause it to deflect in the electric field between the plates, the magnitude of the deflection being a measure of the charge. Analyze the operation and determine the stability of the instrument.

SOLUTION. Let the charge on the fiber be $+e$, the potentials on the plates be $+E$ and $-E$, the deflection be x, and the restoring force on the fiber be kx, as shown in Figure 6.12.

The fiber, with charge $+e$, is at a distance x from the center and $d - x$ from the plate $-E$. The initial step in the solution is to calculate the potential energy of the charged fiber. One component of the energy is due to electric forces and another to the mechanical or restoring force on the fiber. The fiber, being between the two conducting plates, will form a series of electric images, as pictured in the sketch. It has an image of charge $-e$ in plate $-E$ at a distance $d - x$ behind the plate. This charge, in turn, has an image charge $+e$ in plate $+E$ at a distance $3d - x$ behind it or at a distance $4d$ from the fiber, and so on, there being an infinite number of images. As a result of these image charges, there is a

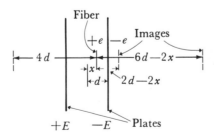

FIGURE 6.12 Schematic diagram for fiber electrometer.

force on the fiber in the direction of increasing x of amount

$$F_1 = \frac{e^2}{(2d - 2x)^2} + \frac{e^2}{(6d - 2x)^2} + \frac{e^2}{(10d - 2x)^2} + \cdots$$

$$+ \frac{e^2}{(4d)^2} + \frac{e^2}{(8d)^2} + \frac{e^2}{(12d)^2} + \cdots$$

where all quantities are in *cgs* electrostatic units.

There is also an infinite series of images starting with the image charge $-e$ in plate $+E$ at a distance $d + x$ behind it. The charges and distances are the same as in the first series except that x is written with a positive sign. Thus, there is a second force in the direction of decreasing x of amount

$$F_2 = - \frac{e^2}{(2d + 2x)^2} - \frac{e^2}{(6d + 2x)^2} - \frac{e^2}{(10d + 2x)^2} - \cdots$$

$$- \frac{e^2}{(4d)^2} - \frac{e^2}{(8d)^2} - \frac{e^2}{(12d)^2} - \cdots$$

Adding F_1 and F_2, the resultant force due to the system of images is

$$F_c = e^2 x d \left[\frac{1}{(d^2 - x^2)^2} + \frac{3}{(9d^2 - x^2)^2} + \cdots + \frac{n}{(n^2 d^2 - x^2)^2} + \cdots \right]$$

To obtain the total force acting on the fiber, the force due to the charge being in a uniform field and that due to the elastic restoring force must be added to F_c. Thus

$$F = \frac{Ee}{d} - kx + F_c$$

The force F_c represents the net force acting on the fiber due to the infinite series of charge images in the conductive plates. That these image forces are small, except when the fiber approaches closely to one of the plates, is evident from the equation for F_c which contains the factor e^2, a very small number compared to e. Thus the electrometer would be expected to have essentially a linear deflection characteristic with increasing charge e except when the fiber approaches one of the plates. For this latter condition the effect of the image force is to increase the deflection and, therefore, the sensitivity of the instrument. However, there arises the possibility of instability which must be considered in a situation of this kind.

For this purpose, it is advantageous to use the principle of virtual work and the concept of potential energy. The potential energy of the system is given by

$$V = - \int_0^x F \, dx = - \frac{Eex}{d} + \tfrac{1}{2}kx^2 - \int_0^x F_c \, dx$$

Equilibrium occurs at a point x_0 defined by

$$\frac{dV}{dx} = -F = -\frac{Ee}{d} + kx - e^2xd\left[\frac{1}{(d^2 - x^2)^2} + \frac{3}{(9d^2 - x^2)^2} + \cdots\right] = 0$$

The stability of the equilibrium is investigated by finding $\dfrac{d^2V}{dx^2}$

$$\frac{d^2V}{dx^2}\bigg]_{x_0} = -\frac{dF}{dx}\bigg]_{x_0} = k - e^2d\left[\frac{1}{(d^2 - x_0^2)^2} + \frac{3}{(9d^2 - x_0^2)^2} + \cdots\right]$$

$$-e^2x_0d\left[\frac{4x_0}{(d^2 - x_0^2)^3} + \frac{12x_0}{(9d^2 - x_0^2)^3} + \cdots\right]$$

The equation $\dfrac{dV}{dx} = 0$ yields

$$e^2d\left[\frac{1}{d^2 - x_0^2} + \frac{3}{(9d^2 - x_0^2)^2}\right] + \cdots = k - \frac{Ee}{x_0d}$$

Substitution of this expression in the equation for $\dfrac{d^2V}{dx^2}$ gives

$$\frac{d^2V}{dx^2}\bigg]_{x_0} = \frac{Ee}{x_0d} + \left(kx_0 - \frac{Ee}{d}\right)\frac{\left[\dfrac{4x_0}{(d^2 - x_0^2)^3} + \cdots\right]}{[(d^2 - x_0^2)^2 + \cdots]}$$

$$= \frac{Ee}{x_0d} + \left(kx_0 - \frac{Ee}{d}\right)f(x_0)$$

where $f(x_0)$ represents the expression in brackets and is positive for all values of x_0 permitted by the problem.

The sign of $\dfrac{d^2V}{dx^2}\bigg]_{x_0}$ may be determined by rearranging the equation to the form

$$\frac{d^2V}{dx^2}\bigg]_{x_0} = \frac{Ee\,f(x_0)}{d}\left[\frac{1}{x_0f(x_0)} + \frac{kx_0d}{Ee} - 1\right]$$

Stability requires that

$$\frac{d^2V}{dx^2}\bigg]_{x_0} > 0$$

and hence, that

$$\frac{1}{x_0f(x_0)} > 1 - \frac{kx_0d}{Ee} > 0$$

since both x_0 and $f(x_0)$ are always positive. Instability will occur if

$$\frac{d^2V}{dx^2}\bigg]_{x_0} < 0$$

that is, for

$$\frac{1}{x_0f(x_0)} < 1 - \frac{kx_0d}{Ee}$$

These results show that attempts to improve the sensitivity of the instrument (which increases with E/k) may lead to instability, that is, the attractive force between the needle and the plates overpowers the restraining force of the fiber suspension, causing the needle to fly to the plates and disrupt the instrument. This example shows that even for relatively simple instruments stable operation may not always be taken for granted.

Constrained Systems. Solution by Lagrange Multiplier Method. Consider a system with coordinates $(x_1, x_2, \ldots, x_f, x_{f+1})$ and a constraint

$$F(x_1, x_2, \ldots, x_f, x_{f+1}) = 0$$

where the x_i are any set of $f + 1$ coordinates and f is the number of degrees of freedom. Let λ be a Lagrangian multiplier. Then, from Chap. 2, the condition for equilibrium is

$$\nabla(V + \lambda F) = 0 \tag{6.38}$$

where the term λF may be thought of as the potential energy of the constraint force. The generalized force associated with a virtual displacement δx_i is

$$f_i = -\frac{\partial}{\partial x_i}(V + \lambda F) = -\frac{\partial V}{\partial x_i} - \lambda \frac{\partial F}{\partial x_i} \tag{6.39}$$

The term $-\dfrac{\partial V}{\partial x_i}$ is the external or applied force, while $-\lambda \dfrac{\partial F}{\partial x_i}$ is the force of reaction due to the constraint. Thus, we see that the Lagrangian multiplier term has the remarkable property of giving the reaction force of the constraint.

Example 5. A particle constrained to move on a circular path is subjected to forces F_1 and F_2, as shown in Figure 6.13a. Apply the principle of virtual work and the Lagrangian multiplier method to find the positions of equilibrium and the forces of reaction.

SOLUTION. Use polar coordinates (r, θ) with origin at 0. The force F_1 is along the x-direction and F_2 along the r-direction. Hence, the potential energy V is

$$V = F_2 r - F_1 x$$

with the constraint

$$f = r - 2a \cos \theta = 0$$

(equation of circle in polar coordinates). Introduce the Lagrangian multiplier λ. Then the condition for equilibrium is

$$\begin{aligned}
\nabla(V + \lambda f) &= \nabla[F_2 r - F_1 r \cos \theta + \lambda(r - 2a \cos \theta)] \\
&= (F_2 - F_1 \cos \theta + \lambda)\delta r + (F_1 r \sin \theta + 2a\lambda \sin \theta)\delta\theta \\
&= 0
\end{aligned}$$

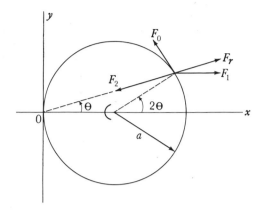

(a) Generalized coordinate θ

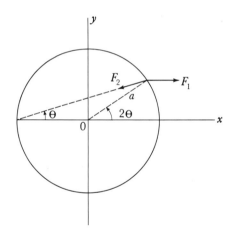

(b) Generalized coordinate 2θ

FIGURE 6.13 Force equilibrium diagrams.

which is equivalent to the scalar equations

$$F_2 - F_1 \cos \theta + \lambda = 0$$
$$(F_1 r + 2a\lambda) \sin \theta = 0$$

To these must be added the constraint equation

$$r - 2a \cos \theta = 0$$

There are three equations and three unknowns (r, θ, λ). Eliminating λ and r yields

$$(2F_1 \cos \theta - F_2) \sin \theta = 0$$

Hence, there are two equilibrium values of θ given by

Case I $\quad \sin \theta = 0$

Case II $\quad \cos \theta = \dfrac{F_2}{2F_1} \leq 1$

To determine the nature of the equilibrium, examine the second derivative of V with respect to θ for both Cases I and II.

$$\frac{\partial^2 V}{\partial \theta^2} = (-2F_1 \sin \theta) \sin \theta + (2F_1 \cos \theta - F_2) \cos \theta = 2F_1(\cos^2 \theta - \sin^2 \theta)$$
$$- F_2 \cos \theta$$

Case I $\quad \left. \dfrac{\partial^2 V}{\partial \theta^2} \right]_{\theta=0} = 2F_1 - F_2$

Stable only if $2F_1 - F_2 > 0$ or $2F_1 > F_2$

Case II $\quad \left. \dfrac{\partial^2 V}{\partial \theta^2} \right]_{\theta = \cos^{-1} \frac{F_2}{2F_2}} = 2F_1(\cos^2 \theta - \sin^2 \theta - \cos^2 \theta)$
$$= -2F_1 \sin^2 \theta$$

which is negative, and hence the point $\theta = \cos^{-1} \dfrac{F_2}{2F_1}$ is unstable.

The forces of reaction may be found by the method of the Lagrangian multiplier as follows:

$$Q_r = F_r = -\frac{\partial}{dr}(V + \lambda f) = F_1 \cos \theta - F_2 - \lambda$$

$$Q_\theta = rF_\theta = -\frac{\partial}{\partial \theta}(V + \lambda f) = -(F_1 r + 2a\lambda) \sin \theta$$

where Q_r and Q_θ are the generalized forces and F_r and F_θ are the forces acting in the directions of increasing r and θ respectively. Thus, we find

$$F_r = (F_1 \cos \theta - F_2) - \lambda$$
$$F_\theta = -F_1 \sin \theta - \frac{2a\lambda \sin \theta}{r}$$
$$= -F_1 \sin \theta - \lambda \tan \theta$$

where $\lambda = F_1 \cos \theta - F_2$ at the equilibrium position.

ALTERNATIVE SOLUTION. Use the angle 2θ (Fig. 6.13b) as a generalized coordinate; then

$$V = F_2 r - F_1 x = 2aF_2 \cos \theta - aF_1 \cos 2\theta$$
$$\nabla V = 2a(-F_2 \sin \theta + F_1 \sin 2\theta) = 0$$

or

$$(2F_1 \cos \theta - F_2) \sin \theta = 0$$

By inspection there are two equilibrium values of θ

Case I $\quad \sin \theta = 0$

Case II $\quad \cos \theta = \dfrac{F_2}{2F_1}, \quad$ where $\quad \dfrac{F_2}{2F_1} \leq 1$

$$\frac{\partial^2 V}{\partial \theta^2} = 2a(-F_2 \cos \theta + 2F_1 \cos 2\theta)$$

Case I $\quad \dfrac{\partial^2 V}{\partial \theta^2}\bigg]_{\theta=0} = 2a(2F_1 - F_2)$

which is positive and, therefore, stable if $2F_1 > F_2$.

Case II $\quad \dfrac{\partial^2 V}{\partial \theta^2}\bigg]_{\theta=\theta_1} = -4aF_1 \sin^2 \theta_1$

where $\theta_1 = \cos^{-1} \dfrac{F_2}{2F_1}$

The equilibrium in this case is unstable.

Fourier's Inequality. The usual formulation of the principle of virtual work is valid only for reversible displacements. This limitation requires that the equilibrium point be inside but not on the boundaries of the configuration space, because, at a boundary, the virtual displacement can occur only inward or along the surface of the configuration space, not outward, as this would lead out of the permitted region. As an example, consider a ball inside a spherical shell or on a table top. The ball can move along or away from, but not through, the bounding surface.

Fourier called attention to this limitation and pointed out that the conventional formulation of the principle of virtual work is restricted to reversible displacements. For irreversible displacements, the principle must be stated as an inequality

$$\delta W \leq 0 \tag{6.40}$$

or, if a potential energy function exists, as

$$\delta V \geq 0 \tag{6.41}$$

A system is in equilibrium if the work done by the forces is zero or negative (in which case energy is supplied to the system) for all permissible virtual displacements. Thus, a system which does not come to equilibrium within the configuration space will move to the boundary and be in equilibrium there. Notice that at the boundary, equilibrium does not require a stationary point of the potential V.

6.5 HAMILTON'S CANONICAL EQUATIONS; PHASE SPACE

Although historically Lagrange derived his equations of motion long before Hamilton's principle was formulated, in this book we have taken Hamilton's principle as a basic postulate and derived Lagrange's equations as a consequence of Hamilton's principle.

Lagrange had achieved a high degree of perfection and mathematical elegance by reducing the equations of motion to a universal form which is the same for all coordinate systems and which has no extraneous variables. The Lagrangian formalism is in terms of generalized coordinates and generalized velocities. The equations of motion are second-order ordinary differential equations, and the motion of the system is specified in terms of the initial values of the coordinates and velocities.

Hamilton conceived and carried out the fundamental idea of transforming Lagrange's f second-order differential equations in the generalized coordinate variables q_i to $2f$ first-order differential equations in the variables q_i and p_i, where the f quantities p_i are the generalized momenta for the system. It is interesting to observe that Hamilton's development of the canonical equations, as the set of $2f$ first-order differential equations are known because of their concise symmetry, was the forerunner not only of many modern developments in physics such as quantum mechanics and statistical mechanics, but also of the concept of state variables, which is proving so useful in the formulation and computer-aided solution of many engineering problems.

Although Hamilton's canonical equations may be derived from Lagrange's equations, it is advantageous to give a separate proof. The first step in the derivation is to define the Hamiltonian function H as follows:

$$H(\mathbf{q}, \mathbf{p}, t) = \mathbf{p} \cdot \dot{\mathbf{q}} - L(\mathbf{q}, \dot{\mathbf{q}}, t) \tag{6.42}$$

where $\mathbf{q}, \dot{\mathbf{q}}, \mathbf{p}$ are the generalized position, velocity, and momenta vectors, with p_i given by

$$p_i = \frac{\partial L}{\partial \dot{q}_i} \tag{6.43}$$

Solving for L from (6.42) and substituting in Hamilton's principle gives

$$\delta \int_{t_0}^{t_1} L \, dt = \delta \int_{t_0}^{t_1} [\mathbf{p} \cdot \dot{\mathbf{q}} - H(\mathbf{q}, \mathbf{p}, t)] \, dt = 0$$

Since the end-point times t_0 and t_1 are not varied, the variation may be carried out inside the integral sign yielding

$$\int_{t_0}^{t_1} \sum_{i=1}^{f} \left(p_i \, \delta \dot{q}_i + \dot{q}_i \, \delta p_i - \frac{\partial H}{\partial q_i} \delta q_i - \frac{\partial H}{\partial p_i} \delta p_i \right) dt = 0 \tag{6.44}$$

The variation and differentiation operations may be interchanged as follows:

$$\delta \dot{q}_i = \delta\left(\frac{dq_i}{dt}\right) = \frac{d}{dt}(\delta q_i)$$

Making this interchange in the first term in integral (6.44) and integrating by parts gives

$$\int_{t_0}^{t_1} \sum_{i=1}^{f} p_i \, \delta \dot{q}_i = \int_{t_0}^{t_1} \sum_{i=1}^{f} p_i \frac{d}{dt}(\delta q_i) \, dt$$

$$= \int_{t_0}^{t_1} \sum_{i=1}^{f} \left[\frac{d}{dt}(p_i \, \delta q_i) - \dot{p}_i \, \delta q_i\right] dt$$

$$= \sum_{i=1}^{f} (p_i \, \delta q_i)\Big]_{t_0}^{t_1} - \int_{t_0}^{t_1} \sum_{i=1}^{f} (\dot{p}_i \, \delta q_i) \, dt$$

The integrated term vanishes because the quantities $\delta q_i = 0$ at t_0 and t_1. Hence, Eq. (6.44) reduces to the form

$$\int_{t_0}^{t_1} \sum_{i=1}^{f} \left[\left(-\dot{p}_i - \frac{\partial H}{\partial q_i}\right)\delta q_i + \left(\dot{q}_i - \frac{\partial H}{\partial p_i}\right)\delta p_i\right] dt = 0$$

Because of the relations $p_i = \dfrac{\partial L}{\partial \dot{q}_i}$ it is possible that the δp_i and δq_i may not be independent. However, it follows from the expression

$$\frac{\partial H}{\partial p_i} = \frac{\partial}{\partial p_i}\left[\sum_{i=1}^{f} p_i \dot{q}_i - L\right] = \dot{q}_i + \sum_{i=1}^{f}\left[p_i \frac{\partial \dot{q}_i}{\partial p_i} - \frac{\partial L}{\partial \dot{q}_i}\frac{\partial \dot{q}_i}{\partial p_i}\right]$$

that

$$\frac{\partial H}{\partial p_i} = \dot{q}_i$$

since the summation terms on the right cancel out. Therefore, the δq_i and δp_i can be chosen arbitrarily and the conditions for the integral immediately above to vanish are expressed by

$$\left.\begin{aligned}
\frac{\partial H}{\partial p_i} &= \dot{q}_i, & i = 1, 2, \ldots, f \\[2mm]
\frac{\partial H}{\partial q_i} &= -\dot{p}_i, & i = 1, 2, \ldots, f
\end{aligned}\right\} \tag{6.45}$$

Equations (6.45) are known as Hamilton's or Jacobi's canonical equations. They comprise a set of $2f$ first-order differential equations in the canonical variables $(q_1, q_2, \ldots, q_f, p_1, p_2, \ldots, p_f)$ which, together with the boundary conditions, provide a complete set of differential equations for the dynamical problem. The general solution contains $2f$

arbitrary constants. As in the case of Lagrange's equations, it is necessary to determine the values of these constants in order to specify uniquely the motion of the system. This requires a knowledge of the state of the system, that is, the values of all the generalized coordinates and generalized momenta at some given instant of time, which is usually taken as the time t_0 at the start of motion.

The distinction from Lagrange's equations is that the q_i's and p_i's are independent variables in Hamilton's canonical equations and have equal status in determining the motion of the system, whereas in Lagrange's equations only the q_i's are taken as independent variables. The state of a dynamical system at any time t is completely determined by the canonical variables $(q_1, q_2, \ldots, q_f, p_1, p_2, \ldots, p_f)$. When the values of these variables are known for some given time, then the state of the system for all later times is known from the dynamical laws. The canonical variables are, therefore, a set of state variables in the sense introduced in Chap. 4. If $\mathbf{q} = [q_1, q_2, \ldots, q_f]$ and $\mathbf{p} = [p_1, p_2, \ldots, p_f]$, then $\mathbf{s} = \lceil \mathbf{q}, \mathbf{p} \rceil$ is a state vector.

One of the fundamental and useful properties of Lagrange's equations is that they are invariant, i.e., do not change their form under point transformations (relations of the type shown in Eq. (6.9)). It may be shown that Hamilton's canonical equations are invariant, not only under point transformations, but also under the much more general transformations represented by Eq. (6.46):

$$\left. \begin{aligned} q_i' &= q_i'(\mathbf{q}, \mathbf{p}), & i = 1, 2, \ldots, f \\ p_i' &= p_i'(\mathbf{q}, \mathbf{p}), & i = 1, 2, \ldots, f \end{aligned} \right\} \tag{6.46}$$

The reader is referred to Ref. 10, p. 209, for proof of the validity of the invariance of Hamilton's equations under contact transformations. This invariance is useful for solving certain problems and also for advanced theory.

Conservation Law. The Hamiltonian function H satisfies a conservation law of the form

$$H = \sum_{i=1}^{f} p_i \dot{q}_i - L = E' = \text{constant} \tag{6.47}$$

if the Lagrangian L does not contain time t explicitly. We note that Eq. (6.47) is essentially a restatement of the result previously obtained in Eq. (6.23), but with p_i substituted for $\dfrac{\partial L}{\partial \dot{q}_i}$. This conservation law reduces to the law of conservation of energy only for the special case of a system with fixed constraints and a conservative potential function $V = V(\mathbf{q})$. For this case,

$$E' = E = T + V = \text{total energy of system} \tag{6.48}$$

Cyclic Coordinates. The Hamiltonian procedure is especially effective for the treatment of cyclic coordinates. A cyclic coordinate is defined as one which does not appear in the Lagrangian or the Hamiltonian. To show this, we note that if coordinate q_f is cyclic, then

$$\dot{p}_f = \frac{\partial H}{\partial q_f} = 0 \tag{6.49}$$

and $p_f = \text{constant} = \alpha$. Then the Hamiltonian takes the form

$$H = H(q_1, q_2, \ldots, q_{f-1};\ p_1, p_2, \ldots, p_{f-1}, \alpha, t) \tag{6.50}$$

In effect, the Hamiltonian now describes a problem involving only $f - 1$ coordinates. The α is a constant of integration, to be determined from the initial conditions. The behavior of the cyclic coordinate itself with time is then found by integrating the equation of motion (6.51)

$$\dot{q}_f = \frac{\partial H}{\partial \alpha} \tag{6.51}$$

Example I. Simple Harmonic Oscillator. The problem of the simple harmonic oscillator has already been treated using Lagrange's equations (see p. 182). As a matter of comparison, we treat the same problem by Hamilton's equations. The Hamiltonian function H for the oscillator is by definition

$$H = p\dot{q} - L = p\dot{q} - (\tfrac{1}{2}m\dot{q}^2 - \tfrac{1}{2}kq^2)$$

where

$$p = \frac{\partial L}{\partial \dot{q}} = m\dot{q}$$

Substitution of $\dot{q} = \dfrac{p}{m}$ in the equation for H gives

$$H(p, q) = \frac{p^2}{2m} + \tfrac{1}{2}kq^2$$

The equations of motion are

$$\frac{\partial H}{\partial p} = \frac{p}{m} = \dot{q}$$

$$\frac{\partial H}{\partial q} = kq = -\dot{p}$$

These may be solved analytically and lead to the result already found on p. 183. Note that the equations of motion in Hamilton's form can be readily programmed for numerical solution by computer once the initial conditions are specified. This can be done, also, for more complex problems where analytical solutions are impossible.

We wish here, however, to show another aspect of the Hamiltonian approach through the use of a contact transformation. By an appropriate choice of a contact transformation, it may be possible to convert to a cyclic coordinate for which the solution is known. There is an extensive theory of contact transformations. For our present purpose, it will suffice to use the transformation

$$q = \sqrt{\frac{2p'}{m\omega}} \sin q'$$

$$p = \sqrt{2m\omega p'} \cos q'$$

where

$$\omega^2 = \frac{k}{m}$$

The Hamiltonian is transformed

$$H'(q', p') = \frac{1}{2m} (2m\omega p' \cos^2 q') + \frac{m\omega^2}{2} \left(\frac{2p'}{m\omega} \sin^2 q'\right)$$

$$= \omega p'$$

Using this transformed Hamiltonian gives equations of motion as follows:

$$\dot{q}' = \frac{\partial H'}{\partial p'} = \omega$$

$$-\dot{p}' = \frac{\partial H'}{\partial q'} = 0$$

Integration yields

$$q' = \omega t + \beta \qquad p' = \alpha$$

where α and β are constants. Putting these in the equations for q and p we obtain

$$q = \sqrt{\frac{2\alpha}{m\omega}} \sin (\omega t + \beta)$$

$$p = \sqrt{2m\omega\alpha} \cos (\omega t + \beta)$$

which are the familiar equations of simple harmonic motion.

6.6 CLASSIFICATION OF DYNAMICAL SYSTEMS

Dynamical systems which possess a finite number of degrees of freedom may be further classified on the basis of three fundamental distinctions as follows:

Scleronomic or Rheonomic. These terms were used by Boltzmann to distinguish between systems whose energy functions and constraints do not depend explicitly on time (scleronomic), and those which do (rheo-nomic). Both terms are derived from Greek roots: *sclero* = hard, fixed, rigid; *rheo* = flow, current, fluid, changing.

Scleronomic Systems. The configuration of the system is fully deter-mined when the values of a set of generalized coordinates $(q_1, q_2, \ldots, q_f)$ are assigned. This implies that all constraints as well as the energy functions depend only on the coordinates and the velocities, and not on the time. In mathematical language

$$\frac{\partial L}{\partial t} = 0 \qquad \text{(6.52)}$$

Scleronomic systems, from a corollary of the calculus of variations (Chap. 3), satisfy a conservation law of the form

$$\sum_{i=1}^{f} \dot{q}_i \frac{\partial L}{\partial \dot{q}_i} - L = \text{const.} = E' \qquad \text{(6.53)}$$

This reduces to the law of conservation of energy only for the special case where V is a function of only the coordinates q_i and T is a homogeneous quadratic function of the $\dot{q}_i$'s.

Rheonomic Systems. In contrast to scleronomic systems, the con-figuration of a rheonomic system requires the assignment of time t in addition to assignment of a set of generalized coordinates. Configuration is a function of $(q_1, q_2, \ldots, q_f, t)$. A rheonomic system is one which has moving constraints or energy functions which contain time t explicitly. Rheonomic systems do not satisfy a conservation law of the type (6.53).

Conservative or Nonconservative. A conservative system is defined as a system which has the following characteristics: (1) the law of conser-vation of energy is obeyed; (2) the generalized forces Q_i are derivable from a potential energy function V; (3) the Lagrangian L does not contain t explicitly; and (4) T is a homogeneous quadratic form. Mathematically,

(1) $\qquad\qquad T + V = E$

(2) $\qquad\qquad Q_i = -\dfrac{\partial V}{\partial q_i}, \; (i = 1, 2, \ldots, f)$

(3) $\qquad\qquad \dfrac{\partial L}{\partial t} = 0$ $\qquad\qquad\qquad\qquad$ **(6.54)**

(4) $\qquad\qquad T = \tfrac{1}{2} \displaystyle\sum_{i=1}^{f} \sum_{j=1}^{f} a_{ij} \dot{q}_i \dot{q}_j$

Conservation law (6.53) reduces to the form

$$\sum_{i=1}^{f} \dot{q}_i \frac{\partial L}{\partial \dot{q}_i} - L = T + V = \text{const.} = E \qquad \text{(6.55)}$$

Systems which do not satisfy these conditions are nonconservative.

For example, systems with sliding friction or systems with moving constraints are nonconservative.

Holonomic or Anholonomic. The terms holonomic and anholonomic were introduced by Hertz to distinguish between constraints which can be expressed in integral form $f(q_1, q_2, \ldots, q_n) = 0$ and those which can be expressed only in nonintegrable form

$$A_1 \, \delta q_1 + A_2 \, \delta q_2 + \cdots + A_n \, \delta q_n = 0 \qquad (6.56)$$

In general, a system subject to r anholonomic constraints has f degrees of freedom in finite motion, but only $f - r$ degrees of freedom in infinitesimal motion. In a holonomic system, arbitrary independent variations can be made in the generalized coordinates without violating the constraints, whereas in a nonholonomic system this cannot be done.

Simple Systems. Systems which are scleronomic, conservative, and holonomic are the simplest to deal with and are, therefore, appropriately referred to as simple systems. At the other extreme, systems which are rheonomic, nonconservative, and nonholonomic are, in general, the most difficult to deal with.

BIBLIOGRAPHY

1. S. Banach, *Mechanics*. Warsaw: 1951. (Translated from the Polish by E. J. Scott.)
2. F. Bouny, *Leçons de Mécanique Rationnelle*, Vols. I and II. Librairie Scientifique. Paris: A. Blanchard, 1929.
3. H. Goldstein, *Classical Mechanics*. Cambridge: Addison-Wesley, 1950.
4. J. L. Synge and B. A. Griffith, *Principles of Mechanics*, Third Edition. New York: McGraw-Hill, 1959.
5. H. C. Corben and P. Stehle, *Classical Mechanics*, Second Edition. New York: Wiley, 1960.
6. G. W. Kilmister, *Hamiltonian Dynamics*. New York: Wiley, 1964.
7. C. Lanczos, *The Variational Principles of Mechanics*. Second Edition. Toronto: University of Toronto Press, 1962.
8. L. D. Landau and E. M. Lifschitz, *Mechanics*. Oxford: Pergamon Press, 1960. (Translated from the Russian by J. B. Sykes and J. S. Bell.)
9. C. Inglis, *Applied Mechanics for Engineers*. New York: Dover, 1963.
10. A. Sommerfield, *Mechanics*. New York: Academic Press, 1952. (Translated from the German by M. O. Stern.)
11. J. J. Thompson, *Application of Dynamics to Physics and Chemistry*. Cambridge: Cambridge Univ. Press, 1888.
12. A. G. Webster, *The Dynamics of Particles*, Second Edition. New York: Dover, 1959.
13. P. Appel, *Traité de mécanique rationnelle*, Sixth Edition. Paris: Gauthier-Villars, Tome I, 1941, Tome II, 1953.
14. E. T. Whittaker, *Analytical Dynamics*. New York: Dover, 1944.
15. L. A. Pars, *A Treatise on Analytical Dynamics*. London: Heinemann, 1965.
16. B. R. Gossick, *Hamilton's Principle and Physical Systems*. New York: Academic Press, 1967.
17. R. S. Schechter, *The Variational Method in Engineering*. New York: McGraw-Hill, 1967.
18. B. L. Moiseiwitsch, *Variational Principles*. New York: Interscience, 1966.
19. P. Naslin, *The Dynamics of Linear and Nonlinear Systems*. New York: Gordon and Breach, 1965.

7

ELECTROMAGNETIC FIELDS
AND ENERGY RELATIONS

Mechanics and electromagnetism are the basic sciences which underlie most of the technologies derived from classical physics. The methods of generalized mechanics, based on Hamilton's principle as a fundamental postulate, provide a very general approach for the analysis of mechanical systems. It remains, however, to inquire as to the extent to which Hamilton's principle can be applied to electromagnetic phenomena.

Although it turns out that such application can be made on a very general basis (see Ref. 3 of preceding chapter), our interest in this book is primarily in applications to electric networks, electromechanical energy conversion, and feedback systems. These involve quasi-static or slowly-varying electric and magnetic fields for which it is possible to reduce complex electromagnetic field problems to far simpler circuit problems. Problems of this type are characterized by a finite number of degrees of freedom and by energy expressions which can be identified with the kinetic and potential energies of Hamilton's principle. It may be concluded, therefore, that the corresponding systems are dynamical and come under the purview of generalized mechanics.

Although it is possible to introduce elements of electromagnetic theory on an ad hoc basis as needed in the treatment of electrical systems, this approach is disconnected and leaves much to be desired in logical development. On the other hand, a reasonable mastery of electromagnetic theory may require one or two years' study at the senior and graduate level, a preparation exceeding that needed for analysis of physical systems covered in this book. The alternative, which is the purpose of the present chapter, is to introduce the reader to the essential aspects of electromagnetic theory necessary to cope with system applications.

The starting point is Maxwell's equations and the Lorentz law of force, which in electromagnetism compare to Newton's laws in mechanics. This is followed by a study of energy relations for electromagnetic fields, a classification of electromagnetic fields in terms of their interactions and radiation effects, and development of energy equations for electrostatic and magnetostatic fields (valid also for the quasi-static case), terminating with energy relations for moving conductors. The results thus obtained are used in later chapters devoted to applications.

7.1 MAXWELL'S EQUATIONS*

The existence of two field vectors **E** and **B** in Euclidean space (E_3) is postulated; **E** is called the electric field intensity and **B** the magnetic induction. The source of an electromagnetic field is a distribution of electric charge and current, denoted by the charge density ρ and the vector current density **J**. It is further postulated that the vectors satisfy Maxwell's equations. The field vectors are assumed to be differentiable functions of space and time at all ordinary points and to be of finite magnitude throughout the entire field.

Maxwell's equations, which may be regarded as a generalization of experiment and experience in electromagnetism, comprise the set of four partial differential equations (7.1a) through (7.1d), valid in an inertial frame of reference.

(a) $$\nabla \times \mathbf{E} = -\frac{\partial \mathbf{B}}{\partial t}$$

(b) $$\nabla \times \mathbf{B} = \frac{1}{\varepsilon_0 c^2}\left(\mathbf{J} + \varepsilon_0 \frac{\partial \mathbf{E}}{\partial t}\right)$$

(c) $$\nabla \cdot \mathbf{E} = \frac{\rho}{\varepsilon_0}$$

(d) $$\nabla \cdot \mathbf{B} = 0$$

(7.1)

To these equations are added the law of conservation of charge, Eq. (7.2), and Lorentz' law of force, Eq. (7.3)

$$\nabla \cdot \mathbf{J} + \frac{\partial \rho}{\partial t} = 0 \tag{7.2}$$

$$\mathbf{F} = q(\mathbf{E} + \mathbf{v} \times \mathbf{B}) \tag{7.3}$$

---

* MKS units are used throughout this chapter.
  c = velocity of light = 3×10^8 meters per second
  ε_0 = permittivity of free space = 8.854×10^{-12} farad per meter

The conservation of charge is not an independent relation, but may be derived from Eqs. (7.1). Lorentz' law states that the electric force **F** on a charge has two components: one, $q\mathbf{E}$, is in the direction of **E** and independent of the motion of the charge; the second, $q\mathbf{v} \times \mathbf{B}$, exists only for a moving charge and is perpendicular both to **B** and to **v**, the velocity of the charge.

Lorentz' law (7.3) provides the basis for experimental measurement of the field vectors **E** and **B** of (7.1) and, in this sense, defines these vectors, since **F** and **v** are known mechanical quantities, and q may be taken as a test charge.

Maxwell called the term $\varepsilon_0 \dfrac{\partial \mathbf{E}}{\partial t} = \mathbf{J}'$ the displacement current density, because the field **B** produced by it is exactly the same as would be produced by a conduction current of density $\mathbf{J}'$. The displacement current is usually very small for low-frequency fields. For example, the displacement currents encountered in electric power systems can, because of the low frequencies used, be neglected in nearly all cases except for large capacitors and for transmission lines of several hundred miles or more.

Maxwell's equations describe electromagnetic phenomena in terms of field quantities set up by charges and the effect of these fields on charges. An alternate course is to treat the laws of electrodynamics in terms of the forces between charges, as for example Coulomb's inverse square law of force between stationary point charges. The distinction is between forces acting at a distance between the charges, and the transmission of the forces through a medium. In the nineteenth century, attempts were made to construct a hypothetical medium, called the *ether*, through which forces and disturbances could be transmitted, somewhat in analogy to the transmission of wave motion through an elastic medium. Today these attempts have been abandoned because of the failure of the Michelson-Morley ether-drift experiments and the contradictions with relativity theory. The existence of electromagnetic fields is therefore postulated without reference to any hypothetical medium.

Although the law of force between stationary charges is simple, the force relationships for charges in motion are complicated by geometrical effects and by time delays. As a result, it is much simpler and more productive to cast the laws of electrodynamics in terms of fields. One of the most important properties of the field equations is that they are linear; therefore, the *principle of superposition* of fields applies, i.e., the total field at a given point is just the vector sum of the individual fields.

Maxwell's equations (7.1) and the Lorentz law of force (7.3) constitute the fundamental laws of electrodynamics. These, together with just one more law, namely Newton's law of motion $\mathbf{F} = \dot{\mathbf{p}}$, provide the necessary foundation for much of classical physics and for the technologies stemming therefrom.

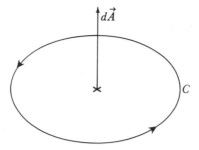

FIGURE 7.1 Convention for positive direction of normal relative to contour of integration in Stokes' theorem.

Integral Form of the Field Equations. The field equations may be transformed to integral relations by using Stokes' theorem and the divergence theorem of vector analysis. For details of these transformations the reader is referred to any of the standard works on electromagnetic theory listed in the bibliography.

The resulting equations (7.4a) through (7.4d) are given here for later use and for reference purposes. In the first two equations, the integrals on the left are taken around a closed contour C, and the integrals on the right over the surface A bounded by the contour C. The convention used for the positive direction of traverse around the contour is shown in Figure 7.1. The integrals on the left of the second two equations are taken over a closed regular surface S, and on the right over the volume τ bounded by the closed surface S. The quantity Φ is called the magnetic flux, and the quantity ψ the electric flux.

(a) *
$$\int_C \mathbf{E} \cdot d\mathbf{s} = -\frac{\partial}{\partial t}\int_A \mathbf{B} \cdot d\mathbf{A} = -\frac{\partial \Phi}{\partial t}$$

(b) *
$$\int_C \varepsilon_0 c^2 \mathbf{B} \cdot d\mathbf{s} = I + \frac{\partial}{\partial t}\int_A \varepsilon_0 \mathbf{E} \cdot d\mathbf{A} = I + \frac{\partial \psi}{\partial t}$$

(c)
$$\int_S \varepsilon_0 \mathbf{E} \cdot d\mathbf{S} = \int_\tau \rho \, d\tau = q$$

(d)
$$\int_S \mathbf{B} \cdot d\mathbf{S} = 0$$

(7.4)

The remaining two relations for the conservation of charge and the Lorentz law, when written in integral form, are

(a)
$$\int_S \mathbf{J} \cdot d\mathbf{S} = -\frac{\partial}{\partial t}\int_\tau \rho \, d\tau = -\frac{\partial q}{\partial t}$$

and

(b)
$$\mathbf{F} = \int_\tau (\rho \mathbf{E} + \mathbf{J} \times \mathbf{B}) \, d\tau$$

(7.5)

---

* The partial derivatives with respect to time in these equations imply variable flux densities threading a fixed contour C. The case of a moving and deforming contour is discussed on p. 232.

7.2 ENERGY RELATIONS FOR ELECTROMAGNETIC FIELDS

Consider an electromagnetic system of currents and charges in a finite region of space of volume τ bounded by a closed surface Σ. These charges and currents are the sources of an electromagnetic field which also exists in the bounded region. Assume that external mechanical forces act on and deliver energy to or from the system. These may be, for example, the forces which move electrical charge on the insulating belt of a Van de Graaff machine, or torques which rotate shafts in dynamos and generators. Finally, assume also that energy may enter or leave the system through the surface Σ via conductors or in the form of electromagnetic radiation.

In addition to the postulates for the electromagnetic field already stated, the law of conservation of energy is postulated. Let u represent the energy per unit volume of the field, $\mathbf{S}$ the energy flux or rate of flow of energy per unit area through Σ, and $\dot{Q}$ the rate at which work is done on matter in the system. $\mathbf{S}$ is called the Poynting vector, after the originator of this analysis. Then, the corresponding energies for the system are

$$\int_{\tau} u \, d\tau = \text{total energy of the field inside } \Sigma$$

$$\int_{\Sigma} \mathbf{S} \cdot d\mathbf{a} = \text{rate at which energy flows out of } \Sigma$$

Since the total energy is conserved, these quantities are related by the equation

$$\frac{\partial}{\partial t} \int_{\tau} u \, d\tau + \int_{\Sigma} \mathbf{S} \cdot d\mathbf{a} + \dot{Q} = 0 \qquad (7.6)$$

The term $\dot{Q}$, representing the time rate that work is done on matter in the system, may be evaluated from the definition of work and the Lorentz force equation (7.3). Then

$$\dot{Q} = \int_{\tau} \mathbf{F} \cdot \mathbf{v} \, d\tau = \int_{\tau} Nq(\mathbf{E} + \mathbf{v} \times \mathbf{B}) \cdot \mathbf{v} \, d\tau$$

$$= \int_{\tau} Nq\mathbf{E} \cdot \mathbf{v} \, d\tau = \int_{\tau} \mathbf{E} \cdot \mathbf{J} \, d\tau$$

where N is the number of charges per unit volume, and $Nq\mathbf{v} = \mathbf{J}$ is the current density. Hence, (7.6) can be written as

$$\frac{\partial}{\partial t} \int_{\tau} u \, d\tau + \int_{\Sigma} \mathbf{S} \cdot d\mathbf{a} + \int_{\tau} \mathbf{E} \cdot \mathbf{J} \, d\tau = 0 \qquad (7.7)$$

Converting the surface integral to a volume integral by the divergence theorem gives

$$\int_{\Sigma} \mathbf{S} \cdot d\mathbf{a} = \int_{\tau} \mathbf{\nabla} \cdot \mathbf{S} \, d\tau$$

and (7.6) becomes

$$\int_{\tau} \left[\mathbf{E} \cdot \mathbf{J} + \mathbf{\nabla} \cdot \mathbf{S} + \frac{\partial u}{\partial t} \right] d\tau = 0$$

where $\dfrac{\partial}{\partial t}$ can be moved inside the integral sign since τ is fixed in space. This result is true for all volumes τ and, hence, the integrand itself must vanish, i.e.,

$$\mathbf{E} \cdot \mathbf{J} + \mathbf{\nabla} \cdot \mathbf{S} + \frac{\partial u}{\partial t} = 0 \tag{7.8}$$

which is the differential form of the law of conservation of energy for the electromagnetic field.

To complete the derivation, it is necessary to express the energy flux $\mathbf{S}$ and the energy density u in terms of the fields $\mathbf{E}$ and $\mathbf{B}$. A method for doing this was worked out by Poynting in 1884, and is based on rewriting $\mathbf{E} \cdot \mathbf{J}$ in terms of field quantities. The procedure follows. First, substitute Maxwell's equation (7.1b) for $\mathbf{J}$ and obtain

$$\mathbf{E} \cdot \mathbf{J} = \mathbf{E} \cdot \left[\varepsilon_0 c^2 \mathbf{\nabla} \times \mathbf{B} - \varepsilon_0 \frac{\partial \mathbf{E}}{\partial t} \right]$$

$$= \varepsilon_0 c^2 \mathbf{E} \cdot (\mathbf{\nabla} \times \mathbf{B}) - \varepsilon_0 \mathbf{E} \cdot \frac{\partial \mathbf{E}}{\partial t}$$

Expanding the first term on the right according to the rules of vector analysis and (7.1a) gives

$$\mathbf{E} \cdot (\mathbf{\nabla} \times \mathbf{B}) = \mathbf{\nabla} \cdot (\mathbf{B} \times \mathbf{E}) + \mathbf{B} \cdot (\mathbf{\nabla} \times \mathbf{E})$$

$$= \mathbf{\nabla} \cdot (\mathbf{B} \times \mathbf{E}) + \mathbf{B} \cdot \left(-\frac{\partial \mathbf{B}}{\partial t} \right)$$

which, when substituted in the expression for $\mathbf{E} \cdot \mathbf{J}$, leads to

$$\mathbf{E} \cdot \mathbf{J} - \varepsilon_0 c^2 \left[\mathbf{\nabla} \cdot (\mathbf{B} \times \mathbf{E}) - \mathbf{B} \cdot \frac{\partial \mathbf{B}}{\partial t} \right] + \varepsilon_0 \mathbf{E} \cdot \frac{\partial \mathbf{E}}{\partial t} = 0$$

or

$$\mathbf{E} \cdot \mathbf{J} + \varepsilon_0 c^2 \mathbf{\nabla} \cdot (\mathbf{E} \times \mathbf{B}) + \frac{\partial}{\partial t} \left[\frac{\varepsilon_0 c^2 B^2}{2} + \frac{\varepsilon_0 E^2}{2} \right] = 0 \tag{7.9}$$

Comparison of (7.9) with (7.8) shows that energy flux **S** and energy density u may be identified as follows:

$$u = \frac{\varepsilon_0 c^2 B^2}{2} + \frac{\varepsilon_0 E^2}{2} \tag{7.10}$$

$$\mathbf{S} = \varepsilon_0 c^2 \mathbf{E} \times \mathbf{B} \tag{7.11}$$

These identifications clearly are not unique, but do give results consistent with experiment and, therefore, are useful. The first term of (7.10) represents the stored energy of the magnetic field and the second the stored energy of the electric field. These energies are denoted by the symbols U_m and U_e respectively, as follows:

$$U_m = \int_\tau \frac{\varepsilon_0 c^2 B^2}{2} \, d\tau \tag{7.12}$$

$$U_e = \int_\tau \frac{\varepsilon_0 E^2}{2} \, d\tau \tag{7.13}$$

Equation (7.11) may be interpreted as the electromagnetic energy radiated through the bounding surface Σ.

7.3   CLASSIFICATION OF ELECTROMAGNETIC FIELDS

It is convenient in practice to classify electromagnetic fields into several categories as follows:

Stationary or Time-Independent Fields. These are often called *static* fields, and are defined by the condition that the time derivatives in Maxwell's equations vanish. For such fields Maxwell's equations reduce to two independent pairs, one for the electric field and a second for the magnetic field.

$$\nabla \cdot \mathbf{E} = \frac{\rho}{\varepsilon_0} \qquad \nabla \times \mathbf{E} = 0$$

$$\nabla \times \mathbf{B} = \frac{\mathbf{J}}{\varepsilon_0 c^2} \qquad \nabla \cdot \mathbf{B} = 0$$

Since no interaction exists between the electric and the magnetic fields, they may be treated separately.

Slowly-Varying or Quasi-Stationary Fields. Fields which vary so slowly that the energy radiated is negligible in comparison with the other energies are called *slowly-varying* or *quasi-stationary* fields. Slowly-varying in this context is to be interpreted in terms of the physical dimensions of the current and charge systems, the time rates of change of the fields, and the propagation velocity of electromagnetic waves. If the system dimension is small, propagation velocity may be considered infinite; thus, in ordinary electric circuit theory, voltages and currents are assumed to be transmitted instantaneously, i.e., effects appear instantaneously throughout the system. On the other hand, if the system dimension and the time rate of change are such that the fields change significantly during the time required for an electromagnetic wave to traverse the system, then appreciable energy may be radiated and the system will not be slowly varying. For slowly-varying fields, the surface integral of the Poynting vector approaches zero as r^{-3}, as may be seen from the consideration that $\mathbf{E} \times \mathbf{B}$ decreases as r^{-5} while Σ increases only as r^2.

The restriction to slowly-varying fields may be viewed in another light by considering sinusoidal variations. An electromagnetic wave of frequency f has a wavelength in free space of $\lambda = c/f$, where c is the velocity of light. The restriction to slowly-varying fields can be met by requiring the maximum dimension L of the system to be small compared with λ, that is,

$$L \ll \lambda$$

When this condition is satisfied, the electromagnetic field will be negligibly small at distances of only a few wavelengths from the circuit in all directions, and thus the fields are confined to the vicinity of the system. Table 7.1 gives examples.

Table 7.1 Examples of limitations on the maximum dimensions or lengths of electromagnetic equipment in order that the quasi-static approximation be satisfied, based on $L = 0.1\lambda$.

Application	Frequency	Wavelength λ	Maximum System dimension L
a-c power	60 cps	5×10^6 meters	5×10^5 meters (310 miles)
Audio frequency	10 kc	3×10^4 meters	3000 meters (1.9 miles)
Radio broadcast	1 mc	300 meters	30 meters
TV, FM	100 mc	3 meters	30 cm
Microwave (*S-band radar*)	3 gc	10 cm	1 cm

General Case. The fields vary sufficiently rapidly, or the fields extend over sufficiently large regions of space, so that Maxwell's equations must be used in their complete form. For this case $\mathbf{E} \times \mathbf{B}$ decreases not as r^{-5} but as r^{-2}, and hence, the surface integral of the Poynting vector approaches a constant value as r increases. Fields of this type are called *radiation fields* and are propagated through space with the velocity of light c. The energy associated with radiation fields is called *radiation* or *radiant energy.*

7.4 ELECTROSTATIC ENERGY

Before deriving expressions for the energy of an electrostatic system, it is necessary to introduce and quantitatively define the scalar potential for an electrostatic field. The relation $\mathbf{\nabla} \times \mathbf{E} = 0$ is sufficient to insure the existence of a scalar potential function φ. Existence of this function greatly facilitates the solution of electrostatic problems because the field problem is thereby reduced to a much simpler scalar problem. The vector identity

$$\mathbf{\nabla} \times (\mathbf{\nabla}\varphi) = 0$$

shows that $\mathbf{E}$ may be expressed as the gradient of a scalar point function φ, i.e.,

$$\mathbf{E} = -\mathbf{\nabla}\varphi \tag{7.14}$$

The negative sign is arbitrary, but is chosen to make $\mathbf{E}$ positive for decreasing values of φ.

Substitution of Eq. (7.14) in (7.1c) leads to the partial differential equation for φ known as Poisson's equation

$$\mathbf{\nabla} \cdot \mathbf{E} = \mathbf{\nabla} \cdot \mathbf{\nabla}(-\varphi) = -\nabla^2\varphi = \frac{\rho}{\varepsilon_0}$$

or

$$\nabla^2\varphi = -\frac{\rho}{\varepsilon_0} \tag{7.15}$$

In a charge-free region, $\rho = 0$, and (7.15) reduces to Laplace's equation

$$\nabla^2\varphi = 0 \tag{7.16}$$

Poisson's equation for a finite distribution of charge has a general solution of the form

$$\varphi = \int_\sigma \frac{\rho \, d\sigma}{4\pi\varepsilon_0 r} + \varphi_0 \tag{7.17}$$

where r is the distance from the element of charge $\rho \, d\sigma$ to the point where φ is calculated. The constant of integration φ_0 is generally chosen to satisfy the condition that φ vanishes at infinity.

However, the charge distribution usually is not known, and it is then necessary to solve Poisson's equation or Laplace's equation as a boundary value problem in partial differential equations.

Potential Energy. The work required to bring a system of charges together from an infinite distance is called the potential energy of the system. For a volume distribution of charge it may be calculated in the following way. Consider a process whereby the charge density ρ is increased by an infinitesimal amount $d\rho$ at each point. The work dW necessary to effect this increase is

$$dW = \int_\tau \varphi \, d\rho \, d\tau$$

Now

$$\varphi \, d\rho = \varphi d[\varepsilon_0 \boldsymbol{\nabla} \cdot \mathbf{E}] = \varepsilon_0 \varphi \boldsymbol{\nabla} \cdot (d\mathbf{E})$$

which, from the vector identity for $\boldsymbol{\nabla} \cdot (\varphi \, d\mathbf{E})$, can be written

$$\varphi \, d\rho = \varepsilon_0 [\boldsymbol{\nabla} \cdot (\varphi \, d\mathbf{E}) - (\boldsymbol{\nabla}\varphi) \cdot d\mathbf{E}]$$

Hence

$$dW = \int_\tau \varepsilon_0 [\boldsymbol{\nabla} \cdot (\varphi \, d\mathbf{E}) - (\boldsymbol{\nabla}\varphi) \cdot d\mathbf{E}] \, d\tau$$

The first term of the integral drops out by Gauss' divergence theorem, and in the second term $-\boldsymbol{\nabla}\varphi = \mathbf{E}$, so that

$$dW = \int_\tau \varepsilon_0 \mathbf{E} \cdot d\mathbf{E} \, d\tau = \int_\tau \varepsilon_0 \, d(\tfrac{1}{2}E^2) \, d\tau$$

Integration with τ constant and writing U_e for W gives

$$U_e = \frac{1}{2}\int_\tau \varepsilon_0 E^2 \, d\tau \qquad (7.18)$$

which agrees with the result obtained for the dynamic field in Eq. (7.13). Equation (7.18) may be interpreted as a volume energy distribution of density $\tfrac{1}{2}\varepsilon_0 E^2$. This concept is illustrated in Ex. 1. The presence of dielectric materials influences the electric field distribution through molecular polarization or displacement of bound charges in the dielectric.

For most dielectrics encountered in practice the macroscopic effects of internal or polarization fields may be represented in Maxwell's equations by using a quantity ε in place of ε_0. The ratio of ε to ε_0 is a dimensionless number called the relative dielectric constant of the material. For most solid and liquid dielectrics, the dielectric constant lies in the range between about two and about ten, and is usually temperature-dependent. For our purposes, it is sufficient to treat dielectric media in this manner.

The potential energy may be expressed also in terms of φ and ρ by using Eq. (7.17)

$$dU_e = dW = \int_\tau \varphi \, d\rho \, d\tau = \int_\tau \int_\sigma \frac{\rho \, d\rho \, d\tau \, d\sigma}{4\pi\varepsilon_0 r}$$

Hence

$$U_e = \int_0^\rho \int_\tau \int_\sigma \frac{\rho \, d\rho \, d\tau \, d\sigma}{4\pi\varepsilon_0 r} = \frac{1}{2}\int_\tau \int_\sigma \frac{\rho^2 \, d\tau \, d\sigma}{4\pi\varepsilon_0 r}$$

$$= \frac{1}{2}\int_\tau \varphi\rho \, d\tau \qquad (7.19)$$

For a system of n point charges, the potential energy may be approximated by

$$U_e = \frac{1}{2} \int_\tau \varphi \rho \, d\tau = \frac{1}{2} \int_\tau \sum_{i=1}^{n} \varphi_i \rho_i \, d\tau$$

where the charge on each particle is assumed to be distributed over a small but finite volume with a charge density ρ_i. In the limit, where the charge distributions approach point charges of magnitude q_i, the energy U_e becomes

$$U_e = \frac{1}{2} \sum_{i=1}^{n} \varphi_i \int_\tau \rho_i \, d\tau = \frac{1}{2} \sum_{i=1}^{n} \varphi_i q_i \qquad (7.20)$$

Systems of Conductors. The electrostatic energy of a system of n conductors follows directly from Eq. (7.19). Denote the potentials of the conductors by $\varphi_1, \varphi_2, \ldots, \varphi_n$, and the corresponding charge distributions by $\rho_1, \rho_2, \ldots, \rho_n$. Then, since the potential is constant through each conductor

$$U_e = \frac{1}{2} \int_\tau \varphi \rho \, d\tau = \frac{1}{2} \sum_{i=1}^{n} \varphi_i \int_\tau \rho_i \, d\tau$$

$$= \frac{1}{2} \sum_{i=1}^{n} \varphi_i q_i = \tfrac{1}{2} \boldsymbol{\varphi} \cdot \mathbf{q} \qquad (7.21)$$

where $\mathbf{q}$ and $\boldsymbol{\varphi}$ are vectors defined by

$$\mathbf{q} = [q_1, q_2, \ldots, q_n]$$
$$\boldsymbol{\varphi} = [\varphi_1, \varphi_2, \ldots, \varphi_n]$$

and $q_1, q_2, \ldots, q_n$ are the charges on the n conductors.

It is clear from Eq. (7.17) that the potential of each conductor of the system depends on the charge distribution and on the geometry of the system. Moreover, since the potential of any given conductor varies linearly with ρ in Eq. (7.17), it follows that the principle of superposition applies and the potential of the conductor can be calculated by superposition. To carry out this calculation, assume that the n conductors of the system are fixed in position and initially uncharged. Let a unit charge be placed on conductor (1), the other conductors remaining uncharged. Then the potential and field distribution can be found by solving Laplace's equation as a boundary value problem.

Suppose the result is to produce potential p_{11} on conductor (1), potential p_{12} on conductor (2), and so on for all n conductors. Then the result of placing charge q_1 on conductor (1) and leaving the other conductors uncharged will be to produce potentials $p_{11}q_1, p_{12}q_1, \ldots, p_{1n}q_1$. If now charges $q_1, q_2, \ldots, q_n$ are placed on the corresponding conductors simultaneously, the net result, again by the superposition principle, will be to produce potentials

$$\varphi_1 = p_{11}q_1 + p_{21}q_2 + \cdots + p_{n1}q_n$$
$$\varphi_2 = p_{12}q_1 + p_{22}q_2 + \cdots + p_{n2}q_n$$
$$\cdot \quad \cdot \quad \cdot \quad \cdot \quad \cdot \quad \cdot \quad \cdot \quad \cdot \quad \cdot \quad \cdot \quad \cdot$$
$$\varphi_n = p_{1n}q_1 + p_{2n}q_2 + \cdots + p_{nn}q_n$$

where the coefficients p_{ij} are purely geometric quantities which depend on the size, shape, and position of the different conductors. These equations are equivalent to the matrix equation

$$\boldsymbol{\varphi} = P\mathbf{q} \tag{7.22}$$

where the matrix $P = [\, p_{ij}]$, and the coefficients p_{ij} are called the potential coefficients.

The inverse relation for $\mathbf{q}$ is obtained by inversion of (7.22)

$$\mathbf{q} = P^{-1}\boldsymbol{\varphi} = C\boldsymbol{\varphi} \tag{7.23}$$

where the coefficients of c_{ij} of matrix C are called capacity coefficients if $i = j$ and electrostatic induction coefficients if $i \neq j$.

The electrostatic energy of the system of conductors may be exhibited as a quadratic form in the potentials by substitution of (7.23) in (7.21)

$$U_e = \frac{1}{2}\sum_{i=1}^{n}\sum_{j=1}^{n} c_{ij}\varphi_i\varphi_j = \tfrac{1}{2}\boldsymbol{\varphi}C\boldsymbol{\varphi}' \tag{7.24}$$

or, in terms of the charges,

$$U_e = \frac{1}{2}\sum_{i=1}^{n}\sum_{i=1}^{n} P_{ij}q_iq_j = \tfrac{1}{2}\mathbf{q}P\mathbf{q}' \tag{7.25}$$

Both of these quadratic forms are positive definite, since U_e by definition is positive for all positive nonzero values of the charges and potentials, and hence

$$c_{ii} > 0 \qquad c_{ii}c_{jj} > c_{ij}^2 \qquad c_{ij} = c_{ji}$$

with similar relations for the potential coefficients p_{ij}.

The significance of the capacity and potential coefficients is illustrated in Exs. 2 and 3 of this section.

Equilibrium Forces on Charged Conductors. The forces may be found by use of the principle of virtual work. Denote the configuration of the conductors by a set of generalized coordinates

$$\mathbf{x} = [x_1, x_2, \ldots, x_f]$$

and the corresponding external forces necessary for equilibrium by

$$\mathbf{F} = [f_1, f_2, \ldots, f_f]$$

Let the conductors be given a virtual displacement $\delta \mathbf{x}$. This displacement will not affect the charges, but will change the potential coefficients and therefore the energy U_e. Since the system is in equilibrium under the electric forces and the external mechanical forces imposed on the conductors, it follows from the principle of virtual work that

$$\delta W = \mathbf{F} \cdot \delta \mathbf{x} - \delta U_e = 0$$

Note that the sign of δU_e must be taken as negative in this expression because the work done by the electric forces is negative. Using the relation

$$\delta U_e = \sum_{i=1}^{f} \frac{\partial U_e}{\partial x_i} \, \delta x_i = (\mathbf{\nabla} U_e) \cdot \delta \mathbf{x}$$

leads to the result

$$\delta W = (\mathbf{F} - \mathbf{\nabla} U_e) \cdot \delta \mathbf{x} = 0$$

from which

$$\mathbf{F} = \mathbf{\nabla} U_e \quad \text{or} \quad f_i = \frac{\partial U_e}{\partial x_i}$$

From Newton's law of action and reaction, the electric forces $\mathbf{F}_e$ are the negative of $\mathbf{F}$, or

$$\mathbf{F}_e = -\mathbf{F} = -\mathbf{\nabla} U_e \tag{7.26}$$

Example 1. Energy Stored in a Parallel Plate Capacitor. Compute the electrostatic energy stored in a parallel plate condenser of area A, spacing b, and potential difference V. Neglect edge effects and assume the plates are in a vacuum.

SOLUTION

From Eq. (7.18) the energy U_e is given by

$$U_e = \frac{1}{2} \int_\tau \varepsilon_0 E^2 \, d\tau = \frac{1}{2} \varepsilon_0 \int_0^b \left(\frac{V}{b}\right)^2 A \, dx$$

$$= \frac{\varepsilon_0 A V^2}{2b}$$

It is of interest to estimate the maximum energy density which can be achieved in practice. Experience shows that under the best conditions in a vacuum, $E_{max} \cong 10^7$ volts/cm $= 10^9$ volts/meter. Hence

$$\frac{U_{max}}{\text{Unit Volume}} = \tfrac{1}{2}\varepsilon_0 E_{max}^2 = \tfrac{1}{2} 8.85 \times 10^{-12} \times (10^9)^2$$

$$= 4.43 \times 10^6 \text{ joules per meter}^3$$

$$= 4.43 \text{ joules per cm}^3$$

Example 2. Capacity Coefficients for a Sphere. Find the capacity coefficient for an isolated sphere in free space.

SOLUTION

The potential φ_1 from Eq. (7.17) is

$$\varphi_1 = \int_\tau \frac{\rho \, d\tau}{4\pi\varepsilon_0 r} = \frac{1}{4\pi\varepsilon_0 a} \int_\tau \rho \, d\tau = \frac{q_1}{4\pi\varepsilon_0 a}$$

where a is the radius of the sphere. Hence

$$c_1 = \frac{q_1}{\varphi_1} = 4\pi\varepsilon_0 a$$

which is the well-known result for the capacity of a sphere in free space.

Example 3. Mutual Capacity of Two Conductors. Find the mutual capacity c of two conductors, with charges $\perp q$, in terms of the coefficients c_{ij}.

SOLUTION

The mutual capacity c is defined as

$$c = \frac{q}{\varphi_2 - \varphi_1}$$

From (7.21)

$$U_e = \tfrac{1}{2}\mathbf{q} \cdot \boldsymbol{\varphi} = \tfrac{1}{2}(-q\varphi_1 + q\varphi_2)$$

$$= \tfrac{1}{2}q(\varphi_2 - \varphi_1) = \frac{q^2}{2c}$$

and from (7.25)

$$U_e = \tfrac{1}{2}\mathbf{q}P\mathbf{q}'$$

$$= \frac{q^2}{2}(p_{11} - 2p_{12} + p_{22})$$

Hence

$$\frac{1}{c} = p_{11} - 2p_{12} + p_{22}$$

But

$$P = C^{-1} = \begin{bmatrix} c_{11} & c_{12} \\ c_{21} & c_{22} \end{bmatrix}^{-1} = \frac{1}{\det C}\begin{bmatrix} c_{22} & -c_{12} \\ -c_{21} & c_{11} \end{bmatrix} = \begin{bmatrix} p_{11} & p_{12} \\ p_{21} & p_{22} \end{bmatrix}$$

so that

$$\frac{1}{c} = \frac{c_{22} + 2c_{12} + c_{11}}{c_{11}c_{22} - c_{12}^2}$$

Example 4. Electrostatic Force Between Charged Plates. Find the force of attraction between the two plates of a parallel plate condenser charged to a potential difference V and placed in a vacuum.

SOLUTION

From Ex. 1, the electrostatic energy is

$$U_e = \frac{\varepsilon_0 A V^2}{2b}$$

To express V in terms of q, apply Gauss' theorem; then

$$q = \int_A \varepsilon_0 \mathbf{E} \cdot d\mathbf{A} = \int_A \frac{\varepsilon_0 V}{b} \, dA = \frac{\varepsilon_0 V A}{b}$$

Substituting $V = \dfrac{bq}{\varepsilon_0 A}$ in the equation for U_e gives

$$U_e = \frac{q^2 b}{2\varepsilon_0 A}$$

from which the force F_e is

$$F_e = -\frac{\partial U_e}{\partial b} = -\frac{q^2}{2\varepsilon_0 A} = -\frac{\varepsilon_0 V^2 A}{2b^2} = -\frac{\varepsilon_0 E^2 A}{2}$$

From Ex. 1, it follows that the maximum electrostatic force which can be achieved in practice is about

$$(F_e)_{\max} = \tfrac{1}{2}\varepsilon_0 E_{\max}^2 A = 4.43 \times 10^6 \text{ newtons}$$

for $E_{\max} = 10^9$ volts/meter and $A = 1$ square meter.

7.5 MAGNETIC ENERGY

The stored energy of the magnetic field for the dynamic case is given by Eq. (7.12), developed as a part of the energy relations for the electromagnetic field.

$$U_m = \frac{1}{2} \int_\tau \varepsilon_0 c^2 B^2 \, d\tau$$

This relation is also valid for the static case.*

For ferromagnetic materials, the stored energy depends on the shape of the magnetization curve for a given material under given conditions. In general, the magnetization curve must be found by laboratory test methods. Assuming that the magnetization curve is known, the stored magnetic energy is given by

$$U_m = \int_\tau \left\{ \int_0^B \varepsilon_0 c^2 \mathbf{H} \cdot d\mathbf{B} \right\} d\tau \tag{7.27}$$

---

* Note that strictly speaking it is impossible to establish a static field without going through a dynamic phase. Even for very slow changes in **B**, the effects of electromagnetic induction are cumulative and must be taken into account in computing the magnetic energy. Thus, energy calculations based on the static situation alone must introduce some artifice to arrive at the correct expression.

In this expression

$$\mathbf{B} = \mu\mathbf{H}$$

where μ is the relative permeability of the material and $\mathbf{H}$ is called the magnetizing field.

In the absence of ferromagnetic material, $\mathbf{B} = \mathbf{H}$ and $\mu = 1$, i.e., the magnetizing field is just the magnetic field $\mathbf{B}$ for a vacuum or for a nonmagnetic substance.

Magnetic Vector Potential. Although there is no scalar counterpart in the magnetic case for Poisson's equation of electrostatics, there does exist an analogous equation in terms of a vector potential. This follows from the well-known theorem of vector analysis that if $\nabla \cdot \mathbf{B} = 0$, then can be expressed as the curl of another vector $\mathbf{A}$, i.e.,

$$\mathbf{B} = \nabla \times \mathbf{A} \tag{7.28}$$

Substitution of Eq. (7.28) in (7.1b), with $\dfrac{\partial \mathbf{E}}{\partial t} = 0$, gives

$$\nabla \times (\nabla \times \mathbf{A}) = \frac{\mathbf{J}}{\varepsilon_0 c^2} \tag{7.29}$$

Expanding the vector cross product leads to

$$\nabla \times (\nabla \times \mathbf{A}) = \nabla(\nabla \cdot \mathbf{A}) - \nabla^2\mathbf{A} = -\nabla^2\mathbf{A}$$

where $\nabla \cdot \mathbf{A}$ is taken as zero, which is permissible since it does not affect the value of $\mathbf{B}$. Equation (7.29) is thus transformed into Poisson's equation for a vector potential $\mathbf{A}$

$$\nabla^2\mathbf{A} = -\frac{\mathbf{J}}{\varepsilon_0 c^2} \tag{7.30}$$

For any physical system of currents, the current density $\mathbf{J}$ is everywhere finite and confined to a finite region of space. Therefore, using a well-known result of vector analysis, the solution of Poisson's equation (7.30) is

$$\mathbf{A} = \int_\tau \frac{\mathbf{J}\,d\tau}{4\pi\varepsilon_0 c^2 r} \tag{7.31}$$

where the volume of integration τ includes all space for which $\mathbf{J} \neq 0$, and r is the distance from the current element $\mathbf{J}\,d\tau$ to the point where $\mathbf{A}$ is calculated.

In most practical current systems the currents flow in wires, i.e., conductors whose transverse dimensions are small compared with their lengths. Such currents are said to be *filamentary*. The current element for a filamentary conductor may be expressed as

$$\mathbf{J}\,d\tau = I\,d\mathbf{s}$$

and (7.31) becomes

$$A = \int_s \frac{I \, ds}{4\pi\varepsilon_0 c^2 r}$$

where I is the total current in the wire and ds is a line element along the current path. The field **B** is found by substitution of the expression immediately above for **A** in Eq. (7.28), giving

$$\mathbf{B} = \nabla \times \mathbf{A} = \int_s \frac{I}{4\pi\varepsilon_0 c^2} \nabla \times \left(\frac{ds}{r}\right)$$

The law of Biot and Savart for calculating the magnetic field of a current circuit follows directly from the vector identity

$$\nabla \times \frac{ds}{r} = \nabla \frac{I}{r} \times ds + \frac{1}{r} \nabla \times ds = \frac{ds \times \mathbf{e}_r}{r^2}$$

where $\mathbf{e}_r$ is a unit vector in the direction of r and $\nabla \times ds$ is identically zero. Using this in the preceding expression for **B** gives the law of Biot and Savart as follows

$$\mathbf{B} = \int_s \frac{I \, ds \times \mathbf{e}_r}{4\pi\varepsilon_0 c^2 r^2} \tag{7.32}$$

By a procedure similar to that used for developing Eq. (7.19) for the energy of the electrostatic field, it is possible to express the magnetic energy in terms of the source currents, giving the result

$$U_m = \frac{1}{2} \int_\tau \mathbf{J} \cdot \mathbf{A} \, d\tau \tag{7.33}$$

Energy of Magnetically Coupled Circuits. A system of current-carrying conductors with mutual magnetic energy is said to be magnetically coupled. The conductors may be extended media or, more frequently, wires. In this section the magnetic energy of a system of currents in wires (filamentary currents) is calculated. Let there be n filamentary circuits carrying currents $I_1, I_2, \ldots, I_n$. For a filamentary conductor the element of current is given by

$$\mathbf{J} \, d\tau = I \, ds$$

where I is the current in the wire and ds is a line element of the wire. With this substitution Eq. (7.33) becomes

$$U_m = \frac{1}{2} \int_s I \, ds \cdot \mathbf{A} = \frac{1}{2} \sum_{k=1}^n I_k \int_{s_k} \mathbf{A} \cdot ds_k$$

But from Stokes' theorem

$$\int_{s_k} \mathbf{A} \cdot ds_k = \int_{S_k} (\nabla \times \mathbf{A}) \cdot d\mathbf{S}_k = \int_{S_k} \mathbf{B} \cdot d\mathbf{S}_k = \Phi_k \tag{7.34}$$

where Φ_k is the total magnetic flux linking circuit k. Hence

$$U_m = \frac{1}{2} \sum_{k=1}^{n} I_k \Phi_k = \tfrac{1}{2} \mathbf{I} \cdot \boldsymbol{\Phi} \tag{7.35}$$

where

$$\mathbf{I} = [I_1, I_2, \ldots, I_m]$$

and

$$\boldsymbol{\Phi} = [\Phi_1, \Phi_2, \ldots, \Phi_n]$$

The magnetic energy may be written in terms of the currents and the configuration of the wire conductors. For this purpose, it is necessary to express the flux linkages Φ_k as functions of the currents. The general solution for the magnetic vector potential $\mathbf{A}$ due to a current distribution in space represented by a current density $\mathbf{J}$ is, from Eq. (7.31)

$$\mathbf{A} = \frac{1}{4\pi} \int_\tau \frac{\boldsymbol{\nabla} \times \mathbf{B}}{r} \, d\tau = \frac{1}{4\pi\varepsilon_0 c^2} \int_\tau \frac{\mathbf{J}}{r} \, d\tau$$

For a system of filamentary currents this takes the form

$$\mathbf{A} = \frac{1}{4\pi\varepsilon_0 c^2} \sum_{i=1}^{n} \int_{s_i} \frac{I_i \, d\mathbf{s}_i}{r_{ik}}$$

where r_{ik} is the distance between the element $d\mathbf{s}_i$ and the point for which $\mathbf{A}$ is calculated. If this point is at the position of $d\mathbf{s}_k$, Eq. (7.34) can be written as

$$\begin{aligned}
\Phi_k &= \int_{s_k} \left[\frac{1}{4\pi\varepsilon_0 c^2} \sum_{i=1}^{n} \int_{s_i} \frac{I_i \, d\mathbf{s}_i \cdot d\mathbf{s}_k}{r_{ik}} \right] \\
&= \sum_{i=1}^{n} I_i \left[\frac{1}{4\pi\varepsilon_0 c^2} \int_{s_i}\int_{s_k} \frac{d\mathbf{s}_i \cdot d\mathbf{s}_k}{r_{ik}} \right] \\
&= \sum_{i=1}^{n} I_i L_{ik}
\end{aligned} \tag{7.36}$$

where

$$L_{ik} = \frac{1}{4\pi\varepsilon_0 c^2} \int_{s_i}\int_{s_k} \frac{d\mathbf{s}_i \cdot d\mathbf{s}_k}{r_{ik}} \tag{7.37}$$

is a geometric quantity depending only on the configuration of the conductors. L_{ik} is called the mutual inductance between circuits i and k if $i \neq k$, and the self inductance of circuit i if $i = k$. Equation (7.37) is known as Neumann's formula.

Using relation (7.36) for Φ_k in Eq. (7.35) gives for the magnetic energy of the system

$$U_m = \frac{1}{2} \sum_{i=1}^{n} \sum_{k=1}^{n} L_{ik} I_i I_k$$

The energy may also be written in vector notation as follows

$$U_m = \tfrac{1}{2} \mathbf{I} L \mathbf{I}' \tag{7.38}$$

where $L = [L_{ik}]$ is the inductance matrix for the system. The analogy with the similar expression (7.25) for the electrostatic energy of a system of charged conductors is evident.

Equilibrium Forces on Current-Carrying Conductors. A system of current-carrying conductors can be maintained in equilibrium only by application of external forces to counteract the internal magnetic forces between conductors. To derive these forces, denote the configuration of the conductors of the system by a set of generalized coordinates

$$\mathbf{x} = [x_1, x_2, \ldots, x_f]$$

with corresponding external forces

$$\mathbf{F} = [f_1, f_2, \ldots, f_f]$$

Let the conductors be given a virtual displacement $\delta\mathbf{x}$ at infinitesimally small velocity so that the kinetic energy of the conductors is also infinitesimal, under the condition that the currents remain constant, additional energy being supplied by external emf sources as needed. Maxwell's investigations showed that the energy of a system of current-carrying conductors is to be regarded as kinetic energy and is equal to the magnetic energy. Hence, we must use Lagrange's equations with $T = U_m$. Thus

$$f_j = -\frac{\partial T}{\partial x_j} = -\frac{\partial U_m}{\partial x_j} \qquad i = 1, 2, \ldots, f \tag{7.39}$$

or

$$\mathbf{F} = -\nabla U_m$$

Since the magnetic forces between the conductors are equal and opposite to the external forces, we have

$$\mathbf{F}_m = -\mathbf{F} = \nabla U_m = \left[\frac{\partial U_m}{\partial x_1}, \frac{\partial U_m}{\partial x_2}, \ldots, \frac{\partial U_m}{\partial x_f}\right]$$

where $\mathbf{F}_m$ represents the magnetic forces and

$$\frac{\partial U_m}{\partial x_j} = \frac{1}{2}\sum_{i=1}^{n}\sum_{k=1}^{n} I_i I_k \frac{\partial L_{ik}}{\partial x_j} \tag{7.40}$$

Example 1. Force Between Circular Currents. Given two parallel circular one-turn coils, each 6 inches in radius, and spaced one inch apart, and one coil carrying 5 amperes, the other 200 amperes, find the force between the coils.

SOLUTION

The force is given by Eq. (7.40) which yields

$$f_y = I_1 I_2 \frac{\partial L_{12}}{\partial y}$$

where f_y is the force along the axis of the coils, taken here as the y-axis. The mutual inductance L_{12} is a geometric quantity given by the integral of Eq. (7.37) and depends only on the configuration of the conductors. In general, calculation of L_{12} leads to rather complicated integrals. Even for the present case, which is relatively simple, the integrals lead to elliptic functions. However, when the coils are close together, L_{12} can be approximated by the expression

$$L_{12} = \frac{4\pi a}{4\pi\varepsilon_0 c^2} \ln \frac{8a}{y}$$

where a is the radius of the coils and y their separation. Hence

$$f_y = I_1 I_2 \frac{\partial}{\partial y} \left(\frac{4\pi a}{4\pi\varepsilon_0 c^2} \ln \frac{8a}{y} \right)$$

$$= -\frac{a I_1 I_2}{\varepsilon_0 c^2 y} = -4\pi \times 10^{-7} I_1 I_2 \frac{a}{y}$$

$$= -4\pi \times 10^{-7} \times 5 \times 200 \times \tfrac{6}{1} = -7.55 \times 10^{-3} \text{ newton}$$

and is attractive if the currents flow in the same direction in the coils.

7.6 ENERGY RELATIONS FOR MOVING CONDUCTORS

Electrical charges and currents always exhibit forces of interaction in accord with the Lorentz' law of force. The concept of a purely electrical system in which forces are omitted or ignored is obviously a simplification, for convenience or expediency, to facilitate analysis. For example, it is conventional in discussing an electrostatic system of charges to ignore the fact that equilibrium requires external mechanical forces to balance the electrical forces. A similar situation exists with respect to magnetostatic systems.

Lorentz' law also implies that moving charges involve an interplay of forces accompanied by energy exchanges. By this means energy can be transformed reversibly between electromagnetic and mechanical forms. The motion of a charged conductor in an electric field can be used to convert mechanical to electrical energy or electrical to mechanical energy. Similarly, the motion of a current-carrying conductor through a magnetic field provides a basis for energy conversion between electrical and mechanical forms.

Consider a system of electrical and mechanical elements contained in a closed region R as shown in the diagram of Figure 7.2. Energy transformations between electromagnetic and mechanical forms occur within the system, while in addition energy may be delivered to and from the system by outside electrical and mechanical forces. Applying the

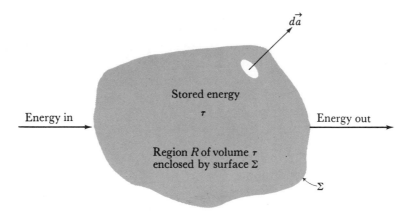

FIGURE 7.2   Energy diagram for a system containing electrical and mechanical elements.

law of conservation of energy as developed in Chap. 6 and expressed by Eq. (6.32), we write

$$\frac{d}{dt}(T + V) = \mathbf{Q}_d \cdot \dot{\mathbf{q}} + \mathbf{Q}_e \cdot \dot{\mathbf{q}} \qquad (7.41)$$

where $T + V$ is total energy stored in the system R, $\mathbf{Q}_d \cdot \dot{\mathbf{q}}$ is the rate at which energy is dissipated in R, and $\mathbf{Q}_e \cdot \dot{\mathbf{q}}$ is the rate at which energy enters or leaves R due to the external forces. The sign convention adopted is that $\mathbf{Q}_e \cdot \dot{\mathbf{q}} > 0$ corresponds to energy flow into the system, and vice versa. It is understood that in the present context the energy functions include both mechanical and electromagnetic energies, and that energy may be dissipated, i.e., converted to heat, by both mechanical and electrical action.

The mechanical energy components of Eq. (7.41) belong to ordinary dynamics and introduce no new features. The electromagnetic energies and their interactions with the mechanical elements of the system do, however, bring in new factors which require consideration. Energy relations for the electromagnetic field were derived earlier in this chapter (p. 215) and lead to the following equation:

$$\frac{\partial U}{\partial t} + \int_\tau \mathbf{E} \cdot \mathbf{J}\, d\tau + \int_\Sigma \mathbf{S} \cdot d\mathbf{a} = 0 \qquad (7.42)$$

where τ is the volume and Σ is the closed surface which bounds the system R. The other quantities are identified as follows: U is the electromagnetic energy of the fields $\mathbf{E}$ and $\mathbf{B}$ in the volume τ; the integral of $\mathbf{E} \cdot \mathbf{J}$ is the work done on matter in the system by the electromagnetic forces; and $\mathbf{S}$ is the energy flux or rate at which energy flows out of a unit area of Σ. The total energy of the field in R, from Eqs. (7.12) and (7.13), is given by

$$U = U_m + U_e = \int_\tau \frac{\varepsilon_0 c^2 B^2}{2}\, d\tau + \int_\tau \frac{\varepsilon_0 E^2}{2}\, d\tau \qquad (7.43)$$

To evaluate the integral of $\mathbf{E} \cdot \mathbf{J}$ in Eq. (7.42), it is assumed that the currents in the system are carried by ohmic conductors of conductivity σ, and that the conductors are in motion. Using Ohm's law and the Lorentz law of force, the current density $\mathbf{J}$ in an element of volume $d\tau$ is

$$\mathbf{J} = \sigma(\mathbf{E} + \mathbf{v} \times \mathbf{B}) \tag{7.44}$$

where $\mathbf{E}$ is the electric field in the conductor, $\mathbf{v}$ is the velocity of the element $d\tau$ of the conductor, and $\mathbf{B}$ is the magnetic field in the inertial frame in which $\mathbf{E}$ and $\mathbf{B}$ are defined. The question as to the possible effect of the motion on $\mathbf{J}$ for a moving conductor is answered by Einstein's special theory of relativity which shows that $\mathbf{J}$ is not significantly affected when $v \ll c$.

Using (7.44) it is found that

$$\int_\tau \mathbf{E} \cdot \mathbf{J}\, d\tau = \int_\tau \frac{J^2}{\sigma}\, d\tau - \int_\tau (\mathbf{v} \times \mathbf{B}) \cdot \mathbf{J}\, d\tau \tag{7.45}$$

The first integral on the right of (7.45) is always positive and gives the ohmic heat loss in the conductors.

The second integral represents the rate at which mechanical energy is converted to electromagnetic energy, as can be seen by writing the vector identity

$$-(\mathbf{v} \times \mathbf{B}) \cdot \mathbf{J} = (\mathbf{J} \times \mathbf{B}) \cdot \mathbf{v}$$

and noting that $\mathbf{J} \times \mathbf{B} = \mathbf{F}$, the force per unit volume of conductor. Then the relation between mechanical and electromagnetic power is given by

$$\int_\tau \mathbf{F} \cdot \mathbf{v}\, d\tau = -\int_\tau (\mathbf{v} \times \mathbf{B}) \cdot \mathbf{J}\, d\tau \tag{7.46}$$

Thus, the mechanical power used to move the conductors is just equal to the electromagnetic power produced. It is also evident from the reversible nature of the process that if, on the other hand, electromagnetic power is fed into the conductors, that power will appear in mechanical form as a force applied by the conductors in moving some mechanical element.

If the conductors are also assumed to be filamentary (wires), then $\mathbf{J}\, d\tau = I\, d\mathbf{s}$ where I is the total current in a wire and $d\mathbf{s}$ is a line element of the wire. Substitution of $I\, d\mathbf{s}$ in the energy expression (7.45) gives the following:

(1) the ohmic or heat loss energy

$$\int_\tau \frac{J^2}{\sigma}\, d\tau = \mathbf{I}R\mathbf{I}' \tag{7.47}$$

where R is the resistance matrix and $\mathbf{I}$ is the current vector for the system conductors; and

(2) the motional energy

$$\int_\tau (\mathbf{v} \times \mathbf{B}) \cdot \mathbf{J}\, d\tau = \int_\tau I(\mathbf{v} \times \mathbf{B}) \cdot d\mathbf{s} \tag{7.48}$$

Since for the slowly-varying currents considered here the radiated energy is negligible, the electromagnetic energy emerging from the system is that carried by the conductors through the bounding surface Σ. Therefore, the energy flux term in Eq. (7.42) becomes

$$\int_\Sigma \mathbf{S} \cdot d\mathbf{a} = \mathbf{V} \cdot \mathbf{I} \qquad (7.49)$$

where $\mathbf{V}$ represents the voltages at the system terminals.

Induced emf in a Moving Conductor. The emf ε induced in a closed moving conductor circuit may be calculated directly from the Lorentz law of force. By definition, the emf is the line integral of the force along a given path. Hence

$$\varepsilon = \int_s (\mathbf{E} + \mathbf{v} \times \mathbf{B}) \cdot d\mathbf{s} = \int_s \mathbf{E} \cdot d\mathbf{s} + \int_s (\mathbf{v} \times \mathbf{B}) \cdot d\mathbf{s}$$

The first integral, from Maxwell's equation (7.4a), has the value

$$\int_s \mathbf{E} \cdot d\mathbf{s} = -\frac{\partial}{\partial t} \int_A \mathbf{B} \cdot d\mathbf{A} = -\left(\frac{\partial \Phi}{\partial t}\right)_{v=0}$$

and gives the induced emf due to changes in the magnetic field.

Using the vector identity

$$(\mathbf{v} \times \mathbf{B}) \cdot d\mathbf{s} = -\mathbf{B} \cdot (\mathbf{v} \times d\mathbf{s})$$

in the second integral gives

$$\int_s (\mathbf{v} \times \mathbf{B}) \cdot d\mathbf{s} = -\int_s \mathbf{B} \cdot (\mathbf{v} \times d\mathbf{s})$$

But $\mathbf{v} \times d\mathbf{s}$ represents the rate at which $d\mathbf{s}$ is moving in a direction perpendicular to itself, so that $\mathbf{B} \cdot (\mathbf{v} \times d\mathbf{s})$ is the rate that element $d\mathbf{s}$ cuts magnetic flux due to the motion of the conductor. Thus

$$-\int_s \mathbf{B} \cdot (\mathbf{v} \times d\mathbf{s}) = -\left(\frac{\partial \Phi}{\partial t}\right)_{B=\text{const.}}$$

which is the induced emf due to the motion of the conductor.

Adding both emf's gives the total emf

$$\varepsilon = -\left(\frac{\partial \Phi}{\partial t}\right)_{v=0} - \left(\frac{\partial \Phi}{\partial t}\right)_{B=\text{const.}} = -\frac{d\Phi}{dt} = -\dot{\Phi} \qquad (7.50)$$

Equation (7.50) is observed to be Faraday's law of electromagnetic induction.

Conservation of Energy for Electromagnetic Field with Moving Conductors. Substitution of Eqs. (7.47) through (7.50) in (7.42) gives the law of conservation of energy in a form appropriate for a system of moving conductors:

$$\frac{\partial U}{\partial t} + IRI' - \boldsymbol{\epsilon} \cdot \mathbf{I} - \left(\frac{\partial \Phi}{\partial t}\right)_{v=0} \cdot \mathbf{I} + \mathbf{V} \cdot \mathbf{I} = 0$$

This expression can be simplified by writing $U = U_m + U_e$ and substituting for U_m from Eq. (7.39). Thus

$$\frac{\partial U}{\partial t} = \frac{\partial U_m}{\partial t} + \frac{\partial U_e}{\partial t}$$

and

$$\frac{\partial U_m}{\partial t} = \frac{\partial}{\partial t}(\tfrac{1}{2}\mathbf{I}L\mathbf{I}') = \mathbf{I} \cdot \frac{\partial}{\partial t}(L\mathbf{I}')$$

$$= \mathbf{I} \cdot \left(\frac{\partial \Phi}{\partial t}\right)_{v=0}$$

Hence, the conservation law reduces to

$$\frac{\partial U_e}{\partial t} + IRI' + \mathbf{V} \cdot \mathbf{I} = \boldsymbol{\epsilon} \cdot \mathbf{I} \qquad (7.51)$$

where $\boldsymbol{\epsilon}$ is the emf vector and includes both the emf's induced by changes in the magnetic field and the emf's induced by the motions of the conductors.

Lagrange's Equations for Systems with Moving Conductors. The energy functions for the electromagnetic part of an electromechanical system can be identified from Eq. (7.51). To these must be added the terms for the mechanical energies, however, which, as noted earlier, present no special problem. The generalized coordinates for the system include both the mechanical and the electrical coordinates. In the present treatment, the latter are to be taken as circuit charges, so that the $\dot{q}_i$ are the currents. The terminal voltages are obviously applied forces, and IRI' clearly represents the ohmic energy loss. Since the electrostatic energy U_e is that stored in the capacitors of the system, it follows from Eq. (7.25) that

$$U_e = \tfrac{1}{2}\mathbf{q}C^{-1}\mathbf{q}' \qquad (7.52)$$

where $\mathbf{q}$ represents the electric charges of the system and C is the capacitance matrix.

The emf vector $\boldsymbol{\epsilon}$ represents the emf's produced in the system, both by changing magnetic fluxes and by moving conductors. Faraday's law of electromagnetic induction

$$\boldsymbol{\epsilon} = -\dot{\boldsymbol{\Phi}} = -\frac{d}{dt}(L\mathbf{I}')$$

is the same as would be derived from Lagrange's equations by considering the magnetic field energy U_m to be kinetic energy. This can be made evident by the following argument. The magnetic energy U_m from Eq. (7.38) is

$$U_m = \tfrac{1}{2}ILI' = \tfrac{1}{2}\dot{q}L\dot{q}' \tag{7.53}$$

Write the time derivative component of Lagrange's equation for component $\dot{q}_k$ as follows

$$\frac{d}{dt}\left(\frac{\partial U_m}{\partial \dot{q}_k}\right) = \frac{d}{dt}\left\{\frac{\partial}{\partial \dot{q}_k}\left[\frac{1}{2}\sum_{i=1}^{n}\sum_{k=1}^{n}L_{ik}\dot{q}_i\dot{q}_k\right]\right\}$$

$$= \frac{d}{dt}\left\{\sum_{i=1}^{n}L_{ik}\dot{q}_i\right\} = \frac{d\Phi_k}{dt} = -\varepsilon_k \tag{7.54}$$

Hence, it follows that the term $\boldsymbol{\varepsilon}$ can be derived by using U_m as a kinetic energy in Lagrange's equations.

In case the magnetic flux is produced by permanent magnets, the induced emf's are due entirely to the motions of the conductors and U_m has the form

$$U_m = \mathbf{I} \cdot \boldsymbol{\Phi} = \dot{\mathbf{q}} \cdot \boldsymbol{\Phi} \tag{7.55}$$

Substitution in Lagrange's equation for component $\dot{q}_j$ gives

$$\frac{d}{dt}\left(\frac{\partial U_m}{\partial \dot{q}_j}\right) = \frac{d\Phi_j}{dt} = -\varepsilon_j \tag{7.56}$$

In the general case, there may be magnetic energies of both types, as represented by Eqs. (7.53) and (7.55).

BIBLIOGRAPHY

1. J. H. Jeans, *Electricity and Magnetism*, Fifth Edition. London: Cambridge University Press, 1925.
2. L. D. Landau and E. M. Lifshitz, *Electrodynamics of Continuous Media*. London: Pergamon Press, 1960.
3. W. K. H. Panofsky and M. Phillips, *Classical Electricity and Magnetism*, Second Edition. Reading, Mass.: Addison-Wesley, 1964.
4. J. C. Maxwell, *A Treatise on Electricity and Magnetism*, Third Edition. London: Oxford University Press, 1892.
5. R. W. Pohl, *Electricity and Magnetism*. New York: Van Nostrand, 1930.
6. J. A. Stratton, *Electromagnetic Theory*. New York: McGraw-Hill, 1941.
7. W. R. Smythe, *Static and Dynamic Electricity*. New York: McGraw-Hill, 1939.
8. R. P. Feynman et al., *The Feynman Lectures on Physics*, Vol. II. Reading, Mass.: Addison-Wesley, 1964.
9. O. Heaviside, *Electromagnetic Theory*. New York: Dover, 1950. (Unabridged edition Vols. I, II, III, originally published 1891–1912.)
10. J. R. Reitz and F. J. Milford, *Foundations of Electromagnetic Theory*. Reading, Mass.: Addison-Wesley, 1960.
11. E. Weber, *Electromagnetic Fields*, Vol. I. New York: Wiley, 1950.

Part III

Applications

8

LINEAR SYSTEMS

Any system which obeys Hamilton's principle and whose behavior may be described mathematically by linear algebraic or linear differential equations may be called a *linear dynamical system* or, more briefly, a linear system. Such systems are of great practical importance because (1) mathematical methods exist for the general solution of linear equations, and (2) many engineering and physical systems can be adequately described by linear equations. Some of the fields of widest application are electric networks, structural and elastic systems, thermal systems, hydraulic systems, acoustical and electroacoustical systems, electro-mechanical energy conversion systems, vibrational systems, and feedback control systems.

Although in practice separate theories have been developed for each area of application of linear systems, the need for separate theories can be reduced by a unified, abstract theory, which then can be applied to all real systems of this type. Trends toward more general treatment of dynamical systems are well established in several fields, as for example the generalization of electric network theory, the utilization of dynamical analogies, and the broadening of automatic and optimal control theory. These steps toward generality capitalize on existing knowledge and familiarity with some specialized field. The best known illustration is probably that of electric network theory, where the great amount of effort which has been directed toward the solution of electric circuit problems and the familiarity of many engineers with electric circuits provide strong motivation to apply this knowledge to other areas.

The approach used in this book, however, is to develop an abstract theory of linear dynamic systems based on the methods of generalized mechanics and, then, to demonstrate its applicability to some of the many

fields where it is useful. All such systems may be grouped under the name
generalized linear dynamical systems. In this chapter, applications to electric
networks, electric systems, electromechanical systems, and structures are
studied.

8.1 GENERAL THEORY OF LINEAR DYNAMICAL SYSTEMS

The general theory is developed starting with Lagrange's equations
for systems with energy dissipation and with energy inputs and outputs.
Lagrange's equations for such systems, from Chap. 6, Eq. (6.31), have the
form

$$\frac{d}{dt}\left(\frac{\partial L}{\partial \dot{q}_r}\right) - \frac{\partial L}{\partial q_r} + \frac{\partial(F - P)}{\partial \dot{q}_r} = 0 \qquad r = 1, 2, \ldots, f \qquad (8.1)$$

where the energy functions are

$$L = T - V = \tfrac{1}{2}\dot{\mathbf{q}}'M\dot{\mathbf{q}} - \tfrac{1}{2}\mathbf{q}'K\mathbf{q} \qquad (8.2)$$

and

$$F - P = \tfrac{1}{2}\dot{\mathbf{q}}'R\dot{\mathbf{q}} - \mathbf{Q} \cdot \dot{\mathbf{q}} \qquad (8.3)$$

In these equations, the generalized coordinates $\mathbf{q}$ and the generalized
forces $\mathbf{Q}$ are column vectors as follows*

$$\mathbf{q} = \begin{bmatrix} q_1 \\ q_2 \\ \cdot \\ \cdot \\ \cdot \\ q_f \end{bmatrix} \qquad \mathbf{Q} = \begin{bmatrix} Q_1 \\ Q_2 \\ \cdot \\ \cdot \\ \cdot \\ Q_f \end{bmatrix}$$

The matrices M, K, and R are the generalized mass, elastance, and resistance matrices, respectively.

The kinetic energy T, the potential energy V, and the Rayleigh
dissipation function F are positive definite quadratic forms with constant
coefficients, and P, which represents the power input or power output of
the system, is a linear function of the generalized velocities $\dot{q}_r$. The
generalized force components Q_r are constants or given functions of time,
but do not depend on either the $\dot{q}_r$ or the q_r.

Carrying out the indicated differentiations in Lagrange's equations
(8.1) leads to the differential equation

$$(MD^2 + RD + K)\mathbf{q} = \mathbf{Q} \qquad (8.4)$$

where D represents the differentiation operator $\dfrac{d}{dt}$. For convenience we
define

$$A = MD^2 + RD + K \qquad (8.5)$$

---

* It is sometimes convenient to write vectors in the form of column vectors as is done
here for $\vec{q}$ and $\vec{Q}$. Clearly $\vec{q}'$ and $\vec{Q}'$ are then the corresponding row vectors.

and write (8.4) in abbreviated form

$$A\mathbf{q} = \mathbf{Q} \qquad (8.6)$$

Here A is a linear differential operator matrix with elements

$$A_{rs} = M_{rs}\frac{d^2}{dt^2} + R_{rs}\frac{d}{dt} + K_{rs} \qquad (8.7)$$

Premultiplication of (8.6) by A^{-1} yields the symbolic solution

$$\mathbf{q} = A^{-1}\mathbf{Q} \qquad (8.8)$$

Equation (8.8) will be an actual solution if the inverse matrix operator A^{-1} can be properly interpreted.

Characteristic Matrix, Transfer Matrix, and Transfer Functions. It is useful to interpret $\mathbf{Q}$ as the input vector to the system, A as the characteristic matrix of the system (because A characterizes or represents the system structure), and $\mathbf{q}$ as the output of the system. Associated with the characteristic matrix A is the transfer matrix A^{-1} which changes $\mathbf{Q}$ to $\mathbf{q}$. Mathematically, this amounts to considering A^{-1} as a "black box" operator which transforms an input $\mathbf{Q}$ to an output $\mathbf{q}$. This symbolic language, derived largely from concepts in wide use in the theory of electrical and acoustical networks, is useful in a broader context and may be used in an abstract setting.

The elements A_{rs}^{-1} of the transfer matrix A^{-1} are called transfer functions, and are of special importance in dealing with systems which have a single input and a single output. For such systems, and where A has been appropriately interpreted, the relation between q_r and Q_s may be written directly from the matrix expression (8.8) in the form

$$q_r = \frac{\Delta_{rs}Q_s}{|A|} \qquad (8.9)$$

where Δ_{rs} is the cofactor of A'_{rs} in det $[A']$.

For many applications the generalized velocity $\dot{\mathbf{q}}$ is the quantity of basic interest, and for such applications we can write

$$\dot{\mathbf{q}} = D\mathbf{q} = DA^{-1}\mathbf{Q} \qquad (8.10)$$

where D is the differentiation operator d/dt. Denoting the transfer matrix DA^{-1} by $Y = Z^{-1}$, Eq. (8.10) may be expressed as

$$\dot{\mathbf{q}} = Z^{-1}\mathbf{Q} = Y\mathbf{Q} \qquad (8.11)$$

where, in correspondence with electric network usage, Z may be called the impedance matrix and Y the admittance matrix. This terminology is convenient and can be used in a more general and abstract context without introducing ambiguities. The elements of Y and Z are related in the following manner

$$Y_{rs} = \frac{\delta_{rs}}{|Z|} \tag{8.12}$$

where δ_{rs} is the cofactor of Z'_{rs} in det $[Z']$.

Linear systems may be classified in terms of the characteristic matrix A and the input or driving force $\mathbf{Q}$. For example, a system with no input and no dissipation would be classified as conservative, whereas a system with only dissipative elements and with a constant force $\mathbf{Q}$ would, in electric-network terminology, be called a direct current or d-c network. A more complete classification based on this approach is given in Table 8.1.

Table 8.1 Classification of Linear Systems

System type	System Structure A	System Input $\mathbf{Q}$	Typical Applications
Conservative	$MD^2 + K$	None	Theory of small vibrations Specific heats of crystals Critical frequencies of structures Electric wave filters
Static	K	$\mathbf{Q} = \mathbf{Q}_0 = $ const.	Equilibrium of structures Electrostatics
Resistive	RD	$\mathbf{Q} = \mathbf{Q}(t)$	Resistance networks Heat flow and diffusion
Periodically excited	$MD^2 + RD + K$	$\mathbf{Q} = \mathbf{Q}_p e^{i\omega t}$	Communication systems Power systems a-c electric networks Forced vibrations, resonance Filter theory
Transient	$MD^2 + RD + K$	$\mathbf{Q} = \mathbf{Q}(t)$	Automatic control Electric machinery Electronic circuits

Solution of the differential equations for the system reduces to interpretation of the inverse matrix operator A^{-1} acting on the input vector $\mathbf{Q}$ of Eq. (8.8). Although it is possible to proceed directly to quite general solutions of (8.8), it is more useful here to identify A^{-1}, first, for simpler systems and, then, for systems of increasing complexity.

The student familiar with the theory of differential equations will recognize the inverse operator A^{-1} as nothing more than the Green's function for the given operator. In this context it is clear that the Green's function is determined by the operator alone. Thus, once we have found the Green's function for the operator A, we may, in principle, solve the system for any input vector $\mathbf{Q}$. For further information on the method of Green's function see Ref. 6.

Static Systems. For these systems, the characteristic matrix A reduces to K, the input vector $\mathbf{Q}$ is constant, and Eq. (8.8) becomes

$$\mathbf{q} = K^{-1}\mathbf{Q} \tag{8.13}$$

Resistive Systems. These are the simplest nonstatic systems inasmuch as they involve only resistive elements and linear algebraic equations. For such systems,

$$A\mathbf{q} = R\dot{\mathbf{q}} = \mathbf{Q}$$

and

$$\dot{\mathbf{q}} = R^{-1}\mathbf{Q} \tag{8.14}$$

where R^{-1} is a constant matrix.

Lagrange's equations for resistive systems yield a useful result in the form of a minimal principle. This principle follows immediately from the condition that T and V in Eq. (8.2) are both zero (no stored energy), so that Lagrange's equations take the form

$$\frac{\partial(F - P)}{\partial \dot{q}_r} = 0 \qquad r = 1, 2, \ldots, f$$

which may be expressed as the gradient in $\dot{\mathbf{q}}$-space

$$\nabla(F - P) = \left[\frac{\partial(F - P)}{\partial \dot{q}_1}, \frac{\partial(F - P)}{\partial \dot{q}_2}, \ldots, \frac{\partial(F - P)}{\partial \dot{q}_f} \right] = 0 \tag{8.15}$$

Equation (8.15) is recognized as the condition for a stationary point of $F - P$.

To determine if the stationary point is also an extremum, find the matrix of the coefficients $\dfrac{\partial^2(F - P)}{\partial \dot{q}_r \, \partial \dot{q}_s}$. This is easily shown to be just the matrix R, and since R is the matrix of a positive definite quadratic form, it follows that the stationary point is also a minimum of $F - P$.

Systems with Periodic Excitation. The theory of systems of this class focuses, in its simplest form, on the steady-state response of systems energized by sinusoidal sources of a single fixed frequency. It is assumed that the systems are dissipative, which is sufficient to insure that starting transients decay asymptotically with time. A slight extension of the primitive theory is sufficient to include the response of systems energized by periodic sources of arbitrary waveform which can be represented by Fourier series. A further extension, without introducing any basically new principle, suffices to include the transient case where the excitation can be represented by a Fourier integral.

It is convenient to represent sinusoidal sources by exponential functions of imaginary argument as follows

$$\mathbf{Q} = \mathbf{Q}_p e^{i\omega t} \tag{8.16}$$

where $\mathbf{Q}_p$ is the complex amplitude and $\omega(= 2\pi f)$ is the angular frequency. Since

$$De^{i\omega t} = i\omega e^{i\omega t}$$

the operator D may be replaced by $i\omega$ in the matrix A. Then, inasmuch as $\mathbf{q}$ must be of the form

$$\mathbf{q} = \mathbf{q}_p e^{i\omega t} = A^{-1}\mathbf{Q}_p e^{i\omega t}$$

it follows that

$$\mathbf{q}_p = A^{-1}\mathbf{Q}_p \tag{8.17}$$

where the elements A_{rs} of A are given by

$$A_{rs} = M_{rs}(i\omega)^2 + R_{rs}i\omega + K_{rs}$$
$$= K_{rs} - M_{rs}\omega^2 + iR_{rs}\omega$$

The generalized velocity $\dot{\mathbf{q}}$ is given by

$$\dot{\mathbf{q}} = i\omega\mathbf{q}_p e^{i\omega t} = \dot{\mathbf{q}}_p e^{i\omega t}$$

Then, since $e^{i\omega t} \neq 0$, it follows that

$$\dot{\mathbf{q}}_p = i\omega\mathbf{q}_p = i\omega A^{-1}\mathbf{Q}_p = Y\mathbf{Q}_p = Z^{-1}\mathbf{Q}_p \tag{8.18}$$

where $\mathbf{q}_p$ is the complex amplitude of $\mathbf{q}$.

The impedance matrix Z has elements Z_{rs} given by

$$Z_{rs} = \frac{A_{rs}}{i\omega} = R_{rs} + i\left(M_{rs}\omega - \frac{K_{rs}}{\omega}\right) \tag{8.19}$$

Systems with Transient Excitation. We consider first quiescent systems (no initial stored energy) excited by transient forces

$$\mathbf{Q} = \mathbf{Q}(t)$$

and use the Laplace transform method of analysis.* Since

$$\mathscr{L}(D\mathbf{q}) = \mathscr{L}(\dot{\mathbf{q}}) = s\bar{\mathbf{q}}$$

where s is the Laplace transform parameter, the operator D may be replaced by s in the matrix A. Then, taking the Laplace transform of Eq. (8.6) yields

$$\mathscr{L}(A\mathbf{q}) = A\bar{\mathbf{q}} = \mathscr{L}(\mathbf{Q}) = \bar{\mathbf{Q}}$$

from which it follows that

$$\bar{\mathbf{q}} = A^{-1}\bar{\mathbf{Q}} \tag{8.20}$$

where the elements A_{rs} in A are given by

$$A_{rs} = M_{rs}s^2 + R_{rs}s + K_{rs} \tag{8.21}$$

---

* The bar symbol is used to denote the Laplace transform of a function, i.e., $\mathscr{L}[f(t)] = \bar{f}$.

To complete the solution it is necessary to determine the inverse transform, i.e.,

$$\mathbf{q} = \mathscr{L}^{-1}(\bar{\mathbf{q}}) = \mathscr{L}^{-1}(A^{-1}\bar{\mathbf{Q}}) \tag{8.22}$$

Calculation of $\mathscr{L}^{-1}(A^{-1}\bar{\mathbf{Q}})$ may be carried out by the method of partial fractions, by use of tables of Laplace transforms, or by the inversion theorem for Laplace transforms which involves complex variable theory. The first two methods suffice for many problems.

Moving now to systems with initial stored energy, we find

$$\mathscr{L}(A\mathbf{q}) = A\bar{\mathbf{q}} - (Ms + R)\mathbf{q}_0 + M\dot{\mathbf{q}}_0 = \bar{\mathbf{Q}} \tag{8.23}$$

where

$$\mathbf{q}_0 = \mathbf{q}(0) \qquad \dot{\mathbf{q}}_0 = \dot{\mathbf{q}}(0)$$

are the initial values of $\mathbf{q}$ and $\dot{\mathbf{q}}$. Taking the inverse transform of (8.23) gives the solution

$$\mathbf{q} = \mathscr{L}^{-1}\{A^{-1}[\bar{\mathbf{Q}} + (Ms + R)\mathbf{q}_0 + M\dot{\mathbf{q}}_0]\} \tag{8.24}$$

In addition to the generalized coordinates $\mathbf{q}$, there are numerous applications for which the generalized velocities $\dot{\mathbf{q}}$ are required. These may be determined by direct differentiation of $\mathbf{q}$, but it is usually more expeditious to use the Laplace transform relation

$$\mathscr{L}(\dot{\mathbf{q}}) = s\bar{\mathbf{q}} - \mathbf{q}_0$$

which when substituted in Eq. (8.24) gives the desired result as follows:

$$\mathbf{q} = \mathscr{L}^{-1}\{A^{-1}[s\bar{\mathbf{Q}} + (Ms^2 + Rs)\mathbf{q}_0 + Ms\mathbf{q}_0 - A\mathbf{q}_0]\}$$

But

$$Ms^2 + Rs - A = Ms^2 + Rs - Ms^2 - Rs - K = -K$$

so that

$$\dot{\mathbf{q}} = \mathscr{L}^{-1}\{A^{-1}[s\bar{\mathbf{Q}} + Ms\dot{\mathbf{q}}_0 - K\mathbf{q}_0]\} \tag{8.25}$$

If there is no initial stored energy in the system, $\mathbf{q}_0 = \dot{\mathbf{q}}_0 = 0$, and (8.25) simplifies to

$$\dot{\mathbf{q}} = \mathscr{L}^{-1}\{sA^{-1}\bar{\mathbf{Q}}\} \tag{8.26}$$

Linear Systems Theorems.   Several theorems often found useful in applications follow directly from the linear and symmetry properties of Eqs. (8.6) and (8.8). These theorems masquerade under various historical names such as superposition, reciprocity, and Thevenin's theorems in electric network theory, Maxwell's reciprocity theorem in the theory of structures, and many others; but all are implicit in the symmetry and linear characteristics of the systems studied. The more important of these theorems are discussed in this section.

Superposition Theorem.   This theorem, first stated by Daniel Bernoulli in 1755 under the name of the principle of coexistence of small oscillations, states that the total response of a given generalized coordinate q_i due to all the forces acting on the system is the sum of the response produced by the individual forces acting separately. Proof follows immediately from the linear properties of Eq. (8.8).

Reciprocity Theorem. Maxwell proved this theorem for linear resistance networks and for systems of conductors in electrostatic theory. It has been generalized since then by Lord Rayleigh and others. The theorem is usually stated for static, resistive, and steady-state sinusoidally excited systems in the form "a force Q_r produces the same response on coordinate q_s as a force $Q_s = Q_r$ produces on coordinate q_r." To prove this result, we note that from Eq. (8.8)

$$q_r = A_{rs}^{-1}Q_s \text{ for a single force } Q_s$$

and

$$q_s = A_{sr}^{-1}Q_r \text{ for a single force } Q_r$$

But $A_{rs}^{-1} = A_{sr}^{-1}$ because of symmetry, and therefore

$$q_r = q_s \quad \text{if} \quad Q_r = Q_s$$

which completes the proof.

Analogy Between Static and Resistive Systems. It is evident that static and resistive systems are analogous in the sense that the governing equations

$$\mathbf{q} = K^{-1}\mathbf{Q} \text{ (static systems)}$$

and

$$\dot{\mathbf{q}} = R^{-1}\mathbf{Q} \text{ (resistive systems)}$$

have the same form. Moreover, since both K^{-1} and R^{-1} are constant matrices for given systems, it is possible to study solutions $\mathbf{q}$ for static systems in terms of corresponding solutions $\dot{\mathbf{q}}$ for the resistive systems. For example, the deflections $\mathbf{q}$ of an elastic structure, which often are difficult to measure experimentally, may be simulated by easily measured currents $\dot{\mathbf{q}}$ in an electric network of resistors and batteries. Such systems are called *analogs* and find wide use in science and engineering. Analogs may also be established for nonphysical systems, as for example transportation and distribution systems in economics.

Dynamical Analogies. The abstract theory of linear dynamical systems may be used to establish the basis for the concept of dynamical analogies. In this sense, any two systems which obey the same dynamic laws, but are comprised of physically different kinds of elements, may be called dynamical analogs. Dynamical analogies can thus be established between electrical, mechanical, acoustical, and (incompletely) thermal systems. It is also possible to extend the concept of dynamical analogies to nonphysical systems, as has been done, for example, by Forrester (Ref. 4) to certain economic systems. Abstract analogs are also commonly formulated for computer simulation, using either analog or digital computers. Dynamical analogs are not unique since it is always possible to devise different embodiments of the same analogous system.

The utility of analogs depends largely on the background, education, experience, and philosophy of the person using the analogs. Maxwell long ago developed the dynamical theory of currents and used mechanical analogs to demonstrate electrical circuit phenomena (Ref. 1). Lord Rayleigh, in his classic nineteenth century work "The Theory of Sound," showed how electrical circuit vibrations could be brought within the scope of acoustics, the electrical vibrations being treated in terms of acoustical analogs (Ref. 2, Vol. I, Ch. X_R). Today, it is far more common to use electric circuits as analogs for mechanical and acoustical systems because of the great amount of effort which has gone into the development of electric network theory and the large number of engineers familiar with network theory.

For reference purposes, the abstract quantities appearing in Eqs. (8.4) and (8.5) are identified in Table 8.2 for electrical, mechanical, and acoustical systems. It should be noted, however, that although the symbols and names shown for the various quantities are typical, there is in fact no generally accepted usage except for the electrical case where usage is nearly standardized.

8.2 STATE VARIABLE METHOD OF SOLUTION

First-order differential equations are of particular importance in application not only because of their comparative simplicity but also because special methods can be used which cannot be applied directly to systems of higher order. Further, as shown in Chap. 4, any system of general order differential equations can always be reduced to a system of first order by suitable choice of new variables. These new variables constitute a set of state variables for the system, since it is known from the theory of differential equations that the future behavior of the system may be predicted from a knowledge of the initial values of the state variables. That is, the solutions exist, are unique, and depend continuously on the initial values of the variables. The advantages of first-order or state equations are especially evident in the application of digital and analog computers to systems problems.

A dynamical system of f degrees of freedom clearly has, in the general case, $2f$ state variables. These may be chosen in many ways. Hamilton's canonical variables $(q_1, q_2, \ldots, q_f; p_1, p_2, \ldots, p_f)$, from which the concept of state variables evolved, represent one choice. Another choice is

Table 8.2 Systems Quantities

Abstract Quantity	Electrical		Mechanical Translation		Mechanical Rotation		Acoustical	
	Symbol	Name	Symbol	Name	Symbol	Name	Symbol	Name
M_{rs}	L_{rs}	Inductance	m_{rs}	Mass	J_{rs}	Moment of inertia	M_{rs}	Inertance
R_{rs}	R_{rs}	Resistance	R_{rs}	Mechanical resistance	R_{rs}	Rotational resistance	R_{rs}	Acoustical resistance
K^{-1}_{rs}	C_{rs}	Capacitance	K^{-1}_{rs}	Mechanical compliance	K^{-1}_{rs}	Rotational compliance	C_{rs}	Acoustical capacitance
q_r	q_r	Charge	x_r	Linear displacement	θ_r	Angular displacement	X_r	Volume displacement
$\dot{q}_r$	$\dot{q}_r = I_r$	Current	$\dot{x}_r$	Velocity	$\dot{\theta}_r = \omega_r$	Angular velocity	$\dot{X}_r$	Volume current
Q_r	ε_r	Electromotive force	F_r	Force	M_r	Torque	P_r	Pressure
P	P	Power	P	Power	P	Power	P	Power

the set of generalized coordinates and generalized velocities $(q_1, q_2, \ldots, q_f;$ $\dot{q}_1, \dot{q}_2, \ldots, \dot{q}_f)$ for the system. Many other choices are possible. For example, in an electric network the mesh currents and node voltages may be used.

Although any system of general order differential equations, whether linear or nonlinear, can be reduced to first order, we are in this section interested in reducing the second-order linear dynamical equation (8.4), i.e.,

$$(MD^2 + RD + K)\mathbf{q} = \mathbf{Q}$$

to first order and in obtaining solutions for the system in this form. In essence, this is the strategy followed in the state variable approach.

To reduce Eq. (8.4) to first order, multiply by M^{-1} and solve for $D^2\mathbf{q}$ as follows:

$$D^2\mathbf{q} = -M^{-1}RD\mathbf{q} - M^{-1}K\mathbf{q} + M^{-1}\mathbf{Q} \qquad (8.27)$$

where it is assumed that $\det [M] \neq 0$, so that M^{-1} exists. Introducing the new vector variable $\mathbf{x}$ defined by

$$\mathbf{x} = \begin{bmatrix} x_1 \\ x_2 \\ \cdot \\ \cdot \\ \cdot \\ \cdot \\ \cdot \\ \cdot \\ \cdot \\ \cdot \\ \cdot \\ \cdot \\ \cdot \\ x_n \end{bmatrix} = \begin{bmatrix} q_1 \\ q_2 \\ \cdot \\ \cdot \\ \cdot \\ q_f \\ \dot{q}_1 \\ \dot{q}_2 \\ \cdot \\ \cdot \\ \cdot \\ \dot{q}_f \end{bmatrix} = \begin{bmatrix} \mathbf{q} \\ \dot{\mathbf{q}} \end{bmatrix} \qquad (8.28)$$

and combining with (8.27), gives the normal or state variable form

$$D\mathbf{x} = B\mathbf{x} + \mathbf{b} \qquad (8.29)$$

The system matrix B has the structure

$$B = \begin{bmatrix} 0 & I_f \\ -M^{-1}K & -M^{-1}R \end{bmatrix} \qquad (8.30)$$

where I_f is the identity matrix of order f and $-M^{-1}K$, $-M^{-1}R$ are also of order f. Hence, B is a square matrix of order $2f$ and is in general nonsymmetric. The vector $\mathbf{b}$, from Eqs. (8.27) and (8.28), has the form

$$\mathbf{b} = \begin{bmatrix} 0 \\ M^{-1}\mathbf{Q} \end{bmatrix} \qquad (8.31)$$

Several special cases of state variable equation (8.29) may occur in practice and require consideration. It may happen that M is singular, so that M^{-1} does not exist. In that case, the number of variables is reduced. This follows from the readily proved result that if M is of rank $r < f$, then there are only

$$n = 2f - (f - r) = f + r$$

variables. In the extreme case of $M = 0$, it is evident that $n = f$ and Eq. (8.4) is of first order as it stands. Finally, if both $M = 0$ and $R = 0$, Eq. (8.4) is reduced to the static case, that is, to a set of algebraic equations.

Solution of State Equation.   There are several methods for solving the state equation (8.29). We consider first the method based on reducing the system matrix B to its simplest form. Introduce the transformation

$$\mathbf{x} = P\mathbf{y} \tag{8.32}$$

where P is a nonsingular $n \times n$ matrix. Then Eq. (8.29) is transformed as follows:

$$D\mathbf{x} = \dot{\mathbf{x}} = P\dot{\mathbf{y}} = BP\mathbf{y} + \mathbf{b}$$

Multiplying through by P^{-1} gives

$$P^{-1}P\dot{\mathbf{y}} = \dot{\mathbf{y}} = P^{-1}BP\mathbf{y} + P^{-1}\mathbf{b} \tag{8.33}$$

which is of the same form as (8.29) but in the new variable $\mathbf{y}$.

By properly choosing the transformation matrix P, it may be possible to reduce matrix B to a simpler form and thus simplify the solution of (8.33). Since a diagonal matrix is the simplest form to deal with, we seek a matrix P such that

$$P^{-1}BP = \Lambda \tag{8.34}$$

where Λ is a diagonal matrix. From linear algebra (Chap. 1, p. 35), this is always possible if the eigenvalues of B are distinct. Thus, for this case we can write (8.33) in the form

$$\dot{\mathbf{y}} = P^{-1}BP\mathbf{y} + P^{-1}\mathbf{b}$$
$$= \Lambda\mathbf{y} + P^{-1}\mathbf{b} \tag{8.35}$$

where

$$\Lambda = \begin{bmatrix} \lambda_1 & & & \\ & \lambda_2 & & 0 \\ & & \cdot & \\ & & & \cdot \\ 0 & & & \cdot \\ & & & & \lambda_n \end{bmatrix} \tag{8.36}$$

and $\lambda_1, \lambda_2, \ldots, \lambda_n$ are the eigenvalues (distinct) of B.

The complete solution of (8.35) consists of the homogeneous solution plus a particular solution. The homogeneous solution can be seen by inspection to be

$$y_j = y_j(0)e^{\lambda_j t} \qquad j = 1, 2, \ldots, n$$

where $y_j(0)$ is the initial value of y_j. These n equations can be expressed in matrix form

$$\mathbf{y} = e^{\Lambda t}\mathbf{y_0} \tag{8.37}$$

where $e^{\Lambda t}$ is the matrix exponential

$$e^{\Lambda t} = \begin{bmatrix} e^{\lambda_1 t} & & & & \\ & e^{\lambda_2 t} & & 0 & \\ & & \cdot & & \\ & & & \cdot & \\ & 0 & & \cdot & \\ & & & & e^{\lambda_n t} \end{bmatrix} \tag{8.38}$$

Substituting $\mathbf{x} = P\mathbf{y}$ and $\mathbf{y_0} = P^{-1}\mathbf{x_0}$ in (8.37) gives the homogeneous solution in terms of $\mathbf{x}$

$$\mathbf{x} = P\mathbf{y} = Pe^{\Lambda t}P^{-1}\mathbf{x_0}$$

To find the complete solution, we use the identity

$$\frac{d}{dt}(e^{-\Lambda t}\mathbf{y}) = e^{-\Lambda t}(\dot{\mathbf{y}} - \Lambda \mathbf{y})$$

in Eq. (8.35) and obtain

$$\frac{d}{dt}(e^{-\Lambda t}\mathbf{y}) = e^{-\Lambda t}P^{-1}\mathbf{b}$$

Denoting time by τ and integrating over the interval 0 to t, gives the result

$$\int_0^t \frac{d}{d\tau}(e^{-\Lambda \tau}\mathbf{y})\, d\tau = e^{-\Lambda t}\mathbf{y} - \mathbf{y_0} = \int_0^t e^{-\Lambda \tau}P^{-1}\mathbf{b}(\tau)\, d\tau$$

from which

$$\mathbf{y} = e^{\Lambda t}\mathbf{y_0} + e^{\Lambda t}\int_0^t e^{-\Lambda \tau}P^{-1}\mathbf{b}(\tau)\, d\tau$$

Transforming back to $\mathbf{x}$ then yields the solution

$$\mathbf{x} = P\mathbf{y} = Pe^{\Lambda t}P^{-1}\mathbf{x_0} + \int_0^t Pe^{\Lambda(t-\tau)}P^{-1}\mathbf{b}(\tau)\, d\tau \tag{8.39}$$

Equation (8.39) is the complete solution, the first term on the right being the homogeneous solution and the second or integral term a particular solution.

There remains the question of finding the transformation matrix P.

This can be accomplished by using the basic properties of similar matrices (cf. Chap. 1, p. 35). Let λ_i be an eigenvalue of B, and C_i the corresponding eigenvector. Then

$$BC_i = \lambda_i C_i \qquad i = 1, 2, \ldots, n \tag{8.40}$$

and

$$B[C_1, C_2, \ldots, C_n] = [\lambda_1 C_1, \lambda_2 C_2, \ldots, \lambda_n C_n]$$

$$= [C_1, C_2, \ldots, C_n]
\begin{bmatrix}
\lambda_1 & & & \\
 & \lambda_2 & & 0 \\
 & & \cdot & \\
 & & & \cdot \\
0 & & & & \cdot \\
 & & & & & \lambda_n
\end{bmatrix}$$

This last equation may be rewritten as

$$BC = C\Lambda \tag{8.41}$$

where the matrix $C = [C_1, C_2, \ldots, C_n]$
Premultiplying (8.41) by C^{-1} gives

$$C^{-1}BC = \Lambda \tag{8.42}$$

Comparing (8.42) with (8.34) shows that the transformation matrix P can be taken as C, i.e., we can choose

$$P = C = [C_1, C_2, \ldots, C_n] \tag{8.43}$$

where $C_1, C_2, \ldots, C_n$ is a set of eigenvectors associated with B. Since these eigenvectors are linearly independent, P is nonsingular and P^{-1} therefore exists.

Solution (8.39) for the state equation exists when B has distinct eigenvalues. The solution for the case of repeated eigenvalues is discussed in Chap. 4. Therein it is shown that if there is an eigenvalue λ_j of multiplicity m, the corresponding part of the solution of the homogeneous equation is given by

$$x_j = (C_j + C_{j+1}t + C_{j+2}t^2 + \cdots + C_{j+m-1}t^{m-1})e^{\lambda_j t} \tag{8.44}$$

where $C_j, C_{j+1}, C_{j+2}, \ldots, C_{j+m-1}$ is a set of m linearly independent n-dimensional vectors which can be calculated from the eigenvalue λ_j. The general solution of the homogeneous equation is the sum of terms of the form (8.44). The case of a distinct eigenvalue, i.e., $m = 1$, reduces to

$$x_j = C_j e^{\lambda_j t}$$

corresponding to the result found previously in (8.39). It is evident that since the total number of eigenvalues is n, the multiplicities must obey the relation

$$m_1 + m_2 + \cdots + m_r = n$$

where m_1 is the multiplicity of λ_1, and so on.

Example 1. Solve the following dynamical system using state variable methods:

$$(M_{11}D^2 + R_{11}D + K_{11})q_1 = Q_1 = \text{const.}, \ t \geqslant 0 \tag{1}$$
$$q_1(0) = q_0 \qquad \dot{q}_1(0) = \dot{q}_0$$

SOLUTION

Choose the state variables

$$\mathbf{x} = \begin{bmatrix} x_1 \\ x_2 \end{bmatrix} = \begin{bmatrix} q_1 \\ \dot{q}_1 \end{bmatrix} \qquad \mathbf{x_0} = \begin{bmatrix} q_0 \\ \dot{q}_0 \end{bmatrix}$$

Then, from Eq. (8.29), the dynamical equation becomes

$$D\mathbf{x} = B\mathbf{x} + \mathbf{b}$$

$$= \begin{bmatrix} 0 & 1 \\ -\dfrac{K_{11}}{M_{11}} & -\dfrac{R_{11}}{M_{11}} \end{bmatrix} \begin{bmatrix} x_1 \\ x_2 \end{bmatrix} + \begin{bmatrix} 0 \\ \dfrac{Q_1}{M_{11}} \end{bmatrix} \tag{2}$$

The eigenvalues of B are the roots of the characteristic equation

$$\begin{vmatrix} \lambda & 1 \\ -\dfrac{K_{11}}{M_{11}} & -\dfrac{R_{11}}{M_{11}} - \lambda \end{vmatrix} = \lambda^2 + \dfrac{R_{11}}{M_{11}}\lambda + \dfrac{K_{11}}{M_{11}} = 0$$

These are

$$\lambda_1, \lambda_2 = -\dfrac{R_{11}}{2M_{11}} \pm i\sqrt{\dfrac{K_{11}}{M_{11}} - \left(\dfrac{R_{11}}{2M_{11}}\right)^2} = -\alpha \pm \beta i \tag{3}$$

where it is assumed that

$$\dfrac{K_{11}}{M_{11}} > \left(\dfrac{R_{11}}{2M_{11}}\right)^2$$

The corresponding eigenvectors are determined by the relations

$$\begin{bmatrix} -\lambda_1 & 1 \\ -\dfrac{K_{11}}{M_{11}} & -\dfrac{R_{11}}{M_{11}} - \lambda_1 \end{bmatrix} \begin{bmatrix} C_{11} \\ C_{21} \end{bmatrix} = 0 \qquad \begin{bmatrix} -\lambda_2 & 1 \\ -\dfrac{K_{11}}{M_{11}} & -\dfrac{R_{11}}{M_{11}} - \lambda_2 \end{bmatrix} \begin{bmatrix} C_{12} \\ C_{22} \end{bmatrix} = 0$$

which yield the transformation matrix

$$P = [\mathbf{C_1}, \mathbf{C_2}] = \begin{bmatrix} C_{11} & C_{12} \\ \lambda_1 C_{11} & \lambda_2 C_{12} \end{bmatrix}$$

Since C_{11} and C_{12} are arbitrary (but not zero), we let $C_{11} = C_{12} = 1$. Then

$$P = \begin{bmatrix} 1 & 1 \\ \lambda_1 & \lambda_2 \end{bmatrix} \quad \text{and} \quad P^{-1} = \dfrac{1}{\lambda_2 - \lambda_1}\begin{bmatrix} \lambda_2 & -1 \\ -\lambda_1 & 1 \end{bmatrix} \tag{4}$$

The solution of (2) may be obtained from Eq. (8.39). Letting

$$e^{At} = Pe^{\Lambda t}P^{-1}$$

$$= \begin{bmatrix} 1 & 1 \\ \lambda_1 & \lambda_2 \end{bmatrix} \begin{bmatrix} e^{\lambda_1 t} & 0 \\ 0 & e^{\lambda_2 t} \end{bmatrix} \left(\frac{1}{\lambda_2 - \lambda_1}\right) \begin{bmatrix} \lambda_2 & -1 \\ -\lambda_1 & 1 \end{bmatrix}$$

and substituting in (8.39), leads to

$$\mathbf{x} = e^{At} \begin{bmatrix} q_0 \\ \dot{q}_0 \end{bmatrix} + \int_0^t e^{A(t-\tau)} \begin{bmatrix} 0 \\ Q_1 \\ M_{11} \end{bmatrix} d\tau \tag{5}$$

Calculation gives

$$e^{At} = \begin{bmatrix} e^{-\alpha t}\left(\dfrac{\alpha}{\beta}\sin\beta t + \cos\beta t\right) & \dfrac{1}{\beta}e^{-\alpha t}\sin\beta t \\[2ex] \dfrac{\alpha^2 + \beta^2}{-\beta} \cdot e^{-\alpha t}\sin\beta t & -e^{-\alpha t}\left(\dfrac{\alpha}{\beta}\sin\beta t - \cos\beta t\right) \end{bmatrix} \tag{6}$$

Substituting (6) in (5) and carrying out the indicated calculations, there is found

$$\mathbf{x} = \begin{bmatrix} q_0 e^{-\alpha t}\left(\dfrac{\alpha}{\beta}\sin\beta t + \cos\beta t\right) + \dfrac{\dot{q}_0}{\beta}e^{-\alpha t}\sin\beta t \\[2ex] q_0\left(\dfrac{\alpha^2 + \beta^2}{-\beta}\right)e^{-\alpha t}\sin\beta t - \dot{q}_0 e^{-\alpha t}\left(\dfrac{\alpha}{\beta}\sin\beta t - \cos\beta t\right) \end{bmatrix}$$

$$+ \begin{bmatrix} \dfrac{Q_1}{K_{11}}\left\{1 - e^{-\alpha t}\left(\cos\beta t + \dfrac{\alpha}{\beta}\sin\beta t\right)\right\} \\[2ex] \dfrac{Q_1}{M_{11}} \cdot \dfrac{e^{-\alpha t}\sin\beta t}{\beta} \end{bmatrix}$$

Laplace Transform Solution. Another approach to the solution of state equation (8.29) is to use Laplace transform methods. The merit of the Laplace transform is that it can be used to transform linear differential equations with constant coefficients and also transcendental functions into algebraic functions. Laplace transform methods are applicable to matrix as well as to scalar equations.

Taking transforms of (8.29), we have

$$\mathscr{L}(D\mathbf{x}) = \mathscr{L}(B\mathbf{x} + \mathbf{b})$$

which gives

$$s\bar{\mathbf{x}} - \mathbf{x}_0 = B\bar{\mathbf{x}} + \bar{\mathbf{b}}$$

Solving for $\bar{\mathbf{x}}$, we obtain

$$\bar{\mathbf{x}} = (sI - B)^{-1}(\mathbf{x}_0 + \bar{\mathbf{b}}) \tag{8.45}$$

To solve (8.45) for $\mathbf{x}(t)$, we may use partial fraction decomposition as follows:
Write

$$\bar{\mathbf{x}} = (sI - B)^{-1}(\mathbf{x}_0 + \bar{\mathbf{b}}) = \frac{\text{adj } [sI - B](\mathbf{x}_0 + \bar{\mathbf{b}})}{\det [sI - B]} = \frac{\mathbf{y}(s)}{\det [sI - B]} \tag{8.46}$$

and let $s_1, s_2, \ldots, s_r$ be the roots of

$$\det [sI - B] = f(s) = 0$$

where $f(s)$ is the characteristic polynomial of B. Then

$$f(s) = (s - s_1)^{m_1}(s - s_2)^{m_2} \cdots (s - s_r)^{m_r} \tag{8.47}$$

where $m_1, m_2, \ldots, m_r$ are the multiplicities of the roots, and $m_1 + m_2 + \cdots + m_r = n$. Expanding (8.46) in partial fractions gives

$$\bar{\mathbf{x}} = \frac{\mathbf{y}(s)}{f(s)} = \frac{\mathbf{A}_1}{(s - s_1)^{m_1}} + \frac{\mathbf{A}_2}{(s - s_1)^{m_1-1}} + \cdots + \frac{\mathbf{A}_{m_1}}{s - s_1}$$

$$+ \frac{\mathbf{B}_1}{(s - s_2)^{m_2}} + \frac{\mathbf{B}_2}{(s - s_2)^{m_2-1}} + \cdots + \frac{\mathbf{B}_{m_2}}{s - s_2} \tag{8.48}$$

$$+ \cdots \cdots \cdots \cdots \cdots$$

where, with $j = 1, 2, \ldots, m_1$,

$$\mathbf{A}_j = \frac{1}{(j-1)!} \frac{d^{j-1}}{ds^{j-1}} \left[(s - s_1)^{m_1} \frac{\mathbf{y}(s)}{f(s)} \right]_{s=s_1} \tag{8.49}$$

Similar expressions hold for $\mathbf{B}_j$, $\mathbf{C}_j$, and so forth.

In the case of simple roots

$$\mathbf{A} = \frac{\mathbf{y}(s)}{f'(s)} \bigg]_{s=s_1} \tag{8.50}$$

Complex roots are difficult to handle by the preceding formulas; the following formulas are more satisfactory. Let $\alpha \pm \beta i$ be a complex pair of simple roots of $f(s) = 0$, and also let

$$\mathbf{B} - \mathbf{C}i = \frac{\mathbf{y}(s)}{f'(s)} \bigg]_{s=\alpha+\beta i} \tag{8.51}$$

Then the partial fraction derived from the complex pair of roots $\alpha \pm \beta i$ is

$$\frac{2(s - \alpha)\mathbf{B} + 2\beta\mathbf{C}}{(s - \alpha)^2 + \beta^2} \tag{8.52}$$

which has the inverse transform

$$2e^{\alpha t}(\mathbf{B} \cos \beta t + \mathbf{C} \sin \beta t) \tag{8.53}$$

Example 2.

Solve the state equation

$$D\begin{bmatrix} x_1 \\ x_2 \\ x_3 \end{bmatrix} = \begin{bmatrix} 0 & 1 & 0 \\ 0 & 0 & 1 \\ 1 & 0 & 0 \end{bmatrix}\begin{bmatrix} x_1 \\ x_2 \\ x_3 \end{bmatrix} \quad \mathbf{x_0} \text{ arbitrary}$$

by the Laplace transform method.

SOLUTION

The Laplace transform equation is

$$\overline{\mathbf{x}} = (sI - B)^{-1}\mathbf{x_0} \tag{1}$$

where

$$(sI - B) = \begin{bmatrix} s & -1 & 0 \\ 0 & s & -1 \\ -1 & 0 & s \end{bmatrix} \tag{2}$$

The inverse matrix is

$$(sI - B)^{-1} = \frac{1}{s^3 - 1}\begin{bmatrix} s^2 & s & 1 \\ 1 & s^2 & s \\ s & 1 & s^2 \end{bmatrix} = \frac{\text{adj}\,[sI - B]}{f(s)}$$

Hence

$$\mathbf{y}(s) = \text{adj}\,[sI - B]\mathbf{x_0}$$

$$= \begin{bmatrix} s^2 & s & 1 \\ 1 & s^2 & s \\ s & 1 & s^2 \end{bmatrix}\begin{bmatrix} x_{10} \\ x_{20} \\ x_{30} \end{bmatrix} = \begin{bmatrix} s^2x_{10} + sx_{20} + x_{30} \\ x_{10} + s^2x_{20} + sx_{30} \\ sx_{10} + x_{20} + s^2x_{30} \end{bmatrix} \tag{3}$$

The roots of the characteristic equation

$$f(s) = \det\,[sI - B] = s^3 - 1 = (s - 1)(s^2 + s + 1) = 0$$

are found to be

$$s_1 = 1 \qquad s_2, s_3 = -\frac{1}{2} \pm \frac{\sqrt{3}}{2}i = \alpha \pm \beta i \tag{4}$$

The vector **A** corresponding to the simple root $s = 1$ is calculated from (8.50)

$$\mathbf{A} = \frac{\mathbf{y}(s)}{f'(s)}\bigg]_{s=1} = \frac{1}{3}\begin{bmatrix} x_{10} + x_{20} + x_{30} \\ x_{10} + x_{20} + x_{30} \\ x_{10} + x_{20} + x_{30} \end{bmatrix} \tag{5}$$

The vectors **B** and **C** are found from Eq. (8.51),

$$\mathbf{B} - \mathbf{C}i = \frac{\mathbf{y}(s)}{f'(s)}\Bigg]_{s=s_2} = \frac{1}{3s_2^2}\begin{bmatrix} s_2^2 x_{10} + s_2 x_{20} + x_{30} \\ x_{10} + s_2^2 x_{20} + s_2 x_{30} \\ s_2 x_{10} + x_{20} + s_2^2 x_{30} \end{bmatrix}$$

Carrying out the indicated calculations gives for **B** and **C** the values

$$\mathbf{B} = \frac{1}{3}\begin{bmatrix} x_{10} - \tfrac{1}{2}x_{20} - \tfrac{1}{2}x_{30} \\ -\tfrac{1}{2}x_{10} + x_{20} - \tfrac{1}{2}x_{30} \\ -\tfrac{1}{2}x_{10} - \tfrac{1}{2}x_{20} - x_{30} \end{bmatrix} \tag{6}$$

$$\mathbf{C} = \frac{1}{3}\begin{bmatrix} 0 & + \frac{\sqrt{3}}{2}x_{20} & - \frac{\sqrt{3}}{2}x_{30} \\ -\frac{\sqrt{3}}{2}x_{10} + & 0 & + \frac{\sqrt{3}}{2}x_{30} \\ \frac{\sqrt{3}}{2}x_{10} + & \frac{\sqrt{3}}{2}x_{20} + & 0 \end{bmatrix} \tag{7}$$

Substituting in (8.48) gives

$$\bar{\mathbf{x}} = \frac{\mathbf{A}}{s - 1} + \frac{2(s + \tfrac{1}{2})\mathbf{B} + \sqrt{3}\mathbf{C}}{(s + \tfrac{1}{2})^2 + \tfrac{3}{4}} \tag{8}$$

The solution may now be written by inspection

$$\mathbf{x}(t) = \mathbf{A}e^t + 2e^{-t/2}\left(\mathbf{B}\cos\frac{\sqrt{3}}{2}t + \mathbf{C}\sin\frac{\sqrt{3}}{2}t\right) \tag{9}$$

where **A**, **B**, **C** have the values given by (5), (6), and (7) above.

8.3 APPLICATION TO ELECTRIC NETWORKS

It is customary to use the term *electric network* as a synonym for an electric system which can be represented pictorially by means of a circuit diagram. Electric networks are a generalization of the circuit concepts originated by Kirchhoff, Maxwell, Rayleigh, and others.

General Properties of Networks

I. Network Definitions. The following definitions are used in network theory:

(1) Passive elements: resistors, inductances, capacitors;
(2) Source elements: energy or emf sources;

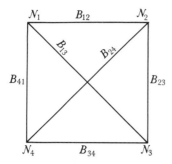

FIGURE 8.1 Topological diagram for an electric network.

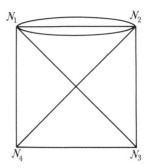

FIGURE 8.2 Multiply connected network.

(3) Branch: part of a network, containing one or more elements (at least one of which must be passive) in series, and extending from one node to another;

(4) Node: a·network junction where two or more branches are connected;

(5) Network: a system formed by interconnection of one or more branches to form closed paths for current flow;

(6) Loop or mesh: a single closed path for current flow in a network.

2. Network Structures. The basic structure of a network may be portrayed graphically in schematic form by a topological diagram such as illustrated in Figure 8.1. Branches are denoted by the symbol B_{rs} and nodes by N_r. In the figure there are four nodes interconnected by six branches. Since all nodes are interconnected, this may be called a fully-connected network. A second example of a fully-connected network is shown in Figure 8.2 wherein several branches are connected in parallel between two of the nodes. Branches connected in parallel may be called multiply-connected, in contrast to simply-connected.

The more detailed structure of a network is usually exhibited by means of a schematic circuit diagram in which the network elements are depicted symbolically. Symbols in common use are shown in Figure 8.3, drawn for a single branch B_r of a network. Elements and symbols are identified as follows:

$$R_r = \text{resistance of branch } r$$
$$L_r = \text{inductance of branch } r$$
$$C_r = \text{capacitance of branch } r$$
$$\varepsilon_r = \text{emf source in branch } r$$
$$i_r = \text{current in branch } r$$
$$V_j = \text{potential or voltage of node } j$$
$$I_k = \text{current in mesh } k$$

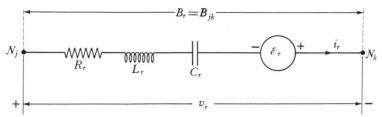

FIGURE 8.3 Branch structure for an electric network.

Our concern here is generally with networks containing a finite number of branches, but it should be mentioned that certain field problems may be usefully represented by networks with an infinite number of branches. A typical example would be a lossless transmission line, as shown in Figure 8.4.

3. Kirchhoff's Laws. These laws provide the basis for the analysis of steady-current networks and are stated here for reference purposes:

(1) At any point of a network system the algebraic sum of all the currents is zero.

(2) In any complete circuit formed by the conductors, the sum of the emf's taken around the circuit is equal to the sum of the products of the current in each conductor multiplied by the resistance of that conductor.

We observe that law (1) is contained in the concept of generalized coordinates, and that law (2) is implicit in Lagrange's equations when applied to electric circuits. Hence, in the context of generalized mechanics, it is unnecessary to state Kirchhoff's laws as separate principles. We therefore regard Kirchhoff's laws as consequences rather than as independent statements.

In electric network theory, Kirchhoff's laws and Ohm's law are sufficient to provide the solution for steady currents in any system of conductors, using only algebraic methods to solve for the unknown currents or emf's. Extension of Kirchhoff's laws to systems with time-varying currents requires justification based on Maxwell's equations. On this basis the first law is correct if the total current (displacement plus conduction currents) is used. The second law is correct if the emf's are

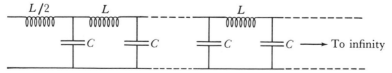

FIGURE 8.4 Infinite network to represent a lossless transmission line.

interpreted by Eq. (8.54), obtained by integrating the field equations about a closed path coincident with the conductors.

$$\int_s \boldsymbol{\epsilon} \cdot d\mathbf{s} - \int_s \frac{\mathbf{J}}{\sigma} \cdot d\mathbf{s} - \int_s \frac{\partial \mathbf{A}}{\partial t} \cdot d\mathbf{s} - \int_s (\boldsymbol{\nabla}\varphi) \cdot d\mathbf{s} = 0 \qquad \textbf{(8.54)}$$

| emf sources | Ohmic voltage drop | Magnetic effects voltage drop | Electrostatic effects voltage drop |

For slowly-varying currents, where the fields may be considered localized, Eq. (8.54) reduces to the form

$$\varepsilon(t) - RI - L\frac{dI}{dt} - \frac{1}{C}\int I\, dt = 0 \qquad \textbf{(8.55)}$$

where $\varepsilon(t)$ is the emf in the circuit, $I(t)$ is the current in the mesh, R is the total resistance, L the total self inductance, and C the capacitance. The network-theory approximations of localized fields and negligible time retardation effects permit the use of inductance and capacitance coefficients as defined for the magnetostatic and electrostatic cases. Note that this result has already been arrived at in Chap. 7 in Sects. 2 and 3 pertaining to quasi-static or slowly-varying currents. The abstract dynamical theory, therefore, gives correct results for the quasi-static approximation assumed in electric network theory.

4. Number of Independent Loops. The number of independent loops for any network may be found from a fundamental theorem of topology. The theorem relates the number of independent loops ℓ to the number of nodes n and the number of branches b, as given in Eq. (8.56).

$$\ell = b - n + 1 \qquad \textbf{(8.56)}$$

Proof of the theorem may be made by induction, starting with the special case of two nodes and two branches, Figure 8.5. Obviously, there is only one loop, L_1. Hence, $\ell = b - n + 1 = 1$. Suppose now that another node N_3 is added in either of the two branches. The number of nodes and branches both increase to three, but ℓ remains unchanged at one, in accord with (8.56). Next, assume that another branch is added,

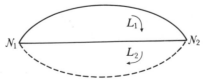

FIGURE 8.5   Diagram for loop and node theorem.

as shown by the dashed lines in Figure (8.5). This adds a second loop L_2, so that again

$$\ell = b - n + 1 = 3 - 2 + 1 = 2$$

The theorem follows by extension of the network to b branches and n nodes.

5. Number of Nodes and Node Pairs. Since any one of the n nodes in a network may be chosen as a reference node, the number of independent nodes or voltages is $n - 1$. From this it follows that the total number of node pairs is $n(n - 1)$, of which only one half are independent. Hence, the total number of independent node pairs is $\frac{1}{2}n(n - 1)$.

6. Types of Networks. Networks may be classified as follows:

(1) Linear constant parameter circuits, or those which may be represented by linear differential equations with constant coefficients;

(2) Linear time-varying circuits, or those which may be represented by linear differential equations with variable coefficients;

(3) Nonlinear circuits, or those which require nonlinear differential equations for their representation;

(4) Passive circuits, that is, circuits which contain no external energy sources;

(5) Active circuits, which do contain external energy sources;

(6) Direct-current (d-c) circuits. This term is equivalent to constant-current or stationary-current circuits;

(7) Alternating-current (a-c) circuits, which are activated by periodic emf sources;

(8) Transient circuits; voltages and currents change with time in a nonrepeating manner;

(9) Electronic circuits; these contain active elements which are electronic devices, or nonlinear magnetic or dielectric devices.

Identification of Electric Systems Quantities. The abstract quantities for linear electric networks or systems are, from the electromagnetic equations of Chap. 7, identified as shown in Table 8.3. Application of

Table 8.3 Electric Systems Quantities

Abstract Element	Electric Quantity Symbol	Name
M	L	Inductance matrix
R	R	Resistance matrix
K	C^{-1}	Inverse of capacitance matrix
$\mathbf{q}$	$\mathbf{q}$	Electric-charge vector
$\dot{\mathbf{q}}$	$\mathbf{I}$	Electric-current vector
Q	ϵ	Electromotive force (emf) vector
P	P	Electric power scalar

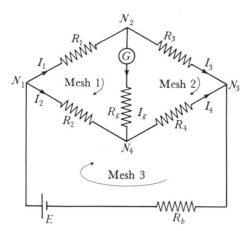

FIGURE 8.6 Wheatstone bridge.

the general theory of linear systems developed in Sects. 8.1 and 8.2, together with the identifications of Table 8.3, provides not only solutions but also general properties for a wide class of electric systems.

Illustrative Examples. Application of generalized linear system theory to a broad cross section of electrical problems is displayed in Examples 1 through 11. These will serve not only to indicate the methods of application, but also to suggest further applications.

Example 1. Wheatstone Bridge. As a first example, we treat the Wheatstone bridge, which, although familiar to students in introductory physics and electrical courses, is not trivial. Figure 8.6 depicts the Wheatstone bridge in the graphical symbolic form usually designated as the schematic circuit diagram. The resistors R_1, R_2, R_3, R_4 form the arms of the bridge; the galvanometer G with internal resistance R_g indicates balance; the battery E with internal resistance R_b provides the power to operate the bridge. An unknown resistor, R_3, can be measured in terms of a standard resistor, R_1, by adjustment of the ratio R_2/R_4.

To find the condition for balance and the effects of unbalance, it is necessary to determine the galvanometer current I_g. The steps of the solution are as follows:

(1) Find the degrees of freedom (number of independent meshes) by use of Eq. (8.56)

$$\ell = b - n + 1 = 6 - 4 + 1 = 3 \text{ independent meshes}$$

(2) Write the expression for the current vector **I** from Eq. (8.14), which applies since this is a resistive system. Thus, we have

$$\dot{\mathbf{q}} = \mathbf{I} = R^{-1}\boldsymbol{\epsilon}$$

In this expression,

$$R = \begin{bmatrix} R_{11} & R_{12} & R_{13} \\ R_{21} & R_{22} & R_{23} \\ R_{31} & R_{32} & R_{33} \end{bmatrix}$$

$$\boldsymbol{\epsilon} = \begin{bmatrix} 0 \\ 0 \\ E \end{bmatrix} \quad \text{and} \quad \mathbf{I} = \begin{bmatrix} I_1 \\ I_2 \\ I_3 \end{bmatrix}$$

(3) Find the elements R_{rs} by inspection of the circuit diagram and calculate **I**.

$R_{11} = $ total resistance in mesh 1 $= R_1 + R_g + R_2$

$R_{12} = $ (resistance common to meshes 1 and 2)* $= -R_g$

and so on for the remaining elements. Hence, we find

$$R = \begin{bmatrix} R_1 + R_g + R_2 & -R_g & -R_2 \\ -R_g & R_3 + R_4 + R_g & -R_4 \\ -R_2 & -R_4 & R_2 + R_4 + R_b \end{bmatrix}$$

Inverting the matrix R by the method of determinants yields

$$R^{-1} = \frac{\Delta}{|R|}$$

where Δ is the matrix of the cofactors of $|R|$ (which is also the adjoint matrix since R is symmetrical). Then

$$\mathbf{I} = \frac{1}{|R|} \begin{bmatrix} \Delta_{11} & \Delta_{12} & \Delta_{13} \\ \Delta_{21} & \Delta_{22} & \Delta_{23} \\ \Delta_{31} & \Delta_{32} & \Delta_{33} \end{bmatrix} \begin{bmatrix} 0 \\ 0 \\ E \end{bmatrix} = E \begin{bmatrix} \Delta_{13} \\ \Delta_{23} \\ \Delta_{33} \end{bmatrix}$$

(4) Find the galvanometer current I_g.

$$I_g = I_1 - I_2 = \frac{E(\Delta_{13} - \Delta_{23})}{|R|}$$

where

$$\Delta_{13} = (-1)^{1+3} \begin{vmatrix} -R_g, & R_3 + R_4 + R_g \\ -R_2, & -R_g \end{vmatrix}$$

and

$$\Delta_{23} = (-1)^{2+3} \begin{vmatrix} R_{11} + R_g + R_2, & -R_g \\ -R_2, & -R_4 \end{vmatrix}$$

---

* The sign is negative if currents I_1 and I_2 flow through the common resistance in opposite directions, otherwise the sign is positive.

so that

$$I_g = \frac{E(\Delta_{13} - \Delta_{23})}{|R|} = \frac{E(R_2 R_3 - R_1 R_4)}{|R|}$$

The condition for balance is

$$R_2 R_3 - R_1 R_4 = 0$$

the well-known result for balance of a Wheatstone bridge. Sensitivity, i.e., the change ΔI_g for a change ΔR_3 in the resistor to be measured, is

$$\Delta I_g \cong \frac{E R_2\, \Delta R_3}{|R|}$$

since the effect of ΔR_3 on $|R|$ is small.

Example 2.   Resistance Network, Three Meshes. In order to show the solution procedure for a numerical problem, a resistance network with three independent meshes and two emf sources, Figure 8.7, is treated next.

SOLUTION

(1) Formulate the resistance matrix R by the same methods as used in Example 1. Thus

$$R = \begin{bmatrix} 10 + 20 & -20 & 0 \\ -20 & 20 + 15 + 5 & -5 \\ 0 & -5 & 5 + 25 \end{bmatrix} = \begin{bmatrix} 30 & -20 & 0 \\ -20 & 40 & -5 \\ 0 & -5 & 30 \end{bmatrix}$$

(2) Find the inverse matrix R^{-1}. This may be done by the method of determinants, by any one of several numerical procedures such as operations on the rows of R, or by means of a computer program. The result as calculated by computer is

$$R^{-1} = \begin{bmatrix} 0.05054 & 0.02581 & 0.00430 \\ 0.02581 & 0.03871 & 0.00645 \\ 0.00430 & 0.00645 & 0.03441 \end{bmatrix}$$

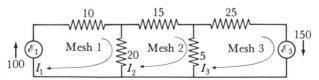

FIGURE 8.7   Three-mesh resistance network.

(3) Calculate the current vector **I**.

$$\mathbf{I} = R^{-1}\boldsymbol{\epsilon} = R^{-1}\begin{bmatrix} 100 \\ 0 \\ 150 \end{bmatrix} = \begin{bmatrix} 5.70 \\ 3.55 \\ 5.59 \end{bmatrix} \text{amperes}$$

All the essential steps necessary to solve resistance networks are illustrated in the two examples just treated. More complex networks of this kind require more arithmetic, but no new principles. Digital computers are necessary to carry out the massive arithmetical computations needed to solve even moderate size problems. The general procedure for a computer-solved network problem is illustrated in Ex. 3.

Example 3. Computer Solution for an Eight-Mesh Resistance Network. It is desired to find the currents and the node voltages for the resistance network shown in Figure 8.8.

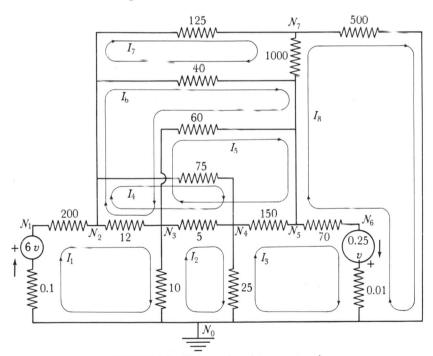

FIGURE 8.8 Eight-mesh resistance network.

SOLUTION

(1) Find the number of independent meshes

$$\ell = b - n + 1 = 15 - 8 + 1 = 8 \text{ independent meshes.}$$

(2) Formulate the resistance matrix

$$R = \begin{bmatrix} 222.1 & -10 & 0 & -12 & 0 & -12 & 0 & 0 \\ -10 & 40 & -25 & -5 & -5 & 0 & 0 & 0 \\ 0 & -25 & 245.01 & 0 & -150 & 0 & 0 & -70 \\ -12 & -5 & 0 & 92 & 5 & 12 & 0 & 0 \\ 0 & -5 & -150 & 5 & 215 & -60 & 0 & 0 \\ -12 & 0 & 0 & 12 & -60 & 112 & -40 & 0 \\ 0 & 0 & 0 & 0 & 0 & -40 & 1165 & -1000 \\ 0 & 0 & -70 & 0 & 0 & 0 & -1000 & 1570.01 \end{bmatrix}$$

(3) Compute the current vector

$$\mathbf{I} = R^{-1}\boldsymbol{\epsilon} = R^{-1} \begin{bmatrix} 6 \\ 0 \\ 0.25 \\ 0 \\ 0 \\ 0 \\ 0 \\ -0.25 \end{bmatrix} = \begin{bmatrix} 0.02808 \\ -0.01220 \\ 0.01805 \\ 0.00314 \\ -0.00482 \\ 0.00488 \\ 0.00072 \\ 0.00058 \end{bmatrix}$$

(4) Determine the node voltages N_0, N_1, etc. as follows:

$$\begin{aligned} N_0 &= 0 & &= 0 \\ N_1 &= 6 - 0.1 I_1 & &= 5.9972 \\ N_2 &= N_1 - 200 I_1 & &= 0.38150 \\ N_3 &= 10(I_1 - I_2) & &= 0.15879 \\ N_4 &= 25(I_2 - I_3) & &= 0.14569 \\ N_5 &= N_4 + 150(I_3 - I_5) & &= 0.15524 \\ N_6 &= -0.25 + 0.01 I_8 & &= -0.24994 \\ N_7 &= 500 I_8 & &= 0.29156 \end{aligned}$$

Example 4.   Minimum Heat Generation.   Show that the principle of minimum heat generation applied to the network of Figure 8.9 leads to results which are in agreement with Kirchhoff's laws.

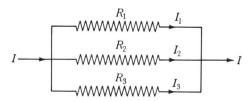

FIGURE 8.9   Resistance network to illustrate minimum heat generation principle.

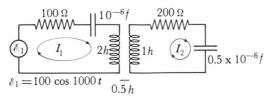

FIGURE 8.10 Inductively coupled *a-c* circuits.

SOLUTION

The heat loss in the resistors per unit time is

$$H = I_1^2 R_1 + I_2^2 R_2 + I_3^2 R_3 \quad \text{where} \quad I_1 + I_2 + I_3 = I = \text{constant.}$$

Use the Lagrange multiplier method to find the minimum of H. Let

$$g = H + \lambda(I_1 + I_2 + I_3 - I)$$
$$\nabla g = [2I_1 R_1 + \lambda, \, 2I_2 R_2 + \lambda, \, 2I_3 R_3 + \lambda] = 0$$

Solving for I_1, I_2, I_3 gives

$$I_1 = \left[\frac{R_2 R_3}{R_1 R_2 + R_2 R_3 + R_3 R_1}\right] I$$

$$I_2 = \left[\frac{R_3 R_1}{R_1 R_2 + R_2 R_3 + R_3 R_1}\right] I$$

$$I_3 = \left[\frac{R_1 R_2}{R_1 R_2 + R_2 R_3 + R_3 R_1}\right] I$$

a result which agrees with Kirchhoff's laws.

Example 5. **Alternating-Current Network.** This problem illustrates the solution for a periodically excited system. The mesh currents and the voltage across the capacitor in the secondary circuit, Figure 8.10, are to be determined.

SOLUTION

(1) Write the impedance matrix Z from Eq. (8.19).

$$Z = \begin{bmatrix} R_1 + i\left(L_1\omega - \dfrac{1}{C_1\omega}\right) & -iL_{12}\omega \\ -iL_{21}\omega & R_2 + i\left(L_2\omega - \dfrac{1}{C_2\omega}\right) \end{bmatrix}$$

$$= \begin{bmatrix} 100 + 1000i & -500i \\ -500i & 200 - 1000i \end{bmatrix}$$

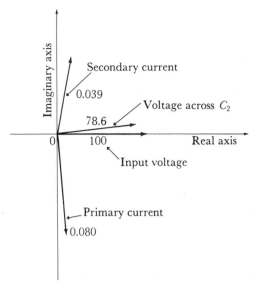

FIGURE 8.11 Argand diagram for currents and voltages of Example 5, Sect. 8.3.

(2) Calculate the inverse matrix Z^{-1}.

$$Z^{-1} = \frac{\begin{bmatrix} Z_{22} & -Z_{21} \\ -Z_{12} & Z_{11} \end{bmatrix}}{|Z|} = \frac{\begin{bmatrix} 200 - 1000i & 500i \\ 500i & 100 + 1000i \end{bmatrix}}{(12.7 + i) \times 10^5}$$

(3) Compute the current vector $\mathbf{I}_p$ from Eq. (8.18), where $\mathbf{I}_p$ represents the peak currents.

$$\mathbf{I}_p = Z^{-1} \begin{bmatrix} 100 \\ 0 \end{bmatrix} = \begin{bmatrix} 0.0095 - 0.0796i \\ 0.0031 + 0.0392i \end{bmatrix} = \begin{bmatrix} 0.080e^{-1.45i} \\ 0.039e^{1.49i} \end{bmatrix}$$

(4) Determine the voltage V_2 across C_2

$$V_2 = \frac{q_2}{C_2} = \frac{I_{p2}}{iC_2\omega} = \frac{0.039e^{1.49i}}{0.5 \times 10^{-3}i}$$

$$= 78.2e^{-0.08i}$$

(5) The phase and the magnitude relations for the several voltages and currents are more clearly shown by plotting on the complex plane (Argand diagram), Figure 8.11.

The preceding examples in this section treat steady-state d-c and a-c networks. It is apparent that the solutions of such problems involve no more than routine procedures which are readily programmed for computers, or for desk calculations in the case of modest size problems.

Solution of networks with transient excitation, although formally not complicated, does require, in general, more mathematical manipulation. The range of problems encountered is also vastly greater because of the many kinds of excitation functions of practical interest, and because of the many initial conditions possible.

Example 6.   Rectangular Voltage Pulse Excitation.   To illustrate the method of solution for a system of this type, consider an L, R series circuit excited by a rectangular pulse of amplitude E and duration $-\tau/2$ to $\tau/2$, as shown in Figure 8.12. It is required to find the current response.

SOLUTION

Using Eq. (8.26), we can write

$$\dot{q} = \mathcal{L}^{-1}[(sA^{-1})\bar{\varepsilon}]$$

where

$$A = Ls^2 + Rs \qquad A^{-1} = \frac{1}{Ls^2 + Rs}$$

and

$$\bar{\varepsilon} = \frac{E}{s}\left[e^{s\tau/2} - e^{-s\tau/2}\right]$$

Since $\bar{\dot{q}} = I = sq$, the expression for $\bar{I}$ is

$$\bar{I} = s\bar{q} = \frac{\bar{\varepsilon}}{Ls + R} = \frac{E[e^{s\tau/2} - e^{-s\tau/2}]}{L\left(s + \dfrac{R}{L}\right)}$$

$$= \frac{E}{R}\left[e^{s\tau/2} - e^{-s\tau/2}\right]\left[\frac{1}{s} - \frac{1}{s + \dfrac{R}{L}}\right]$$

Taking the inverse transform gives the solution*

$$I(t) = \frac{E}{L}\left\{u\left(t + \frac{\tau}{2}\right)[1 - e^{-R(t+\tau/2)/L}] - u\left(t - \frac{\tau}{2}\right)[1 - e^{-R(t-\tau/2)/L}]\right\}$$

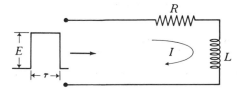

FIGURE 8.12   *L-R* circuit excited by rectangular voltage pulse.

* We use the notation $u(t)$ to denote the Heaviside unit function.

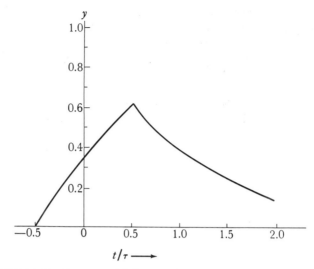

FIGURE 8.13   Response of *L-R* network to a rectangular voltage pulse.

To see the characteristic shape of the current response, it is convenient to introduce the dimensionless variables IR/E and t/τ. Then, we obtain

$$\frac{IR}{E} = y = \left\{ u\!\left(\frac{t}{\tau} + \frac{1}{2}\right)[1 - e^{-R\tau(t/\tau-1/2)/L}] - u\!\left(\frac{t}{\tau} - \frac{1}{2}\right)[1 - e^{-R\tau(t/\tau-1/2)/L}]\right\}$$

which is plotted in Figure 8.13.

Example 7. Alternating-Current Transient.   Alternating emf E sin ωt is applied at $t = 0$ to L, R in series, Figure 8.14. Calculate $I(t)$ for $I(0) = 0$.

Solution

$$\bar{I} = A^{-1}s\bar{\varepsilon}$$

$$A = Ls + R \qquad \bar{\varepsilon} = \frac{E\omega}{s^2 + \omega^2}$$

Hence

$$\bar{I} = \frac{sE\omega}{L\!\left(s + \dfrac{R}{L}\right)(s^2 + \omega^2)}$$

Taking inverse transform, we obtain

$$I(t) = \frac{E}{R^2 + L^2\omega^2}\,(L\omega e^{-(R/L)t} - L\omega \cos \omega t + R \sin \omega t)$$

$$= \frac{E}{\sqrt{R^2 + L^2\omega^2}}\,[e^{-(R/L)t} \sin \theta + \sin (\omega t - \theta)]$$

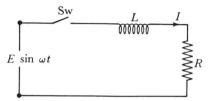

FIGURE 8.14   Circuit diagram of *L-R* network with *a-c* transient excitation.

where

$$\tan \theta = \frac{L\omega}{R}$$

Note that as $t \to \infty$, the solution becomes the steady-state a-c case

$$I = \frac{E}{\sqrt{R^2 + L^2\omega^2}} \sin(\omega t - \theta)$$

Example 8.   Charging Transient for Two-Mesh R-C Circuit.   Find the mesh charges q_1 and q_2 as functions of time for the network of Figure 8.15 when a step emf E is applied to the network at $t = 0$. Assume zero initial charges on the capacitors.

SOLUTION

From Eq. (8.22) we have

$$\mathbf{q} = \mathscr{L}^{-1}[A^{-1}\overline{\mathbf{Q}}]$$

where

$$A = \begin{bmatrix} Rs + \dfrac{1}{C}, & -\dfrac{1}{C} \\[2ex] -\dfrac{1}{C}, & Rs + \dfrac{2}{C} \end{bmatrix}$$

and

$$A^{-1}\overline{\mathbf{Q}} = \frac{1}{|A|} \begin{bmatrix} Rs + \dfrac{2}{C}, & \dfrac{1}{C} \\[2ex] \dfrac{1}{C}, & Rs + \dfrac{1}{C} \end{bmatrix} \begin{bmatrix} \dfrac{E}{s} \\[2ex] 0 \end{bmatrix}$$

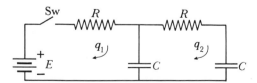

FIGURE 8.15   Two-mesh *R-C* circuit.

The inverse transforms may be found by the method of partial fractions, or more conveniently from a table of transforms, giving

$$
q = \begin{bmatrix} \dfrac{CE}{\sqrt{5}}(e^{-s_1 t} - e^{-s_2 t}) + 2CE\left(1 - \dfrac{s_2 e^{-s_1 t} - s_1 e^{-s_2 t}}{s_2 - s_1}\right) \\[2em] CE\left(1 - \dfrac{s_2 e^{-s_1 t} - s_1 e^{-s_2 t}}{s_2 - s_1}\right) \end{bmatrix}
$$

where

$$
s_1 = \frac{3 - \sqrt{5}}{2RC} \qquad s_2 = \frac{3 + \sqrt{5}}{2RC} \quad \text{are the roots of } |A| = 0
$$

As a partial check of the validity of these results, we note that

$$
q_1(0) = q_2(0) = 0
$$

and

$$
\lim_{t \to \infty} q_1(t) = 2CE \qquad \lim_{t \to \infty} q_2(t) = CE
$$

Example 9.   Electric Wave Filters.   A sequence of coupled circuits in tandem as shown in Figure 8.16 has interesting and useful transmission characteristics for electric signals and is usually referred to as an electric wave filter. An input signal or emf $\varepsilon(t)$ is supplied to the input terminals and, after transmission through the filter, appears in modified form across the output or load impedance Z_L. It is desired to study the transmission characteristics of such an iterative filter structure in terms of the filter impedances Z_1 and Z_2.

SOLUTION

(1) Write the matrix equation

$$
\mathbf{ZI} = \boldsymbol{\epsilon} \tag{1}
$$

where the system matrix Z is given by

$$
Z = \begin{bmatrix}
\frac{1}{2}Z_1 + Z_2, & -Z_2 & 0 & 0 & \cdots & 0 \\
-Z_2, & Z_1 + 2Z_2, & -Z_2 & 0 & \cdots & 0 \\
0 & -Z_2, & Z_1 + 2Z_2, & -Z_2 & \cdots & 0 \\
\cdot & \cdot & \cdot & \cdot & \cdot & \cdot \\
\cdot & \cdot & \cdot & \cdot & \cdot & \cdot \\
\cdot & \cdot & \cdot & \cdot & \cdot & \cdot \\
0 & \cdots, & -Z_2, & Z_1 + 2Z, & -Z_2 \\
0 & \cdots, & 0, & -Z_2, & \frac{1}{2}Z_1 + Z_2 + Z_L
\end{bmatrix} \tag{2}
$$

(2) Note that except for the input and output meshes the mesh equations are of the form

$$
-Z_2 I_{r-1} + (Z_1 + 2Z_2)I_r - Z_2 I_{r+1} = 0 \tag{3}
$$

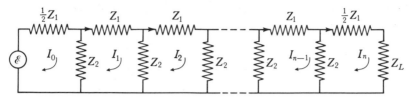

FIGURE 8.16   Chain of coupled electric circuits (*T*-section filter).

where r denotes the rth mesh.   This is seen to be a linear difference equation with constant coefficients in the integral variable r.   By the theory of difference equations (or by analogy with the corresponding linear differential equation), we seek a solution of the type

$$I_r = k e^{r\theta} \tag{4}$$

where k is a constant and θ is a parameter to be determined.

Substitution of (4) in (3) and factoring leads to the condition equation

$$Z_2 k e^{(r-1)\theta}\left[-1 + 2\left(1 + \frac{Z_1}{2Z_2}\right)e^\theta - e^{2\theta}\right] = 0 \tag{5}$$

which has the two roots

$$e^{\theta_1}, e^{\theta_2} = 1 + \frac{Z_1}{2Z_2} \pm \sqrt{\left(1 + \frac{Z_1}{2Z_2}\right)^2 - 1}$$

Hence, the general solution of (3) is

$$I_r = k_1 e^{r\theta_1} + k_2 e^{r\theta_2}$$

This expression may be simplified by observing that

$$(e^{\theta_1})(e^{\theta_2}) = 1 \quad \text{or} \quad e^{\theta_2} = e^{-\theta_1}$$

We may now drop the subscript as no longer necessary and write

$$I_r = k_1 e^{r\theta} + k_2 e^{-r\theta} \tag{6}$$

where

$$\theta = \ln\left[1 + \frac{Z_1}{2Z_2} + \sqrt{\left(1 + \frac{Z_1}{2Z_2}\right)^2 - 1}\right]$$

$$= \cosh^{-1}\left(1 + \frac{Z_1}{2Z_2}\right) \tag{7}$$

Substitution of (6) in (1) gives the relations

$$-k_1 \sinh\theta + k_2 \sinh\theta = \frac{\varepsilon}{Z_2}$$

$$k_1 e^{n\theta}\left(\frac{Z_L}{Z_2} + \sinh\theta\right) + k_2 e^{-n\theta}\left(\frac{Z_L}{Z_2} - \sinh\theta\right) = 0$$

Solving for k_1 and k_2 and substituting in (6) leads to the general solution for I_r in terms ε, Z_L/Z_2 and n

$$I_r = \frac{\varepsilon}{Z_2} \cdot \frac{\sinh \theta \cosh (n - r)\theta + \dfrac{Z_L}{Z_2} \sinh (n - r)\theta}{\sinh \theta \left[\sinh n\theta \sinh \theta + \dfrac{Z_L}{Z_2} \cosh n\theta\right]} \tag{8}$$

For the output section, $r = n$, and (8) reduces to

$$I_n = \frac{\dfrac{\varepsilon}{Z_2}}{\sinh n\theta \sinh \theta + \dfrac{Z_L}{Z_2} \cosh n\theta} \tag{9}$$

The ratio of output to input voltage is

$$\frac{\varepsilon_n}{\varepsilon} = \frac{Z_L I_n}{\varepsilon} = \frac{1}{\cosh n\theta + \dfrac{Z_2}{Z_L} \sinh \theta \sinh n\theta} \tag{10}$$

The theory of electric wave filters will be demonstrated by application to a resistance attenuator, and to band pass filters. These are only two of the many useful applications of this theory.

RESISTANCE ATTENUATOR. The filter reduces to a resistance attenuator if all the impedances are chosen as pure resistors. Using Formula (7) the parameter θ is given by

$$\theta = \cosh^{-1}\left(1 + \frac{R_1}{2R_2}\right)$$

and therefore is real and positive. The voltage attenuation ratio $\varepsilon_n/\varepsilon$ is plotted in Figure 8.17 for several values of R_1/R_2 and with $R_L = R_2$. It is seen that high attenuation ratios are easily achieved with a moderate number of sections and with moderate resistance ratios R_1/R_2. In practice this means that high attenuations are feasible without resorting to abnormally small resistance values which would be difficult to provide and to maintain with sufficient accuracy.

BAND PASS FILTERS. Frequency-sensitive filters find wide application in communication systems. Such filters can be derived by choosing Z_1 and Z_2 to have pure reactive components. To show this, consider first a pure reactance filter with an emf $E \cos \omega t$ applied at the input. Then, from Eq. (7)

$$\cosh \theta = 1 + \frac{Z_1}{2Z_2}$$

we observe that if

$$-1 < \cosh \theta < 1$$

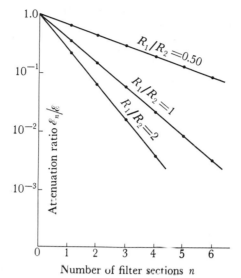

FIGURE 8.17   Resistance attenuator characteristics.

then θ must be pure imaginary and

$$-2 < \frac{Z_1}{2Z_2} < 0$$

If Z_1/Z_2 satisfies this inequality, the hyperbolic functions in Eq. (10) become trigonometric functions and $\varepsilon_n/\varepsilon$ is not attenuated. Thus, all frequencies in the band determined by the inequality are passed by the filter, and all frequencies outside the band are attenuated.

Several types of filters are shown in Table 8.4, together with the frequency ranges or pass bands which are transmitted without attenuation.

There is an apparent time delay as the signal passes through the filter. This may be seen from Eq. (10). As an example, consider a low-pass filter and an angular frequency

$$\omega = \tfrac{1}{2}\omega_0 \qquad \text{where} \quad \omega_0 = \frac{2}{\sqrt{LC}}$$

Then

$$\theta = \cosh^{-1}\left(1 + \frac{Z_1}{2Z_2}\right) = \cosh^{-1}\left(1 - \frac{2\omega^2}{\omega_0^2}\right)$$

$$= \cosh^{-1}\left(\tfrac{1}{2}\right) = 1.047i$$

Assuming $n = 10$ sections, the apparent time delay is of the order of

$$t_d = \frac{10.47}{\omega} = \frac{20.94}{\omega_0}$$

Table 8.4 Band Pass Filters

Filter Type	Filter Section	$\dfrac{Z_1}{4Z_2}$	Pass Band
Low pass		$\dfrac{-LC\omega^2}{4}$	$0 < \omega < \dfrac{2}{\sqrt{LC}}$
High Pass		$\dfrac{-1}{4LC\omega^2}$	$2\sqrt{LC} < \omega < \infty$
Intermediate pass		$\dfrac{1 - L_1C\omega^2}{4L_2C\omega^2}$	$\dfrac{1}{\sqrt{(L_1 - 4L_2)C}} < \omega < \dfrac{1}{\sqrt{L_1C}}$

which for an angular frequency

$$\omega_0 = 2\pi f_0 = 2000\pi (f_0 = 1000 \text{ cps}) \text{ has the value}$$

$$t_d = \frac{20.94}{2000\pi} = 3.34 \times 10^{-3} \text{ sec.}$$

Example 10. Alternating Current Charging for Radar Pulse-Forming Network.[*] High-power modulators for microwave radar transmitters contain pulse-forming networks which supply high-voltage electrical energy to the magnetron or other oscillator. The pulse-forming network must in turn be recharged after each pulse, typically at a frequency of a few hundred pulses per second. One effective and efficient recharging method is known as a-c charging. The equivalent circuit for the charging system is shown in Figure 8.18. In operation, we may regard the switch (which is actually not present in the circuit but is included to simulate operation) as closing at $t = 0$ with initial conditions $q(0) = q_0$, $\dot{q}(0) = I_0$. The problem is to determine (1) the voltage and current at the condenser C and (2) the conditions for periodic charging.

ANALYSIS. The system is linear and is energized by a suddenly applied sinusoidal voltage. Equation (8.4) applies, with $\mathbf{q} = q$ and $\mathbf{Q} = E \cos(\omega t + \varphi)$, i.e.,

$$\left(LD^2 + RD + \frac{1}{C}\right)q = E \cos(\omega t + \varphi) \tag{1}$$

---

[*] From MIT Radiation Laboratory Series, Vol. 5, Pulse Generators, McGraw-Hill, 1948. See Sect. 9.4 by H. J. White on a-c charging networks.

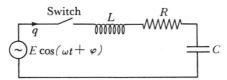

FIGURE 8.18   Schematic circuit for a-c charging of radar pulse network.

The general solution of (1) consists of a transient term q_t plus a steady-state term q_s.

Substitution of

$$q_t = ae^{\lambda t}$$

in the homogeneous equation gives

$$L\lambda^2 + R\lambda + \frac{1}{C} = 0$$

which has the roots

$$\lambda_1, \lambda_2 = -\frac{R}{2L} \pm i\sqrt{\frac{1}{LC} - \frac{R^2}{4L^2}} = -\alpha \pm i\beta$$

where the oscillatory case, $\dfrac{1}{LC} > \dfrac{R^2}{4L^2}$, is assumed.   Hence, the transient solution has the form

$$q_t = a_1 e^{\lambda_1 t} + a_2 e^{\lambda_2 t} \tag{2}$$

where a_1 and a_2 are constants of integration.   Equation (2) may also be written in complex notation as follows:

$$q_t = \text{Re}\,[\bar{a}e^{\lambda_1 t}] \tag{3}$$

where "Re" means "real part of" and usually may be omitted without ambiguity.

Complex notation also expedites finding q_s.   Let

$$E \cos(\omega t + \varphi) = \text{Re}\,[\bar{E}e^{i\omega t}]$$

where

$$\bar{E} = Ee^{i\varphi}$$

Then, since (1) is linear, q_s must be of the form

$$q_s = \bar{b}e^{i\omega t} \tag{4}$$

where $\bar{b}$ is a complex constant determined by the condition

$$\left(-L\omega^2 + iR\omega + \frac{1}{C}\right)\bar{b}e^{i\omega t} = \bar{E}e^{i\omega t}$$

Solving for $\bar{b}$ gives

$$\bar{b} = \frac{C\bar{E}}{1 - LC\omega^2 + iRC\omega} \tag{5}$$

The general solution for q is the sum of (3) and (4), while $\dot{q}$ is found by differentiation

$$q = q_t + q_s = \bar{a}e^{\lambda_1 t} + \bar{b}e^{i\omega t} \Bigg\}$$
$$\dot{q} = \lambda_1 \bar{a}e^{\lambda_1 t} + i\omega \bar{b}e^{i\omega t} \qquad\qquad (6)$$

A necessary condition for stable repeating charging transients is that the magnitude of the current be the same at the beginning and end of each charging period, i.e.,

$$\dot{q}(\tau) \pm \dot{q}(0) = 0$$

which from Eq. (6) may be written as

$$\dot{q}(\tau) \pm \dot{q}(0) = \lambda_1 \bar{a}(e^{\lambda_1 \tau} \pm 1) + i\omega(e^{i\omega t} \pm 1) = 0$$

This condition may be satisfied by choosing

$$\omega\tau = n\pi \qquad \text{where} \quad n = 1, 2, 3, \ldots \qquad (7)$$

and

$$\lambda_1 \bar{a}(e^{\lambda_1 \tau} \pm 1) = 0 \qquad\qquad (8)$$

Equation (8) is transcendental and can be solved numerically. However, in practice it is usually sufficiently accurate to neglect the damping constant and write $\lambda_1 = i\beta$. Then, (8) has the solutions

$$\beta\tau = m\pi \qquad \text{where } m = 1, 2, 3, \ldots \qquad (9)$$

Combining (7) and (9), we find

$$\frac{\omega}{\beta} = \frac{n}{m} \qquad\qquad (10)$$

where n and m are either both even or both odd.

There are many special cases of interest, but resonant charging with a period equal to the period of one cycle of the applied a-c voltage is of greatest interest in practice. For this case

$$\omega\tau = \beta\tau = 2\pi$$

and

$$1 - LC\omega^2 = 0$$

The constant of integration $\bar{a}$ is found from the initial condition

$$q(0) = q_0 = \bar{a} + \bar{b}$$

while $\bar{b}$ is given by (5)

$$\bar{b} = \frac{CEe^{i\varphi}}{iRC\omega} = \frac{E}{R\omega} e^{i[\varphi - (\pi/2)]}$$

Hence, the condenser charge at $t = \tau$ is

$$q(\tau) = q_0 e^{\lambda_1 \tau} + b(e^{i\omega\tau} - e^{-\lambda_1\tau})$$

$$= q_0 e^{-\alpha\tau} \cdot e^{i\omega\tau} + b(e^{i\omega\tau} - e^{-\alpha\tau} \cdot e^{i\omega\tau})$$

$$= \mathrm{Re}\left[q_0 e^{-\alpha\tau} + \frac{E}{R\omega} e^{i(\varphi-\pi/2)}(1 - e^{-\alpha\tau}) \right]$$

$$= q_0 e^{-\alpha\tau} + \frac{E}{R\omega} \sin\varphi\,(1 - e^{-\alpha\tau})$$

Introducing the circuit quality factor

$$Q = \frac{L\omega}{R} = \frac{1}{RC\omega}$$

and noting that

$$\alpha\tau = \frac{R}{2L} \cdot \frac{2\pi}{\omega} = \frac{\pi}{Q}$$

we can write

$$q(\tau) = q_0 e^{-\pi/Q} + QCE(1 - e^{-\pi/Q})\sin\varphi$$

$$= q_0\left(1 - \frac{\pi}{Q}\right) + \pi CE\left(1 - \frac{\pi}{2Q}\right)\sin\varphi \tag{11}$$

For the particular case of zero initial charge on the capacitor and negligible loss, i.e., $Q \gg 0$,

$$\frac{q(\tau)}{C} = \pi E \sin\varphi \tag{12}$$

and the maximum capacitor voltage is π times the peak value E of the applied a-c voltage wave. Practical pulse generators of this type have a voltage multiplication of about 2.7 or 2.8 compared to π for the ideal lossless case. The voltage wave on C as observed by oscilloscope on a typical radar pulser is shown in Figure 8.19.

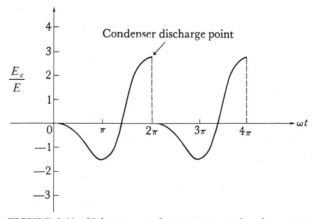

FIGURE 8.19 Voltage wave for *a-c* resonant charging system.

Example II. Heaviside's Expansion Theorem. This famous theorem is one of the foundations of Heaviside's operational calculus; it gives the current flowing in mesh r of a network when a unit step emf is applied in series with mesh s. The theorem is expressed by

$$a_r(t) = \frac{1}{Z_{rs}(0)} + \sum_{k=1}^{2f} \frac{e^{s_k t}}{s_k Z'_{rs}(s_k)} \tag{1}$$

where $a_r(t)$ is the current in mesh r due to application of unit step emf in mesh s, and is called the *indicial admittance.* $Z_{rs}(s)$ is the transfer impedance function and s_k are its roots, all of which are assumed to be simple (non-repeated) roots.

Proof of the theorem follows immediately by taking the Laplace transform of Eq. (8.11) with $\mathbf{Q} = eu(0)$ and $e =$ unit emf

$$\bar{I}_r = \frac{1}{sZ_{rs}} = \bar{a}_r$$

We note that the roots of $Z_{rs}(s) = 0$ are identical with the characteristic roots of $|Z| = 0$. Expand $\bar{a}_r(t)$ in partial fractions and obtain

$$\bar{a}_r = \frac{1}{sZ_{rs}(0)} + \sum_{k=1}^{2f} \frac{1}{s_k Z'_{rs}(s_k)(s - s_k)}$$

Taking the inverse transform gives the final result

$$a_r(t) = \frac{1}{Z_{rs}(0)} + \sum_{k=1}^{2f} \frac{e^{s_k t}}{s_k Z'_{rs}(s_k)}$$

where

$$Z'_{rs}(s_k) = \frac{d}{ds}[Z_{rs}(s)]\big|_{s=s_k}$$

Kron's Mesh Analysis of Electrical Networks. Application of the general theory of linear systems developed in Sect. 8.1 to electric networks corresponds to the mesh analysis method originated by Maxwell (Ref. 1, Vol. 1, Sects. 282b and 347). Finding the branch currents and voltages of a network, after the mesh currents are known, might seem to be a relatively minor problem. Kron, however, recognized the importance of reducing these calculations to an automatic routine. The resulting method of analysis is a direct extension of Maxwell's mesh method, but takes advantage of the superiority of matrix-tensor methods for organizing the calculation of networks and also separates the step of formulating the properties of the branches from that of dealing with problems concerning the interconnections of the branches. In the context of a system, the branches may be thought of as the elements and the interconnections as the interactions. Kron's pioneering work (Ref. 13), which was done before the advent of the digital computer, has provided important logical and procedural foundations for digital computer analysis of large size networks (Ref. 14, p. 1787) and forms the basis for many of these methods.

In this section, we summarize a matrix method for finding the branch voltages and branch currents for a given network. Consider a general network having b branches and n nodes. Using the terminology for the branch structure shown in Figure 8.3, the voltage relations for each branch can be written in the form

$$-\varepsilon_r + i_r Z'_r + \sum i_k Z'_{rk} = v_r \qquad r = 1, 2, \ldots, b \qquad (8.57)$$

where

$$Z'_r = L_r D + R_r + \frac{1}{C_r} D^{-1} = \text{branch impedance}$$

and

$Z'_{rk} = $ coupling impedance between branch r and branch k, $k \neq r$

For inductive coupling,

$$Z'_{rk} = Z'_{kr} = L_{rk} D$$

Asymmetric couplings, $Z'_{rk} \neq Z'_{kr}$, may also be present if active elements such as transistors or vacuum tubes are present. In this section, primed notation is used to designate branch impedances in order to distinguish these from the unprimed notation already used for mesh impedances. The branch equations may be written in vector notation as follows

$$\mathbf{v} + \boldsymbol{\epsilon} = Z' \mathbf{i} \qquad (8.58)$$

Equations (8.57) and (8.58) give the properties of the branches, but do not show the structure of the network. Denoting the mesh currents by $I_1, I_2, \ldots, I_l$, where l is the number of independent meshes given by Eq. (8.56), the branch currents may be related to the mesh currents in the following manner:

$$i_1 = B_{11} I_1 + B_{12} I_2 + \cdots + B_{1l} I_l$$
$$\cdot \qquad \cdot \qquad \cdot \qquad \cdot \qquad \cdot$$
$$\cdot \qquad \cdot \qquad \cdot \qquad \cdot \qquad \cdot$$
$$\cdot \qquad \cdot \qquad \cdot \qquad \cdot \qquad \cdot$$
$$i_b = B_{b1} I_1 + B_{b2} I_2 + \cdots + B_{bl} I_l$$

where

$B_{rs} = +1$ if branch r is contained in mesh s, and the branch and mesh currents are assumed positive in the same direction;

$B_{rs} = -1$ if branch r is contained in mesh s, and the branch and mesh currents are assumed positive in opposite directions;

$B_{rs} = 0$ if branch r is not contained in mesh s.

In matrix-vector notation

$$\mathbf{i} = B \mathbf{I} \qquad (8.59)$$

where B is the $b \times l$ matrix with elements B_{rs} and is called the *branch-node* matrix.

Mesh voltage relations may be expressed as follows:

$$E_{11}v_1 + E_{12}v_2 + \cdots + E_{1b}v_b = 0$$

$$\cdot \qquad \cdot \qquad \cdot \qquad \cdot \qquad \cdot$$
$$\cdot \qquad \cdot \qquad \cdot \qquad \cdot \qquad \cdot$$
$$\cdot \qquad \cdot \qquad \cdot \qquad \cdot \qquad \cdot$$

$$E_{l1}v_1 + E_{l2}v_2 + \cdots + E_{lb}v_b = 0$$

where

$E_{rs} = +1$ if branch r is contained in mesh s, and the voltage reference direction of branch r agrees with the voltage reference direction of mesh s;

$E_{rs} = -1$ if branch r is contained in mesh s, and the voltage reference direction of branch r is opposite to that of mesh s;

$E_{rs} = 0$ if branch r is not contained in mesh s.

The matrix $E = [E_{rs}]$ is called the *branch-mesh* matrix. The definitions of elements E_{rs} show that E is the transpose of B, i.e.,

$$E = B'$$

and hence, that the branch voltage equations may be stated as

$$B'\mathbf{v} = 0 \tag{8.60}$$

The mesh current equations may now be expressed in terms of the branch quantities by combining Eqs. (8.58), (8.59), and (8.60) to obtain

$$\mathbf{v} + \boldsymbol{\epsilon} = Z'\mathbf{i} = Z'B\mathbf{I}$$

Multiplying through by $E = B'$ then gives

$$B'\mathbf{v} + B'\boldsymbol{\epsilon} = B'\boldsymbol{\epsilon} = B'Z'B\mathbf{I}$$

or

$$B'Z'B\mathbf{I} = B'\boldsymbol{\epsilon} \tag{8.61}$$

The branch currents and branch voltages are now readily found. Solving (8.61) for $\mathbf{I}$ gives

$$\mathbf{I} = (B'Z'B)^{-1}B'\boldsymbol{\epsilon} \tag{8.62}$$

where $B'Z'B$ is an $l \times l$ nonsingular matrix. The branch currents are then given by

$$\mathbf{i} = B\mathbf{I} = B(B'Z'B)^{-1}B'\boldsymbol{\epsilon} \tag{8.63}$$

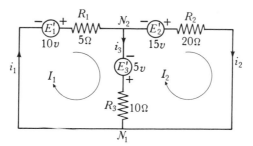

FIGURE 8.20 Two-mesh network to illustrate Kron's method of analysis.

and the branch voltages by

$$\mathbf{v} = Z'\mathbf{i} - \boldsymbol{\epsilon} = Z'B(B'Z'B)^{-1}B'\boldsymbol{\epsilon} - \boldsymbol{\epsilon} \tag{8.64}$$

Remarks

1. It is evident that B is a topological matrix which defines the network structure in terms of the interconnections of the branches.

2. Comparison of Eq. (8.62) with (8.11) shows that $B'Z'B$ can be identified as the mesh impedance matrix Z for the network, i.e.,

$$B'Z'B = Z \tag{8.65}$$

3. The Kron matrix method is highly systematic and routine and therefore can be programmed easily for an automatic computer.

4. The method is also very general, so that it is not necessary to develop a new computer program for each new network.

Example 12. Kron's Mesh Analysis Method. This example for the two-mesh network of Figure 8.20 shows the steps involved in Kron's mesh analysis method. The procedure for handling large size network problems parallels the procedure used for this problem except that the numerical work would be programmed for a digital computer.

SOLUTION

Step 1: Determine the branch-node or connection matrix B for the network.

$$\mathbf{i} = \begin{bmatrix} i_1 \\ i_2 \\ i_3 \end{bmatrix} = \begin{bmatrix} I_1 \\ I_2 \\ I_1 - I_2 \end{bmatrix} = \begin{bmatrix} 1 & 0 \\ 0 & 1 \\ 1 & -1 \end{bmatrix} \begin{bmatrix} I_1 \\ I_2 \end{bmatrix} = B\mathbf{I}$$

Step 2: Write mesh current equations and solve for **I**. From Eq. (8.62),

$$\mathbf{I} = (B'Z'B)^{-1}B'\boldsymbol{\epsilon}$$

Calculation gives

$$
B'Z'B = \begin{bmatrix} 1 & 0 & 1 \\ 0 & 1 & -1 \end{bmatrix} \begin{bmatrix} R_1 & 0 & 0 \\ 0 & R_2 & 0 \\ 0 & 0 & R_3 \end{bmatrix} \begin{bmatrix} 1 & 0 \\ 0 & 1 \\ 1 & -1 \end{bmatrix}
$$

$$
= \begin{bmatrix} R_1 + R_3, & -R_3 \\ -R_3, & R_2 + R_3 \end{bmatrix} = \begin{bmatrix} 15 & -10 \\ -10 & 30 \end{bmatrix}
$$

$$
(B'Z'B)^{-1} = \begin{bmatrix} 15 & -10 \\ -10 & 30 \end{bmatrix}^{-1} = \frac{1}{70} \begin{bmatrix} 6 & 2 \\ 2 & 3 \end{bmatrix}
$$

$$
(B'Z'B)^{-1}B' = \frac{1}{70} \begin{bmatrix} 6 & 2 \\ 2 & 3 \end{bmatrix} \begin{bmatrix} 1 & 0 & 1 \\ 0 & 1 & -1 \end{bmatrix} = \frac{1}{70} \begin{bmatrix} 6 & 2 & 4 \\ 2 & 3 & -1 \end{bmatrix}
$$

$$
\mathbf{I} = \begin{bmatrix} I_1 \\ I_2 \end{bmatrix} = \frac{1}{70} \begin{bmatrix} 6 & 2 & 4 \\ 2 & 3 & -1 \end{bmatrix} \begin{bmatrix} 10 \\ 15 \\ 5 \end{bmatrix} = \begin{bmatrix} \frac{11}{7} \\ \frac{6}{7} \end{bmatrix}
$$

Step 3:   Calculate the branch currents.

$$
\mathbf{i} = B\mathbf{I} = \begin{bmatrix} 1 & 0 \\ 0 & 1 \\ 1 & -1 \end{bmatrix} \begin{bmatrix} \frac{11}{7} \\ \frac{6}{7} \end{bmatrix} = \begin{bmatrix} \frac{11}{7} \\ \frac{6}{7} \\ \frac{5}{7} \end{bmatrix} = \begin{bmatrix} i_1 \\ i_2 \\ i_3 \end{bmatrix}
$$

Step 4:   Determine the branch voltages.

$$
\mathbf{v} = Z'\mathbf{i} - \boldsymbol{\epsilon} = \begin{bmatrix} 5 & 0 & 0 \\ 0 & 20 & 0 \\ 0 & 0 & 10 \end{bmatrix} \begin{bmatrix} \frac{11}{7} \\ \frac{6}{7} \\ \frac{5}{7} \end{bmatrix} - \begin{bmatrix} 10 \\ 15 \\ 5 \end{bmatrix} = \begin{bmatrix} -\frac{15}{7} \\ \frac{15}{7} \\ \frac{15}{7} \end{bmatrix} = \begin{bmatrix} v_1 \\ v_2 \\ v_3 \end{bmatrix}
$$

Step 5:   Check the mesh voltage equations. We find

$$
B'\mathbf{v} = \begin{bmatrix} 1 & 0 & 1 \\ 0 & 1 & -1 \end{bmatrix} \begin{bmatrix} -\frac{15}{7} \\ \frac{15}{7} \\ \frac{15}{7} \end{bmatrix} = \begin{bmatrix} 0 \\ 0 \end{bmatrix} = 0
$$

which checks.

8.4   APPLICATION TO ELECTROMECHANICAL SYSTEMS

It has been shown in Chap. 7 that electromechanical systems can be treated by the methods of generalized mechanics. This is an especially useful area of application because it involves the interface of two separate sciences, i.e., mechanics and electromagnetism. It is obviously advantageous to solve electromechanical problems by a unified method rather than by piecemeal methods based on combining results from the two separate fields.

Systems which combine electrical and mechanical elements are fundamental to modern science and technology and have numerous applications in many fields. Examples are energy conversion in the production and utilization of electric power; electroacoustical devices in the field of electrical communications; recorders, plotters, and meters in instrumentation; and energy converters and controllers in automation and automatic control systems.

In addition to the illustrative examples treated in this section, further illustrations of electromechanical systems are given in Chaps. 10, 11, and 12.

Example I. Energy Conversion by Moving Straight Wire in a Magnetic Field. Find the motion for an electromechanical system consisting of a straight wire moving through a magnetic field, with a weight attached to the wire, as shown in Figure 8.21. The magnetic field is normal to the wire, and it is assumed that the external resistance is much greater than the resistance of the rails and moving wire.

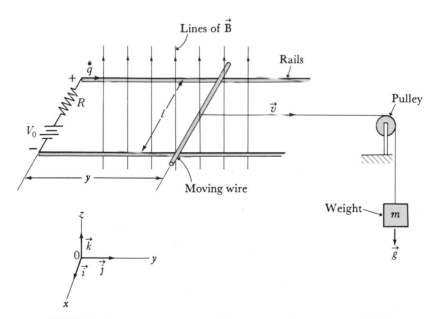

FIGURE 8.21 Energy conversion by a moving wire in a magnetic field.

SOLUTION.   Let q and y denote the two degrees of freedom of the system. Then, the energy functions are

$$L = T - V = \tfrac{1}{2}m\dot{y}^2 + mgy + \dot{q}lBy$$
$$F = \tfrac{1}{2}R\dot{q}^2 \qquad P = V_0\dot{q}$$

Substitution in Lagrange's equations gives the equations of motion

$$\frac{d}{dt}\left(\frac{\partial L}{\partial \dot{q}}\right) - \frac{\partial L}{\partial q} + \frac{\partial(F - P)}{\partial \dot{q}} = lB\dot{y} + R\dot{q} - V_0 = 0$$

$$\frac{d}{dt}\left(\frac{\partial L}{\partial \dot{y}}\right) - \frac{\partial L}{\partial y} + \frac{\partial(F - P)}{\partial \dot{y}} = m\ddot{y} - mg - lB\dot{q} = 0$$

which may be expressed in matrix form as follows:

$$\begin{bmatrix} R & lB \\ -lB & mD \end{bmatrix}\begin{bmatrix} \dot{q} \\ \dot{y} \end{bmatrix} = \begin{bmatrix} V_0 \\ mg \end{bmatrix}$$

To solve these equations, let $\dot{q} = I$ and $\dot{y} = v$, and eliminate I between the equations. This gives

$$\frac{dv}{dt} + \frac{l^2B^2}{mR}v - \left(\frac{lBV_0}{mR} + g\right) = 0$$

Substituting a new variable w defined by

$$w = \frac{l^2B^2}{mR}v - \left(\frac{lBV_0}{mR} + g\right) = bv + c$$

reduces the equation to

$$\frac{dw}{dt} + bw = 0$$

This has the obvious solution

$$w = w_0 e^{-bt}$$

where w_0 is the value of w at $t = 0$.

Hence, in terms of v, the solution is

$$bv + c = (bv_0 + c)e^{-bt}$$

Since b is always positive, it is evident that v approaches a constant value as t approaches infinity, i.e.,

$$\lim_{t \to \infty} v = -\frac{c}{b} = \frac{Rm}{l^2B^2}\left(\frac{lBV_0}{mR} + g\right)$$

The current I is given by

$$I = \frac{V_0 - lBv}{R}$$

and has the limiting value

$$\lim_{t \to \infty} I = \frac{V_0}{R} - \frac{lB}{R} \qquad \lim_{t \to \infty} v = -\frac{mg}{lB}$$

The energy conversion rate is

$$\varepsilon I = lbvI .$$

COMMENT ON MOTIONAL AND GYROSCOPIC FORCES. Note that the Lagrangian for this example contains a term which is linear in the velocity, corresponding to a force proportional to velocity. Thomson and Tait (Ref. 3, Vol. 1, p. 370) introduced the term *motional* to describe forces of this type in contrast to *positional* for forces proportional to displacement Although viscous forces are proportional to velocity, these forces are always dissipative and do not enter into the Lagrangian. Nondissipative motional forces of the type encountered in the present example are called *gyroscopic*. Gyroscopic forces are anti-symmetric as may be seen from the system matrix for this example, i.e.,

$$\begin{bmatrix} R & lB \\ -lB & mD \end{bmatrix}$$

Gyroscopic forces occur here because of the presence of the static magnetic field.

Example 2. Electromechanical Vibration Transducer. Many technical and scientific applications require means for converting acoustical and mechanical vibrations to electric vibrations, and vice versa. Devices for converting energy from one form to another form are called transducers. Examples of vibration transducers are microphones, loudspeakers, vibration testing equipment, and servomechanism control devices. One widely used type of transducer is illustrated in Figure 8.22. A coil mounted on a spring or equivalent elastic element moves axially in the annular air gap of a permanent magnet, and is coupled both to electrical and mechanical elements which serve as inputs and outputs.

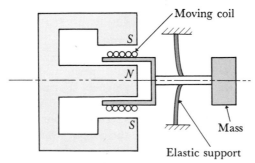

FIGURE 8.22 Electromechanical vibration transducer.

Such a transducer is used in the moving coil microphone, where the coil is attached to a thin metallic diaphragm. Vibrations of the diaphragm caused by impinging sound waves give rise to corresponding emf's generated by the coil as it moves in the magnetic field. These emf's appear across a load resistance in the form of an output signal. Thus, acoustic energy is imaged as electrical energy. The reverse process takes place in a loudspeaker, where electrical energy is imaged as acoustic energy.

The following analysis is made for the basic device. This is followed by application to one specific form, an electromagnetic driver for vibration testing.

ANALYSIS.   Denote the two degrees of freedom for the system by q and x as shown in Figure 8.23. The energy functions may be written by inspection, and are

$$L = T - V = \tfrac{1}{2}L_c\dot{q}^2 + \tfrac{1}{2}m\dot{x}^2 - \tfrac{1}{2}kx^2 + k_1\dot{q}x$$

$$F = \tfrac{1}{2}(R_c + R)\dot{q}^2 + \tfrac{1}{2}\mu\dot{x}^2$$

$$P = V\dot{q} + f\dot{x}$$

In these equations,

$$\tfrac{1}{2}L_c\dot{q}^2 + k_1\dot{q}x = U_m$$

where the parameter k_1 depends on the coil and magnetic field details. For a uniform field **B** in the air gap (if one neglects the slight pitch of the helical coil winding) $k_1 = Bl$, where l is the length of wire in the coil. In practice, B is not strictly uniform and other imperfections will exist, but k_1 will still be approximately constant and can be found by test measurements.

Substitution in Lagrange's equations yields the equations of motion for the system as follows.

$$L_c\ddot{q} + (R_c + R)\dot{q} + k_1\dot{x} - V = 0$$

$$m\ddot{x} + \mu\dot{x} + kx - k_1\dot{q} - f = 0$$

Writing $\dot{q} = I$ and expressing in matrix form leads to the matrix equation

$$\begin{bmatrix} L_cD + R_c + R & k_1 \\ -k_1 & mD + \mu + kD^{-1} \end{bmatrix}\begin{bmatrix} I \\ \dot{x} \end{bmatrix} = \begin{bmatrix} V \\ f \end{bmatrix} \tag{1}$$

Various transducer operations can be performed by choice of driving functions V and f, and of system parameters. For instance, if V is a constant voltage and f represents an impinging sound wave, the device will operate as a microphone with the electrical output signal across the resistor R. On the other hand, if V is the output from an audio amplifier, f is chosen to be zero, and the coil is connected to an appropriate diaphragm, the system becomes a loudspeaker. Still another application is

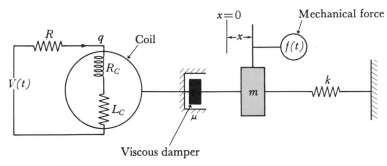

FIGURE 8.23 Schematic diagram of vibration transducer.

to choose V as a servomechanism signal, f as zero, and m as a hydraulic control valve, in which case the transducer functions as a positioning element in an automatic feedback control system. Numerous other applications are evident. To illustrate one of these, the system equations will be solved for the case where the device is used to test the reliability of equipment assemblies under vibration. When used this way, the force function f is zero and $V(t)$ is sinusoidal. Since the steady-state vibrations are of primary interest, we use complex notation to find the solution. Let

$$V(t) = E \cos \omega t = \text{Re} \, [Ee^{i\omega t}]$$

where ω is the angular frequency. Then, because of the linearity of the system, both I and $\dot{x}$ are also sinusoidal, i.e.,

$$\dot{q} = \bar{I}e^{i\omega t} \qquad \dot{x} = \bar{X}e^{i\omega t}$$

The equations become

$$iL_c\omega \bar{I} + (R_c + R)\bar{I} + ik_1\omega\bar{X} = E$$

$$-k_1\bar{I} - m\omega^2\bar{X} + i\mu\omega\bar{X} + k\bar{X} = 0$$

or in matrix form

$$\begin{bmatrix} R_c + R + iL_c\omega & ik_1\omega \\ -k_1 & k - m\omega^2 + i\mu\omega \end{bmatrix} \begin{bmatrix} \bar{I} \\ \bar{X} \end{bmatrix} = \begin{bmatrix} E \\ 0 \end{bmatrix} \tag{2}$$

The matrix solution of (2) is

$$\begin{bmatrix} \bar{I} \\ \bar{X} \end{bmatrix} = A^{-1} \begin{bmatrix} E \\ 0 \end{bmatrix} = \frac{1}{\det A} \begin{bmatrix} \Delta_{11} & E \\ \Delta_{21} & E \end{bmatrix} \tag{3}$$

where A is the matrix and Δ_{11}, Δ_{21} are cofactors of A'.

$$\Delta_{11} = k - m\omega^2 + i\mu\omega$$

$$\Delta_{21} = k_1$$

$$\det [A] = (R_c + R + iL_c\omega)(k - m\omega^2 + i\mu\omega) + ik_1^2\omega$$

$$= (R_c + R)(k - m\omega^2) - L_c\mu\omega^2$$

$$+ i\omega[(k - m\omega^2)L_c + (R_c + R)\mu + k_1^2]$$

Inspection shows that there is only one resonant frequency possible, since the imaginary component of det $[A]$ vanishes for only one real positive value of ω.

Example 3.   Unipolar Induction Machines.   Faraday's law of electromagnetic induction states that the induced emf in a closed circuit is proportional to the time rate of change of magnetic flux threading the circuit, whether due to changes in flux density or to motion of the conductor.   There are situations, however, where the flux rule can not be applied.   Many examples of such exceptions are known in the literature, some given by Faraday himself.   The exceptions occur whenever a continuous conductor, such as a metal sheet, is moved through a magnetic field.   In these cases the magnetic flux linking the circuit remains constant, but the material itself is moved through the field.

Several types of machines have been designed to utilize this effect. One such machine, known as the Faraday disc generator, is illustrated in Figure 8.24.   The fundamental principle can be understood in terms of Lorentz' law

$$\mathbf{F} = q\mathbf{v} \times \mathbf{B}$$

The emf may be found by taking a path of integration along the radial line between the contact brushes

$$\varepsilon = \int_a^b \frac{\mathbf{F} \cdot d\mathbf{s}}{q} = \int_a^b r\omega B \, dr = \tfrac{1}{2}\omega B(b^2 - a^2)$$

With the directions of $\mathbf{B}$ and ω shown in the sketch, the positive terminal will be at the outer contact.

The power relations are

$$\varepsilon I = \frac{\varepsilon^2}{R} = \frac{\omega^2 B^2 (b^2 - a^2)^2}{R} = \tau\omega$$

where τ is the driving torque on the shaft and R is the circuit resistance.

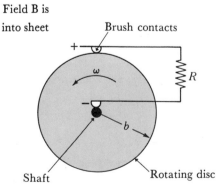

FIGURE 8.24   Faraday disc generator.

Note that an attempt to calculate the emf by the rate of change of flux in the circuit fails because the flux linkage is constant. The Lorentz force, however, gives the correct result.

Faraday or homopolar induction generators are characterized by very low internal resistance and are, therefore, best suited for applications requiring very high currents at relatively low voltages, such as certain metallurgical reduction or melting processes, and generating currents for super strength magnetic fields. Homopolar machines have been built for extremely high currents. For example, one of the largest existing machines (Ref. 5) uses four discs, each disc rated for 240 volts, 1,600,000 amperes in pulse duty. In a typical application this machine produces two-second pulses with peak output of 350 volts and 1,500,000 amperes or 525 megawatts

Example 4. Condenser Microphone. A condenser microphone uses electrostatic forces to convert acoustical vibrations impinging on a diaphragm to electric signals. Figure 8.25 shows the essential elements. Forced vibrations of the diaphragm, caused by the pressure variations of the impinging sound wave, produce an output voltage due to variations in the capacitance of the condenser formed by the diaphragm and back plate. Performance characteristics of the microphone can be calculated by treating it as an electromechanical system.

There are two degrees of freedom, denoted by x for the mechanical displacement of the diaphragm and q for the electric charge on the condenser. Energy functions for the system are given by

$$T = \tfrac{1}{2}m\dot{x}^2 + \tfrac{1}{2}L\dot{q}^2$$

$$V = \tfrac{1}{2}kx^2 + \frac{q^2}{2C}$$

$$F = \tfrac{1}{2}\mu\dot{x}^2 + \tfrac{1}{2}R\dot{q}^2$$

$$P = Q_x\dot{x} + V_0\dot{q}$$

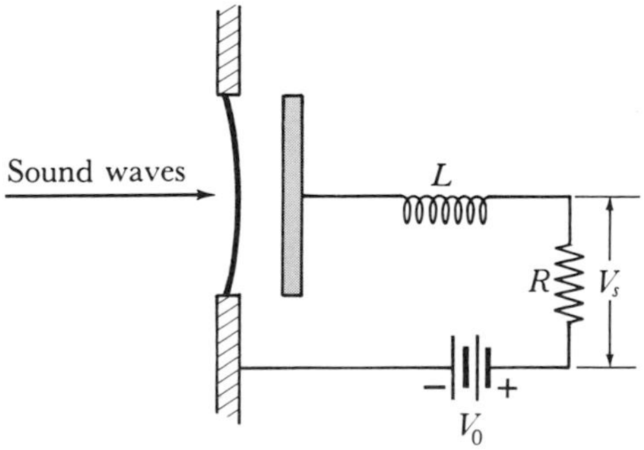

FIGURE 8.25 Condenser microphone.

where m is the equivalent mass of the diaphragm, k its equivalent spring constant, μ the damping constant of the air cushion between the diaphragm and back plate, and $Q_x(t)$ the mechanical driving force of the sound waves along the axis of the microphone.

Substitution in Lagrange's equations gives

$$\left.\begin{aligned} m\ddot{x} + kx + \mu\dot{x} - \frac{q^2}{2C^2}\frac{\partial C}{\partial x} &= Q_x \\[2mm] L\ddot{q} + \frac{q}{C} + R\dot{q} &= V_0 \end{aligned}\right\} \tag{1}$$

The capacitance C varies in a rather complex way with x, but where displacements are small, as they are for such microphones, it is permissible to approximate C by the equation

$$C = \frac{\alpha\varepsilon_0 A_c}{b - x}$$

where α is a dimensionless coefficient of somewhat less than unity and is included to represent the effect of the curvature of the diaphragm. Using this approximation, Eqs. (1) become

$$\left.\begin{aligned} m\ddot{x} + \mu\dot{x} + kx - \frac{q^2}{2\alpha\varepsilon_0 A_c} &= Q_x \\[2mm] L\ddot{q} + R\dot{q} + \frac{q(b - x)}{\alpha\varepsilon_0 A_c} &= V_0 \end{aligned}\right\} \tag{2}$$

Equations (2) can be simplified by noting that at equilibrium (q_0, x_0), we have

$$kx_0 - \frac{q_0^2}{2\alpha\varepsilon_0 A_c} = 0$$

$$\frac{q_0(b - x_0)}{\alpha\varepsilon_0 A} = V_0 = q_0/C_0$$

Measuring the coordinates from their equilibrium values, i.e., letting

$$q_1 = q - q_0$$
$$x_1 = x - x_0$$

and neglecting second order terms, leads to the expressions

$$\left.\begin{aligned} m\ddot{x}_1 + \mu\dot{x}_1 + kx_1 - \frac{q_0 q_1}{\alpha\varepsilon_0 A_c} &= Q_x \\[2mm] L\ddot{q}_1 + R\dot{q}_1 + \frac{q_1}{C_0} - \frac{q_0 x_1}{\alpha\varepsilon_0 A_c} &= 0 \end{aligned}\right\} \tag{3}$$

The output signal voltage is given by

$$V_s = R\dot{q}_1 = RI$$

Since $\dot{q}_1$ can be solved for in terms of Q_x from Eqs. (3), V_s can be expressed as a function of Q_x and the response of the system is thus determined. For steady-state a-c excitation, the solution follows from Eq. (8.18)

$$\dot{\mathbf{q}}_p = i\omega A^{-1}\mathbf{Q}_p = Y\mathbf{Q}_p$$

or, in expanded form

$$\begin{bmatrix} \dot{x}_1 \\ \dot{q}_1 \end{bmatrix} = \begin{bmatrix} \bar{v} \\ \bar{I} \end{bmatrix} = \begin{bmatrix} Y_{11} & Y_{12} \\ Y_{21} & Y_{22} \end{bmatrix}\begin{bmatrix} \bar{Q}_x \\ 0 \end{bmatrix} = \begin{bmatrix} Y_{11} & \bar{Q}_x \\ Y_{21} & \bar{Q}_x \end{bmatrix} \tag{4}$$

Hence

$$\bar{V}_s = RY_{21}\bar{Q}_x \tag{5}$$

where the bar notation indicates complex quantities.

To calculate $\bar{V}_s$, it is noted that

$$Y_{21} = i\omega A_{21}^{-1} = \frac{i\omega \Delta_{12}}{\det A}$$

with

$$A = \begin{bmatrix} (k - m\omega^2) + i\mu\omega & -q_0/\alpha c_0 A_c \\ -q_0/\alpha c_0 A_c & \left(\dfrac{1}{C} - L\omega^2\right) + iR\omega \end{bmatrix}$$

$$\Delta_{12} = q_0/\alpha \varepsilon_0 A_c$$

$$\det A = (k - m\omega^2)\left(\frac{1}{C} - L\omega^2\right) - R\mu\omega^2 - \left(\frac{q_0}{\alpha \varepsilon_0 A_c}\right)^2$$

$$+ i\omega\left[R(k - m\omega^2) + \mu\left(\frac{1}{C} - L\omega^2\right)\right]$$

Resonant frequencies, if they exist, correspond approximately to the vanishing of the real part of det A, i.e.,

$$(k - m\omega^2)\left(\frac{1}{C} - L\omega^2\right) - R\mu\omega^2 - \left(\frac{q_0}{\alpha \varepsilon_0 A_c}\right)^2 = 0$$

This is a quadratic in ω^2 and has real roots only if

$$\left(\frac{k}{m} + \frac{1}{LC} + \frac{R\mu}{Lm}\right)^2 > 4\left[\frac{1}{LC} \cdot \frac{k}{m} - \frac{1}{Lm}\left(\frac{q_0}{\alpha \varepsilon_0 A_c}\right)^2\right]$$

The frequency response of the microphone is given by Eq. (5), and can be evaluated numerically for given sets of microphone design parameters.

Example 5. Van de Graaff Electrostatic Generator. Very high potential differences are produced by transporting electric charges on a moving insulating belt between large-diameter electrodes, as shown in Figure 8.26. Corona points are used to spray charges on the belt; mechanical forces carry these charges to the second electrode where the charges are removed. Potential differences as high as several million volts are achieved, and currents of the order of a few milliamperes can be carried. These machines are used primarily for accelerating ions to high energy for nuclear work.

To indicate the order of magnitude of the forces and energies involved, assume a generator capable of producing 10^6 volts at a current of 10^{-3} ampere. The calculated power between electrodes 1 and 2 is

$$P = \varepsilon I = \int_1^2 \mathbf{E} \cdot (\sigma v \, d\mathbf{s}) = \int_1^2 \mathbf{E} \cdot (\sigma \, d\mathbf{s})v$$

$$= \int_1^2 v \, dF = \text{mechanical power}$$

where σ is the electric charge carried per unit length of the belt. Hence, for this example

$$P = \varepsilon I = 10^6 \times 10^{-3} = 1000 \text{ watts}$$

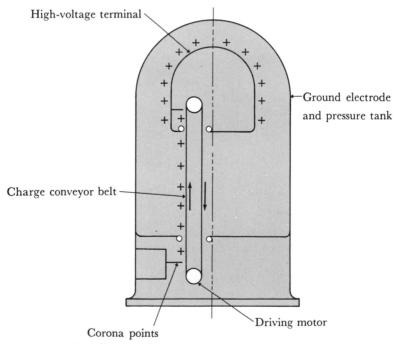

High-voltage terminal

Ground electrode
and pressure tank

Charge conveyor belt

Corona points

Driving motor

FIGURE 8.26 Van de Graaff high-voltage generator.

Assuming a belt velocity v of 3000 ft./min., the mechanical force exerted on the belt to overcome the electrostatic force is

$$F = \frac{P}{v} = 1000 \left/ \frac{3000}{60 \times 3.28} \right. = 65.6 \text{ newtons} = 14.7 \text{ lbs}$$

Note that the moving belt is the source of emf in the Van de Graaff device, and that the concept of emf as the work done in moving unit charge from one electrode to the other is clearly evidenced in the operation of the machine.

Example 6. Direct-Current Generator. The ordinary d-c generator uses a ferromagnetic rotor wound with coils which continuously cut through a magnetic field as shown in Figure 8.27; d-c instead of a-c current is produced by including a commutator system which functions in such a manner that the output always maintains constant polarity independent of rotor position. In practice, this is readily accomplished by using a distributed winding on the rotor, with the segments connected to the commutator bars as indicated in the figure. The orientation of the rotor coils with respect to brush position is such that the direction of the magnetic field of the rotor is always perpendicular to that of the stator. Commutation thus assures that the rotor field is stationary in space, regardless of the fact that the rotor itself turns at high speed. It should be noted that for this type of operation no magnetic coupling exists between the rotor and stator coils.

The equations of motion and the operational characteristics of the d-c generator are determined by means of the Lagrangian procedure. There are two degrees of freedom, θ for the angle of rotation of the armature and q for the electric charge displacement through the rotor. Letting L_a be the armature inductance and R_a the armature resistance,

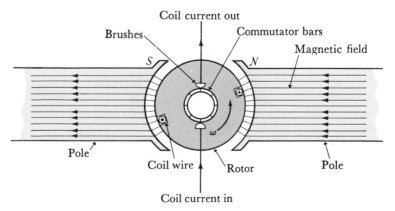

FIGURE 8.27 Schematic diagram for a two-pole d-c generator.

the energy functions are

$$L = T - V = \tfrac{1}{2}L_a\dot{q}^2 + \tfrac{1}{2}J\dot{\theta}^2 + k\dot{q}\theta$$
$$F = \tfrac{1}{2}R_a\dot{q}^2$$
$$P = \tau\dot{\theta} - V_1\dot{q}$$

In these equations J is the moment of inertia of the rotor and τ the torque driving the rotor. The energy term $k\dot{q}\theta$ is calculated as follows: let B be the average magnetic field in the air gap between rotor and stator, n the number of armature turns per unit length of rotor periphery, l the axial length of the rotor exposed to the field, w the peripheral length of the air gap field at one pole, and r_0 the radius of rotor winding. Then

$$U_m = \dot{q}\Phi = \dot{q}(2lwnBr_0)\theta = k\dot{q}\theta$$

Using Lagrange's equations gives

$$\left. \begin{aligned} \frac{d}{dt}\left(\frac{\partial L}{\partial \dot{q}}\right) - \frac{\partial L}{\partial q} + \frac{\partial(F - P)}{\partial \dot{q}} &= L_a\ddot{q} + R_a\dot{q} + V_1 + k\dot{\theta} = 0 \\ \frac{d}{dt}\left(\frac{\partial L}{\partial \dot{\theta}}\right) - \frac{\partial L}{\partial \theta} + \frac{\partial(F - P)}{\partial \dot{\theta}} &= J\ddot{\theta} - \tau - k\dot{q} = 0 \end{aligned} \right\}$$

These may also be written as

$$\begin{bmatrix} L_aD + R_a & k \\ -k & JD \end{bmatrix}\begin{bmatrix} I \\ \omega \end{bmatrix} = \begin{bmatrix} -V_1 \\ \tau \end{bmatrix}$$

where I is the rotor current and ω is the angular velocity of the rotor. Note the presence of gyroscopic forces indicated by the elements k and $-k$ in the system matrix.

For steady-state operation, the controlling equations reduce to

$$\begin{bmatrix} R_a & k \\ -k & 0 \end{bmatrix}\begin{bmatrix} I_0 \\ \omega_0 \end{bmatrix} = \begin{bmatrix} -V_0 \\ \tau \end{bmatrix}$$

where I_0 and ω_0 are the steady-state values of current and angular frequency, respectively. The solution may be written by inspection

$$\begin{bmatrix} I_0 \\ \omega_0 \end{bmatrix} = \frac{1}{k^2}\begin{bmatrix} -k\tau \\ -kV_0 + R_a\tau \end{bmatrix}$$

The emf and torque are given by

$$\varepsilon = -k\omega_0 = V_0 + R_aI_0$$
$$\tau = -kI_0$$

Also for steady-state conditions

$$\text{power input} = \tau\omega_0 = \text{loss} + \text{power output} = R_aI_0^2 + V_0I_0$$

8.5 APPLICATION TO STRUCTURES

In this section, linear systems methods are applied to the analysis of statically loaded structures. The treatment indicates the general approach to these problems and is illustrated by two examples, one of an elementary nature to show the principles, and the other requiring a digital computer for its solution.

For structures subject to static loads, the characteristic matrix of Eq. (8.6) reduces to K which may be called the *structure stiffness matrix*. The input vector $\mathbf{Q}$, representing the applied generalized forces on the structure, is constant. Therefore Eq. (8.13) applies, i.e.,

$$\mathbf{q} = K^{-1}\mathbf{Q}$$

where K^{-1} may be called the *structure flexibility matrix* and $\mathbf{q}$ is the output vector representing joint displacements of the structure.

The general method for formulating the structure stiffness matrix K is demonstrated in terms of the structure shown in Figure 8.28. Diagram 8.28a shows the externally applied moments Q_i and the corresponding rotations q_i. Similarly, diagram 8.28b represents the member end moments M_i and the member end rotations θ_i. It should be noted that the number of coordinates q_i determines the number of degrees of freedom for the structure which, in this example, is two.

The external moments $\mathbf{Q}$ may be expressed in terms of the member end moments $\mathbf{M}$ by applying static equilibrium equations utilizing the free body diagram of Figure 8.28c. This gives

$$Q_1 = M_2 + M_3$$
$$Q_2 = M_4$$

In matrix form

$$\mathbf{Q} = A\mathbf{M} \tag{8.66}$$

where A may be called the *statics matrix* (see Ref. 15).

The internal end rotations $\boldsymbol{\theta}$ can be expressed in terms of joint displacements as follows:

$$\boldsymbol{\theta} = \begin{bmatrix} \theta_1 \\ \theta_2 \\ \theta_3 \\ \theta_4 \end{bmatrix} = \begin{bmatrix} 0 \\ q_1 \\ q_1 \\ q_2 \end{bmatrix}$$

In matrix form,

$$\boldsymbol{\theta} = B\mathbf{q} \tag{8.67}$$

where B may be called the *deformation matrix* (see Ref. 15).

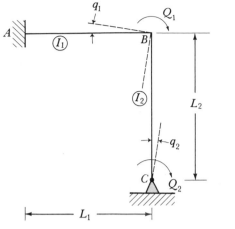

(a) External moments

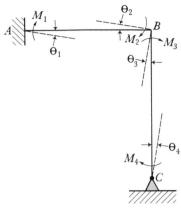

(b) Member end moments

FIGURE 8.28 Diagrams for formulation of structure equations.

(c) Free body diagram

Matrices A and B are seen to be the transpose of each other, i.e.,

$$A = \begin{bmatrix} 0 & 1 & 1 & 0 \\ 0 & 0 & 0 & 1 \end{bmatrix} \qquad B = \begin{bmatrix} 0 & 0 \\ 1 & 0 \\ 1 & 0 \\ 0 & 1 \end{bmatrix} = A'$$

This result can be generalized by using the principle of virtual work, but the proof is outside the intended scope of this book. Those readers interested in the details of the proof are referred to item 15 in the bibliography

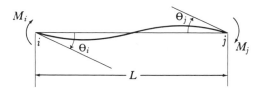

FIGURE 8.29 Diagram for slope-deflection equations.

of this chapter. The result in general form is

$$B = A' \tag{8.68}$$

and therefore, from Eq. (8.67),

$$\boldsymbol{\theta} = A'\mathbf{q} \tag{8.69}$$

For slender prismatic members of a frame with constant moment of inertia, the member end moments can be expressed in terms of internal end rotations $\boldsymbol{\theta}$ by applying the slope-deflection equations (Ref. 16) to span ij shown in Figure 8.29. Thus

$$\left. \begin{aligned} M_i &= \frac{4EI}{L}\,\theta_i + \frac{2EI}{L}\,\theta_j \\[2mm] M_j &= \frac{2EI}{L}\,\theta_i + \frac{4EI}{L}\,\theta_j \end{aligned} \right\} \tag{8.70}$$

where E is the modulus of elasticity and I is the moment of inertia of the member. In the derivation of the above equations, the effects of axial and shear forces are neglected as their influence on the final displacements is generally very minor for frames with slender members. The matrix form of these equations is

$$\mathbf{M} = S\boldsymbol{\theta} \tag{8.71}$$

where S is called the *member stiffness matrix*. Application of the slope-deflection equations to the frame considered produces the following

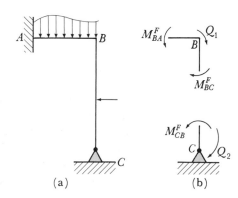

FIGURE 8.30 Frame loading.

member stiffness matrix:

$$
S = \begin{bmatrix}
\dfrac{4EI_1}{L_1} & \dfrac{2EI_1}{L_1} & 0 & 0 \\[2mm]
\dfrac{2EI_1}{L_1} & \dfrac{4EI_1}{L_1} & 0 & 0 \\[2mm]
0 & 0 & \dfrac{4EI_2}{L_2} & \dfrac{2EI_2}{L_2} \\[2mm]
0 & 0 & \dfrac{2EI_2}{L_2} & \dfrac{4EI_2}{L_2}
\end{bmatrix} \tag{8.72}
$$

The structure stiffness matrix K can now be derived. Substituting Eq. (8.69) in Eq. (8.71) gives

$$
\mathbf{M} = SA'\mathbf{q} \tag{8.73}
$$

and substituting Eq. (8.73) in Eq. (8.66) yields

$$
\mathbf{Q} = ASA'\mathbf{q} \tag{8.74}
$$

Comparing (8.74) with (8.13), it is seen that

$$
K = ASA' \tag{8.75}
$$

The input vector $\mathbf{Q}$ for the generalized forces can now be determined in the following manner.

If the frame under study has no freedom in translation (no sidesway), Q_i is found as the algebraic sum of the fixed-end moments at joint i. If at a given joint i, all fixed-end moments sum up to zero, then $Q_i = 0$. For example, if the frame considered in Figure 8.28, which has no translational degrees of freedom, is subject to the loads shown in Figure 8.30a, the generalized forces Q_1 and Q_2 can be easily determined from Figure 8.30b. Thus

$$
\left.\begin{aligned}
Q_1 &= M_{BC}^F - M_{BA}^F \\
Q_2 &= -M_{CB}^F
\end{aligned}\right\} \tag{8.76}
$$

where M_{BC}^F is the fixed-end moment at end B for span BC, and so on. The reader should be w... ned that in the case of frames with translational freedom (sidesway) additional degrees of freedom to account for these must be included in the analysis.

The stiffness matrix $K = ASA'$ and the flexibility matrix K^{-1} are normally computed by means of digital computers, and joint displacements $\mathbf{q}$ are then found from Eq. (8.13). Once $\mathbf{q}$ is determined, the member end moments are easily calculated from Eq. (8.73). Final moments are found as the sum of fixed-end moments and member end moments.

The above theory is illustrated by the following numerical examples.

Example 1. Elementary Structure. Find the moments at A and B, and the rotation at B for the beam shown in Figure 8.31a.

SOLUTION. The (Q, q) and (M, θ) diagrams are shown in Figures 8.31b and 8.31c. The solution procedure follows.

The equilibrium of moments about B gives

$$Q_1 = (0)M_1 + M_2$$

Thus

$$A = [0 \quad 1]$$

and

$$A' = B = \begin{bmatrix} 0 \\ 1 \end{bmatrix}$$

The member stiffness matrix is then written as

$$S = \begin{bmatrix} \dfrac{4EI}{10} & \dfrac{2EI}{10} \\[2mm] \dfrac{2EI}{10} & \dfrac{4EI}{10} \end{bmatrix}$$

The structure stiffness matrix K is determined by

$$K = ASA'$$

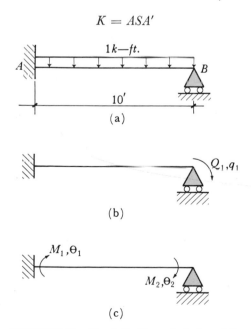

FIGURE 8.31 Sketch for Example 1, Sect. 8.5.

Hence,

$$K = [0 \quad 1] \begin{bmatrix} \dfrac{4EI}{10} & \dfrac{2EI}{10} \\[2mm] \dfrac{2EI}{10} & \dfrac{4EI}{10} \end{bmatrix} \begin{bmatrix} 0 \\ 1 \end{bmatrix} = \begin{bmatrix} \dfrac{2EI}{5} \end{bmatrix}$$

and

$$K^{-1} = \begin{bmatrix} \dfrac{5}{2EI} \end{bmatrix}$$

In this particular example the vector of applied forces **Q** reduces to the fixed-end moment at B

$$M_{AB}^F = M_{BA}^F = \frac{wl^2}{12} = \frac{(1)(10)^2}{12} = \frac{25}{3} \text{ k-ft.}$$

Since the directions of M_{BA}^F and Q_1 are opposite, we have

$$\mathbf{Q} = \begin{bmatrix} -\dfrac{25}{3} \end{bmatrix}$$

The rotation at B can now be determined from the displacement **q** in Eq. (8.13)

$$\mathbf{q} = K^{-1}\mathbf{Q} = \begin{bmatrix} \dfrac{5}{2EI} \end{bmatrix}\begin{bmatrix} -\dfrac{25}{3} \end{bmatrix} = \begin{bmatrix} -\dfrac{125}{6EI} \end{bmatrix}$$

Therefore

$$\theta_B = \theta_2 = -\frac{125}{6EI} \text{ radian}$$

where the negative sign indicates that the actual rotation is counterclockwise or opposite to the selected sense of θ_2.

Member end moments are easily found by Eq. (8.73).

$$\mathbf{M} = SA'\mathbf{q}$$

$$\mathbf{M} = \begin{bmatrix} \dfrac{4EI}{10} & \dfrac{2EI}{10} \\[2mm] \dfrac{2EI}{10} & \dfrac{4EI}{10} \end{bmatrix}\begin{bmatrix} 0 \\ 1 \end{bmatrix}\begin{bmatrix} -\dfrac{125}{6EI} \end{bmatrix} = \begin{bmatrix} -\dfrac{25}{6} \\[2mm] -\dfrac{25}{3} \end{bmatrix}$$

Thus, $M_1 = -\frac{25}{6}$ k-ft., and $M_2 = -\frac{25}{3}$ k-ft. Final moments at A and B are determined by

$$M_A = M_{BA}^F + M_1 = -\tfrac{25}{3} - \tfrac{25}{6} = -12.5 \text{ k-ft.}$$

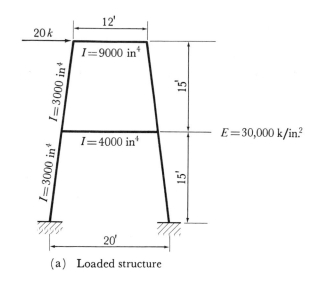

(a) Loaded structure

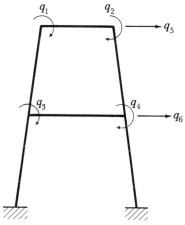

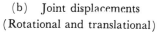

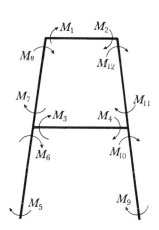

(b) Joint displacements
(Rotational and translational)

$\vec{Q}, \vec{q}$ Diagram

(c) Member end moments

$\vec{M}, \vec{\theta}$ Diagram

FIGURE 8.32

and

$$M_B = M_{BA}^F + M_2 = \tfrac{25}{3} - \tfrac{25}{3} = 0$$

It is clearly seen that $M_A = 12.5$ k-ft. corresponds to the well known result $M_A = \dfrac{wl^2}{8} = \dfrac{(1)(10)^2}{8} = 12.5$ k-ft., and since B is a roller support the moment there should be zero as determined by the preceding solution.

Example 2. Digital Computer Solution of Structure Problem. Given a rigid frame with two degrees of sidesway, as shown in Figure 8.32a, compute joint displacements and bending moments at ends of each member.

SOLUTION. Figure 8.32b shows the rotational (q_1, q_2, q_3, q_4) and the translational (q_5, q_6) degrees of freedom. Member end moments M_i and internal end rotations θ_i, $i = 1, 2, \ldots, 12$, are shown in the $(\mathbf{M}, \boldsymbol{\theta})$-diagram, Figure 8.32c.

Referring to Figures 8.32b and 8.32c, the generalized forces $\mathbf{Q}$ may be expressed in terms of member end moments $\mathbf{M}$ by applying static equilibrium equations. Thus

$$Q_1 = M_1 + M_8$$
$$Q_2 = M_2 + M_{12}$$
$$Q_3 = M_3 + M_6 + M_7$$

Hence,

$$
A =
\begin{bmatrix}
1 & 0 & 0 & 0 & 0 & 0 & 0 & 1 & 0 & 0 & 0 & 0 \\
0 & 1 & 0 & 0 & 0 & 0 & 0 & 0 & 0 & 0 & 0 & 1 \\
0 & 0 & 1 & 0 & 0 & 1 & 0 & 0 & 0 & 0 & 0 & 0 \\
0 & 0 & 0 & 1 & 0 & 0 & 1 & 0 & 0 & 1 & 1 & 0 \\
0.0222 & 0.0222 & 0 & 0 & -0.0667 & 0 & -0.0567 & -0.0667 & 0 & 0 & 0.0667 & -0.0667 \\
0 & 0 & 0.01667 & 0.01667 & -0.0667 & -0.0667 & 0.0667 & 0.0667 & -0.0667 & -0.0667 & 0.0667 & 0.0667
\end{bmatrix}
$$

The member stiffness matrix S, from Eqs. (8.70) and (8.72), is

$$S = \begin{bmatrix}
6.25 & 3.12 & 0 & 0 & 0 & 0 & 0 & 0 & 0 & 0 & 0 & 0 \\
3.12 & 6.25 & 0 & 0 & 0 & 0 & 0 & 0 & 0 & 0 & 0 & 0 \\
0 & 0 & 2.08 & 1.04 & 0 & 0 & 0 & 0 & 0 & 0 & 0 & 0 \\
0 & 0 & 1.04 & 2.08 & 0 & 0 & 0 & 0 & 0 & 0 & 0 & 0 \\
0 & 0 & 0 & 0 & 1.65 & 0.825 & 0 & 0 & 0 & 0 & 0 & 0 \\
0 & 0 & 0 & 0 & 0.825 & 1.65 & 0 & 0 & 0 & 0 & 0 & 0 \\
0 & 0 & 0 & 0 & 0 & 0 & 1.65 & 0.825 & 0 & 0 & 0 & 0 \\
0 & 0 & 0 & 0 & 0 & 0 & 0.825 & 1.65 & 0 & 0 & 0 & 0 \\
0 & 0 & 0 & 0 & 0 & 0 & 0 & 0 & 1.65 & 0.825 & 0 & 0 \\
0 & 0 & 0 & 0 & 0 & 0 & 0 & 0 & 0.825 & 1.65 & 0 & 0 \\
0 & 0 & 0 & 0 & 0 & 0 & 0 & 0 & 0 & 0 & 1.65 & 0.825 \\
0 & 0 & 0 & 0 & 0 & 0 & 0 & 0 & 0 & 0 & 0.825 & 1.65
\end{bmatrix} 10^5 \text{ (k-ft.)}$$

The input vector **Q** is determined by statics. Since all fixed-end moments are zero, we have

$$Q_1 = Q_2 = Q_3 = Q_4 = 0 \qquad Q_5 = 20 \text{ k.}, Q_6 = 0$$

Thus

$$\mathbf{Q} = \begin{bmatrix} 0 \\ 0 \\ 0 \\ 0 \\ 20 \\ 0 \end{bmatrix}$$

The stiffness matrix $K = ASA'$ and the flexibility matrix K^{-1} are calculated by digital computer, and from Eq. (8.13) the joint displacements **q** are found to be

$$\mathbf{q} = \begin{bmatrix} -1.21685 \times 10^{-4} \text{ rad.} \\ -1.21685 \times 10^{-4} \text{ rad.} \\ 2.12667 \times 10^{-4} \text{ rad.} \\ 2.12667 \times 10^{-4} \text{ rad.} \\ 9.13734 \times 10^{-3} \text{ ft.} \\ 4.68147 \times 10^{-3} \text{ ft.} \end{bmatrix}$$

The final end moments are determined from Eq. (8.73)

$$\mathbf{M} = \begin{bmatrix} 76.0916 \\ 76.0916 \\ 90.7005 \\ 90.7005 \\ -59.7379 \\ -42.1929 \\ -48.5076 \\ -76.0916 \\ -59.7379 \\ -42.1929 \\ -48.5076 \\ -76.0916 \end{bmatrix} \text{ k-ft.}$$

Since the fixed-end moments are all zero for the given loading, the member end moments **M** directly represent the final moments.

BIBLIOGRAPHY

1. J. C. Maxwell, *A Treatise on Electricity and Magnetism*, Third Edition. London: Oxford University Press, 1892. (See Vol. II, Chap. VI.)
2. J. W. S. Rayleigh, *The Theory of Sound*. New York: Dover, 1945. (Paperback edition, two volumes.)

3. W. Thomson and P. G. Tait, *Treatise on Natural Philosophy*. Cambridge University Press, 1879.

4. J. W. Forrester, *Industrial Dynamics*. New York: Wiley, 1961.

5. R. J. Stakler, "More about Homopolar Machines." *International Science and Technology*, p. 13, Sept. 1966.

6. C. Lanczos, *Linear Differential Operators*. New York: Van Nostrand, 1961. (See Chap. 5.)

7. E. A. Guillemin, *Theory of Linear Physical Systems*. New York: Wiley, 1963.

8. S. A. Stigant, *Matrix and Tensor Analysis in Electrical Network Theory*. London: Macdonald, 1964.

9. N. L. Schmitz and D. W. Novotny, *Introductory Electromechanics*. New York: Ronald Press, 1965.

10. W. W. Harmon and D. W. Lytle, *Electrical and Mechanical Networks*. New York: McGraw-Hill, 1962.

11. H. F. Olson, *Dynamical Analogies*. New York: Van Nostrand, 1943.

12. W. A. Lynch and J. G. Truxal, "Control System Engineering." *International Science and Technology*, *60*, March, 1966.

13. G. Kron, *Tensor Analysis of Networks*. New York: Wiley, 1939.

14. *Proceedings of the IEEE*, Special Issue on Computer-Aided Design, *55*, Nov. 1967.

15. C. K. Wang, *Matrix Methods of Structural Analysis*. Scranton, Pa.: International Textbook Co., 1966. (See Chap. 7.)

16. C. H. Norris and J. B. Wilbur, *Elementary Structural Mechanics*. New York: McGraw-Hill, 1960. (See Chap. 13.)

9

SATELLITE ORBITS

Satellite orbits provide an excellent illustration of the utility of systems methods in the solution of nonlinear problems of modern interest. Although precise computations of satellite trajectories require elaborate digital computer methods to include the effects of the many perturbations which are present, the results of these numerical integrations provide little insight for general understanding of orbit problems. For the latter purpose, it is much more useful to study ideal central force motions as a foundation and then introduce refinements and extensions. Lagrangian analytical methods yield basic solutions whose general properties can be determined. These, then, can serve as a basis for analysis and design. The purpose of this chapter, therefore, is to apply the methods of generalized mechanics to the study of central force motions and satellite orbits as an illustration of the broad application of these methods.

Planetary and satellite motions were among the earliest problems studied in mechanics and indeed led Newton to his mathematical analysis of the solar system and to the formulation of the law of universal gravitation. The foundations of modern *celestial mechanics* were laid by Euler, Lagrange, and Laplace in the eighteenth century. By 1800 celestial mechanics had been developed to a high degree of refinement. Modern interest in the field has been greatly stimulated by the massive space programs, in particular the successful development of artificial satellites (Sputnik–1957) and space vehicles (Mariner–1962). Calculation of the trajectories of these bodies requires extreme refinements, but no new principles.

The simplest problem in celestial mechanics is to determine the motion of two attracting bodies. Complete solution of the two-body problem requires 12 integrals of motion or constants of integration, six for the positions of the bodies and six more for the velocities. The two-body problem was solved by Newton. Application of the two-body solution to the motions of the planets about the Sun yields results in close, although not complete, agreement with observations. Refinements in the theory mainly take account of the attractions between the planets which produce second-order effects compared to the attraction to the Sun.

If three bodies are considered, 18 integrals or constants of motion are necessary for a complete solution. The conservation laws of mechanics lead to 10 of the required 18 integrals. Poincaré showed that there are no further integrals in closed form. Additional progress is therefore dependent on approximation methods, of which many have been developed.

Since the two-body problem is fundamental to all other work in celestial mechanics, and also is important in atomic physics and its practical applications, we treat this problem in some depth. The solution consists of finding the paths or orbits of the two bodies. The orbits may be closed, in which case the motion is periodic, or they may be open, in which case the motion is infinite. In the usual case, one of the bodies is massive compared with the other body, and if the small body moves in a closed orbit it is called a satellite of the massive body. In the theoretical calculations the two bodies are replaced by particles of the same masses located at the centers of the bodies. Newton proved that this idealization is exact for spheres of homogeneous or radially symmetric mass distribution.

Closed orbits for the two-body problem may be interpreted as a form of vibrations in two dimensions, also called space vibrations. There are, of course, other forms of two-dimensional vibrations in mechanics, as well as in electrical phenomena and other areas. Although the solution of the two-body problem in celestial mechanics is based on Newton's law of gravitation, the most characteristic feature is not the inverse-square law of force but rather that the force on each body is directed toward a common point, called the center of mass, which lies on the line joining the two bodies. In the center-of-mass coordinate system these forces are directed toward a fixed point. For a massive body and satellite, the center of mass sensibly coincides with the center of the massive body. Systems which have this central force characteristic are called central force or central motion systems. For example, a mass moving on a smooth horizontal table and attached by a coil spring to a fixed point is a central force system even though only one body is involved. A second example is "Rutherford scattering" in physics, i.e., the deflection of light nuclei, such as alpha particles, by heavy nuclei such as the nuclei of gold atoms.

9.1 CENTRAL FORCE MOTION

We now define central force motion more precisely as motion in which (1) the particle is acted on by a force directed toward or away from a central point, and (2) the magnitude of the force is a function only of the distance. Several characteristic features of the motion of a particle under a central force do not depend specifically on the law of force.

The first characteristic property of central force motion follows directly from the principle of conservation of angular momentum. Let **r** be the radius vector of the particle measured from the center of force and **p** the linear momentum of the particle. Then, the angular momentum **H** is defined as

$$\mathbf{H} = \mathbf{r} \times \mathbf{p} = \mathbf{r} \times m\dot{\mathbf{r}}$$

Differentiating with respect to time gives

$$\dot{\mathbf{H}} = \dot{\mathbf{r}} \times m\dot{\mathbf{r}} + \mathbf{r} \times m\ddot{\mathbf{r}} = \mathbf{r} \times \mathbf{F}$$

where from Newton's second law of motion $m\ddot{\mathbf{r}} = \mathbf{F}$, the force acting on the particle. Since for central force motion, **F** has the same direction as **r**, it follows that $\dot{\mathbf{H}} = 0$ and therefore

$$\mathbf{H} = \mathbf{C} \tag{9.1}$$

where **C** is a constant, i.e., angular momentum is conserved throughout the motion. Since **H** is perpendicular to **r** and is constant in direction as well as magnitude, we conclude that the radius vector **r** of the particle lies in a plane perpendicular to **H**. Thus, the particle trajectory for central force motion is always planar, and the problem of finding the motion reduces to one involving only two degrees of freedom.

Further properties of the motion may be examined using the Lagrangian written in terms of polar coordinates r, φ lying in the plane of motion. Thus

$$L = T - V = \tfrac{1}{2}\mu(\dot{r}^2 + r^2\dot{\varphi}^2) - V(r) \tag{9.2}$$

where μ is the reduced mass, defined by $\mu = \dfrac{m_1 m_2}{m_1 + m_2}$, and m_1 and m_2 are the masses of the two bodies.

There are two integrals of motion evident, one resulting from the ignorable coordinate φ and the other from the conservation of energy.

(1) Ignorable coordinate φ

$$\frac{\partial L}{\partial \dot{\varphi}} = \mu r^2 \dot{\varphi} = \text{const.} \tag{9.3}$$

The quantity $\tfrac{1}{2} r^2 \dot{\varphi}$ is the rate at which the radius vector of the particle sweeps out area in the plane of motion. Let this area be denoted by A. Then

$$\frac{dA}{dt} = \tfrac{1}{2} r^2 \dot{\varphi} = \text{const.} = \tfrac{1}{2} h \tag{9.4}$$

and is known as the sector velocity. In planetary motion this result is known as Kepler's law of areas. More generally, the quantity $\mu r^2 \dot{\varphi}$ represents the angular momentum of the system, and clearly is constant for central force motions.

(2) Integral of energy

$$T + V = \tfrac{1}{2}\mu(\dot{r}^2 + r^2\dot{\varphi}^2) V(r) = E \tag{9.5}$$

Elimination of $\dot{\varphi}$ by using (9.4) reduces the equation to one in a single dependent variable

$$\tfrac{1}{2}\mu\left(\dot{r}^2 + \frac{h^2}{r^2}\right) + V(r) = E$$

which has the integral

$$t = \int \frac{dr}{\sqrt{\dfrac{2}{\mu}(E - V) - \dfrac{h^2}{r^2}}} + C \tag{9.6}$$

The integral for φ may now be found as follows:

$$\varphi = \int \frac{h}{r^2}\,dt + C \tag{9.7}$$

Equations (9.6) and (9.7) give the orbit in parametric form. Elimination of t between these two equations results in the orbital equation in direct form

$$\varphi = \int \frac{h\,dr}{r^2\sqrt{\dfrac{2}{\mu}(E - V) - \dfrac{h^2}{r^2}}} + C \tag{9.8}$$

We now consider a central force law of the form

$$F(r) = -\alpha r^n \tag{9.9}$$

The force is attractive if $\alpha > 0$ and repulsive if $\alpha < 0$, where α is a parameter and n is a constant. This law includes cases of major interest. For example, the gravitational force in celestial motions as found by Newton is

$$F(r) = -\frac{GMm}{r^2} \tag{9.10}$$

where M and m are the masses of the two bodies and G is the universal constant of gravitation equal to 6.67×10^{-11} in MKS (meter-kilogram-second) units.

Coulomb's law of electrostatic force is similar:

$$F(r) = \frac{1}{4\pi\varepsilon_0} \cdot \frac{q_1 q_2}{r^2} \tag{9.11}$$

where q_1 and q_2 are point charges, and $\varepsilon_0 (= 8.85 \times 10^{-12}$ in rationalized

MKSA units) is called the permittivity of free space. A further example of interest is Hooke's law of force which holds for an ideal linear spring, i.e.,

$$F(r) = -kr \tag{9.12}$$

where k is the spring or elastic constant.

The potential energy corresponding to force law (9.9) is found by integration

$$V(r) = -\int -\alpha r^n \, dr + \text{const} = \frac{\alpha r^{n+1}}{n+1} + C \qquad n \neq -1$$

For the inverse square law, $n = -2$ and

$$V(r) = -\frac{\alpha}{r} + C \tag{9.13}$$

It is customary to choose the constant of integration so that V vanishes at infinity and therefore $C = 0$. Using next Hooke's law of force for which $n = 1$, we find

$$V(r) = \tfrac{1}{2}kr^2 + C \tag{9.14}$$

In this case the constant is usually taken so that V vanishes at $r = 0$, and again $C = 0$.

Equation (9.8) is integrable in terms of trigonometric functions for values of $n = +1, -2, -3$. This may be shown by making the change of variable $u = \dfrac{1}{r}$ so that

$$\varphi = \int \frac{-du}{\sqrt{\dfrac{2u}{\mu h^2}\left[E - \dfrac{\alpha u^{-n-1}}{n+1}\right] - u^2}} + C \tag{9.15}$$

If $n = -4$, for example, the exponent $-n - 1 = 3$, which means that (9.15) is an elliptic integral. A similar result is obtained if $n = 2$. (The case of $n = -1$ is not of practical interest because it corresponds to $V = \alpha \ln r$).

Because the motion is planar, the 12 constants of integration for the general two-body problem reduce to 8, and are further reduced to 4 by use of center-of-mass coordinates.

9.2 KEPLER MOTION

The most important central force motion is that governed by the inverse square law of force which, as we have seen, includes both gravitational and electrostatic forces. Central force motions of this kind are called Kepler motions in recognition of Kepler's fundamental work on planetary orbits.

The orbital equation for Kepler motion is found from Eq. (9.15) by setting $n = -2$.

$$\varphi = \int \frac{-du}{\sqrt{\dfrac{2}{\mu h^2}[E + \alpha u] - u^2}} + C$$

To simplify notation we set $u_0 = \dfrac{\alpha}{\mu h^2}$ and factor u_0^2 from the radical, so that

$$\varphi = \int \frac{-du}{u_0 \sqrt{1 + \dfrac{2E}{\alpha u_0} - \left[\dfrac{u}{u_0} - 1\right]^2}} + C \qquad (9.16)$$

Making the change of variable $w = u/u_0 - 1$ and defining a new parameter e^2 given by

$$e^2 = 1 + \frac{2E}{\alpha u_0} = 1 + \frac{2r_0 E}{\alpha}$$

reduces the integral to the standard from

$$\varphi = \int \frac{-dw}{\sqrt{e^2 - w^2}} + \varphi_0$$

$$= \cos^{-1} \frac{w}{e} + \varphi_0 \qquad -e < w < e \qquad e > 0 \qquad (9.17)$$

Hence, we find

$$w = \frac{u}{u_0} - 1 = \frac{r_0}{r} - 1 = e \cos(\varphi - \varphi_0)$$

or

$$\frac{r_0}{r} = 1 + e \cos(\varphi - \varphi_0) \qquad (9.18)$$

which is the polar form for the equation of a conic section of eccentricity e. The type of conic section depends on e, and it is known from geometry that for

$$\left. \begin{array}{l} e = 0, \text{ orbit is a circle} \\ 0 < e < 1, \text{ orbit is an ellipse} \\ e = 1, \text{ orbit is a parabola} \\ e > 1, \text{ orbit is a hyperbola} \end{array} \right\} \qquad (9.19)$$

For a given two-body system, the character of the motion depends on the total energy E through the relationship

$$E = \frac{\alpha}{2r_0}(e^2 - 1) \tag{9.20}$$

as summarized in Table 9.1.

Table 9.1

Eccentricity	Orbit	Total Energy
$e = 0$	Circle	$E = -\dfrac{\alpha}{2r_0}$
$0 < e < 1$	Ellipse	$-\dfrac{\alpha}{2r_0} < E < 0$
$e = 1$	Parabola	$E = 0$
$e > 1$	Hyperbola	$E > 0$

Existence of negative energy E violates no physical law, but is merely the result of choosing the constant of integration in Eq. (9.13) as zero. This corresponds to zero total energy at $r = \infty$. A body with zero initial velocity at infinity approaching a center of force moves in a parabolic orbit as shown in Table 9.1. If the body has kinetic energy at $r = \infty$ and a component of velocity directed toward the center of force, then $E > 0$ and the path is a hyperbola. On the other hand, if the body starts with a value of kinetic energy at a finite distance from the center of force such that $E < 0$, then the orbit is closed. A special case occurs if the initial velocity is directed toward the center of force, for it is clear that the body would then be on a collision course.

Let us now assume that a body is projected from some point r_e with a velocity $\dot{r}_e = v_e$ directly away from the center of force. Then, if

$$T + V = \tfrac{1}{2}\mu\dot{r}_e^2 - \frac{\alpha}{r_e} = 0$$

it is clear that the body will escape from the center of force, that is, it will move to infinity. The corresponding critical velocity v_e is known as the escape velocity and is given by

$$v_e = \sqrt{\frac{2}{\mu r_e}} \tag{9.21}$$

In elliptical motion, the radial distance r will vary between a maximum value r_2 and a minimum value r_1. These limits may be found from (9.17).

$$-e \le w = \frac{r_0}{r} - 1 \le e$$

or

$$\frac{r_0}{1+e} = r_1 \le r \le r_2 = \frac{r_0}{1-e} \tag{9.22}$$

Hence

$$r_1 = \text{minimum value of } r$$

$$r_2 = \text{maximum value of } r$$

We note that $r_2 \to \infty$ as $e \to 1$, and the motion is unbounded corresponding to the parabolic orbit of Table 9.1. If, on the contrary, $e = 0$ we see that

$$r_0 = r_1 = r_2$$

which is the special case of a circular orbit.

9.3 PLANETARY ORBITS

Planetary orbits for the two-body model of the Sun and a given planet may be found directly from the theory of Kepler motion. It has been established in the preceding section that Kepler motions are planar, although the planes are different for different planets. Since the Sun is many orders of magnitude more massive than any of the planets, some 333,000 times the mass of Earth for example, the Sun may be assumed to be stationary and the planets to revolve about a point at the center of the Sun. This is equivalent to taking the reduced mass μ to be the mass m of a given planet. From Newton's law of gravitation, Eq. (9.10), the constant α has the value

$$\alpha = GMm \tag{9.23}$$

where M is the mass of the Sun and G the gravitation constant.

The orbital equation in polar form is given by (9.18)

$$\frac{r_0}{r} = 1 + e \cos (\varphi - \varphi_0)$$

where $0 \le e < 1$ for a closed (elliptical) orbit. To evaluate the phase angle φ_0, we choose $r = r_2$ for $\varphi = 0$, which is satisfied by $\varphi_0 = \pi$. Hence, the equation is

$$\frac{r_0}{r} = 1 - e \cos \varphi \tag{9.24}$$

The general features of the planetary orbit are depicted in Figure 9.1 for an eccentricity $e = \frac{1}{2}$. There are two foci, f_1 and f_2, with the Sun located at f_1. The point of greatest distance from the Sun is called the aphelion, while the point of closest approach is known as the perihelion. These points correspond to $r = r_2$ and $r = r_1$ in our terminology, and it is clear that $\dot{r} = 0$ at both points. The points $r = r_0$, $\varphi = \pm \dfrac{\pi}{2}$ define the chord (known as the latus rectum) which passes through f_1 perpendicular to the major axis. The major and minor axes of the ellipse are given by

$$\left. \begin{aligned} a &= \frac{r_1 + r_2}{2} = \frac{r_0}{1 - e^2} \\ b &= \sqrt{r_1 r_2} = a\sqrt{1 - e^2} \end{aligned} \right\} \tag{9.25}$$

From these relations it follows that

$$\left. \begin{aligned} r_1 &= a(1 - e) \\ r_2 &= a(1 + e) \\ r_0 &= a(1 - e^2) \end{aligned} \right\} \tag{9.26}$$

The major axis a is also simply related to the total energy E of the motion. In Eq. (9.20) we substitute $\alpha = GMm$, $r_0 = a(1 - e^2)$ and thereby obtain

$$E = \frac{-GMm}{2a} \tag{9.27}$$

Thus, the energy depends only on the major axis and not on the eccentricity.

Application of Kepler's law of areas leads to the direct relationship between the area of the ellipse and the period of revolution of the planet. From Eq. (9.4), this relationship is

$$\frac{dA}{dt} = \tfrac{1}{2}h \qquad A = \tfrac{1}{2}ht$$

For one revolution, $A = \pi ab =$ area of ellipse and $t = \tau$. Hence

$$\tfrac{1}{2}h\tau = \pi ab = \pi a^2 \sqrt{1 - e^2}$$

and

$$\tau = \frac{2\pi a^2 \sqrt{1 - e^2}}{h}$$

Eliminate h by noting that u_0 is defined by

$$u_0 = \frac{GM}{h^2} = \frac{1}{r_0}$$

from which

$$h = \sqrt{GMr_0} = \sqrt{GMa(1 - e^2)}$$

Substitution of this value of h in the equation for τ gives

$$\tau = \frac{2\pi a^{3/2}}{(GM)^{1/2}}$$

or

$$\tau^2 = \frac{4\pi^2}{GM} a^3 \tag{9.28}$$

which is Kepler's third law of planetary motion.

Equations (9.24) through (9.28) are sufficient to calculate all quantities of interest in planetary motion when considered as an idealized two-body problem. These equations are in terms of masses and lengths, but it should be pointed out that the usual scientific units, the meter and kilogram, cannot be employed with sufficient precision for astronomical purposes. This comes about because distances in the solar system cannot be measured directly; only angular quantities are directly observable. But these observations can be made with high precision, a tenth of a second of arc for a single observation up to a thousandth of a second of arc or possibly better in favorable cases where periodic observations can be made. Relative distances in the solar system can be inferred to eight or nine significant figures. It is customary to express relative distances in terms of astronomical units, the mean distance of the Earth to the Sun being taken as unity. Masses of solar system bodies are usually expressed relative to the Sun. For example, the mass of the Earth is 1/333,432 that of the Sun, and the mass of the Earth plus Moon is 1/329,390 on the same basis.

Most of the principal planets have eccentricities only slightly greater than zero, that is, their orbits are nearly circular; for instance, $e = 0.007$ for Venus, 0.0167 for Earth, 0.048 for Jupiter, and 0.008 for Neptune. (Mercury and Pluto are exceptions, with eccentricities of 0.206 and 0.249, respectively.) However, even small eccentricities are reflected in significantly different maximum and minimum distances. For example, for the Earth's orbit, the mean distance to the Sun is 93,200,000 miles, and the maximum and minimum distances are 91,600,000 and 94,800,000 miles, respectively, a significant difference.

9.4 EARTH SATELLITES

Artificial terrestrial satellites were discussed in a serious scientific sense by Newton and others, and the fundamental conditions necessary to launch such satellites successfully have long been known. There was no possibility of making the satellite idea a reality, however, until the technical problem of imparting sufficient momentum and energy to carry a vehicle through the Earth's atmosphere and overcome the Earth's gravitational field could be solved. Development of high energy chemical fuels, coupled with precision automatic control and guidance systems, have been the key factors in the solution of the launching problem.

Artificial satellites are powered by rocket motors. Hot gases are ejected at high velocity from a tail nozzle on the rocket and the rocket is propelled by the so-called reaction force. The thrust may be calculated by application of the principle of conservation of linear momentum. Let $\mathbf{F}$ be the resultant external force acting on the rocket. Then

$$
\begin{aligned}
\mathbf{F} = \dot{\mathbf{p}} &= \left[\frac{d}{dt} m\mathbf{v} + \int (\mathbf{v_0} - \mathbf{v})\, dm \right] \\
&= m\frac{d\mathbf{v}}{dt} - \mathbf{v_0}\frac{dm}{dt}
\end{aligned}
\tag{9.29}
$$

where $m = m(t)$ is the mass of the rocket at time t, $\mathbf{v}$ is the velocity of the rocket, and $\mathbf{v_0}$ is the terminal velocity of the ejected gases measured relative to the rocket velocity. Equation (9.29) may be rearranged in the form

$$
m\frac{d\mathbf{v}}{dt} = \mathbf{F} + \mathbf{v_0}\frac{dm}{dt} = \mathbf{F} + \mathbf{T}
\tag{9.30}
$$

where $\mathbf{T}$ may be thought of as the "thrust" on the rocket due to the ejection of the tail gases. We note that since $\frac{dm}{dt}$ is negative, the thrust $\mathbf{T}$ is opposite in direction to the velocity $\mathbf{v_0}$ of the ejected gas.

For a rocket rising through the atmosphere,

$$
\mathbf{F} = -\mathbf{R}(v, h) - m\dot{\mathbf{g}}
$$

where $\mathbf{R}$ is the atmospheric drag on the rocket and is mainly a function of speed v and altitude h. Hence, (9.30) takes the form

$$
m\frac{d\mathbf{v}}{dt} = -\mathbf{R} - m\dot{\mathbf{g}} + \mathbf{T}
\tag{9.31}
$$

At take-off $\mathbf{R} = 0$, and the condition for the rocket to rise is $\mathbf{T} > m\mathbf{g}$. Equation (9.31) may be solved numerically if $\mathbf{R}$ is known as a function of speed and altitude of the rocket. In practice, rockets have several stages, which are shed at predetermined altitudes to make better use of fuel and to permit increase in payload. It is obvious, of course, that the changes in mass and other parameters in the rocket equation must be taken into account for multistage rockets.

Once clear of the atmosphere, which extends to an elevation of about 100 miles, with very slight effects to about 500 miles, the rocket and satellite are subjected only to gravitational forces. Under this condition,

$\mathbf{R} = 0$ and Eq. (9.31) reduces to

$$m \frac{d\mathbf{v}}{dt} = -m\mathbf{g} + \mathbf{T} \tag{9.32}$$

A further simplification occurs in the idealized case where the rocket is free of all external forces. Then (9.29) becomes

$$\dot{\mathbf{p}} = \frac{d}{dt}\left[m\mathbf{v} + \int (\mathbf{v_0} - \mathbf{v})\, dm \right] = 0$$

If $\mathbf{v}$ and $\mathbf{v_0}$ are collinear, which is the usual case for rockets, the scalar form may be used, and an elementary integration yields the relation

$$v = v_0 \ln \frac{M}{m} \tag{9.33}$$

The constant M is the initial mass of the rocket for $v = 0$. Note that if v_0 is negative with respect to v, the rocket will be decelerated. Similar equations apply for changing the direction of the motion or the orientation of the vehicle. It is evident that small auxiliary jets may be used to make slight corrections in the motions, and that retro-jets may be used to decelerate the vehicle (for re-entry into the Earth's atmosphere, for example).

Following the powered flight through the atmosphere, the final stage of the rocket is fired to give the desired final velocity. This will depend on the purpose of the satellite. For orbiting the Earth, the velocity will be perpendicular to or nearly perpendicular to the Earth's gravitational field. For deep space probes having highly elliptic orbits, the velocity will have a component directed away from the Earth. After separation of the final stage of the rocket, the satellite moves in space subject only to the gravitational field. In the idealized two-body model, the center of force is at the center of the Earth and the inverse square law of force holds.

This model is a close approximation, but certain refinements must be made in order to account for the slight oblateness of the Earth, the effect of the Moon's and Sun's gravitational pull, solar radiation, and so on. These additional forces produce higher-order effects on the motion. These usually can be ignored during a single revolution of a satellite, but can have important cumulative effects over many revolutions.

Let us measure time from the instant of separation of the last stage of the rocket, and let the position vector and velocity of the satellite at $t = 0$ be $\mathbf{r}(0)$ and $\mathbf{v}(0)$, respectively. The orbit of the satellite will then lie in the plane defined by the center of the Earth and the velocity vector of the satellite. Using polar coordinates in this plane with origin at the

$$\left\|\vec{e_1}\right\| = \left\|\vec{e_2}\right\| = 1$$

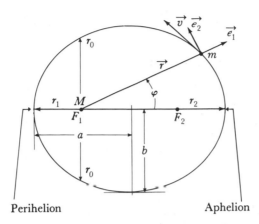

Perihelion   Aphelion

FIGURE 9.1 Diagram for planetary orbits.

center of the Earth, we have (cf. Fig. 9.1)

$$\mathbf{r}(0) = r(0)\mathbf{e_1}, \ \mathbf{v}(0) = \dot{\mathbf{r}}(0) = \dot{r}(0)\mathbf{e_1} + r(0)\dot{\varphi}(0)\mathbf{e_2}$$

The total energy is

$$E = T + V = \tfrac{1}{2}m[\dot{r}^2(0) + r^2(0)\dot{\varphi}^2(0)] - \frac{GMm}{r(0)} \qquad (9.34)$$

From Eq. (9.27), the major axis a of the orbit, which we assume to be elliptical, is given by

$$a = \frac{GMm}{2E} \qquad (9.35)$$

and the eccentricity by

$$e^2 = 1 + \frac{2h^2 E}{G^2 M^2 m} = 1 - \frac{h^2}{GMa} \qquad (9.36)$$

where

$$h = r^2(0)\dot{\varphi}(0)$$

On physical grounds, it is apparent that there is a minimum energy necessary for a stable closed orbit about the Earth to exist. This minimum energy may be calculated from Eq. (9.35) by taking a to be the radius of the smallest permissible circular orbit. Since the radius of the Earth is 3960 miles and the altitude necessary to avoid serious viscous effects of the atmosphere* is somewhat in excess of 100 miles, we may assume the

---

* A satellite which comes below about 90 miles will usually stay in orbit for only a few revolutions; above about 500 miles the effect of atmospheric drag is almost negligible.

minimum value of a to be about 4100 miles; hence, these results for E are

$$E = -\frac{GMm}{2a} = \frac{-3.99 \times 10^{14}m}{2 \times 4100 \times 1610} \quad \text{where} \quad GM = 6.67 \times 10^{-11}$$

$$\times 5.98 \times 10^{24}$$

$$= -3.02 \times 10^7 m \text{ joules (in } MKS \text{ units)}$$

This corresponds to a satellite velocity at altitude 4100 miles of

$$v = \sqrt{2\left(\frac{E}{m} + \frac{GM}{a}\right)} = \sqrt{\frac{GM}{a}} = \sqrt{\frac{-2E}{m}}$$

$$= \sqrt{6.04 \times 10^7} = 7780 \text{ meters/sec.} = 17450 \text{ miles/hr.}$$

and a period

$$\tau = \frac{2\pi a}{v} = \frac{2\pi \times 6.6 \times 10^6}{7.78 \times 10^3} = 5330 \text{ sec.} = 89 \text{ min.}$$

A related quantity of interest is the escape velocity v_e of a body propelled from the Earth's surface. This quantity is given by Eq. (9.21) as follows:

$$v_e = \sqrt{\frac{2}{\mu r_e}} = \sqrt{\frac{2GM}{r_e}} = \sqrt{\frac{2 \times 3.99 \times 10^{14}}{6.37 \times 10^6}}$$

$$= 11,200 \text{ meters/sec.} = 25,100 \text{ miles/hr.}$$

Example I. A satellite is set into orbit at an altitude of 500 miles. Find the velocity required for the orbit to be a circle, and also the period.

SOLUTION

Orbital Eq. (9.24) reduces to the form

$$\frac{r_0}{r} = 1 \text{ for a circular orbit. Hence}$$

$$r = r_0 = r(0) = \frac{h^2}{GM} = \frac{r^2(0)v^2(0)}{GM}$$

Solving for $v(0)$, we obtain in MKS units

$$v(0) = \sqrt{\frac{GM}{r(0)}} = \sqrt{\frac{6.67 \times 10^{-11} \times 5.98 \times 10^{24}}{6.37 \times 10^6 + 500 \times 1610}}$$

$$= 7460 \text{ meters/sec.} = 16,700 \text{ miles/hr.}$$

In this example, $\mathbf{v}(0)$ is perpendicular to $\mathbf{r}(0)$ and, hence, the period τ may be computed from the relation

$$\tau = \frac{2\pi r(0)}{v(0)} = \frac{6.28 \times 7.18 \times 10^6}{7.46 \times 10^3} = 6060 \text{ sec.} = 101 \text{ min.}$$

Example 2. Find the satellite orbit for an initial altitude of 500 miles and an initial velocity perpendicular to r of 20,000 miles/hr.

SOLUTION

From Eq. (9.24)

$$\frac{1}{r(0)} = \frac{1}{r_0}(1 + e) = \frac{GM}{h^2}(1 + e)$$

or

$$1 + e = \frac{r(0)v^2(0)}{GM} = \frac{7.18 \times 10^6 \times (0.893)^2 \times 10^8}{6.66 \times 10^{-11} \times 5.98 \times 10^{24}} = 1.44$$

and hence, $e = 1.44 - 1 = 0.44$.
Also

$$r_0 = (1 + e)r(0) = 1.44 \times 7.18 \times 10^6 = 1.033 \times 10^7 \text{ meters}$$

$$= \frac{1.033 \times 10^7}{1610} = 6420 \text{ miles}$$

This is equivalent to an altitude of

$$6420 - 3960 = 2460 \text{ miles}$$

Next

$$r_1 = \text{minimum value of } r = \frac{r_0}{1 + e} = r(0)$$

so that the minimum altitude is 500 miles, and

$$r_2 = \text{maximum value of } r$$

$$= \frac{r_0}{1 - e} = \frac{6420}{0.56} = 11,500 \text{ miles}$$

$$= 11,500 - 3960 = 7540 \text{ miles altitude}$$

To find the period, we use Kepler's third law

$$\tau^2 = \frac{4\pi^2}{GM}a^3$$

where

$$a = \frac{r_1 + r_2}{2} = \frac{7.18 \times 10^6 + 18.5 \times 10^6}{2}$$

$$= 1.27 \times 10^7 \text{ meters}$$

Hence

$$\tau = 2\pi\sqrt{\frac{(1.27 \times 10^7)^3}{3.99 \times 10^{14}}} = 2\pi\sqrt{\frac{2.05 \times 10^{21}}{3.99 \times 10^{14}}}$$

$$= 14,400 \text{ sec.} = 240 \text{ min.}$$

Example 3. Find the initial conditions for a Syncom satellite.

SOLUTION. It is clear that a Syncom satellite (communications satellite synchronized with the Earth's rotation) must orbit in the Earth's equatorial plane with angular velocity equal to that of the Earth. Furthermore, since the angular velocity is constant, the orbit must be a circle and the radius may be deduced from Kepler's third law:

$$r^3 = a^3 = GM\left(\frac{\tau}{2\pi}\right)^2 = 3.99 \times 10^{14}\left(\frac{8.64 \times 10^4}{6.28}\right)^2$$

$$= 3.99 \times 10^{14} \times 1.89 \times 10^8$$

$$= 7.55 \times 10^{22}$$

and

$$r = a = 4.23 \times 10^7 \text{ meters} = 26{,}300 \text{ miles}$$

which is equivalent to an altitude of

$$26{,}300 - 3960 = 22{,}340 \text{ miles}$$

The velocity is constant at

$$v = \frac{2\pi a}{\tau} = \frac{2\pi \times 26{,}300}{24} = 6900 \text{ miles/hr.}$$

Example 4. Calculate the orbit for a satellite with initial altitude of 2000 miles and initial velocity $\mathbf{v}(0) = 8350\mathbf{e}_1 + 14{,}470\mathbf{e}_2$ miles/hr., where $\mathbf{e}_1$ and $\mathbf{e}_2$ are the unit vectors for polar coordinates.

SOLUTION

$$E = T + V = \tfrac{1}{2}mv^2(0) - \frac{GMm}{r(0)}$$

$$v(0) = \sqrt{8350^2 + 14{,}470^2} = 16{,}700 \text{ miles/hr.}$$

$$= 7460 \text{ meters/sec.}$$

$$r(0) = (3960 + 2000)1610 = 9.60 \times 10^6 \text{ meters}$$

$$GM = 6.67 \times 10^{-11} \times 5.98 \times 10^{24} = 3.99 \times 10^{14} MKS$$

Hence

$$E = \left[\tfrac{1}{2}(7460)^2 - \frac{3.99 \times 10^{14}}{9.60 \times 10^6}\right]m$$

$$= [2.78 \times 10^7 - 4.16 \times 10^7]m = -1.38 \times 10^7 m$$

and

$$a = \frac{-GMm}{2E} = \frac{3.99 \times 10^{14}}{2.76 \times 10^7} = 1.445 \times 10^6 \text{ meters}$$

$$= 8970 \text{ miles}$$

$$e^2 = 1 + \frac{2h^2 E}{G^2 M^2 m}$$

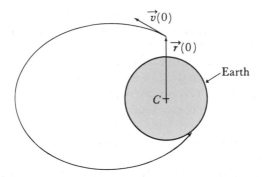

FIGURE 9.2 Satellite orbit illustration for Example 4.

where

$$h = r^2(0)\dot{\varphi}(0) = r(0)v_2(0)$$
$$= (9.60 \times 10^6)(6.47 \times 10^3) = 6.21 \times 10^{10} MKS$$

[Note: $v_2(0)$ = component of $\mathbf{v}(0)$ along $\mathbf{e}_2$]
Thus, we find

$$e^2 = 1 - \frac{2(6.21 \times 10^{10})^2(1.38 \times 10^7 m)}{(3.99 \times 10^{14})^2 m}$$

$$= 1 - \frac{10.63 \times 10^{28}}{15.9 \times 10^{28}} = 1 - 0.669 = 0.331$$

$$e = 0.576$$

$$b = a\sqrt{1 - e^2} = 1.445 \times 10^6\sqrt{0.669}$$
$$= 1.176 \times 10^6 \text{ meters} = 7310 \text{ miles}$$

$$r_2 = a(1 + e) = 8970(1.576) = 14,100 \text{ miles}$$

$$r_1 = a(1 - e) = 8970(0.424) = 3810 \text{ miles}$$

It is evident that this orbit will intersect the Earth, despite the fact that the satellite is launched at the relatively high altitude of 2000 miles. (See Figure 9.2).

9.5 PERTURBATIONS OF SATELLITE ORBITS

The first order theory of satellite orbits assumes a spherically sym-metric Earth, neglects the effects of the Earth's atmosphere, and neglects the gravitational attraction of other bodies in the solar system. Under these circumstances, we have seen that the orbit of a satellite is an ellipse of constant size and shape in a plane whose orientation remains fixed in relation to the stars. The idealized theory gives the main features of the motion, but certain refinements are required to take account of perturbations which occur in practice.

The most important perturbations are those caused by the oblateness of the Earth, atmospheric drag, and the gravitational fields of the Sun and Moon. Solar radiation pressure may also be a factor for balloon-type satellites. Deep space probes and lunar and interplanetary space vehicles present additional problems because the gravitational fields of several bodies of the solar system must be accounted for in the orbit calculations.

Theoretical and practical aspects of these problems have received much attention in recent years because of their importance in space programs. Perturbations due to the figure of the Earth and the gravitational fields of the planets have, of course, long been studied as a part of celestial mechanics, and these have carried over into the space programs. On the other hand, the theory of satellite orbits in the presence of an atmosphere has been largely developed in modern times, motivated by the problems associated with the development of artificial satellites.

The reader is referred to the bibliography of this chapter for treatment of the perturbations of satellite orbits and other more specialized treatments of space mechanics. King-Hele (Ref. 4) gives a basic introduction at an intermediate mathematical level to these problems, covering especially the effects of atmospheric drag. Studies of the kinematics and dynamics of satellite orbits, again at an intermediate mathematical level, are contained in selected reprints published by the American Association of Physics Teachers (Ref. 5). More general aspects of orbital mechanics and space flight are to be found at a fairly elementary level using graphical methods (Ref. 6), and at a more advanced mathematical level (Ref. 7). Some of the basic problems and methods of flight mechanics which enter into exploration of the solar system, including planetary swingby or gravity-assisted trajectory methods that offer possibilities of substantially improved pay loads and lower launching energies, are discussed at a qualitative level in recent articles (see Ref. 8, for example). A general introduction to the space sciences with a minimum of mathematics is available in a summary prepared in cooperation with the National Aeronautics and Space Administration (Ref. 9).

BIBLIOGRAPHY

1. S. W. McCuskey, *An Introduction to Advanced Dynamics*. Reading, Mass.: Addison-Wesley, 1959.
2. *Space Navigation Handbook*—Navpers 92988. Published under the direction of Chief of Naval Personnel. Washington, D.C.: U.S. Government Printing Office, 1962.
3. *NASA University Conference*, Vol. 1 and 2. Proceeding of the Conference held November 1962, Chicago, Ill., Published by the office of Scientific and Technical Information, NASA, Washington, D.C.: December, 1962.

4. D. King-Hele, *Theory of Satellite Orbits in an Atmosphere*. London: Butterworths, 1964.
5. *Kinematics and Dynamics of Satellite Orbits, Selected Reprints*. Published by American Institute of Physics, 335 E. 45 St., New York 10017, 1963.
6. F. A. Heacock, *Graphics in Space Flight*. New York: McGraw-Hill, 1964.
7. D. A. Pogorelov, *Fundamentals of Orbital Mechanics*. San Francisco: Holden-Day, 1964.
8. R. F. Porter, "Trajectories of Unmanned Spacecraft." *Battelle Technical Review, 15* 3, 1966.
9. S. Glasstone, *Sourcebook on the Space Sciences*. Princeton, N.J.: Van Nostrand, 1965.

10

GENERALIZED ENERGY CONVERTERS

The interface between electromagnetic and mechanical systems is identified by the conversion of energy between the two forms. Electromagnetism and mechanics are thus brought together at the most fundamental level. The basic relations required for the systems study of electromechanical energy conversion processes were developed in Chaps. 6 and 7 of this book. In the present chapter, we apply systems methods to the largest and most important class of electromechanical energy converters, that is, those based on the magnetic field component of the Lorentz law of force $\mathbf{F} = q\mathbf{v} \times \mathbf{B}$.

Although many physical embodiments of the Lorentz magnetic energy conversion process exist, the traditional form uses current-carrying conductors moving through a magnetic field. Both linear and rotational motions are used, but most converters are of the rotational type and are generally called *rotating machines*. Such machines have innumerable applications in both the most traditional and the most modern technologies. They are the source of virtually all electric power and provide easily controllable sources of both electrical and mechanical power. They are available in sizes ranging from a fraction of a watt such as are used in control systems to over 1000 megawatt capacity for electric power generation. Automatic feedback control and servosystems depend fundamentally on rotating electrical machines.

Generalized Converters. The systems approach to the theory and design of rotating energy converters stems from the pioneer work of Kron in the 1930's. Kron proposed and developed the concept of the generalized

326

or ancestor energy converter (Ref. 1). The many forms of apparently different energy converters are, in fact, all fundamentally similar and can be derived from a common theoretical model called a *generalized machine*.

Interest in the generalized machine extends beyond theoretical interest or the desire for a neater packaging of knowledge. Extensive application of rotating machines in automatic control systems and automation generally, together with the development of large interconnected electric power systems or grids, have required more general and more powerful methods of analysis. System stability and transient performance characteristics are needed for advanced design of these systems. Earlier design methods, which were concerned primarily with steady-state performance, are no longer equal to the requirements imposed by modern complex many-variable systems.

Kron's model was developed by generalization from the d-c commutator machine. He used the methods of tensor analysis to express the fundamental equations. The same ideas can be expressed in terms of matrices and vectors, as will be done here.* The generalized equations of motion for the machine can be obtained by any one of several analytical techniques; Lagrange's equations are used in the following treatment.

Although various levels of generalization are possible in considering generalized machines, it is sufficient for most system purposes to consider a four-pole machine with one pair of poles along the vertical axis and the other pair along the horizontal axis. These axes are referred to as direct and quadrature axes, respectively. The magnetic fields of the stator and rotor windings are correspondingly called direct and quadrature. A schematic diagram for the generalized machine, together with schematic circuit diagram and magnetic field directions, are shown in Figure 10.1.

Analysis of the generalized machine is based on the assumption that all effects are essentially linear and that the magnetic flux is sinusoidally distributed for both the stator and armature fields. These assumptions are reasonable, and make it possible to carry out calculations for most machines of practical interest. Nonlinear effects can be taken into account by special methods where necessary.

Machines derived from the generalized machine may be classified by groups into commutator, induction, and synchronous forms. Commutator machines are also d-c machines and are the only ones which use actual commutation brushes. Induction and synchronous machines are a-c machines and use slip rings instead of commutators, although hypothetical commutators with brushes may be visualized to permit

---

* Because tensors are limited to essentially geometric applications, we prefer in this book to use matrices and vectors which can be applied to both geometric and nongeometric problems.

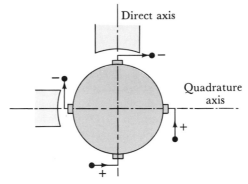

(a) Schematic diagram for machine

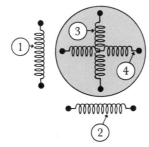

(b) Schematic circuit diagram

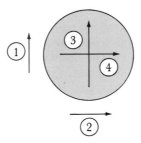

(c) Field directions

FIGURE 10.1

identification of coil polarities and to locate magnetic field directions. In the latter case, slip ring action is simulated by imagining the brushes to rotate at the same speed as the rotor.

    Table 10.1 lists the rotating machines in common use which can be derived from the general machine by choosing specific stator, rotor, and brush arrangements. Reference 3, p. 165, gives a more extensive table of 22 such derivative machines.

Table 10.1 Rotating Machines in Common Use

No.	Machine	a-c or d-c	No. of Phases	Commutator or slip rings
1	Generalized		Theoretical or basic machine	
2	Synchronous motor	a-c	Usually polyphase	Slip rings
3	Synchronous alternator	a-c	Usually polyphase	Slip rings
4	Induction motor	a-c	Usually polyphase	Squirrel cage uses neither; Wound rotor uses slip rings
5	Synchro or selsyn	a-c	Usually 1- or 3-phase	
6	Servomotor	a-c	2-phase	
7	Repulsion motor	a-c	1-phase	Commutator
8	Direct current motor	d-c	—	Commutator
9	Direct current generator	d-c	—	Commutator

10.1 EQUATIONS OF MOTION

Consider a generalized machine of the type described with the armature rotating in a counterclockwise direction as shown schematically in Figure 10.2. Assume that the armature coils are connected via slip rings to external terminals so that the magnetic axes of the armature rotate with the coils. The stator coils and their magnetic axes are of course fixed in space. To carry out the analysis, it is necessary to formulate energy

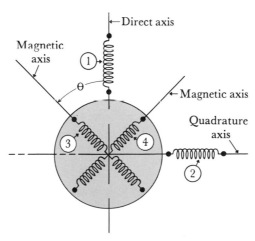

FIGURE 10.2 Schematic diagram of generalized machine showing rotation of armature.

functions for the system. Magnetic energy is stored in the four windings of the machine, kinetic energy is associated with the rotating armature, mechanical energy is exchanged through shaft torque, and, finally, conversion of energy between electromagnetic and mechanical forms occurs as the armature winding rotates through the magnetic field of the stator.

To proceed with the formulation of the energy functions, the four windings of the generalized machine are identified by numbers 1, 2, 3, and 4, as shown in Figure 10.1b. Coils 1 and 2 are called the direct and quadrature field windings respectively, while coils 3 and 4 are the corresponding armature windings. The electric charges q_1, q_2, q_3, q_4 in these windings represent four degrees of freedom, to which must be added the angle of rotation θ of the armature denoted by q_5, making a total of five degrees of freedom for the machine as a whole. Energy functions for the system are then given by

$$\left.\begin{aligned} T &= \tfrac{1}{2}\dot{\mathbf{q}}'M\dot{\mathbf{q}} \\ F &= \tfrac{1}{2}\dot{\mathbf{q}}'R\dot{\mathbf{q}} \\ P &= \mathbf{Q}\cdot\dot{\mathbf{q}} \end{aligned}\right\} \tag{10.1}$$

where R is the resistance matrix, $\mathbf{Q}$ the generalized force vector, and $\dot{\mathbf{q}}$ the generalized velocity vector. Matrix M is the kinetic energy matrix for the system given by

$$M = \left[\begin{array}{cccc|c} & L & & & 0 \\ \hline M_{51} & M_{52} & M_{53} & M_{54} & J \end{array}\right] \tag{10.2}$$

where L is the inductance matrix for the system in the expression for U_m derived from Eq. (7.38).

The self inductances L_{11}, L_{22}, L_{33}, L_{44} for a machine with uniform air gaps do not vary with angular position θ and therefore may be taken as constants. On the other hand, the mutual inductances do in general vary with θ. Consider, for example, the mutual inductance between coils 1 and 3. The maximum value occurs when the magnetic fields of the two coils are aligned, i.e., for $\theta = 0$. Denoting this maximum value by L_{13}, the value of the mutual inductance for any angle θ is given by $L_{13}\cos\theta$ where sinusoidally distributed windings are assumed.

Following this pattern, the inductance coefficients for all the coils are readily expressed.

L_{11}, L_{22}, L_{33}, L_{44} are constants

$$\left.\begin{aligned} L_{12}\cos\frac{\pi}{2} &= 0 \\ L_{21}\cos\left(-\frac{\pi}{2}\right) &= 0 \end{aligned}\right\} \quad \text{since coils 1 and 2 are perpendicular and fixed in position}$$

$$L_{13} \cos \theta = L_{31} \cos \theta \left.\right\}$$
$\qquad$ since coil 3 is aligned with coil 1 for $\theta = 0$

$$L_{14} \cos \left(\theta - \frac{\pi}{2}\right) = L_{14} \sin \theta \atop = L_{41} \sin \theta \left.\right\}$$
$\qquad$ since coil 4 is aligned with coil 1 for $\theta = \pi/2$

$$L_{23} \cos \left(\theta + \frac{\pi}{2}\right) = -L_{23} \sin \theta \atop = -L_{32} \sin \theta \left.\right\}$$
$\qquad$ since coil 3 is aligned with coil 2 for $\theta = -\pi/2$

$$L_{24} \cos \theta = L_{42} \cos \theta \left.\right\}$$
$\qquad$ since coil 2 is aligned with coil 4 for $\theta = 0$

$$L_{34} \cos \left(-\frac{\pi}{2}\right) = 0 \atop L_{43} \cos \left(+\frac{\pi}{2}\right) = 0 \left.\right\}$$
$\qquad$ since coils 3 and 4 are perpendicular for all values of θ

Using these values, the inductance matrix is expressed as a function of θ in Eq. (10.3).

$$L(\theta) = \begin{bmatrix} L_{11} & 0 & L_{13} \cos \theta & L_{14} \sin \theta \\ 0 & L_{22} & -L_{23} \sin \theta & L_{24} \cos \theta \\ L_{31} \cos \theta & -L_{32} \sin \theta & L_{33} & 0 \\ L_{41} \sin \theta & L_{42} \cos \theta & 0 & L_{44} \end{bmatrix} \qquad (10.3)$$

Substituting the expressions for T, F, and P from Eq. (10.1) in Lagrange's equation (6.31), and writing the resulting equations in matrix form yields

$$\frac{d}{dt}(M\dot{q}) + R\dot{q} = M\ddot{q} + \dot{M}\dot{q} + R\dot{q} = Q \qquad (10.4)$$

It is convenient to separate the electrical and mechanical portions of Eq. (10.4) which is readily accomplished because of the forms of matrices M and R.

Using the electrical portion of (10.4) gives

$$\frac{d}{dt}(LI) + RI = L\dot{I} + \dot{L}I + RI = V$$

where I represents the currents $\dot{q}_1$, $\dot{q}_2$, $\dot{q}_3$, $\dot{q}_4$ and V the terminal voltages. This expression may also be written in the form

$$(LD + \dot{L} + R)I = ZI = V \qquad (10.5)$$

where Z may be thought of as the operator impedance matrix for the rotating machine.

Matrix elements M_{51}, M_{52}, M_{53}, M_{54} are associated with the torque equation for the system. Rather than evaluate these elements individually, it is simpler to determine the torque equation directly by using the result obtained in Eq. (7.40), which when applied to the present case gives

$$\tau_m = \frac{\partial U_m}{\partial \theta}$$

where U_m is the magnetic energy of the system and τ_m is the reaction torque due to the magnetic forces acting on the armature windings. Therefore

$$\tau_m = \frac{\partial U_m}{\partial \theta} = \tfrac{1}{2}\mathbf{I}' \frac{\partial L}{\partial \theta} \mathbf{I} \tag{10.6}$$

The term $\dfrac{\partial L}{\partial \theta}$ may be evaluated by differentiation of Eq. (10.3) which yields

$$\frac{\partial L}{\partial \theta} = \begin{bmatrix} 0 & 0 & -L_{13}\sin\theta & L_{14}\cos\theta \\ 0 & 0 & -L_{23}\cos\theta & -L_{24}\sin\theta \\ -L_{31}\sin\theta & -L_{32}\cos\theta & 0 & 0 \\ L_{41}\cos\theta & -L_{42}\sin\theta & 0 & 0 \end{bmatrix} \tag{10.7}$$

Substituting (10.7) in (10.6) and carrying out the matrix multiplication gives the following scalar equation for τ_m

$$\tau_m = -(L_{23}I_2I_3 - L_{14}I_1I_4)\cos\theta - (L_{13}I_1I_3 + L_{24}I_2I_4)\sin\theta \tag{10.8}$$

The torque equation for the system may now be calculated in the form

$$\tau_m + J\ddot{\theta} = \tau \tag{10.9}$$

In summary, the system equations for the generalized machine may be expressed as follows:

$$(MD + \dot{M} + R)\dot{\mathbf{q}} = \mathbf{Q} \tag{10.10}$$

or in terms of electrical and mechanical portions

$$\left. \begin{aligned} (LD + \dot{L} + R)\mathbf{I} = Z\mathbf{I} = \mathbf{V} \\ \tau_m + J\ddot{\theta} = \tau \end{aligned} \right\} \tag{10.11}$$

In these equations

$$\dot{\mathbf{q}} = \begin{bmatrix} I_1 \\ I_2 \\ I_3 \\ I_4 \\ \dot{\theta} \end{bmatrix} \qquad \mathbf{Q} = \begin{bmatrix} V_1 \\ V_2 \\ V_3 \\ V_4 \\ \tau \end{bmatrix}$$

τ_m is given by Eq. (10.8), and the elements of the operator impedance matrix Z have the values listed in (10.12).

$$\left.\begin{aligned}
Z_{11} &= L_{11}D + R_{11} \\
Z_{22} &= L_{22}D + R_{22} \\
Z_{33} &= L_{33}D + R_{33} \\
Z_{44} &= L_{44}D + R_{44} \\
Z_{12} &= Z_{21} = 0 \\
Z_{13} &= Z_{31} = L_{13}\,(\cos\theta D - \sin\theta\,\dot\theta) \\
Z_{14} &= Z_{41} = L_{14}\,(\sin\theta D + \cos\theta\,\dot\theta) \\
Z_{23} &= Z_{32} = -L_{23}\,(\sin\theta D + \cos\theta\,\dot\theta) \\
Z_{24} &= Z_{42} = L_{24}\,(\cos\theta D - \sin\theta\,\dot\theta) \\
Z_{34} &= Z_{43} = 0
\end{aligned}\right\} \qquad \textbf{(10.12)}$$

The sign convention for $\mathbf{Q}$ is determined by Eq. (6.33) in which $\mathbf{Q}_\theta \cdot \dot{\mathbf{q}} > 0$ corresponds to energy flow into the machine. It follows, therefore, that

(1) For a generator, $\tau > 0 \qquad \dot\theta > 0 \qquad V_i > 0 \qquad I_i < 0$

(2) For a motor, $\qquad \tau < 0 \qquad \dot\theta > 0 \qquad V_i > 0 \qquad I_i > 0$

10.2 COMMUTATOR MACHINES

The basic generalized machine depicted in Figure 10.2 uses slip rings and is therefore fundamentally adapted to a-c operation. Direct-current or d-c operation can be obtained by replacing the slip rings with a commutator as shown in Figure 10.1. The physical effect of commutation is to maintain the magnetic axes of the rotor coils stationary in space, even though the rotor itself turns at high speed. With the commutator brushes located as shown in Figure 10.1a, the magnetic axes of the rotor are aligned with those of the stator. Under these conditions it is evident that motional emf's are produced in coil 4 due to the fields of coils 1 and 3 and in coil 3 due to the fields of coils 2 and 4.

To apply the analytical results obtained for the generalized machine with slip rings and rotating armature fields to commutator or d-c machines, it is necessary to refer the armature currents to the fixed direct and quadrature axes. This is readily accomplished by operating on $\mathbf{I}$ with the rotation matrix R_θ^{-1} which rotates I_3 and I_4 through an angle $-\theta$, thus just offsetting the displacement angle θ. The matrix R_θ is defined by

$$R_\theta = \begin{bmatrix} 1 & 0 & 0 & 0 \\ 0 & 1 & 0 & 0 \\ 0 & 0 & \cos\theta & -\sin\theta \\ 0 & 0 & \sin\theta & \cos\theta \end{bmatrix} \qquad \textbf{(10.13)}$$

Denoting the current vector taken with respect to the fixed axes by $\mathbf{I}_f$, we have

$$\mathbf{I}_f = R_\theta^{-1}\mathbf{I} \tag{10.14}$$

Since power flow (because of energy conservation) must have the same value whether expressed in terms of the rotating or fixed axes, the following equality must hold

$$\mathbf{V}' \cdot \mathbf{I} = \mathbf{V}'_f \cdot \mathbf{I}_f = (\mathbf{V}'_f R_\theta^{-1}) \cdot \mathbf{I}$$

Hence

$$\mathbf{V}' = \mathbf{V}'_f R_\theta^{-1}$$

and, since R_θ is an orthogonal matrix,

$$\mathbf{V}_f = R_\theta'\mathbf{V} = R_\theta^{-1}\mathbf{V} \tag{10.15}$$

Substituting (10.5) in (10.15) gives the further relation

$$\mathbf{V}_f = R_\theta^{-1}\mathbf{V} = R_\theta^{-1}Z\mathbf{I} = R_\theta^{-1}ZR_\theta\mathbf{I}_f = Z_f\mathbf{I}_f$$

Hence

$$Z_f = R_\theta^{-1}ZR_\theta = R_\theta^{-1}(R + LD + \dot{L})R_\theta$$

$$= R_\theta^{-1}RR_\theta + R_\theta^{-1}LR_\theta D + R_\theta^{-1}D(LR_\theta) \tag{10.16}$$

and therefore

$$\left.\begin{aligned} R_f &= R_\theta^{-1}RR_\theta \\ L_f &= R_\theta^{-1}LR_\theta \\ G\dot{\theta} &= R_\theta^{-1}D(LR_\theta) \end{aligned}\right\} \tag{10.17}$$

The quantities R_f and L_f are clearly the resistance and inductance matrices referred to fixed axes. However, $G\dot{\theta}$ is a new type of quantity which is associated with the motion of the armature and is known as the motional impedance.

In many applications of interest it is sufficient to assume that the stator coils are identical, that the rotor coils are also identical, and that the nonzero mutual inductances are the same for all coil pairs. With these assumptions, we can write

$$R_{11} = R_{22} = R_s = \text{resistance of stator coils}$$

$$R_{33} = R_{44} = R_a = \text{resistance of armature coils}$$

$$L_{11} = L_{22} = L_s = \text{inductance of stator coils}$$

$$L_{33} = L_{44} = L_a = \text{inductance of armature coils}$$

$$L_{as} = \text{mutual inductances}$$

Calculation gives

$$R_f = R = \begin{bmatrix} R_s & & & \\ & R_s & & 0 \\ & & R_a & \\ 0 & & & R_a \end{bmatrix} \tag{10.18}$$

$$L_f = \begin{bmatrix} L_s & 0 & L_{as} & 0 \\ 0 & L_s & 0 & L_{as} \\ L_{as} & 0 & L_a & 0 \\ 0 & L_{as} & 0 & L_a \end{bmatrix} \tag{10.19}$$

$$G = \begin{bmatrix} 0 & 0 & 0 & 0 \\ 0 & 0 & 0 & 0 \\ 0 & -L_{as} & 0 & -L_a \\ L_{as} & 0 & L_a & 0 \end{bmatrix} \tag{10.20}$$

Thus the system parameters, in contrast to the results for rotating axes, all reduce to constants when referred to the fixed axes. Moreover, the new term $G\dot{\theta}$, which is proportional to the angular speed of rotation of the armature $\dot{\theta}$, appears. It is evident that $G\dot{\theta}$ is associated with the emf generated by the motion of the armature conductors through the magnetic field of the stator and also through its own magnetic field (since the armature field is held stationary in space due to the action of the commutator).

10.3 MACHINES DERIVED FROM THE GENERALIZED MACHINE

Reference has been made (Table 10.1) to the various classes of machines which may be derived from the generalized machine by appropriate specialization. These include both synchronous and induction a-c machines, as well as numerous kinds of d-c machines. In general, any given type of machine can be operated either as a motor (by feeding in electrical power) or as an electric generator (by feeding in mechanical power). All polyphase a-c machines can be reduced to equivalent two-phase machines, so that from a fundamental viewpoint separate analyses of the various possible polyphase machines is unnecessary.

The purpose of this section is to illustrate application of the theory to several important types of machines derivable from the generalized case. These examples will suggest how extensively the theory can be applied, as well as demonstrating the procedures used in applying the theory.

Example I. Direct-Current Motor. There is given a shunt-wound d-c motor driving a load with torque proportional to angular velocity (such as a centrifugal fan or a centrifugal water pump); it is assumed that the armature is connected to the d-c power line at $t = 0$, so that starting transients occur. The stator (shunt winding) is separately excited, and the stator magnetic field is assumed to be fully established when the armature circuit is closed.

This type of motor may be derived from the generalized machine by deleting coils 2 and 3 and the quadrature axis brushes. The basic structure of the motor is similar to that of the d-c generator, Figure 8.27. A circuit diagram for the motor is shown in Figure 10.3.

EQUATIONS OF MOTION. These follow from Eqs. (10.11) through (10.20), with the parameter identifications given as follows:

$$L_{11} = L_s = \text{inductance of stator winding}$$

$$R_{11} = R_s = \text{resistance of stator winding}$$

$$L_{44} = L_a = \text{inductance of armature winding}$$

$$R_{22} = R_a = \text{resistance of armature winding}$$

$$V_1 = V_0 = \text{applied d-c voltage}$$

$$\tau = -\mu\dot{\theta} = \text{load torque, where } \mu \text{ is a load}$$
$$\text{parameter which may be assumed}$$
$$\text{constant for a given case}$$

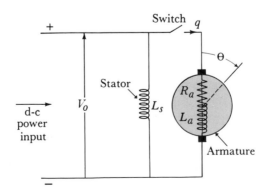

FIGURE 10.3 Circuit diagram for shunt-wound d-c motor.

Thus, the equations of motion are

$$L_s \dot{I}_s + R_s I_s = V_0$$
$$L_a \dot{I}_a + R_a I_a + L_{as} \dot{\theta} I_s = V_0$$
$$J\ddot{\theta} - L_{as} I_a I_s = -\mu\dot{\theta}$$

The steady-state stator current is given by

$$I_s = V_{0/R_s} = \text{const.}$$

Denoting $L_{as} I_s$ by k and $\dot{\theta}$ by ω_a, the remaining equations of motion may be expressed

$$L_a \dot{I}_a + R_a I_a + k\omega_a = V_0$$
$$-k I_a + J\dot{\omega}_a + \mu\omega_a = 0$$

For the steady-state condition,

$$\dot{I}_a = 0 \qquad \dot{\omega}_a = 0 \qquad I_a = I_0 \qquad \omega_a = \omega_0$$

and these equations reduce to

$$R_a I_0 + k\omega_0 = V_0$$
$$-k I_0 + \mu\omega_0 = 0$$

Solving for μ and k gives these parameters in terms of easily measured experimental quantities as follows:

$$\mu = \frac{V_0 I_0 - R_a I_0^2}{\omega_0^2}$$

$$k = \frac{V_0 - R_a I_0}{\omega_0}$$

The transient solution may be found by the Laplace transform method.

$$\begin{bmatrix} L_a s + R_a & k \\ -k & Js + \mu \end{bmatrix} \begin{bmatrix} \bar{I} \\ \bar{\omega}_a \end{bmatrix} = \begin{bmatrix} \dfrac{V_0}{s} \\ 0 \end{bmatrix}$$

$$\begin{bmatrix} \bar{I} \\ \bar{\omega}_a \end{bmatrix} = \begin{bmatrix} \dfrac{V_0(Js + \mu)}{s(L_a s + R_a)(Js + \mu) + k^2} \\ \dfrac{V_0 k}{s(L_a s + R_a)(Js + \mu) + k^2} \end{bmatrix}$$

Here $\bar{I}$ and $\bar{\omega}_a$ stand for the Laplace transforms of I and ω_a respectively.

Simplifying and taking the inverse transform yields the solution

$$\bar{I} = \frac{V_0\mu}{\mu R_a + k^2}\left[1 + \frac{s_2 e^{-s_1 t} - s_1 e^{-s_2 t}}{s_1 - s_2}\right] + \frac{V_0}{L_a}\left[\frac{e^{-s_1 t} - e^{-s_2 t}}{s_2 - s_1}\right]$$

$$\bar{\omega}_a = \frac{V_0 k}{\mu R_a + k^2}\left[1 + \frac{s_2 e^{-s_1 t} - s_1 e^{-s_2 t}}{s_1 - s_2}\right]$$

where

$$s_1, s_2 = \frac{1}{2}\left(\frac{\mu}{J} + \frac{R_a}{L_a}\right) \pm \sqrt{\frac{1}{4}\left(\frac{\mu}{J} + \frac{R_a}{L_a}\right)^2 - \frac{\mu R_a + k^2}{JL_a}}$$

It is of interest to note that s_1, s_2 can be complex roots, in which case the starting transients will involve damped sine waves. On the other hand, if the roots are real, I_a and ω_a will merely increase asymptotically to their steady-state values.

Power relations for the system follow immediately from Eq. (6.33)

$$\frac{d}{dt}(T + V) + 2F = \mathbf{Q}_e \cdot \dot{\mathbf{q}} = P$$

Primary interest is in the steady-state condition for which the equation reduces to

$$k\omega_0 I_0 + R_a I_0^2 = V_0 I_0$$

It is evident that $k\omega_0 I_0$ represents power converted from electrical to mechanical form, $R_a I_0^2$ is the resistance loss in the armature winding, and $V_0 I_0$ is the input power from the power line.

The equations of motion can be represented by an equivalent electric network. For this purpose it is noted that $k\dot{\theta}$ has the dimensions of voltage, while $\dfrac{J\ddot{\theta}}{k}$ and $\dfrac{\mu\dot{\theta}}{k}$ have the dimensions of current. The first equation may therefore be regarded as a loop or mesh equation, and the second, after division by k, as the node equation

$$\frac{J\ddot{\theta}}{k} + \frac{\mu\dot{\theta}}{k} - \dot{q} = 0$$

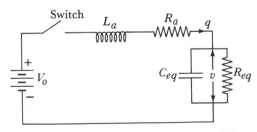

FIGURE 10.4   Equivalent electric circuit for shunt-wound d-c motor with load.

The corresponding network is drawn in Figure 10.4, where

$$R_{eq} = \frac{v}{I_R} = \frac{k\dot{\theta}}{\dfrac{\mu\dot{\theta}}{k}} = \frac{k^2}{\mu}$$

$$C_{eq} = \frac{I_c}{\dfrac{dv}{dt}} = \frac{J\ddot{\theta}}{\dfrac{k}{k\ddot{\theta}}} = \frac{J}{k^2}$$

Note that the inertia of the rotor leads to a capacitor term and that oscillations may occur under certain conditions.

Example 2. Two-Phase Synchronous Machines. Two-phase synchronous machines have a single winding on one magnetic member and a two-phase winding on the other magnetic member. In practice, the single winding is usually the rotor which is energized by d-c power through slip rings. In theory, it is a matter of indifference whether the rotor carries the single winding or the two-phase winding, but since the latter arrangement corresponds to the generalized machine it will be used for the analysis. The resulting arrangement is depicted schematically in Figure 10.5, and is derived from Figure 10.2 by deleting the stator coil on the quadrature axis. Since the machine is synchronous, the angular a-c frequency is equal to the angular rotational speed of the armature, i.e., $\omega = \omega_a$.

Equations of motion for the system may be derived from Eqs. (10.11)

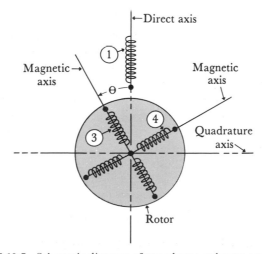

FIGURE 10.5 Schematic diagram of two-phase synchronous machine.

by introducing the conditions

$V_1 = \text{const.} = V_s$

V_2 and I_2 do not exist (because the quadrature stator winding is deleted).

$\theta = \omega_a t + \varphi$, where φ is a phase angle which may be taken as zero without loss of generality.

The volt-ampere relations may then be written in the form

$$
\begin{bmatrix} V_s \\ V_3 \\ V_4 \end{bmatrix} = \begin{bmatrix} L_s D + R_s & L_{as}(\cos\theta D - \sin\theta\omega_a) & L_{as}(\sin\theta D + \cos\theta\omega_a) \\ L_{as}(\cos\theta D - \sin\theta\omega_a) & L_a D + R_a & 0 \\ L_{as}(\sin\theta D + \cos\theta\omega_a) & 0 & L_a D + R_a \end{bmatrix} \begin{bmatrix} I_1 \\ -I_3 \\ -I_4 \end{bmatrix}
\tag{1}
$$

where the mutual inductance coefficients are all equal and denoted by L_{as}, the armature inductances and resistances are indicated by the subscript a, and the negative signs for the rotor currents I_3 and I_4 signify generator action.

For open-circuit, steady-state conditions, $I_3 = I_4 = 0$ and Eqs. (1) simplify to

$$
\begin{bmatrix} V_s \\ V_3 \\ V_4 \end{bmatrix} = \begin{bmatrix} I_1 R_s \\ -I_1 L_{as}\omega_a \sin\omega_a t \\ I_1 L_{as}\omega_a \cos\omega_a t \end{bmatrix} = \begin{bmatrix} V_s \\ \varepsilon_3 \\ \varepsilon_4 \end{bmatrix}
\tag{2}
$$

where ε_3 and ε_4 are the motional emf's of the machine. Thus, the synchronous machine, when operated as a generator, produces two-phase sine-wave emf's in accord with Eq. (2). These emf's are equal in magnitude and differ by 90° or $\pi/2$ radians in phase angle.

Equation (1) for the general operation of the synchronous machine is observed to be nonlinear even for constant inductance parameters. Although it is necessary to use Eq. (1) for transient operating conditions of the synchronous machine, these equations can be greatly simplified for balanced load, steady-state conditions which are of most interest in practice. Under these conditions, $\omega_a = \dot\theta = \text{const.}$, and the generator output currents for linear load conditions are sinusoidal and can be expressed as follows (together with I_1)

$$
\begin{bmatrix} I_1 \\ I_3 \\ I_4 \end{bmatrix} = \begin{bmatrix} V_s/R_s \\ -I_a \sin(\omega_a t + \gamma) \\ I_a \cos(\omega_a t + \gamma) \end{bmatrix}
\tag{3}
$$

where I_a is the amplitude and γ is the phase angle of I_3 and I_4 relative to V_3 and V_4, respectively.

The voltage-current relations for balanced a-c conditions may now be determined by substituting (2) and (3) in (1) and carrying out the matrix multiplication, which gives

$$
\begin{bmatrix} V_s \\ V_3 \\ V_4 \end{bmatrix} = \begin{bmatrix} R_s I_1 \\ -I_1 L_{as}\omega_a \sin \omega_a t + (R_a + L_a D)I_a \sin (\omega_a t + \gamma) \\ I_1 L_{as}\omega_a \cos \omega_a t - (R_a + L_a D)I_a \cos (\omega_a t + \gamma) \end{bmatrix} \tag{4}
$$

Voltages V_3 and V_4 may also be represented in complex notation as follows:

$$
\begin{bmatrix} V_s \\ \bar{V}_3 \\ \bar{V}_4 \end{bmatrix} = \begin{bmatrix} I_1 R_s \\ -i[\bar{\varepsilon}_a - (R_a + iL_a\omega_a)\bar{I}_a] \\ \bar{\varepsilon}_a - (R_a + iL_a\omega_a)\bar{I}_a \end{bmatrix} = \begin{bmatrix} I_1 R_s \\ -i(\bar{\varepsilon}_a - Z_a\bar{I}_a) \\ \bar{\varepsilon}_a - Z_a\bar{I}_a \end{bmatrix} \tag{5}
$$

where

$\bar{\varepsilon}_a = \varepsilon_a = $ motional emf of rotor coil 4

$\bar{I}_a = I_a e^{i\gamma} = $ current in circuit of rotor coil 4

$Z_a = R_a + iL_a\omega_a = $ complex impedance of rotor coils

$D = i\omega_a$

TORQUE. Reaction torque due to interactions of stator and rotor magnetic fields is found by substitution of Eq. (3) in Eq. (10.8).

$$
\tau_m = L_{as} I_1 I_a \lfloor \cos (\omega_a t + \gamma) \cos \omega_a t + \sin (\omega_a t + \gamma) \sin \omega_a t]
$$
$$
= L_{as} I_1 I_a \cos \gamma \tag{6}
$$

Thus, the very significant result is obtained that the torque is constant despite the fact that the currents vary sinusoidally with time. Moreover, since γ is the phase angle between the voltages V_3 and V_4 at the machine terminals and the corresponding currents I_3 and I_4, it is evident that maximum torque occurs for zero phase angle, and minimum or zero torque for $\pm\pi/2$ phase angle.

A further important result follows from the observation that changing the phase angle of currents I_3 and I_4 by π radians, which is equivalent to reversing the directions of these currents, also will reverse the sign of τ_m and corresponds to operating the machine as a motor. Thus, we deduce that

$$
\tau_m > 0 \quad \text{for} \quad -\frac{\pi}{2} < \gamma < \frac{\pi}{2} \text{ (machine acts as a generator)}
$$

$$
\tau_m < 0 \quad \text{for} \quad \frac{\pi}{2} < \gamma < \frac{3\pi}{2} \text{ (machine acts as a motor)}
$$

Rotor Magnetic Field. We now prove the fundamental result that the net magnetic field produced by balanced a-c currents flowing in the rotor coils of a synchronous machine is constant in magnitude and fixed in space (relative to the fixed *d-q* axes). To prove this result, first assume that the rotor coils are fixed in space and aligned with the *d-q* axes, as shown in Figure 10.1. Then, the magnetic field distribution can be represented by the vector **B** as shown in Figure 10.6, where the unit vector **i** is taken along the quadrature axis and the unit vector **j** along the direct axis. Since we are still assuming that a-c currents I_3 and I_4 flow in the rotor coils, the flux vector **B** for any time *t* is given by

$$\mathbf{B} = \mathbf{B}_4 + \mathbf{B}_3$$
$$= B_a[\mathbf{i} \cos (\omega_a t + \gamma) - \mathbf{j} \sin (\omega_a t + \gamma)] \tag{7}$$

But the expression in brackets represents a unit vector which rotates in a clockwise direction with angular speed ω_a. Hence, the flux vector **B** has a constant magnitude B_a and rotates clockwise with an angular speed ω_a. If now the rotor is given a counterclockwise rotation with angular speed ω_a (as in the synchronous machine), it follows that **B** takes the value

$$\mathbf{B} = B_a(\mathbf{i} \cos \gamma - \mathbf{j} \sin \gamma)$$

where the rotor coils are aligned with the *d-q* axes at $t = 0$. Thus, the net effect of the clockwise rotation of the field due to the a-c currents in the rotor and the counterclockwise rotation of the rotor itself is to freeze **B** in space at a clockwise angle $\gamma + \pi/2$ with respect to the **j** or fixed axis of the stator winding. Moreover, it is noted that the rotor voltages and currents are in phase for $\gamma = 0$ (maximum generator torque), and are

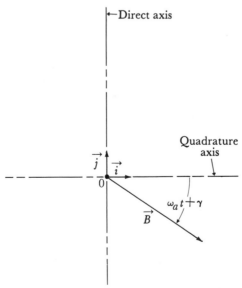

FIGURE 10.6 Vector diagram for rotor magnetic field of synchronous machine assuming fixed rotor aligned with *d-q* axes and with balanced a-c currents in rotor coils.

180° out of phase for $\gamma = \pi$ (maximum motor torque). Zero torque occurs for $\gamma = \pm\dfrac{\pi}{2}$, i.e., when the rotor d-q axes are aligned or counter-aligned with the fixed d-q axes.

Example 3. Two-Phase Induction Motor. Induction machines are distinguished by two characteristics: first, the stator windings are energized with alternating currents in such a manner that the resulting magnetic field rotates at an angular velocity integrally related to the angular frequency of the alternating current; and second, the rotor windings are short-circuited either at their terminals as in squirrel-cage machines, or by external resistors as in wound-rotor machines. Squirrel-cage machines require neither a commutator nor slip rings and therefore can be made exceptionally rugged and durable. Wound-rotor machines use slip rings to make the external connections to the resistors. Induction machines may be used either as motors or as generators, but the former are far more important in practice.

The rotating magnetic field produced by the stator of the induction motor induces electromotive forces and currents in the rotor windings by transformer action. Lorentz forces arising from the interaction between the rotating magnetic field and the rotor currents create a torque on the rotor. This torque is used to drive an external load, and the machine, therefore, operates as a motor.

In a two-phase induction motor the stator winding arrangement is the same as that for the generalized machine shown in Figure 10.2. Winding 1 is energized with an alternating current $I_s \cos(\omega t - \varphi)$ and winding 2 with an alternating current $I_s \cos\left(\omega t - \varphi + \dfrac{\pi}{2}\right)$. By analogy with Ex. 2, Eq. (7), the net effect of the balanced a-c currents in the two fixed windings is to produce a magnetic field distribution which is constant in magnitude but which rotates in this case counterclockwise with an angular speed ω.

Consider now the rotor windings. The rotating field produced by the stator induces emf's in the rotor coils, and these emf's cause relatively large currents to flow because of the very low resistance of the short-circuited windings. In the limit, with vanishing resistance and no load torque, the rotor must rotate in synchronism with the stator field. This conclusion follows from the Lorentz law of force applied to the conduction electrons in the rotor conductors. For a perfect conductor, the force must be zero, otherwise infinite currents would flow. Thus

$$\mathbf{F} = q\mathbf{v} \times \mathbf{B} = 0$$

which is possible only if the rotor turns synchronously with the rotating magnetic field of the stator. On the other hand, if the motor is driving a load, the rotor must "slip" or turn more slowly than the rotating field in order to create magnetic forces on the rotor conductors.

These qualitative observations may be quantified by applying the equations of motion (10.11) for the generalized machine to the coil arrangement of the induction motor, i.e.,

$$\mathbf{V} = (\dot{L} + R + LD)\mathbf{I} \tag{1}$$

$$\tau = \tau_m + J\ddot{\theta} \tag{2}$$

where

$$\mathbf{V} = \begin{bmatrix} V_1 \\ V_2 \\ V_3 \\ V_4 \end{bmatrix} = \begin{bmatrix} V_s \cos \omega t \\ V_s \cos (\omega t + \pi/2) \\ 0 \\ 0 \end{bmatrix} \tag{3}$$

The volt-ampere relations for the induction motor may be obtained by substituting (3) in (1). The resulting equations are in general nonlinear even though the motor inductance coefficients are assumed to be constant. It is usual in order to facilitate analysis to assume that motor speed is constant for a given load condition. This limitation is not unduly restrictive in practice because the speed of response of the relatively inert mechanical part of the system is inherently much slower than the electrical response. Electrical transients of several cycles' duration can therefore occur before motor speed changes perceptibly. Operation with constant rotational speed but nonsteady electrical conditions may be called quasi-stationary.

The volt-ampere relations, obtained by substituting (3) in (1) and writing in expanded matrix form, are given by expression (4).

These equations are general in form and may be specialized for various modes of operation for the induction motor as follows:

QUASI-STATIONARY. Voltages applied to windings 1 and 2 are two-phase a-c of angular frequency ω; armature turns with a constant angular speed $\dot{\theta} = \omega_a < \omega$.

STEADY-STATE. This type of operation is the same as (1) except that frequency and amplitudes of the input a-c voltages are also constant. If the input voltages are also equal in amplitude, the operation is called balanced steady-state. An extension of the concept of steady-state operation can be made to include relatively slow changes in mechanical loading of the motor such that transient effects reflected in the a-c voltages and currents are negligible over any one-cycle time period.

Steady-state operation is of most fundamental interest in practice and is also simplest to analyze. Transient phenomena can be treated by solving the general equations (4).

$$
\begin{bmatrix} V_1 \\ V_2 \\ 0 \\ 0 \end{bmatrix} =
\begin{bmatrix}
R_s + L_s D & 0 & L_{as}(\cos\theta D - \sin\theta\dot{\theta}) & L_{as}(\sin\theta D + \cos\theta\dot{\theta}) \\
0 & R_s + L_s D & -L_{as}(\sin\theta D + \cos\theta\dot{\theta}) & L_{as}(\cos\theta D - \sin\theta\dot{\theta}) \\
L_{as}(\cos\theta D - \sin\theta\dot{\theta}) & -L_{as}(\sin\theta D + \cos\theta\dot{\theta}) & R_a + L_a D & 0 \\
L_{as}(\sin\theta D + \cos\theta\dot{\theta}) & L_{as}(\cos\theta D - \sin\theta\dot{\theta}) & 0 & R_a + L_a D
\end{bmatrix}
\begin{bmatrix} I_1 \\ I_2 \\ I_3 \\ I_4 \end{bmatrix}
\tag{4}
$$

Balanced Steady-state Operation. The voltage-current equations may be found by using $\mathbf{V}$ from (3) and setting $\dot{\theta} = \omega_a$ in the matrix expression (4). However, rather than solving the resulting equations directly, it is instructive as well as expeditious to take advantage of the known properties of the generalized machine by referring the rotating currents of the armature to the fixed d-q axes (cf. Fig. 10.2). This is equivalent to assuming that a commutator with a set of four fixed brushes aligned with the d-q axes exists and that armature currents and voltages are measured with respect to the brushes as in a d-c machine.

Since the net effect of the commutator and fixed brushes is to counteract the rotation θ of the armature, it is evident that the inverse rotation matrix R_θ^{-1} of Eq. (10.14) is the appropriate transformation to accomplish this. The voltage-current equations referred to fixed axes may then be expressed in matrix form as

$$\mathbf{V}_f = Z_f \mathbf{I}_f = (R_f + L_f D + G\dot{\theta})\mathbf{I}_f \tag{5}$$

where R_f, L_f, and G are given by Eqs. (10.18) to (10.20). Writing (5) in expanded form we obtain

$$
\begin{bmatrix} V_1 \\ V_2 \\ 0 \\ 0 \end{bmatrix} =
\begin{bmatrix}
R_s + L_s D & 0 & L_{as} D & 0 \\
0 & R_s + L_s D & 0 & L_{as} D \\
L_{as} D & -L_{as}\omega_a & R_a + L_a D & -L_a\omega_a \\
L_{as}\omega_a & L_{as} D & L_a\omega_a & R_a + L_a D
\end{bmatrix}
\begin{bmatrix} I_1 \\ I_2 \\ I_3 \\ I_4 \end{bmatrix} \tag{6}
$$

Using complex notation, and noting that the currents as well as the voltages must be balanced, allows (6) to be expressed in the form

$$
\begin{bmatrix} V_s e^{i\omega t} \\ iV_s e^{i\omega t} \\ 0 \\ 0 \end{bmatrix} =
\begin{bmatrix}
R_s + iL_s\omega & 0 & iL_{as}\omega & 0 \\
0 & R_s + iL_s\omega & 0 & iL_{as}\omega \\
iL_{as}\omega & -L_{as}\omega_a & R_a + iL_a\omega & -L_a\omega_a \\
L_{as}\omega & iL_{as}\omega & L_a\omega_a & R_a + iL_a\omega
\end{bmatrix}
\begin{bmatrix} I_s e^{i\omega t} \\ iI_s e^{i\omega t} \\ I_a e^{i\omega t} \\ iI_a e^{i\omega t} \end{bmatrix} \tag{7}
$$

The four equations represented by (7) reduce to two independent equations, as may be seen by writing out the equations and observing that the second is equivalent to the first, while the fourth is equivalent to the third. Thus, the voltage-current relations for balanced steady-state operation can be written as follows

$$
\left.
\begin{aligned}
V_s &= (R_s + iL_s\omega)I_s + iL_{as}\omega I_a \\
0 &= iL_{as}(\omega - \omega_a)I_s + [R_a + iL_a(\omega - \omega_a)]I_a
\end{aligned}
\right\} \tag{8}
$$

These may be reduced to more convenient form for some purposes by introducing the concept of slip defined by the relation

$$s = \frac{\omega - \omega_a}{\omega} = \text{slip} \tag{9}$$

For a blocked rotor $s = 1$, while for synchronous speed $s = 0$. In terms of s, Eqs. (8) become

$$V_s = (R_s + iL_s\omega)\bar{I}_s + iL_a\omega\bar{I}_a \left.\begin{array}{c}\\\\\\\end{array}\right\}$$
$$0 = iL_{as}\omega\bar{I}_s + \left(\frac{R_a}{s} + iL_a\omega\right)\bar{I}_a \left.\begin{array}{c}\\\\\\\end{array}\right\} \tag{10}$$

The pair of equations (10) may be represented by the equivalent circuit of Figure 10.7 which is useful in picturing the operation of the motor.

TORQUE. Reaction torque between the armature and stator may be calculated in terms of the currents referred to the d-q axes by noting that $\mathbf{I}'_f \mathbf{V}_f$ is the system power and, hence, that

$$\tau_m \dot{\theta} = \mathbf{I}'_f (G\dot{\theta})\mathbf{I}_f$$

from which

$$\tau_m = \mathbf{I}'_f G \mathbf{I}_f$$

Calculation of the matrix products, using Eq. (10.20) for G, gives

$$\tau_m = L_{as}(I_1 I_4 - I_2 I_3)$$
$$= I'_{us}\left[I_s I_a \cos(\omega t - \varphi) \cos\left(\omega t - \psi + \frac{\pi}{2}\right)\right.$$
$$\left. - I_s I_a \cos\left(\omega t - \varphi + \frac{\pi}{2}\right) \cos(\omega t - \psi)\right]$$
$$- L_{as}I_s I_u \sin(\psi - \varphi)$$

where φ is the phase angle of the stator currents and ψ that of the armature currents.

Since φ can be taken as zero without loss of generality, it follows that τ_m can be written as

$$\tau_m = L_{as}I_s I_a \sin\psi \tag{11}$$

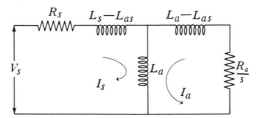

FIGURE 10.7 Equivalent circuit for balanced operation of two-phase induction motor. (Circuit is shown for one phase; an identical circuit exists for the other phase with voltages and currents advanced or retarded 90°.)

where ψ now represents the phase angle difference between the stator and rotor currents. Maximum torque occurs when $\psi = \pm\pi/2$, corresponding to quadrature phase difference between the stator and rotor currents and also their associated magnetic fields. Negative values of τ_m indicate operation of the machine as a motor, while positive values indicate operation as a generator. As in the case of the two-phase synchronous machine, the two-phase induction machine gives constant torque despite the fact that a-c is used.

Torque may also be calculated by referring to the equivalent circuit for the induction motor (Fig. 10.7). From conservation of energy, and keeping in mind that there are two phases, we can write

$$\tau_m\omega_a + 2I_a^2\left(\frac{R_a}{s} - R_a\right) = \tau_m\omega_a + 2I_a^2R_a\left(\frac{1-s}{s}\right) = 0$$

Solving for τ_m gives the result

$$\tau_m = \frac{-2I_a^2R_a}{\omega}\left(\frac{1-s}{s^2}\right) \tag{12}$$

Rotating Magnetic Fields. By analogy with the results of Ex. 2, the stator currents in the two-phase induction machine produce a magnetic field rotating at an angular speed ω, where ω is the angular frequency of the stator voltages. Since a-c currents flow in the rotor, a rotating magnetic field is also set up by the rotor currents and has an angular speed relative to the rotor of $\omega - \omega_a$. Adding to this value the angular speed ω_a of the rotor itself gives a net speed relative to the stator of ω, i.e., the same as the angular speed of the rotating field of the stator. Thus, the stator field and armature field turn at the same speed, differing only in phase angle, as already proved in the calculation of the torque τ_m.

Note on Linear Induction Motors (Ref. 12). Motors of this type have been designed to provide traction for modern railway transportation in Great Britain. These motors have wound stator poles attached to the vehicle and a linear armature consisting of a continuous plate of conducting material attached to the roadbed. Traction does not depend on rail adhesion, but is supplied directly by the magnetic reaction forces between the stator and armature. The principle of operation of the linear induction motor does not differ from that of the conventional rotary induction motor; only the geometrical configuration is different for the two types.

BIBLIOGRAPHY

1. G. Kron, "Non-Riemannian Dynamics of Rotating Electrical Machinery." *Journal of Mathematics and Physics, 13* 103, 1934.
2. D. C. White and H. W. Woodson, *Electromechanical Energy Conversion.* New York: Wiley, 1959.

3. L. V. Bewley, *Tensor Analysis of Electric Circuits and Machines*. New York: Ronald Press, 1961.
4. V. Gourishanker, *Electromechanical Energy Conversion*. Scranton, Pa.: International Textbook Company, 1965.
5. H. K. Messerle, *Dynamic Circuit Theory*. Oxford: Pergamon Press, 1966.
6. R. P. Feynman, et al., *The Feynman Lectures on Physics*, Vol. II. Reading, Mass.: Addison-Wesley, 1964.
7. D. L. Bobroff, "The Classical Dynamics Approach to Connected Electromechanical Systems." *Institute of Electrical and Electronics Engineers, Transactions on Education, E-8*, 109, 1965.
8. H. K. Messerle, "Electrical Energy Conversion." *Institute of Electrical and Electronics Engineers, Student Journal*, p. 27, July, 1966.
9. B. Lehnert, *Dynamics of Charged Particles*. Amsterdam: North-Holland Publishing Co., 1964. (Distributed in U.S. by Interscience Publishers.)
10. "Electromechanical Energy Conversion," a series of papers published by American Institute of Electrical Engineers (now IEEE). Bulletin S-128, April 1961.
11. L. D. Landau and E. M. Lifshitz, *Electrodynamics of Continuous Media*. London: Pergamon Press, 1960. (See p. 205 for discussion of moving conductors.)
12. E. R. Laithewaite, *Induction Machines for Special Purposes*. London: George Newnes, 1966.
13. B. Adkins, *The General Theory of Electrical Machines*. London: Chapman and Hall, 1964.

11

THEORY OF VIBRATIONS

Vibration may be defined as the oscillatory motion of the elements of a system about a position of equilibrium. Vibrations of strings, elastic solids, fluids, and the pendulum are familiar examples for physical systems. Electrical oscillations are fundamental to telephone and radio communications, while acoustic vibrations produced by physiological, mechanical, or electromechanical means provide the basis for speech and music. The concept of vibratory motion is usually applied to physical systems, but also may be generalized to include nonphysical systems.

The field of vibrations pervades most areas of science and engineering. Some of the fundamental applications in these fields are listed as follows for purposes of orientation.

Physics: theory of solids; acoustics; physical optics; molecular spectra; and plasma oscillations.

Acoustical engineering: sound production, recording, and transmission.

Instrumentation: design of automatic recorders, balances, gyroscopes, meters, and timing devices.

Civil and structural engineering: vibrations of bridges, buildings, and other structures.

Mechanical engineering: control of vibrations in machinery; and stability of moving vehicles.

Electrical and electronics engineering: automatic control systems, stability of large power generation and distribution systems; high-frequency oscillators; communications; and filters.

Vibration theory in this book is developed using the methods of generalized mechanics and of matrix calculus. This approach provides the generality and power needed for diverse applications in a variety of fields.

350

The theory of vibrations was treated by Lagrange and others in the eighteenth and nineteenth centuries. Modern work in the field stems from the basic advances made by Poincaré in the latter part of the nineteenth century. Leadership in the first half of this century centered in the Russian school, as evidenced by the fundamental research at the Institute of Oscillations founded in Moscow about 1935. Stability and oscillation problems, which loom so large in the development of automatic control systems and processes and, more generally, in the design and operation of many other complex systems, are responsible for much of today's interest in vibration theory.

Several terms and concepts which are needed in the theory and application of vibrating systems are defined:

Undamped vibrations: the vibrations or natural oscillations which continue indefinitely in a conservative system disturbed initially from a state of stable equilibrium.

Damped vibrations: free vibrations which decrease in amplitude with time owing to frictional dissipation of energy or to withdrawal of energy from the system.

Forced vibrations: vibrations produced by application of external forces to a system.

Linear vibrations: vibrations which may be adequately described by linear differential equations.

Nonlinear vibrations: those which require nonlinear differential equations for their description.

Self-excited vibrations: oscillations which occur in certain physical systems owing to internal excitation or instability. Familiar examples are electronic oscillators, clock mechanisms, and parasitic oscillations which may occur, for example, in automatic control systems or in aircraft wings under flight conditions.

11.1 SMALL VIBRATIONS OF CONSERVATIVE SYSTEMS

The theory of small vibrations about a state of stable equilibrium is of basic importance for the vibrations of structures, behavior of elastic solids, theory of specific heat of solids, vibrations of machinery, theory of electric networks, and other fundamental problems. Although the vibrations of real systems are nonlinear it is usually possible, for sufficiently small displacements from equilibrium, to approximate the vibrations by linear equations. Most systems of interest permit expansions of the energy functions T and V in Taylor's series. It is then a matter of the type of system and the degree of precision required that determines the region over which the linear approximation is satisfactory.

Consider a dynamical system with f degrees of freedom which possesses a potential function $V = V(q_1, q_2, \ldots, q_f) = V(\mathbf{q})$. If a configuration of stable equilibrium exists, it is determined by the condition that V have a local minimum at the given point. From the theory of the extrema of functions, Chap. 2, Sect. 3, the following conditions are necessary and sufficient for a local minimum of V:

$$(1) \qquad \nabla V = \left[\frac{\partial V}{\partial q_1}, \frac{\partial V}{\partial q_2}, \ldots, \frac{\partial V}{\partial q_f}\right] = 0$$

which defines the stationary points of V; and

(2) the matrix K with elements

$$K_{rs} = K_{sr} = \left[\frac{\partial^2 V}{\partial q_r\, \partial q_s}\right]_{\mathbf{q}=\mathbf{q}_0}$$

must be positive definite, where $\mathbf{q} = \mathbf{q}_0$ is a stationary point of V.

Let $\mathbf{x}' = [x_1, x_2, \ldots, x_f]$ be a vector which represents small deviations from the equilibrium point $\mathbf{q}_0$. Expanding V in a Taylor's series about $\mathbf{q}_0$ gives

$$V = V(\mathbf{q}_0) + \frac{1}{2}\sum_{r=1}^{f}\sum_{s=1}^{f}\left[\frac{\partial^2 V}{\partial q_r\, \partial q_s}\right]_{\mathbf{q}_0} x_r x_s + \cdots \qquad (11.1)$$

where the first order terms vanish because of condition (1). For sufficiently small deviations, $V - V(\mathbf{q}_0)$ may be represented by the second order term of the expansion. Furthermore, since $V(\mathbf{q}_0)$ may without loss of generality be set equal to zero, the approximation for V may be written as

$$V = \frac{1}{2}\sum_{r=1}^{f}\sum_{s=1}^{f}\left[\frac{\partial^2 V}{\partial q_r\, \partial q_s}\right]_{\mathbf{q}_0} x_r x_s = \frac{1}{2}\,\mathbf{x}'K\mathbf{x} \qquad (11.2)$$

The kinetic energy of the system is given by

$$T = \frac{1}{2}\sum_{r=1}^{f}\sum_{s=1}^{f} M_{rs}(\mathbf{q}_0 + \mathbf{x})\dot{q}_r\dot{q}_s$$

In many cases the coefficients M_{rs} are constants. If not, they may be expanded in a Taylor's series and the zero order terms $M_{rs}(\mathbf{q}_0)$ retained. The velocities $\dot{q}_r$ require no approximation, since $\dot{q}_r = \dot{x}_r$. Thus, T is a positive definite quadratic form in the velocities $\dot{x}_r, r = 1, 2, \ldots, f$ as follows:

$$T = \frac{1}{2}\sum_{r=1}^{f}\sum_{s=1}^{f} M_{rs}(\mathbf{q}_0)\dot{x}_r\dot{x}_s = \frac{1}{2}\,\dot{\mathbf{x}}'M\dot{\mathbf{x}} \qquad (11.3)$$

The Lagrangian function L may now be expressed in terms of $\mathbf{x}$ and $\dot{\mathbf{x}}$ in the form

$$L = T - V = \tfrac{1}{2}\dot{\mathbf{x}}'M\dot{\mathbf{x}} - \tfrac{1}{2}\mathbf{x}'K\mathbf{x} \qquad (11.4)$$

Lagrange's equations of motion follow immediately

$$M\ddot{\mathbf{x}} + K\mathbf{x} = (MD^2 + K)\mathbf{x} = 0 \qquad (11.5)$$

Matrices M and K are often referred to as the inertia matrix and stiffness matrix respectively. An alternate expression for (11.5) is obtained by multiplying by M^{-1} which yields

$$\ddot{\mathbf{x}} + M^{-1}K\mathbf{x} = 0 \qquad \det[M] \neq 0 \qquad (11.6)$$

where $M^{-1}K$ is called the dynamic matrix.

The general solution of Eq. (11.5) or its equivalent (11.6) may be found either by the exponential-substitution method or by the method of normal coordinates. Both methods are given here because each has wide application to many problems of this kind.

Exponential Substitution Method. (cf. Chap. 4) Let

$$\mathbf{x} = \mathbf{C}e^{\lambda t}$$

Substitution in (11.5) gives

$$(MD^2 + K)(\mathbf{C}e^{\lambda t}) = (M\lambda^2 + K)\mathbf{C}e^{\lambda t} = 0$$

Cancellation of $e^{\lambda t} \neq 0$ leads to the equation

$$(M\lambda^2 + K)\mathbf{C} = 0 \qquad (11.7)$$

which has a nontrivial solution only if

$$\det(M\lambda^2 + K) = 0 \qquad (11.8)$$

Expansion of the determinant yields a polynomial equation of order f in λ^2 which has f negative real roots $-\lambda_1^2, -\lambda_2^2, \ldots, -\lambda_f^2$. This result follows from the fact that both M and K are positive definite matrices.

The general solution (cf. Chap. 4) of Eq. (11.5), assuming that all roots are distinct, can be written in complex notation as

$$\mathbf{x} = \mathbf{C}_1 e^{i\lambda_1 t} + \mathbf{C}_2 e^{i\lambda_2 t} + \ldots + \mathbf{C}_f e^{i\lambda_f t} \qquad (11.9)$$

where $\mathbf{C}_1, \mathbf{C}_2, \ldots, \mathbf{C}_f$ are complex. Finally, taking the real part of (11.9) there is obtained

$$x_r = \sum_{s=1}^{f} B_{rs} \cos(\lambda_s t + \theta_s) \qquad r = 1, 2, \ldots, f \qquad (11.10)$$

where B_{rs} is the magnitude and θ_s is the argument of the element C_{rs} of the column vector $\mathbf{C}_r$.

Normal Coordinates and Normal Vibrations. The equation of motion (11.6) may be greatly simplified by reducing the matrix $M^{-1}K$ to diagonal form, which is always possible because both M and K are positive definite. We use Property Q-9 of Chap. 1, p. 35 which states that if A is a real symmetric matrix with eigenvalues $\lambda_1, \lambda_2, \ldots, \lambda_n$, then there exists a real orthogonal matrix P such that $P^{-1}AP$ is orthogonal. Although not essential, we shall for simplicity assume that the eigenvalues are distinct. The case of repeated roots can be dealt with by a limiting process in which the repeated roots are first assumed to differ slightly in value and then to approach the same value in the limit. There exists, therefore, a real nonsingular linear transformation

$$\mathbf{x} = P\mathbf{y} \qquad (11.11)$$

which transforms $M^{-1}K$ to diagonal form

$$\Lambda = \begin{bmatrix} \mu_1 & & & & \\ & \mu_2 & & & 0 \\ & & \cdot & & \\ & & & \cdot & \\ 0 & & & & \cdot \\ & & & & & \mu_f \end{bmatrix}$$

where $\mu_1, \mu_2, \ldots, \mu_f$ are the eigenvalues of the matrix $M^{-1}K$, that is, the roots of the characteristic equation

$$\det [M^{-1}K - \mu I] = 0$$

or

$$\det [K - M\mu] = 0 \qquad (11.12)$$

Using this transformation reduces Eq. (11.6) to the form

$$\ddot{\mathbf{y}} + \Lambda\mathbf{y} = 0 \qquad (11.13)$$

for which the general solution, assuming distinct eigenvalues and writing $\lambda_r^2 = \mu_r, r = 1, 2, \ldots, f$, is

$$\mathbf{y} = \begin{bmatrix} y_1 \\ y_2 \\ \vdots \\ y_f \end{bmatrix} = \begin{bmatrix} b_1 \cos (\lambda_1 t + \theta_1) \\ b_2 \cos (\lambda_2 t + \theta_2) \\ \cdots \cdots \cdots \cdots \\ b_f \cos (\lambda_f t + \theta_f) \end{bmatrix} \qquad (11.14)$$

where the λ_r are the normal or eigenfrequencies, and the b_r and θ_r are $2f$ constants of integration. Note that the eigenvalues $\mu_1, \mu_2, \ldots, \mu_f$ are all positive, since both M and K are positive definite matrices.

In terms of the coordinates $\mathbf{x}$, the vibrations are linear combinations of the simple harmonic motions of Eq. (11.14). Thus

$$\mathbf{x} = P\mathbf{y} = \begin{bmatrix} p_{11}y_1 + p_{12}y_2 + \cdots + p_{1f}y_f \\ p_{21}y_1 + p_{22}y_2 + \cdots + p_{2f}y_f \\ \cdots \cdots \cdots \cdots \cdots \cdots \cdots \\ p_{f1}y_1 + p_{f2}y_2 + \cdots + p_{ff}y_f \end{bmatrix} \qquad (11.15)$$

or

$$x_r = \sum_{s=1}^{f} p_{rs} b_s \cos{(\lambda_s t + \theta_s)} \qquad r = 1, 2, \ldots, f \qquad \textbf{(11.16)}$$

The transformation matrix P is constructed as follows. Let $\mathbf{p}_1$, $\mathbf{p}_2, \ldots, \mathbf{p}_f$ be an associated set of eigenvectors normalized so that

$$\mathbf{p}_r \cdot \mathbf{p}_s = \delta_{rs}$$

Next, let P be the matrix formed by using the $\mathbf{p}_r$ as columns, that is,

$$P = [\mathbf{p}_1, \mathbf{p}_2, \ldots, \mathbf{p}_f] \qquad \textbf{(11.17)}$$

Then, since the $\mathbf{p}_r$ are eigenvectors of $M^{-1}K$, we have

$$M^{-1}KP = [\mu_1 \mathbf{p}_1, \mu_2 \mathbf{p}_2, \ldots, \mu_f \mathbf{p}_f]$$

and therefore

$$P'(M^{-1}K)P = \begin{bmatrix} \mathbf{p}_1 \\ \mathbf{p}_2 \\ \cdot \\ \cdot \\ \cdot \\ \mathbf{p}_f \end{bmatrix} [\mu_1 \mathbf{p}_1, \mu_2 \mathbf{p}_2, \ldots, \mu_f \mathbf{p}_f]$$

$$= \begin{bmatrix} \mu_1 & & & \\ & \mu_2 & & 0 \\ & & \cdot & \\ 0 & & & \cdot \\ & & & \mu_f \end{bmatrix} = \Lambda \qquad \textbf{(11.18)}$$

This shows that P is the desired transformation.

The elements of P may therefore be found from the normalized eigenvectors $\mathbf{p}_1, \mathbf{p}_2, \ldots, \mathbf{p}_f$ of $M^{-1}K$, i.e.,

$$(M^{-1}K - \mu_r I)\mathbf{p}_r = 0 \qquad \mathbf{p}_r \cdot \mathbf{p}_r = 1 \qquad \textbf{(11.19)}$$

where

$$\mathbf{p}_r = \begin{bmatrix} p_{1r} \\ p_{2r} \\ \cdot \\ \cdot \\ \cdot \\ p_{fr} \end{bmatrix} \qquad r = 1, 2, \ldots, f$$

Since Eq. (11.19) is the compatibility condition for a set of f homogeneous equations in the f unknowns $p_{1r}, p_{2r}, \ldots, p_{fr}$, it follows that only one of the unknowns can be chosen independently. It is convenient to consider p_{1r}

as the independent quantity which, when chosen, determines $p_{2r}, p_{3r}, \ldots,$ p_{fr} through relation (11.19).

Referring again to Eq. (11.16), it is evident that of the constants $p_{rs}b_s$, $r, s = 1, 2, \ldots, f$, only $p_{11}b_1, p_{12}b_2, \ldots, p_{1f}b_f$ can be chosen independently. These, together with $\theta_1, \theta_2, \ldots, \theta_f$, comprise $2f$ constants of integration which can be found from the initial state of the system given by $x_1(0), x_2(0), \ldots, x_f(0)$; $\dot{x}_1(0), \dot{x}_2(0), \ldots, \dot{x}_f(0)$.

Systems with One Degree of Freedom. Applications to systems with one degree of freedom will be covered first as these are simpler mathematically, yet serve to bring out some of the basic features. The solution for this case, from Eq. (11.10) or (11.16), reduces to a single term

$$x = x_0 \cos (\lambda t + \theta)$$

where x_0 is the amplitude, $\lambda = \sqrt{\dfrac{K_{11}}{M_{11}}}$ is the angular frequency, and θ the phase angle of the vibration. Several familiar physical embodiments are shown in Figure 11.1. Solutions for these are discussed in the first four examples which follow.

Example 1.   Elastically Suspended Weight (Fig. 11.1a)

The Lagrangian is

$$L = T - V = \tfrac{1}{2}m\dot{q}^2 - (\tfrac{1}{2}kq^2 - mgq)$$

The equilibrium position is given by

$$\frac{\partial V}{\partial q} = kq - mg = 0 \quad \text{so that} \quad q_0 = \frac{mg}{k}$$

and the coefficient K_{11} by

$$K_{11} = \left[\frac{\partial^2 V}{\partial q^2}\right]_{q=q_0} = k$$

Hence, angular frequency is

$$\lambda = \sqrt{\frac{K_{11}}{M_{11}}} = \sqrt{\frac{k}{m}}$$

Example 2.   Electric Circuit Oscillator (Fig. 11.1b)

Solution steps for the eigenfrequency are:

(1)
$$L = T - V = \tfrac{1}{2}L\dot{q}^2 - \frac{1}{2}\left(\frac{q^2}{C}\right)$$

(2)
$$\frac{\partial V}{\partial q} = \frac{q}{C} = 0 \quad \text{so that} \quad q_0 = 0$$

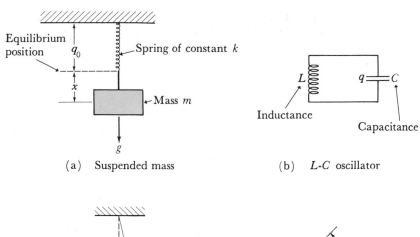

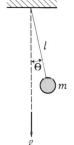

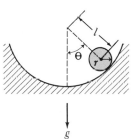

(a)   Suspended mass   (b)   *L-C* oscillator

(c)   Simple pendulum   (d)   Sphere rolling on a spherical surface

FIGURE 11.1   Examples of vibrators with one degree of freedom.

(3)
$$K_{11} = \left[\frac{\partial^2 V}{\partial q^2}\right]_{q=q_0} = \frac{1}{C}$$

(4)
$$\lambda = \sqrt{\frac{K_{11}}{M_{11}}} = \sqrt{\frac{1}{LC}} = \text{eigenfrequency}$$

Example 3.   Simple Pendulum (Fig. 11.1c)

Following the same procedure, we have

$$L = T - V = \tfrac{1}{2}m\ell^2\dot{\theta}^2 - (-mg\ell\cos\theta)$$

$$\frac{\partial V}{\partial \theta} = mg\ell\sin\theta = 0 \quad \text{so that} \quad \theta_0 = 0, \pi, 2\pi, \ldots$$

$$K_{11} = \left[\frac{\partial^2 V}{\partial \theta^2}\right]_{\theta=\theta_0} = mg\ell$$

Note that only the value $\theta_0 = 0$ is significant. The values $\theta_0 = \pi$, $3\pi, \ldots$, correspond to unstable equilibrium, while $\theta_0 = 2\pi, 4\pi, \ldots$, are merely repetitive positions of $\theta_0 = 0$.

The eigenfrequency is then

$$\lambda = \sqrt{\frac{K_{11}}{M_{11}}} = \sqrt{\frac{mg\ell}{m\ell^2}} = \sqrt{\frac{g}{\ell}}$$

Example 4. Sphere Rolling on a Spherical Surface (Fig. 11.1d)

Assume that there is no slipping and that the motion is restricted to a vertical plane. Then, the Lagrangian is

$$L = T - V = \tfrac{1}{2}m\ell^2\dot{\theta}^2 + \tfrac{1}{2}J\omega^2 + mg\ell\cos\theta$$

where

$J = $ moment of inertia of sphere about a diameter

$\quad = \tfrac{2}{5}mr^2$ (assuming a homogeneous sphere)

and

$\omega = $ angular velocity of sphere

$$= \frac{\ell\dot{\theta}}{r}$$

Hence

$$L = \tfrac{1}{2}m\ell^2\dot{\theta}^2 + \tfrac{1}{5}m\ell^2\dot{\theta}^2 + mg\ell\cos\theta$$
$$= \tfrac{7}{10}m\ell^2\dot{\theta}^2 + mg\ell\cos\theta$$

By analogy with the simple pendulum

$$\lambda = \sqrt{\frac{K_{11}}{M_{11}}} = \sqrt{\frac{mg\ell}{\tfrac{7}{5}m\ell^2}} = \sqrt{\frac{5}{7}\left(\frac{g}{\ell}\right)}$$

Example 5. Mass Suspended from a Beam (Fig. 11.2). A mass m is suspended from the end of a cantilever steel beam of length ℓ, modulus of elasticity E, and cross-sectional moment of inertia J as depicted in

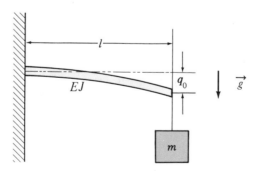

FIGURE 11.2 Small vibrations of a loaded beam.

Figure 11.2. Find the position of equilibrium and the frequency of small oscillations about this position. Neglect the mass of the beam.

SOLUTION

Potential energy of the beam plus the mass m is

$$V = -mgq + \frac{3}{2}\frac{EJ}{\ell^3}q^2$$

where

$$E = \text{modulus of elasticity} = \frac{\text{stress}}{\text{strain}}$$

and it is assumed that the stress does not exceed the elastic limit. The equilibrium position q_0 is given by

$$\frac{\partial V}{\partial q} = -mg + \frac{3EJ}{\ell^3}q = 0$$

Hence

$$q_0 = \frac{mg}{\dfrac{3EJ}{\ell^3}} = \frac{mg\ell^3}{3EJ}$$

$$\frac{\partial^2 V}{\partial q^2}\bigg]_{q=q_0} = \frac{3EJ}{\ell^3} = k > 0$$

$$\lambda = \sqrt{\frac{k}{m}} = \sqrt{\frac{3EJ}{m\ell^3}}$$

$$f = \text{frequency} = \frac{1}{2\pi}\sqrt{\frac{3EJ}{m\ell^3}}$$

NUMERICAL VALUES. Let

$$W = mg = 100 \text{ lbs.}$$

$$\ell = 5 \text{ ft.}$$

$$EJ = 1{,}000{,}000 \text{ lb.-in.}^2 = 7960 \text{ lb.-ft.}^2$$

Then

$$q_0 = \frac{mg\ell^3}{3EJ} = \frac{100 \times (5)^3}{3 \times 7960} = 0.52 \text{ ft.}$$

$$f = \frac{1}{2\pi}\sqrt{\frac{3EJ}{m\ell^3}} = \frac{1}{2\pi}\sqrt{\frac{g}{q_0}} = \frac{1}{6.28}\sqrt{\frac{32.2}{0.524}}$$

$$= 1.25 \text{ oscillations per second}$$

Example 6.   Vibrational Frequencies of Diatomic Molecules

Given a diatomic molecule with atoms of mass m_1 and m_2 and potential energy

$$V = -\frac{a}{q^6} + \frac{b}{q^{12}} \, , a > 0, b > 0$$

where q is the distance between the atoms, find the period of small vibrations about equilibrium.

SOLUTION

The energy for the system is

$$E = \tfrac{1}{2}m_1\dot{q}_1^2 + \tfrac{1}{2}m_2\dot{q}_2^2 + V(q)$$

Choose the origin at the center of mass (Fig. 11.3) defined by

$$-m_1 q_1 + m_2 q_2 = 0 \qquad q_1 + q_2 = q$$

These two equations give

$$q_1 = \frac{m_2 q}{m_1 + m_2} \qquad q_2 = \frac{m_1 q}{m_1 + m_2}$$

Substitution in E gives

$$E = \tfrac{1}{2}m\dot{q}^2 + V(q)$$

where

$$m = \frac{m_1 m_2}{m_1 + m_2} = \text{reduced mass}$$

To find the equilibrium value q_0, we have

$$\frac{\partial V}{\partial q} = \frac{6a}{q^7} - \frac{12b}{q^{13}} = \frac{6aq^6 - 12b}{q^{13}} = 0$$

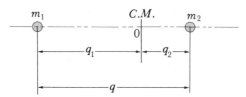

FIGURE 11.3   Center of mass coordinates for diatomic molecule.

from which

$$q_0^0 = \frac{2b}{a}$$

$$K = \left.\frac{\partial^2 V}{\partial q^2}\right]_{q_0} = \frac{-42aq_0^6 + 156b}{q_0^{14}} = \frac{72b}{q_0^{14}} > 0$$

$$\text{Period} = \tau = \frac{2\pi}{\beta_0} = \frac{2\pi}{\sqrt{\dfrac{72b}{mq_0^{14}}}} = \frac{\pi q_0}{3a}\sqrt{2bm}$$

Example 7. Electromechanical Oscillator. A nonlinear electro-mechanical oscillator may be built by suspending a flat metal plate from a spring as shown in Figure 11.4. The plate of area A and mass m is separated a distance x from a fixed conductive ground plane, and is charged to a potential E_0 by means of external battery. The spring is linear, with a spring constant k, and has an effective relaxed length of zero. Investigate the small vibrations of the oscillator, assuming that it is mounted in a vacuum chamber and that edge effects of the capacitor are negligible.

ANALYSIS

(1) *Find the position of stable equilibrium.*
     The potential energy of the system, V, is given by

$$V = \tfrac{1}{2}k(\ell - x)^2 - \frac{AE_0^2}{8\pi\varepsilon_0 x} + mgx \tag{1}$$

The electrostatic energy is taken as negative since it is essentially zero for large values of x and decreases as x is decreased. For stable equilibrium

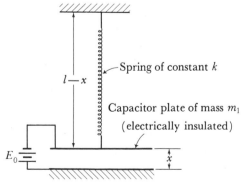

FIGURE 11.4 Electromechanical oscillator.

it is necessary that

$$\frac{\partial V}{\partial x} = -k(\ell - x) + \frac{AE_0^2}{8\pi\varepsilon_0 x^2} + mg = 0 \tag{2}$$

$$\frac{\partial^2 V}{\partial x^2}\bigg]_{x_0} = k - \frac{AE_0^2}{4\pi\varepsilon_0 x_0^3} > 0 \tag{3}$$

where x_0 is a root of Eq. (2). A small calculation shows that these conditions are equivalent to the following inequalities

$$0 < \frac{2}{3}\left(\ell - \frac{mg}{k}\right) < x_0 < \ell \tag{4}$$

(2) *Write the expression for* λ.

$$\lambda = \sqrt{\frac{K_{11}}{M_{11}}} = \sqrt{\frac{k}{m}\left(1 - \frac{AE_0^2}{4\pi\varepsilon_0 k x_0^3}\right)} \tag{5}$$

We note that the eigenfrequency λ reduces to that for a mass suspended from a linear spring when $E_0 = 0$, i.e., to the case of no electric force. Also the system has a region of stability defined by

$$0 \leq \frac{AE_0^2}{4\pi\varepsilon_0 k x_0^3} < 1 \tag{6}$$

Thus, with fixed dimensions and parameters, the field E_0 cannot exceed the value given by inequality (6). Otherwise, the system will be unstable.

Systems with Several Degrees of Freedom. The general case of systems with several or many degrees of freedom is treated in this section. Physical embodiments for a number of two-degree-of-freedom systems are depicted in Figure 11.5. The solutions are given by Eq. (11.16) with $f = 2$, i.e., by (11.20) below

$$\left.\begin{aligned} x_1 &= p_{11}b_1 \cos(\lambda_1 t + \theta_1) + p_{12}b_2 \cos(\lambda_2 t + \theta_2) \\ x_2 &= p_{21}b_1 \cos(\lambda_1 t + \theta_1) + p_{22}b_2 \cos(\lambda_2 t + \theta_2) \end{aligned}\right\} \tag{11.20}$$

where $p_{11}b_1$, $p_{12}b_2$, $p_{21}b_1$, $p_{22}b_2$ are amplitudes, and θ_1 and θ_2 are phase angles. The four constants of integration b_1, b_2, θ_1, and θ_2 are determined by the initial conditions $x_1(0)$, $x_2(0)$, $\dot{x}_1(0)$, and $\dot{x}_2(0)$.

Example 8. Dual Suspended Masses (Fig. 11.5a)
The Lagrangian is

$$L = \tfrac{1}{2}m\dot{q}_1^2 + \tfrac{1}{2}m\dot{q}_2^2 - [\tfrac{1}{2}kq_1^2 + \tfrac{1}{2}k(q_2 - q_1)^2 - mgq_1 - mgq_2]$$

The equilibrium position is defined by

$$\begin{aligned} \nabla V &= \left[\frac{\partial V}{\partial q_1}, \frac{\partial V}{\partial q_2}\right] \\ &= [kq_1 - k(q_2 - q_1) - mg,\ k(q_2 - q_1) - mg] = 0 \end{aligned}$$

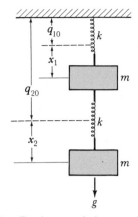

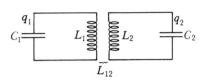

(b)   Coupled L-C oscillators

(a)   Dual suspended masses

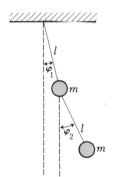

FIGURE 11.5   Vibrating systems
with two degrees of freedom.

(c)   Double pendulum

which gives

$$q_{10} = \frac{2mg}{k} \qquad q_{20} = \frac{3mg}{k}$$

Elements of matrix K are given by

$$K_{11} = \left[\frac{\partial^2 V}{\partial q_1^2}\right]_{q_1=q_{10}} = 2k$$

$$K_{22} = \left[\frac{\partial^2 V}{\partial q_2^2}\right]_{q_2=q_{20}} = k$$

$$K_{12} = K_{21} = \left[\frac{\partial^2 V}{\partial q_1\,\partial q_2}\right]_{\substack{q_1=q_{10}\\q_2=q_{20}}} = -k$$

Hence

$$K = \begin{bmatrix} 2k & -k \\ -k & k \end{bmatrix}$$

Matrix M can be written by inspection from the expression for the kinetic energy, giving

$$M = \begin{bmatrix} m & 0 \\ 0 & m \end{bmatrix}$$

The characteristic equation is

$$\det [K - M\lambda^2] = \begin{vmatrix} 2k - m\lambda^2, & -k \\ -k, & k - m\lambda^2 \end{vmatrix} = 0$$

which has the eigenvalues

$$\lambda_{1,2}^2 = \frac{3 \pm \sqrt{5}}{2}\left(\frac{k}{m}\right)$$

Therefore, the eigenfrequencies are

$$\lambda_1 = \left(\frac{3 + \sqrt{5}}{2} \cdot \frac{k}{m}\right)^{1/2} = 0.618\sqrt{\frac{k}{m}}$$

$$\lambda_2 = \left(\frac{3 - \sqrt{5}}{2} \cdot \frac{m}{k}\right)^{1/2} = 1.62\sqrt{\frac{k}{m}}$$

Using these eigenfrequencies, the general solution is

$$x_1 = p_{11}b_1 \cos (\lambda_1 t + \theta_1) + p_{12}b_2 \cos (\lambda_2 t + \theta_2)$$
$$x_2 = p_{21}b_1 \cos (\lambda_1 t + \theta_1) + p_{22}b_2 \cos (\lambda_2 t + \theta_2)$$

To find the matrix P, we use (11.19) which gives

$$(2k - m\lambda_1^2)p_{11} - kp_{21} = 0$$
$$(2k - m\lambda_2^2)p_{12} - kp_{22} = 0$$

or

$$\frac{p_{21}}{p_{11}} = \frac{2k - m\lambda_1^2}{k} = \frac{1 - \sqrt{5}}{2} = -0.62$$

$$\frac{p_{22}}{p_{12}} = \frac{2k - m\lambda_2^2}{k} = \frac{1 + \sqrt{5}}{2} = 1.62$$

Normalizing the vectors $\mathbf{p}_1$ and $\mathbf{p}_2$ gives the matrix P

$$P = \begin{bmatrix} p_{11} & p_{12} \\ p_{21} & p_{22} \end{bmatrix} = \begin{bmatrix} -0.527, & 0.852 \\ 0.852, & 0.527 \end{bmatrix}$$

Substituting in (11.16) yields the solution

$$x_1 = -0.527b_1 \cos (\lambda_1 t + \theta_1) + 0.852b_2 \cos (\lambda_2 t + \theta_2)$$
$$x_2 = 0.852b_1 \cos (\lambda_1 t + \theta_1) + 0.527b_2 \cos (\lambda_2 t + \theta_2)$$

where the constants of integration b_1, b_2, θ_1, θ_2 are to be determined from the initial conditions.

Example 9. Coupled L-C Oscillators (Fig. 11.5b). Determine the normal vibrations (eigenfrequencies) for the electric oscillations of two inductively coupled L-C circuits.

SOLUTION. Let the circuit parameters be L_1, C_1, L_2, C_2, and L_{12} the mutual inductance. The required eigenvalues are given by the condition

$$\det [C^{-1} - L\lambda^2] = 0$$

Calculation gives

$$\begin{vmatrix} -L_1\lambda^2 + \dfrac{1}{C_1}, & -L_{12}\lambda^2 \\[2mm] -L_{21}\lambda^2, & -L_2\lambda^2 + \dfrac{1}{C_2} \end{vmatrix}$$

$$= (L_1 L_2 - L_{12}^2)\lambda^4 \cdot \left(\frac{L_1}{C_2} + \frac{L_2}{C_1}\right)\lambda^2 + \frac{1}{C_1 C_2} = 0$$

which has the roots

$$\lambda_{1,2}^2 = \frac{(L_1 C_1 + L_2 C_2) \pm \sqrt{(L_1 C_1 - L_2 C_2)^2 + 4L_{12}^2 C_1 C_2}}{2(L_1 L_2 - L_{12}^2)C_1 C_2}$$

The eigenfrequencies are given by $\sqrt{\lambda_{12}^2}$, and are both real. It is noted that if $L_{12} \to 0$, then λ_1 and λ_2 tend to $1/\sqrt{L_1 C_1}$ and $1/\sqrt{L_2 C_2}$, which are the eigenfrequencies of the two circuits taken separately.

Example 10. Double Pendulum (Fig. 11.5c). Find the normal vibrations. The Lagrangian is

$$L = T - V$$
$$T = \tfrac{1}{2}m\ell^2\dot{\varphi}_1^2 + \tfrac{1}{2}m(\ell\dot{\varphi}_1 + \ell\dot{\varphi}_2)^2$$
$$= m\ell^2\dot{\varphi}_1^2 + m\ell^2\dot{\varphi}_1\dot{\varphi}_2 + \tfrac{1}{2}m\ell^2\dot{\varphi}_2^2$$

where quantities of higher order than the second are neglected.

$$V = -mg\ell \cos \varphi_1 - mg\ell \cos \varphi_1 - mg\ell \cos \varphi_2$$
$$= -2mg\ell \cos \varphi_1 - mg\ell \cos \varphi_2$$

The equilibrium position is given by

$$\nabla V = [2mg\ell \sin \varphi_1, \; mg\ell \sin \varphi_2] = 0$$

from which

$$\varphi_{10} = 0 \qquad \varphi_{20} = 0$$

Elements of matrix K are

$$K_{11} = \frac{\partial^2 V}{\partial \varphi_1^2}\bigg]_{\varphi_1=0} = 2mg\ell \qquad K_{22} = \frac{\partial^2 V}{\partial \varphi_2^2}\bigg]_{\varphi_2=0} = mg\ell$$

$$K_{22} = K_{21} = \frac{\partial^2 V}{\partial \varphi_1\, \partial \varphi_2}\bigg]_{\substack{\varphi_1=0 \\ \varphi_2=0}} = 0$$

$$K = \begin{bmatrix} 2mg\ell & 0 \\ 0 & mg\ell \end{bmatrix}$$

By inspection

$$M = \begin{bmatrix} 2m\ell^2 & m\ell^2 \\ m\ell^2 & m\ell^2 \end{bmatrix}$$

The characteristic equation is

$$\det [K - M\lambda^2] = \begin{vmatrix} -2m\ell^2\lambda^2 + 2mg\ell, & -m\ell^2\lambda^2 \\ -m\ell^2\lambda^2, & -m\ell^2\lambda^2 + mg \end{vmatrix} = 0$$

which has the roots

$$\lambda_1^2 = \frac{\sqrt{2}}{\sqrt{2}-1} \cdot \frac{g}{\ell} \qquad \lambda_2^2 = \frac{\sqrt{2}}{\sqrt{2}+1} \cdot \frac{g}{\ell}$$

Hence, the eigenfrequencies are

$$\lambda_1 = \sqrt{\lambda_1^2} = \left[(2 + \sqrt{2})\frac{g}{\ell} \right]^{1/2} = 1.85 \sqrt{\frac{g}{\ell}}$$

$$\lambda_2 = \sqrt{\lambda_2^2} = \left[(2 - \sqrt{2})\frac{g}{\ell} \right]^{1/2} = 0.766 \sqrt{\frac{g}{\ell}}$$

Again, we find one frequency higher and the other lower than the frequency of a simple pendulum of the same length.

Example 11. Coupled Pendulum. This problem involves two equal pendulums coupled by a spring, as depicted in Figure 11.6.

SOLUTION

$$T = \tfrac{1}{2}m\ell^2\dot{\theta}_1^2 + \tfrac{1}{2}m\ell^2\dot{\theta}_2^2$$
$$V = mg\ell(1 - \cos\theta_1) + mg\ell(1 - \cos\theta_2)$$
$$+ \tfrac{1}{2}k(b - a\sin\theta_1 + a\sin\theta_2)^2$$

The position of equilibrium is determined by the conditions

$$\frac{\partial V}{\partial \theta_1} = mg\ell\sin\theta_1 + k(b - a\sin\theta_1 + a\sin\theta_2)(-a\cos\theta_1) = 0$$

$$\frac{\partial V}{\partial \theta_2} = mg\ell\sin\theta_2 + k(b - a\sin\theta_1 + a\sin\theta_2)(a\cos\theta_2) = 0$$

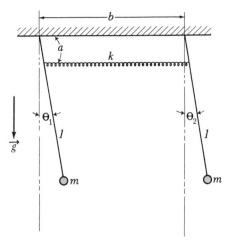

FIGURE 11.6   Coupled pendulums.

For small displacements, we approximate $\sin \theta$ by θ and $\cos \theta$ by 1. Then, we obtain

$$\frac{\partial V}{\partial \theta_1} = mg\ell\theta_1 + k(b - a\theta_1 + a\theta_2)(-a) = 0$$

$$\frac{\partial V}{\partial \theta_2} = mg\ell\theta_2 + k(b - a\theta_1 + a\theta_2)(a) = 0$$

which gives the equilibrium values

$$\theta_{10} = -\theta_{20} = \frac{kab}{mg\ell + 2ka^2}$$

The eigenfrequencies are found from the characteristic equation

$$\begin{vmatrix} mg\ell + ka^2 - m\ell^2\lambda^2 & -ka^2 \\ -ka^2 & mg\ell + ka^2 - m\ell^2\lambda^2 \end{vmatrix} = 0$$

which has the roots

$$\lambda_1^2 = \frac{g}{\ell} \qquad \lambda_2^2 = \frac{g}{\ell} + \frac{2ka^2}{m\ell^2}$$

corresponding to the eigenfrequencies

$$\lambda_1 = \sqrt{\frac{g}{\ell}} \qquad \lambda_2 = \sqrt{\frac{g}{\ell} + \frac{2ka^2}{m\ell^2}}$$

The eigenvectors are determined by

$$(-m\ell^2\lambda_1^2 + mg\ell + ka^2)\,p_{11} - ka^2 p_{21} = 0$$

$$(-m\ell^2\lambda_2^2 + mg\ell + ka^2)\,p_{12} - ka^2 p_{22} = 0$$

from which

$$\frac{p_{21}}{p_{11}} = 1 \qquad \frac{p_{12}}{p_{22}} = -1$$

Hence, matrix P is

$$P = \begin{bmatrix} \dfrac{\sqrt{2}}{2}, & \dfrac{\sqrt{2}}{2} \\[2mm] \dfrac{\sqrt{2}}{2}, & -\dfrac{\sqrt{2}}{2} \end{bmatrix} = \begin{bmatrix} 0.707, & 0.707 \\[2mm] 0.707, & -0.707 \end{bmatrix}$$

and the solution is

$$\theta_1 = 0.707b_1 \cos(\lambda_1 t + \varphi_1) + 0.707b_2 \cos(\lambda_2 t + \varphi_2)$$
$$\theta_2 = 0.707b_1 \cos(\lambda_1 t + \varphi_1) - 0.707b_2 \cos(\lambda_2 t + \varphi_2)$$

There are two normal modes of vibration. For mode λ_1, the two pendulums vibrate together with a frequency equal to that of the uncoupled frequency. On the other hand, for mode λ_2, the two pendulums vibrate in opposite directions with a frequency larger than the uncoupled frequency. The general oscillation is a superposition of these two modes.

PHENOMENON OF BEATS

Suppose that weak coupling exists between the two pendulums such that

$$\lambda_2 = \sqrt{\frac{g}{\ell} + \frac{2ka^2}{m\ell^2}} = \lambda_1(1 + \varepsilon)$$

where $\varepsilon\lambda_1$ is small compared to λ_1. Choosing initial conditions

$$\theta_1(0) = \theta_{10} \qquad \dot{\theta}_1(0) = 0$$
$$\theta_2(0) = 0 \qquad \dot{\theta}_2(0) = 0$$

and solving for b_1, b_2, φ_1, φ_2 yields

$$0.707b_1 = 0.707b_2 = \tfrac{1}{2}\theta_{10} \qquad \varphi_1 = \varphi_2 = 0$$

Thus, we can write

$$\theta_1 = \left[\frac{\theta_{10}}{2} \cos \lambda_1 t + \cos(1 + \varepsilon)\lambda_1 t)\right]$$

$$= \theta_{10} \cos\left(1 + \frac{\varepsilon}{2}\right)\lambda_1 t \cos\frac{\varepsilon\lambda_1}{2} t \qquad (1)$$

$$\theta_2 = \left[\frac{\theta_{10}}{2} \cos \lambda_1 t - \cos(1 + \varepsilon)\lambda_1 t\right]$$

$$= \theta_{10} \sin\left(1 + \frac{\varepsilon}{2}\right)\lambda_1 t \sin\frac{\varepsilon\lambda_1}{2} t \qquad (2)$$

Equations (1) and (2) represent simple harmonic vibrations of angular frequency

$$\left(1 + \frac{\varepsilon}{2}\right)\lambda_1 \cong \lambda_1$$

with amplitudes modulated by sinusoids of angular frequency $\dfrac{\varepsilon\lambda_1}{2}$. If, for example,

$$\lambda_1 = 2\pi \qquad \varepsilon = 0.02$$

then

$$\theta_1 = \theta_{10} \cos 2\pi t \cos 0.01\pi t$$
$$\theta_2 = \theta_{10} \sin 2\pi t \sin 0.01\pi t$$

Initially, when $t \cong 0$, pendulum (1) vibrates at maximum amplitude θ_{10} and pendulum (2) moves imperceptibly. Later when $t \cong 50$, the situation is reversed, with pendulum (1) moving imperceptibly and pendulum (2) vibrating at maximum amplitude. Thus, the system energy transfers back and forth between the two pendulums, producing *beats* of frequency $\dfrac{\varepsilon\lambda_1}{4\pi}$. In the limit, as $\varepsilon \to 0$, the coupling disappears and the two pendulums vibrate independently.

Example 12. Small Vibrations of Periodic Structures. The vibrational characteristics of periodic structures are fundamental in physics and engineering. Many physical systems, such as crystals and electric and acoustic filters, exhibit periodic structure and have important applications in science and technology. Lagrange's investigation of the vibrations of a series of elastically mounted particles, considered in this example, may be regarded as the prototype for all later work in the field (most of which differs only in detail, not in principle, from Lagrange's results).

Consider a system of n particles of equal mass m mounted at equal distances a on an elastic string as shown in Figure 11.7. The string is sufficiently taut that gravity is not a factor. For small displacements, the energy functions are

$$T = \tfrac{1}{2}m\dot{q}_1^2 + \tfrac{1}{2}m\dot{q}_2^2 + \cdots + \tfrac{1}{2}m\dot{q}_n^2$$

$$V = \frac{\tau}{2a}\left[q_1^2 + (q_2 - q_1)^2 + \cdots + (q_n - q_{n-1})^2 + q_n^2\right]$$

where τ is the string tension. Lagrange's equations give

$$m\ddot{q}_1 + \frac{\tau}{a}q_1 - \frac{\tau}{a}(q_2 - q_1) = 0$$

$$m\ddot{q}_2 + \frac{\tau}{a}(q_2 - q_1) - \frac{\tau}{a}(q_3 - q_2) = 0$$

$$\cdot \quad \cdot \quad \cdot \quad \cdot \quad \cdot \quad \cdot \quad \cdot \quad \cdot \quad \cdot \quad \cdot \quad \cdot \quad \cdot \quad \cdot$$
$$\cdot \quad \cdot \quad \cdot \quad \cdot \quad \cdot \quad \cdot \quad \cdot \quad \cdot \quad \cdot \quad \cdot \quad \cdot \quad \cdot$$

$$m\ddot{q}_n + \frac{\tau}{a}(q_n - q_{n-1}) + \frac{\tau}{a}q_n = 0$$

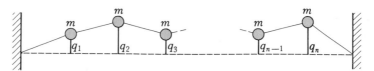

(a)   Loaded string; equidistant particles mounted on taut string; transverse vibrations

(b)   Mass and spring analog; longitudinal vibrations

(c)   Massive discs on compliant shaft; rotational vibrations

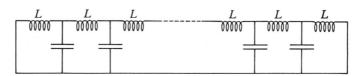

(d)   Electrical analog

FIGURE 11.7   Loaded string vibrator and some of its analogs.

Rearranging and recording in matrix form leads to the matrix* equation

$$
\begin{bmatrix}
mD^2 + \dfrac{2\tau}{a} & -\dfrac{\tau}{a} & 0 & \cdots & 0 \\[2ex]
-\dfrac{\tau}{a} & mD^2 + \dfrac{2\tau}{a} & -\dfrac{\tau}{a} & \cdots & 0 \\[2ex]
0 & -\dfrac{\tau}{a} & mD^2 + \dfrac{2\tau}{a} & \cdots & 0 \\[1ex]
\cdots & \cdots & \cdots & \cdots & \cdots \\
\cdots & \cdots & \cdots & \cdots & \cdots \\
0 & 0 & 0 & -\dfrac{\tau}{a}, & mD^2 + \dfrac{2\tau}{a}
\end{bmatrix}
\begin{bmatrix}
q_1 \\[2ex] q_2 \\[2ex] q_3 \\[1ex] \cdot \\ \cdot \\ \cdot \\ q_n
\end{bmatrix} = 0
$$

---

* A matrix having this general form is known as a *Jacobi matrix*.

The eigenvalues are given by the roots of

$$\det [K - M\lambda^2] = 0$$

which after division by τ/a and putting

$$-\frac{ma}{\tau}\lambda^2 + 2 = C$$

becomes

$$\begin{vmatrix} C & -1 & 0 & 0 & \cdots & 0 \\ -1 & C & -1 & 0 & \cdots & 0 \\ 0 & -1 & C & -1 & \cdots & 0 \\ \cdot & \cdot & \cdot & \cdot & \cdot & 0 \\ \cdot & \cdot & \cdot & \cdot & \cdot & -1 \\ 0 & 0 & 0 & 0 & -1 & C \end{vmatrix} = 0$$

For small values of n, the determinant is easily calculated by direct expansion in minors, and the roots of the characteristic equation are thus found. However, for large values of n this approach becomes difficult and an easier method is needed if one can be found. The recurrent structure of the determinant suggests that some recurrence relation may exist. To find such a relation, denote the determinant by Δ_n and its first minors by $\Delta_{n-1}, \Delta_{n-2}, \ldots, \Delta_1$. Then, it is easily shown that

$$\Delta_n = C\Delta_{n-1} - \Delta_{n-2}$$

which holds for all $n \geq 2$, where

$$\Delta_0 = 1 \qquad \Delta_1 = C$$

If we put $C = 2 \cos \theta$, the recurrence equation is satisfied by

$$\Delta_n = b \sin (n + 1)\theta = b[2 \cos \theta \sin (n - 1)\theta - \sin (n - 2)\theta]$$

where b is a constant to be determined. Letting $n = 1$ gives

$$\Delta_1 = b \sin 2\theta = C = 2 \cos \theta \qquad \therefore b = \frac{1}{\sin \theta}$$

Thus, the characteristic equation becomes

$$\Delta_n = \frac{\sin (n + 1)\theta}{\sin \theta} = 0$$

which is satisfied by

$$(n + 1)\theta = s\pi \qquad s = 1, 2, \ldots, n$$

The corresponding values of λ_s^2 are

$$\lambda_s^2 = \frac{\tau}{ma}(2 - C) = \frac{2\tau}{ma}\left(1 - \cos\frac{s\pi}{n+1}\right)$$

$$= \frac{4\tau}{ma}\sin^2\frac{s\pi}{2(n+1)}   s = 1, 2, \ldots, n$$

The vibrations are linear combinations of the n normal modes of vibration of angular frequency λ_s, $s = 1, 2, \ldots, n$. From Eq. (11.16) we can write the solution

$$q_r = \sum_{s=1}^{n} p_{rs}b_s \cos(\lambda_s t + \theta_s)   r = 1, 2, \ldots, n$$

where the b_s and θ_s are constants of integration.

It is instructive to compare this example with the electric wave filter studied in Chap. 8. The determinantal equations for the two are basically identical if the input and output to the filter are suppressed. Solution of the filter problem in terms of hyperbolic functions is not intrinsically different, because of the close relation between the hyperbolic and the circular (trigonometric) functions. The method of finding the eigenvalues in the present example differs only slightly from the previous method.

11.2   DAMPED VIBRATIONS

The free vibrations of dissipative systems decrease in amplitude with time, owing to frictional dissipation of energy. Lagrange's equations for dissipative systems, from Eq. (6.27) with $\mathbf{Q}_e = 0$ are

$$\frac{d}{dt}\left(\frac{\partial L}{\partial \dot{q}_r}\right) - \frac{\partial L}{\partial q_r} = Q_{dr}   r = 1, 2, \ldots, f \qquad \textbf{(11.21)}$$

The rate of loss of stored energy in the system is given by the expression

$$\frac{d}{dt}(T + V) = \mathbf{Q}_d \cdot \dot{\mathbf{q}} \qquad \textbf{(11.22)}$$

Equations (11.21) and (11.22) provide the basis for the analysis of any dissipative system for which T, V, and $\mathbf{Q}_d$ are known.

Again, linear systems are the simplest to deal with and also are of greatest usefulness in practice. Solutions can be determined in general form. For a linear system, $\mathbf{Q}_d = -R\dot{\mathbf{q}}$ and the equations of motion can be written in matrix form directly from relation (8.4) with $\mathbf{Q} = 0$ as follows

$$(MD^2 + RD + K)\mathbf{q} = 0 \qquad \textbf{(11.23)}$$

where M, R, and K are the system matrices. The canonical form, obtained by multiplying (11.23) by M^{-1}, is

$$(D^2 + M^{-1}RD + M^{-1}K)\mathbf{q} = 0 \tag{11.24}$$

Matrix equation (11.24) cannot be solved by the method of normal coordinates because it is generally not possible to simultaneously diagonalize the matrices. However, it can be solved by the method of exponential substitution. Proceeding therefore, as in the case of small vibrations (p. 353), let

$$\mathbf{q} = \mathbf{C}e^{\lambda t}$$

and substitute in Eq. (11.24). This gives the condition equation

$$(I\lambda^2 + M^{-1}R\lambda + M^{-1}K)\mathbf{C} = 0 \tag{11.25}$$

which has a nontrivial solution only if

$$\det(I\lambda^2 + M^{-1}R\lambda + M^{-1}K) = 0 \tag{11.26}$$

Expansion of the determinant yields a polynomial equation of degree $2f$ in λ which in the general case has f pairs of complex conjugate roots $-\alpha_r \pm i\beta_r$. It is known from physical considerations that the real parts are always negative or, in the limiting case of no damping, zero. Using complex notation, the general solution of (11.24) is therefore

$$\mathbf{q} = \sum_{r=1}^{f} \mathbf{C}_r e^{\lambda_r t} \tag{11.27}$$

where the $\mathbf{C}_r$ are f complex amplitude vectors and the λ_r are the f complex roots $-\alpha_r + i\beta_r$. Taking the real part of (11.27), there is obtained the general solution

$$q_r = \sum_{s=1}^{f} B_{rs}e^{-\alpha_s t} \cos(\beta_s t + \theta_s) \qquad r = 1, 2, \ldots, f \tag{11.28}$$

where B_{rs} is the magnitude and θ_s the argument of C_{rs}.

Equation (11.28) differs from the undamped solution, Eq. (11.10), only by the presence of the damping terms $e^{-\alpha_s t}$. The motion consists of a combination of damped sinusoids. In the limit of no damping, the $\alpha_s \to 0$, and (11.28) reduces to the undamped case.

Critical Damping. It may happen that some or all the $\beta_s = 0$, in which case there is a repeated root of Eq. (11.26) for each $\beta_s = 0$. The solution terms corresponding to these roots are (cf. Chap. 4)

$$B_{rs}e^{-\alpha_s t}(1 + C_s t) \tag{11.29}$$

where B_{rs} and C_s are the two integration constants associated with the repeated roots. Introduction of these terms then gives the general solution.

Overdamping. If some or all the roots of (11.26) are real, the corresponding solution terms are said to be overdamped. In this case, $\lambda_s = -\alpha_s \pm \beta_s < 0$, and the solution term is

$$B_{rs} e^{-\alpha_s t} \cosh (\beta_s t + \theta_s) \tag{11.30}$$

The net effect is to replace cosine terms by hyperbolic cosine terms. Clearly neither the overdamped nor the critically damped terms are oscillatory.

One Degree of Freedom. The canonical equation (11.26) reduces in this case to

$$\lambda_2 + \frac{R_{11}}{M_{11}} \lambda + \frac{R_{11}}{M_{11}} = 0$$

The roots are

$$\lambda_{1,2} = -\frac{R_{11}}{2M_{11}} \pm \sqrt{\left(\frac{R_{11}}{2M_{11}}\right)^2 - \frac{K_{11}}{M_{11}}} = -\alpha \pm i\beta$$

where α and β are defined by

$$\alpha = \frac{R_{11}}{2M_{11}} \qquad \beta = \sqrt{\frac{K_{11}}{M_{11}} - \left(\frac{R_{11}}{2M_{11}}\right)^2}$$

Hence, the solution, from (11.28), is

$$q = Be^{-\alpha t} \cos (\beta t + \theta) \tag{11.31}$$

and represents damped harmonic motion. It is worth noting that although the solution is oscillatory it is not periodic, since it does not satisfy the requirement that a nonzero number τ exists such that $q(t) = q(t + \tau)$ for every value of t.

The damping factor α in Eq. (11.31) is a measure of the decay time. The time for $q(t)$ to decrease by a factor $\frac{1}{e}$, or to 0.37 of its initial amplitude, is $\frac{1}{\alpha}$ and is called the time constant. It is evident that $\frac{1}{\alpha}$ has the physical dimensions of inverse time. A dimensionless measure of the damping is obtained by comparing values of $q(t)$ for successive waves or pseudo periods τ. The logarithm of this ratio is called the logarithmic decrement d, defined by

$$d = \ln \frac{q(t)}{q(t + \tau)} = \alpha\tau \tag{11.32}$$

Not only is d dimensionless, but $\frac{1}{d}$ is the number of oscillations for the amplitude to decrease to $\frac{1}{e}$ of its initial value.

In the modern literature on vibrations, it is customary to express the damping properties of a dissipative system in terms of a quality factor Q defined by the equation

$$Q = \frac{M_{11}\beta}{R_{11}}$$

Typical Q-factors found in practice are 200 for good radio-frequency coils, 2000 for tuning forks, and 200,000 for piezo-electric quartz-crystal resonators. Even larger Q-factors are encountered in the resonators for atomic clocks. Compare also the example of the radar pulse-forming network discussed in Chap. 8.

Applications. There are many practical applications of linear dissipative systems. These include such diverse cases as electric circuits, mechanical vibrations, subsidence oscillations in automatic control devices, and readjustments in economic and production cycles. The many physical applications stem from the fact that energy losses can often be represented, or at least reasonably well approximated, by viscous damping forces.

Example I. Radar Pulse Network. Damped oscillations of coupled L-C circuits are used to generate pulses in high-power radar transmitters. The network considered here is designed to produce approximately square-wave pulses of one microsecond duration. The energy is stored in capacitors and discharged through a high-power switching element to a resistance-type load as shown in Figure 11.8. Initial conditions are: capacitors are charged to potential V_0 at $t = 0$; and currents I_1, I_2, I_3 are zero at $t = 0$. Find the current pulse through the load resistance R_3 when the switch is closed, thus discharging the network. Numerical values of elements are

$$L_{22} = L_{33} = 4.13 \times 10^{-6} \text{ henry} \qquad C = 1.67 \times 10^{-9} \text{ farad}$$

$$L_{11} = 4.80 \times 10^{-6} \text{ henry} \qquad R_3 = 50 \text{ ohms}$$

$$L_{12} = 0.07 \times 10^{-6} \text{ henry} \qquad V_0 = 6000 \text{ volts}$$

$$L_{23} = 0.68 \times 10^{-6} \text{ henry}$$

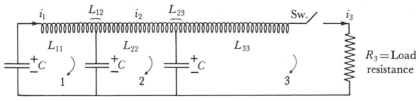

FIGURE 11.8 Radar pulse network.

SOLUTION

(1) Since $\boldsymbol{\epsilon} = 0$, $\mathbf{I}_0 = \dot{\mathbf{q}}_0 = 0$, the Laplace transform for the current $\mathbf{I}$ is, from Eq. (8.25), given by

$$\bar{\mathbf{I}} = -A^{-1}C^{-1}\mathbf{q}_0$$

where

$$\mathbf{I} = \dot{\mathbf{q}} \qquad C^{-1} = K$$

$$C^{-1}\mathbf{q}_0 = \begin{bmatrix} \dfrac{2}{C} & -\dfrac{1}{C} & 0 \\[2ex] -\dfrac{1}{C} & \dfrac{2}{C} & -\dfrac{1}{C} \\[2ex] 0 & -\dfrac{1}{C} & \dfrac{1}{C} \end{bmatrix} \begin{bmatrix} CV_0 \\[2ex] CV_0 \\[2ex] CV_0 \end{bmatrix} = \begin{bmatrix} V_0 \\[2ex] 0 \\[2ex] 0 \end{bmatrix}$$

and the system matrix A is

$$A = \begin{bmatrix} s^2 L_{11} + \dfrac{2}{C} & s^2 L_{12} - \dfrac{1}{C} & 0 \\[2ex] s^2 L_{12} - \dfrac{1}{C} & s^2 L_{22} - \dfrac{2}{C} & s^2 L_{23} - \dfrac{1}{C} \\[2ex] 0 & s^2 L_{23} - \dfrac{1}{C} & s^2 L_{33} + sR + \dfrac{1}{C} \end{bmatrix}$$

(2) The current I_3 through the load resistance R_3 is then

$$I_3 = \mathscr{L}^{-1}(-A_{31}^{-1}V_0) = \mathscr{L}^{-1}\left(\frac{-\Delta_{31}V_0}{|A|}\right)$$

where Δ_{31}, the cofactor of A_{31} in $|A|$, is

$$\Delta_{31} = \left(L_{12}s^2 - \frac{1}{C}\right)\left(L_{23}s^2 - \frac{1}{C}\right)$$

(3) To complete the solution it is necessary to substitute the numerical values for the circuit parameters, find the roots of $|A| = 0$, expand $\bar{I}_3$ in partial fractions, and then find the inverse transform of $\bar{I}_3$. It is seen from the expression for A that $|A|$ is a sixth order polynomial in s and therefore has six roots. The roots of $|A| = 0$ as found by computer are as follows:

$$s_1, s_2 = (-0.312 \pm 0.424i)10^7$$
$$s_3, s_4 = (-0.202 \pm 1.33i)10^7$$
$$s_5, s_6 = (-0.110 \pm 2.34i)10^7$$

(4) These three pairs of complex roots correspond to three damped sinusoids, the sum of which gives the expression for $I_3(t)$. The nature of

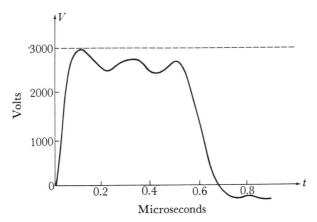

FIGURE 11.9 Pulse shape for radar network.

the solution is indicated graphically in Figure 11.9 which is a plot of $V = I_3 R_3$, the voltage pulse as seen across the load resistor R_3.

Example 2. "Negative Resistance" Feedback Oscillator. It is possible, under certain conditions where energy is supplied to a system from an external source, to develop oscillations of increasing amplitude. If such a system can be approximated sufficiently well by a linear differential equation with constant coefficients, we can still use a damping factor α, but with a negative sign so that the differential equation becomes

$$\ddot{q} - 2\alpha\dot{q} + \beta_0^2 q = 0 \qquad \beta_0^2 = \alpha^2 + \beta^2 \qquad (1)$$

This equation has the solution

$$q = e^{\alpha t}\left[q_0 \cos \beta t + \frac{\dot{q}_0 - \alpha q_0}{\beta} \sin \beta t \right] \qquad (2)$$

which represents an oscillation with increasing amplitude as illustrated in Figure 11.10.

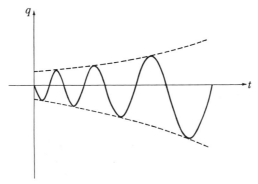

FIGURE 11.10 "Negatively damped" oscillation.

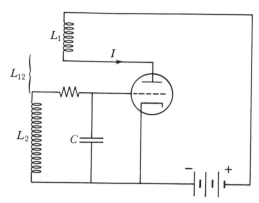

FIGURE 11.11 "Negative resistance" oscillator.

A practical example of a "negative resistance" oscillator is provided by a vacuum tube system with positive feedback, Figure 11.11. The differential equation for this circuit is

$$L_2 C\ddot{v} + RC\dot{v} + v = L_{12}\frac{dI}{dt} \tag{3}$$

where v is the grid voltage and I is the plate current as shown in the circuit diagram.

In general, the plate current I is a complicated function of the plate and grid voltages. For simplicity we neglect the effect of the plate voltage, which is small, and consider the grid voltage alone. Under these assumptions, we write

$$I = f(v) \quad \text{and} \quad \frac{dI}{dt} = \frac{df}{dv}\dot{v} = g\dot{v}$$

where g is called the mutual conductance of the tube and is a variable depending on v in a complicated manner.

For small values of v, g is nearly constant and we may write

$$LC\ddot{v} + (RC - L_{12}g_0)\dot{v} + v = 0 \tag{4}$$

where

$$g_0 = \left(\frac{\partial f}{\partial v}\right)_{v=v_0} = \text{mutual conductance at the operating point}$$

By making L_{12} positive, and of such magnitude that

$$RC - L_{12}g_0 < 0$$

α becomes negative, and the oscillations will grow, rather than attenuate. In practice, the amplitude of the oscillations cannot of course increase indefinitely but, rather, will be limited by nonlinear effects to some finite value.

11.3 FORCED VIBRATIONS

Forced vibrations occur when a periodic force is applied to a system capable of executing natural vibrations. The energy equation for forced vibrations is similar to that for damped vibrations, but with the forcing terms added, i.e.,

$$\frac{d}{dt}(T - V) = \mathbf{Q}_d \cdot \mathbf{q} + \mathbf{Q}_e \cdot \mathbf{q}$$

where $\mathbf{Q}_e$ represents a generalized periodic driving force. Solutions will be derived for the basic case where $\mathbf{Q}_e$ is sinusoidal, with the understanding that solutions for other wave forms can be obtained by the Fourier series method.

As in the case of damped vibrations, linear systems are of most interest in practice and are simplest to deal with. Equations of motion for such systems follow from Eq. (8.4),

$$(MD^2 + RD + K)\mathbf{q} = \mathbf{Q}_p e^{i\omega t} \tag{11.33}$$

where $\mathbf{Q}_p$ represents the amplitudes of the force terms, and the sinusoid is represented in complex notation.

The steady-state solution $\mathbf{q}_s$ of (11.33) may be written from Eq. (8.17) in the form

$$\mathbf{q}_s = \mathbf{q}_p e^{i\omega t} = A^{-1}\mathbf{Q}_p e^{i\omega t} \tag{11.34}$$

where

$$A = -\omega^2 M + i\omega R + K$$

The general solution $\mathbf{q}$ is the sum of $\mathbf{q}_s$ and the transient solution $\mathbf{q}_{tr}$ which was derived for damped vibrations and is given by Eq. (11.27). Hence

$$\mathbf{q} = \mathbf{q}_s + \mathbf{q}_{tr} \tag{11.35}$$

For most applications, the steady-state solution is the only one of interest, because even for light damping the transient terms decay rather rapidly and the solution reduces to the steady-state case. From Eq. (11.34), the steady-state solution for $\mathbf{q}_p$ in complex notation is

$$\mathbf{q}_p = A^{-1}\mathbf{Q}_p \tag{11.36}$$

The corresponding solution for the generalized velocity $\dot{\mathbf{q}}_p$ was developed in Eq. (8.18) as follows:

$$\dot{\mathbf{q}}_p = i\omega\mathbf{q}_p = i\omega A^{-1}\mathbf{Q}_p$$

The matrix $i\omega A^{-1}$ is called the admittance matrix and is denoted by Y or by Z^{-1}, where Z is called the impedance matrix. Hence, $\dot{\mathbf{q}}_p$ may be written

$$\dot{\mathbf{q}}_p = Y\mathbf{Q}_p = Z^{-1}\mathbf{Q}_p \tag{11.37}$$

Resonance. When the applied frequency ω is near or equal to a natural frequency of the system, the response is in general magnified and may rise to dangerous values in lightly damped systems, causing plastic distortion or even rupture in mechanical systems due to excessive stresses; a similar effect may occur in electrical systems where excessive voltages may cause insulation failure or breakdown. On the other hand, resonance may be useful, as for example in the tuned circuits of communication systems, or in certain types of vibration dampers.

A *resonant frequency* ω_r is defined as a value of ω which maximizes the response of the system. There will be in general a resonant frequency for each local maximum of the frequency response characteristic. In most cases of practical interest the damping terms are relatively small and A in Eq. (11.36) can be approximated by

$$A \cong K - \omega^2 M$$

The resonant frequencies are then found with close approximation from the condition

$$\det [A] = \det (K - \omega^2 M) = 0 \qquad \textbf{(11.38)}$$

But this is the same condition as Eq. (11.12) for the eigenfrequencies of the undamped system. Hence, the resonant frequencies are, to a close approximation, equal to the eigenfrequencies, i.e.,

$$\omega_1 = \lambda_1 \qquad \omega_2 = \lambda_2, \ldots \qquad \omega_f = \lambda_f \qquad \textbf{(11.39)}$$

Example I. Resonance for a System of One Degree of Freedom
The system response q_p is given by

$$q_p = A^{-1} Q_p = \frac{Q_p}{K_{11} - M_{11}\omega^2 + iR_{11}\omega}$$

which has the magnitude

$$|q_p| = \frac{Q_p}{[(K_{11} - M_{11}\omega^2)^2 + R_{11}^2 \omega^2]^{1/2}}$$

Resonance corresponds to the value of ω which maximizes $|q_p|$. This is equivalent to finding the value of ω which minimizes the above bracketed expression. Hence, using the theory of extrema, we find

$$\frac{d}{d\omega} [(K_{11} - M_{11}\omega^2)^2 + R_{11}^2 \omega^2] = 0$$

which gives the condition

$$2(K_{11} - M_{11}\omega^2)(-2M_{11}\omega) + 2R_{11}^2 \omega = 0$$

The root $\omega = 0$ corresponds to static equilibrium and is therefore of no interest here. Thus, the resonant frequency is given by

$$\omega_r^2 = \frac{K_{11}}{M_{11}} - \frac{R_{11}^2}{2M_{11}^2} = \beta_0^2 - 2\alpha^2 = \beta^2 - \alpha^2$$

and the maximum amplitude of q, after simplification, is found to be

$$\text{Max }|q_p| = \frac{Q_p}{R_{11}\sqrt{\beta_0^2 - \alpha^2}}$$

Since for light damping $\beta_0^2 \gg \alpha^2$, this result differs only very slightly from the approximate answer given by

$$\text{Max }|q_p| = \frac{Q_p}{R_{11}\beta_0}$$

Thus, for many purposes it is satisfactory to define resonance for a sinusoidally-excited system as a frequency for which the reactance vanishes.

Example 2. Resonance for Inductively Coupled Circuits (Two Degrees of Freedom). Find the resonant frequencies and response of the inductively coupled circuits shown in Figure 11.12, where R_{11} and R_{22} are relatively small and $L_{11}C_{11} = L_{22}C_{22}$.

ANALYSIS

The system matrix is

$$A = \begin{bmatrix} \dfrac{1}{C_{11}} - L_{11}\omega^2 + iR_{11}\omega & -L_{12}\omega^2 \\[2ex] -L_{12}\omega^2 & \dfrac{1}{C_{22}} - L_{22}\omega^2 + iR_{22}\omega \end{bmatrix}$$

Resonant frequencies are, to a close approximation, given by

$$\begin{vmatrix} \dfrac{1}{C_{11}} - L_{11}\omega^2 & -L_{12}\omega^2 \\[2ex] -L_{12}\omega^2 & \dfrac{1}{C_{22}} - L_{22}\omega^2 \end{vmatrix} = 0$$

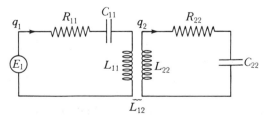

FIGURE 11.12 Inductively coupled oscillator circuits.

which has the roots

$$\omega_{1,2}^2 = \frac{(L_{11}C_{11} + L_{22}C_{22}) \pm \sqrt{(L_{11}C_{11} - L_{22}C_{22})^2 + 4L_{12}^2 C_{11}C_{22}}}{2(L_{11}L_{22} - L_{12}^2)C_{11}C_{22}}$$

But since $L_{11}C_{11} = L_{22}C_{22} = LC$, this expression reduces to

$$\omega_{1,2}^2 = \frac{1}{(1-k)LC}, \frac{1}{(1+k)LC}, \quad k^2 = \frac{L_{12}^2}{L_{11}L_{22}}$$

The two resonant frequencies obviously move closer together as the coupling L_{12} decreases and, in the limit, approach the same value, i.e.,

$$\lim_{L_{12} \to 0} \omega_1^2 = \lim_{L_{12} \to 0} \omega_2^2 = \omega_0^2 = \frac{1}{LC}$$

The system response at resonance for $k \ll 1$ is given by

$$q_p(\omega_1) = A^{-1}(\omega_1) \begin{bmatrix} E_1 \\ 0 \end{bmatrix} = \begin{bmatrix} A_{11}^{-1} & E_1 \\ A_{21}^{-1} & E_1 \end{bmatrix}$$

where

$$A_{11}^{-1} = \frac{A_{22}}{\det A} \cong \frac{\dfrac{1}{C} - L\omega_1^2 + i\omega_1 R_{22}}{-\omega_1^2 R_{11}R_{22}} \cong \frac{1}{iR_{11}\omega_1}$$

$$A_{21}^{-1} = \frac{-A_{12}}{\det A} \cong -\frac{L_{12}}{R_{11}R_{22}}$$

Hence

$$\mathbf{q}_p(\omega_1) = \begin{bmatrix} q_{p1} \\ \\ q_{p2} \end{bmatrix} \cong \begin{bmatrix} -\dfrac{iE_1}{R_{11}\omega_1} \\ \\ -\dfrac{L_{12}E_1}{R_{11}R_{22}} \end{bmatrix}$$

and, correspondingly,

$$\mathbf{q}_p(\omega_2) \cong \begin{bmatrix} -\dfrac{iE_1}{R_{11}\omega_2} \\ \\ -\dfrac{L_{12}E_1}{R_{11}R_{22}} \end{bmatrix}$$

Example 3. Instrument Protection.   In a certain system, it is necessary to mount a delicate instrument on the frame of an electric motor. Since perfect balance of the motor cannot be assured, some method for protection of the instrument against motor vibration must be provided. One commonly used method for accomplishing the isolation is to mount the instrument on a spring support with a dashpot damping device as shown in Figure 11.13. Analyze the protection characteristics for such a device.

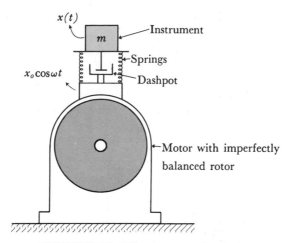

FIGURE 11.13 Vibration protection device.

ANALYSIS

There is one degree of freedom, denoted by x, the vertical displacement of the instrument. The imperfectly balanced motor gives rise to an oscillating motion of the base of the protection device; this motion is represented by $x_0 \cos \omega t$, where x_0 is the amplitude and ω the angular frequency of the motor vibration. It is clear that the system, consisting of the instrument and its mounting, is subjected to a sinusoidal driving force, but that this force is not directly known, and, in fact, cannot be determined until the motion of the system has been calculated. The problem posed is characteristic of a large class of problems which crop up in applications.

The solution, however, is straightforward by Lagrange's equations and proceeds as follows. Energy functions are

$$L = T - V = \tfrac{1}{2}m\dot{x}^2 - \tfrac{1}{2}k(x - x_0 \cos \omega t)^2$$
$$\delta W = R(\dot{x} + \omega x_0 \sin \omega t)\, \delta x$$

where R is the damping resistance of the dashpot. From these functions, the equation of motion is written

$$m\ddot{x} + k(x - x_0 \cos \omega t) = -R(\dot{x} + x_0 \omega \sin \omega t)$$

or

$$m\ddot{x} + R\dot{x} + kx = x_0(k \cos \omega t - R\omega \sin \omega t) = x_0\sqrt{k^2 + R^2\omega^2} \cos (t\omega + \gamma)$$

where

$$\gamma = \tan^{-1} \frac{R\omega}{k}$$

By analogy with Ex. 1, the steady-state solution for x in complex notation is

$$\bar{x} = \frac{(x_0\sqrt{k^2 + R^2\omega^2})e^{i\gamma}}{k - m\omega^2 + iR\omega}$$

A measure or figure of merit of the vibration protection provided by the device is the amplitude reduction ratio r, defined as

$$r = \left|\frac{\bar{x}}{x_0}\right| = \left[\frac{k^2 + R^2\omega^2}{(k - m\omega^2)^2 + R^2\omega^2}\right]^{1/2}$$

$$= \left[\frac{1 + \dfrac{R^2\omega^2}{k^2}}{\left(1 - \dfrac{\omega^2}{\omega_r^2}\right)^2 + \dfrac{R^2\omega^2}{k^2}}\right]^{1/2}$$

where $\omega_r^2 = k/m$ is the undamped resonant frequency.

Since damping is small, the terms containing R may be neglected except near resonance. In frequency ranges away from resonance, the damping terms may therefore be dropped and the expression for r approximated by

$$r = \frac{1}{\left|1 - \dfrac{\omega^2}{\omega_r^2}\right|}$$

To reduce the vibration amplitude ratio to less than one, it is necessary that

$$\left[1 - \frac{\omega^2}{\omega_r^2}\right]^2 > 1, \quad \text{which means that} \quad \frac{\omega^2}{\omega_r^2} > 2$$

Hence, the resonant frequency ω_r of the instrument-mounting system must be chosen such that

$$\omega_r < \frac{\omega}{\sqrt{2}}$$

For effective protection it is necessary that

$$r << 1 \quad \text{i.e.,} \quad \frac{\omega^2}{\omega_r^2} >> 2$$

Thus, for example, a protection ratio of $r = 0.10$, or 10 percent, requires that

$$\frac{\omega^2}{\omega_r^2} = 11 \quad \text{or} \quad w_r = \frac{\omega}{\sqrt{11}}$$

Example 4. Critical Speed of Rotating Machine. An elementary
rotating machine may be represented by a model consisting of a disc of
mass M mounted on a shaft of equivalent spring constant k. In general,
such a body will be subjected to fluctuating moments, that is, to moments
and forces which rotate with the body. These rotating moments and
forces vanish if, and only if, the following conditions are satisfied: (1)
the rotating body must be balanced so that the axis of rotation is a
principal axis of inertia; (2) the center of mass of the body must lie on
the axis of rotation. These conditions are known as the principle of
dynamic balancing.

The effects of unbalance are evident, for example, in the familiar
situation of the front wheels of an automobile with unevenly worn tires.
At certain driving speeds the rotation frequency coincides with the natural
frequency of the wheel and suspension spring. Violent vibrations, known
as "shimmying," then tend to occur. In rotating machines such vibrations
may be severe enough to tear the machine apart.

In the analysis of the elementary model depicted in Figure 11.14, we
neglect damping forces since they have little effect on the critical frequency. Let **r** be the displacement vector in the x, y plane for the center
C of the flywheel, and let **a** be the vector position of the center of mass of
the wheel measured from C. Note that **a** rotates with the wheel.

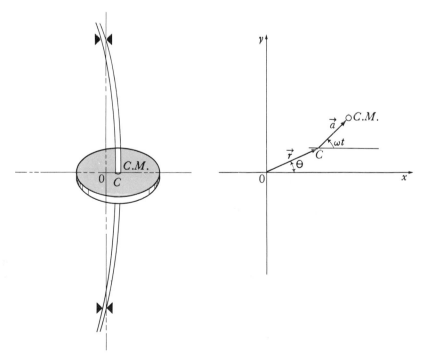

FIGURE 11.14 Schematic diagram for critical speed of rotating machine.

The Lagrangian for the system is

$$L = \tfrac{1}{2}m(\dot{\mathbf{r}} + \dot{\mathbf{a}})^2 - \tfrac{1}{2}kr^2$$

$$= \tfrac{1}{2}m(\dot{r}^2 + 2\dot{\mathbf{r}} \cdot \dot{\mathbf{a}} + \dot{a}^2) - \tfrac{1}{2}kr^2$$

$$\frac{d}{dt}\left(\frac{\partial L}{\partial \dot{r}}\right) - \frac{\partial L}{\partial r} = \frac{d}{dt}(m\dot{r} + m\mathbf{e} \cdot \dot{\mathbf{a}}) + kr = 0$$

where $\mathbf{e}$ is a unit vector in the direction of $\mathbf{r}$. Carrying out the differentiation gives

$$m\ddot{r} + m\mathbf{e} \cdot \ddot{\mathbf{a}} + kr = 0$$

But

$$\ddot{\mathbf{a}} = -\omega^2 \mathbf{a}$$

and

$$\mathbf{e} \cdot \mathbf{a} = a \cos(\omega t - \theta)$$

Therefore, the equation of motion is

$$m\ddot{r} - ma\omega^2 \cos(\omega t - \theta) + kr = 0$$

or

$$\ddot{r} + \frac{k}{m}r = a\omega^2 \cos(\omega t - \theta)$$

The steady-state solution is

$$r = \frac{a\omega^2}{\beta_0^2 - \omega^2} \cos(\omega t - \theta)$$

where $\beta_0 = \sqrt{k/m}$ is the natural frequency of the shaft and wheel. It is obvious that if ω approaches β_0 the vibration amplitude will become relatively large. On the other hand, to minimize the vibration it is necessary to make $\beta_0^2 \gg \omega^2$, that is, the operating frequency ω should be much below the natural frequency β_0.

Example 5. Speedometer Mechanism. One commonly used type of speedometer is depicted in Figure 11.15. The underlying principle of the instrument is to develop a torque proportional to the rotational speed; this torque in turn deflects an indicator needle through an angle proportional to the torque. The conversion from input angular velocity to deflecting torque is achieved through a viscous drag or reaction damper element as shown in the figure.

---

* Adapted from F. D. Cribbins and C. S. Stultz, "Analyze the Torsional Vibration of a Damped, Coupled System." *General Motors Engineering Journal*, Vol. 12, No. 3, 51, 1965.

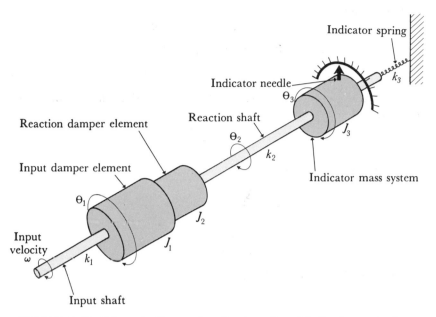

FIGURE 11.15 Schematic diagram for vibrations of speed-indicating mechanism.

ANALYSIS. There are three degrees of freedom, denoted by θ_1, θ_2, and θ_3. Energy functions for the systems are

$$L = T - V = \tfrac{1}{2}J_1\dot{\theta}_1^2 + \tfrac{1}{2}J_2\dot{\theta}_2^2 + \tfrac{1}{2}J_3\dot{\theta}_3^2$$
$$-\tfrac{1}{2}k_1(\theta_1 - \omega t)^2 - \tfrac{1}{2}k_2(\theta_2 - \theta_3)^2 - \tfrac{1}{2}k_3\theta_3^2$$
$$F = \tfrac{1}{2}\mu(\dot{\theta}_1 - \dot{\theta}_2)^2$$

where J_1, J_2, J_3 are the moments of inertia, k_1, k_2, k_3 the torsional spring constants, μ the damping parameter, and ω the angular velocity of the drive shaft.

Substitution in Lagrange's equations in the form (8.1) yields

$$J\ddot{\theta}_1 + k_1(\theta_1 - \omega t) + \mu(\dot{\theta}_1 - \dot{\theta}_2) = 0$$
$$J_2\ddot{\theta}_2 + k_2(\theta_2 - \theta_3) - \mu(\dot{\theta}_1 - \dot{\theta}_2) = 0$$
$$J_3\ddot{\theta}_3 - k_2(\theta_2 - \theta_3) + k_3\theta_3 = 0$$

Rearranging and writing in matrix form gives

$$\begin{bmatrix} J_1D^2 + \mu D + k_1 & -\mu D & 0 \\ -\mu D & J_2D^2 + \mu D + k_2 & -k_2 \\ 0 & -k_2 & J_3D^2 + k_2 + k_3 \end{bmatrix} \begin{bmatrix} \theta_1 \\ \theta_2 \\ \theta_3 \end{bmatrix} = \begin{bmatrix} k_1\omega t \\ 0 \\ 0 \end{bmatrix}$$

or more compactly

$$\boldsymbol{\theta} = A^{-1}\mathbf{Q} = A^{-1}\begin{bmatrix} k_1\omega t \\ 0 \\ 0 \end{bmatrix}$$

We are interested primarily in θ_3, the deflection angle of the needle indicator. Let

$$A^{-1} = \frac{\Delta}{\det A}$$

where Δ is the adjoint matrix of A. Then

$$\theta_3 = \frac{\Delta_{31}k_1\omega t}{\det A} = \frac{k_2 k_1 \mu D\omega t}{\det A}$$

$$= \frac{k_1 k_2 \mu\omega}{\det A}$$

For the steady-state case, $\det A$ reduces to

$$\begin{vmatrix} k_1 & 0 & 0 \\ 0 & k_2 & -k_2 \\ 0 & -k_2 & k_2 + k_3 \end{vmatrix} = k_1[k_2(k_2 + k_3) - k_2^2]$$

$$= k_1 k_2 k_3$$

Therefore, the steady-state deflection is

$$\theta_3 = \frac{k_2 k_1 \mu\omega}{k_1 k_2 k_3} = \frac{\mu\omega}{k_3}$$

Transient analysis for this system is straightforward, using the Laplace transform or other equivalent method. The determinantal equation is obviously of order six and must be solved by numerical means. Properties of the solution can be studied by varying the critical parameters. In practice, changes in the torsional constant k_1 are of most interest. The root locus method, which is widely used in automatic feedback control studies, is useful for this purpose.

11.4 NONLINEAR VIBRATIONS

Most vibration problems in physical and engineering systems are nonlinear. The nonlinearities may arise from nonlinear inertia, resistance, or stiffness elements of a system, or from nonlinear coupling in the system itself. For example, electric circuits with ferromagnetic-core inductances are nonlinear owing to the variation of the inductance coefficients with current. Damping forces which are not proportional to the velocity also cause nonlinearity as, for instance, the vibratory motion of a body in a resistive medium where the resistive force is proportional to the square of the velocity. In general, advanced and special mathematical methods are required to cope with nonlinear vibrations. If, however, the vibrations are sufficiently small, it is often possible to linearize a problem by the method used for small vibrations covered earlier in this chapter. If the linear approximation is not acceptable, then higher order approximations may be possible. Beyond this it is necessary to deal directly with the nonlinear equations of motion.

Since no general theories exist for the solution of nonlinear differential equations, it is necessary to turn to special methods and techniques. Solution of the pendulum problem, for example, can be obtained in terms of elliptic functions Other special functions, of which some 2000 are treated in the literature, are also useful on a selected basis. Some problems such as the Kepler problem for satellite and planetary orbits can even be solved in terms of elementary functions. Various other techniques are available for investigating nonlinear vibrations. Among these may be mentioned perturbation methods, the phase plane method of Poincaré (cf. Chap. 4), the method of isoclines (Ref. 1), and certain methods of generalized mechanics. Digital and analog computer methods play a dominant role in finding numerical solutions of nonlinear vibration problems.

The equations of motion for the vibrations of systems can be obtained by the methods of generalized mechanics. These methods are also useful in solving the equations of motion. For example, it may be possible to choose the generalized coordinates for a system in such a way that one or more of the coordinates are ignorable, thus simplifying the problem. This approach is effective in the solution of the Kepler problem. As a second example, the integral of energy not infrequently may be used to simplify problems. One integral of motion can always be obtained in this way for the vibrations of a conservative system. If, further, the system has only one degree of freedom, the general solution can always be obtained in the form of an integral. The pendulum problem, for example, can be solved by this method.

Conservative Systems. We consider first systems with one degree of freedom. Oscillatory motion can occur only in the neighborhood of a point of minimum potential, i.e., in a *potential well* as illustrated in Figure 11.16. The nature of the vibration is determined by the shape of the potential well and by the total energy of the system. Let $q(t)$ denote the configuration of the system at any time t. Then it is evident that the motion ranges between the points q_1 and q_2 defined by

$$V(q_1) = V(q_2) = E \qquad (11.40)$$

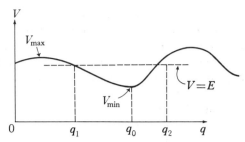

FIGURE 11.16   Potential well for one degree of freedom.

where E is the total energy of the system. The positive and negative excursions are given by $q_2 - q_0$ and $q_0 - q_1$, respectively. In general, these are not equal, and the oscillations are asymmetrical. It is also clear that the range of oscillation increases as the total energy of the system is increased, though it cannot exceed the limit set by the condition $E < V_{max}$ as illustrated in Figure 11.16.

The equation of motion can be derived from the law of conservation of energy

$$T + V = \tfrac{1}{2}m\dot{q}^2 + V(q) = E$$

from which $\dot{q}$ may be written

$$\dot{q} = \frac{dq}{dt} = \sqrt{\frac{2(E - V)}{m}}$$

Solving for dt and integrating gives

$$t = \int \sqrt{\frac{m}{2(E - V)}}\, dq + t_0 \qquad\qquad \textbf{(11.41)}$$

Equation (11.41) is the complete solution for the case of one degree of freedom, since it contains the two constants of motion E and t_0. The constant t_0 merely specifies the initial time and can usually be taken as zero without loss of generality. Thus, the motion for a given system depends on the energy constant E as a parameter.

The general method of applying these results is illustrated in the following examples.

Example I.   Charged-Particle Oscillator.   A charged particle of mass m and charge q moves on the x-axis between two point charges Q located at $x = \pm a$. Determine the motion, assuming that the electromagnetic radiation due to acceleration of the charged particle can be neglected.

SOLUTION

Write the energy equation

$$T + V = \tfrac{1}{2}m\dot{x}^2 + \frac{Qq}{\varepsilon_0(a - x)} + \frac{Qq}{\varepsilon_0(a + x)} = E$$

and test for equilibrium

$$\frac{\partial V}{\partial x} = \frac{Qq}{\varepsilon_0}\left[\frac{1}{(a-x)^2} - \frac{1}{(a+x)^2}\right] = 0$$

Note that V has a stationary point at $x = 0$, and since

$$\left[\frac{\partial^2 V}{\partial x^2}\right]_{x=0} = \frac{Qq}{\varepsilon_0}\left(\frac{2}{a^3} + \frac{2}{a^3}\right) = \frac{4Qq}{\varepsilon_0 a^3} > 0$$

$x = 0$ is a point of stable equilibrium about which vibrations may occur, in accord with Eq. (11.41), i.e.,

$$t = \int\sqrt{\frac{m}{2(E-V)}}\, dx$$

Let x_0 be the amplitude of the vibration. Then, we can express $E - V$ in the form

$$E - V = \frac{2aQq}{\varepsilon_0}\left[\frac{1}{a^2 - x_0^2} - \frac{1}{a^2 - x^2}\right] = \frac{2aQq}{\varepsilon_0(a^2 - x_0^2)}\left[\frac{x_0^2 - x^2}{a^2 - x^2}\right]$$

and integral (11.41) becomes

$$t = \sqrt{\frac{m(a^2 - x_0^2)\varepsilon_0}{4aQq}}\int\sqrt{\frac{a^2 - x^2}{x_0^2 - x^2}}\, dx + \text{const.}$$

This is an elliptic integral which can be reduced to Legendre's normal form for elliptic integrals of the second kind by the substitution

$$x = x_0 \sin\varphi \qquad x_0 = ak$$

Making this substitution and choosing $x = 0$ when $t = 0$, we obtain

$$\int_0^x\sqrt{\frac{a^2 - x^2}{x_0^2 - x^2}}\, dx = \int_0^\varphi\sqrt{1 - k^2\sin^2\varphi}\, d\varphi = E(k, \varphi)$$

where $E(k, \varphi)$ is the normal form for an elliptic integral of the second kind as noted above. Using this result we can write

$$t = \sqrt{\frac{ma^3(1 - k^2)}{4Qq}}\, E(k, \varphi)$$

The period τ for the complete vibration is given by

$$\tau = 4\sqrt{\frac{ma^3(1 - k^2)}{4Qq}}\, E\left(k, \frac{\pi}{2}\right)$$

which follows from the consideration that one-quarter of a swing corresponds to $x = x_0$ or $\varphi = \pi/2$.

The period of the electric vibrator decreases with increasing amplitude, in contrast to the plane pendulum (treated in the next example)

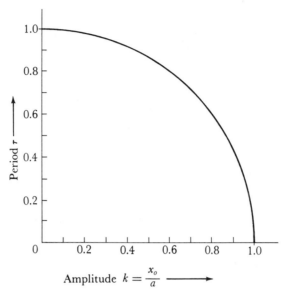

Period $\tau \longrightarrow$

Amplitude $k = \dfrac{x_o}{a} \longrightarrow$

FIGURE 11.17   Period versus amplitude for charged particle oscillator.

where the period increases with amplitude. Compare Figures 11.17 and 11.19.

Example 2. Plane Pendulum. Study of the motion of a plane pendulum, Figure 11.18. Let ℓ be the length, m the mass, and θ the angle of deflection. Then

$$T + V = \tfrac{1}{2}m\ell^2\dot{\theta}^2 + mg\ell(1 - \cos\theta) = E = mg\ell(1 - \cos\theta_0)$$

where θ_0 is the maximum deflection.

The solution written directly from Eq. (11.41) is

$$t = \sqrt{\frac{\ell}{2g}}\int_0^\theta \frac{d\theta}{\sqrt{\cos\theta - \cos\theta_0}}$$

where $\theta = 0$ when $t = 0$.

To evaluate this integral, make a change of variable as follows:

$$\sin\frac{\theta}{2} = k\sin\varphi \quad \text{where} \quad k = \sin\frac{\theta_0}{2}$$

so that

$$t = \sqrt{\frac{\ell}{g}}\int_0^\varphi \frac{d\varphi}{\sqrt{1 - k^2\sin^2\varphi}}$$

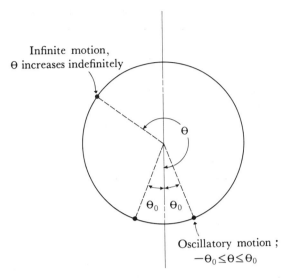

Infinite motion,
θ increases indefinitely

θ

θ_0 θ_0

Oscillatory motion ;
$-\theta_0 \leq \theta \leq \theta_0$

FIGURE 11.18   Oscillatory and infinite motions of a plane pendulum.

which defines the elliptic function

$$\varphi = am \sqrt{\frac{g}{\ell}} t$$

The period τ for a complete oscillation of the pendulum is found by noting that the motion from $\theta = 0$ to $\theta = \theta_0$ corresponds to one-quarter cycle, during which φ increases from 0 to $\frac{\pi}{2}$.   Hence

$$\tau = 4\sqrt{\frac{\ell}{g}} \int_0^{\pi/2} \frac{d\varphi}{\sqrt{1 - k^2 \sin^2 \varphi}} = 4\sqrt{\frac{\ell}{g}} K(k)$$

where

$$K(k) = \int_0^{\pi/2} \frac{d\varphi}{\sqrt{1 - k^2 \sin^2 \varphi}}$$

is the complete elliptic integral of the first kind.   For small vibrations, the expression for τ may be expanded in the series

$$\tau = 2\pi \sqrt{\frac{\ell}{g}} \left(1 + \frac{\theta_0^2}{16} + \cdots \right)$$

Numerical Illustration.   Plot τ as a function of the amplitude θ_0 for a pendulum of length ℓ such that the period for small vibrations is one second.   This gives

$$\sqrt{\frac{\ell}{g}} = \frac{1}{2\pi}$$

Table 11.1 Numerical Values for Pendulum

θ_0	$\dfrac{\theta_0}{2}$	$\sin \dfrac{\theta_0}{2}$	K	$\dfrac{\tau}{\text{sec.}}$
0	0	0	$\pi/2$	1
30	15°	0.2588	1.598	1.02
60	30	0.5000	1.686	1.075
90	45	0.7070	1.854	1.18
120	60	0.866	2.157	1.37
150	75	0.966	2.768	1.76
170	85	0.966	3.832	2.44
180	90	1.000	∞	∞

Hence

$$\tau = \frac{2}{\pi} K(k) = \frac{2}{\pi} K\left(\sin \frac{\theta_0}{2}\right)$$

See Table 11.1 and Figure 11.19 for $\tau = \tau(\theta_0)$, with $0 \le \theta_0 \le \pi$.

For small vibrations the motion is approximately simple harmonic, and is given by

$$\theta = \theta_0 \sin \sqrt{\frac{g}{\ell}}\, t$$

where $\theta = 0$ for $t = 0$, which agrees with Ex. 3, Sect. 11.1.

The potential energy V for the simple pendulum is plotted in Figure 11.20. It is evident that the motion is oscillatory for $\theta_0 < \pi$. For larger values, the deflection angle θ increases indefinitely, i.e., the motion is infinite. Near the origin the curve is parabolic, $V = \frac{1}{2}mg\ell\theta^2$, and the motion is simple harmonic, analogous to the vibrations of a mass suspended from a spring of constant $k = mg\ell$.

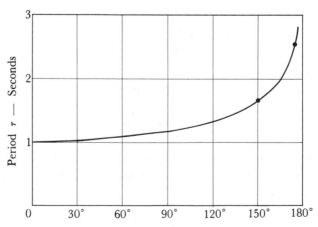

FIGURE 11.19 Period versus amplitude for vibrations of a plane pendulum.

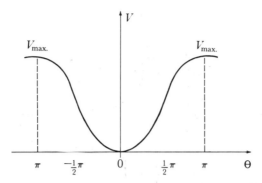

FIGURE 11.20 Potential well for plane pendulum oscillator.

Further insight into the motion is gained by a study of the path lines in the phase plane. To this end we use the Hamiltonian $H = T + V = H(q, p)$, where

$$p = \frac{\partial L}{\partial \dot\theta} = m\ell^2\dot\theta \qquad q = \theta$$

Hence

$$H(q, p) = \frac{p^2}{2m\ell^2} + mg\ell(1 - \cos q) = E$$

and

$$p = 2m\ell^2\left(\frac{g}{\ell}\right)^{1/2}\left(\frac{E}{mg\ell} - 1 + \cos q\right)^{1/2} = p_0\left(\frac{E}{mg\ell} - 1 + \cos q\right)^{1/2}$$

where

$$p_0 = 2m\ell^2\left(\frac{g}{\ell}\right)^{1/2}$$

The streamlines in the phase plane shown in Figure 11.21 are ovals for $E < 2mg\ell$ and sinuous curves for $E > 2mg\ell$. The critical value $E = 2mg\ell$ corresponds to $q_0 = \pi$, a point of unstable equilibrium. Very large values of $E \gg 2mg\ell$ obviously correspond to phase lines which approach straight paths parallel to the q-axis.

Example 3. Soft and Hard Springs. Consider an elastic vibrator with a restoring force $F(q)$ represented by a spring of stiffness S, where S is the ratio of restoring force to displacement. If S is independent of displacement $|q|$, the spring is said to be *linear;* if S decreases with $|q|$ the spring is said to be *soft,* while if S increases with $|q|$ the spring is said to be *hard.* We wish to compare the periods of vibration for springs of the three classes.

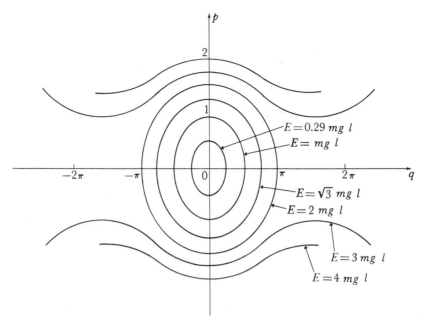

FIGURE 11.21 Streamlines in phase plane for simple pendulum.

For a linear spring, $S = k$, and the period of vibration is given by

$$\tau = 2\int_{-q_0}^{+q_0}\sqrt{\frac{m}{2(E-V)}}\,dq$$

$$= 2\int_{-q_0}^{q_0}\frac{dq}{\sqrt{\dfrac{k}{m}\,(q_0^2 - q^2)}} = 2\sqrt{\frac{m}{k}}\,\sin^{-1}\frac{q}{q_0}\Bigg]_{-q_0}^{+q_0}$$

$$= 2\pi\sqrt{\frac{m}{k}} = \frac{2\pi}{\lambda}$$

where λ is the eigenfrequency.

For a nonlinear spring, it is convenient to write the potential energy in the form

$$V = \int_0^q F(q)\,dq = \int_0^q Sq\,dq$$

Integration by parts gives the relation

$$V = \tfrac{1}{2}Sq^2 - \int_0^q \frac{\partial S}{\partial q}\cdot\left(\frac{q^2}{2}\right)dq$$

Comparing this with the potential energy V_ℓ for a linear spring, we have

$$V - V_\ell = \tfrac{1}{2}Sq^2 - \tfrac{1}{2}S_\ell q^2 + \int_0^q \frac{\partial S}{\partial q}\left(\frac{q^2}{2}\right) dq$$

where $S_\ell = k$ is the stiffness for a linear spring.

In the case of a soft spring,

$$\frac{\partial S}{\partial q} < 0 \qquad S - S_\ell < 0$$

Hence

$$V - V_\ell < 0$$

and it is evident from the expression for τ that

$$\tau_s(q_0) > \tau_\ell(q_0) \qquad \frac{\partial \tau_s}{\partial q_0} > 0$$

where τ_s is the period for the soft spring and τ_ℓ that for the linear spring. For a hard spring, the inequalities are reversed and

$$\tau_h(q_0) < \tau_\ell(q_0) \qquad \frac{\partial \tau_h}{\partial q_0} < 0$$

Summarizing we have

(1) Linear spring: $\tau_\ell(q_0) = $ const.

(2) Soft spring: $\tau_s(q_0) > \tau_\ell(q_0) \qquad \dfrac{\partial \tau_s}{\partial q_0} > 0$

(3) Hard spring: $\tau_h(q_0) < \tau_\ell(q_0) \qquad \dfrac{\partial \tau_h}{\partial q_0} < 0$

The pendulum is an example of a soft spring, while the charged-particle oscillator comes under the category of a hard spring.

When dealing with systems with several degrees of freedom, the integral of energy no longer suffices to find the complete solution. The vibrations of nonlinear conservative systems with several degrees of freedom may sometimes be treated by taking advantage of ignorable coordinates and the integral of energy. As already mentioned, the Kepler problem can be solved in this manner. Other examples are the spherical pendulum and the oscillations of a heavy top. The latter problem is discussed in Ex. 4.

Example 4.   Cyclic Motion of a Top.   Consider the general motion of a top spinning at high speed about an axis of symmetry as shown in Figure 11.22. The top is acted on by Earth's gravitational field, represented by the force $m\mathbf{g}$ at the center of gravity of the top.

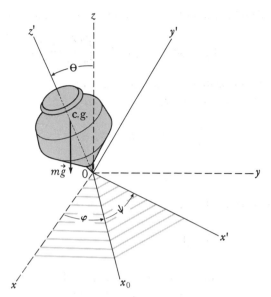

FIGURE 11.22 Motion of top.

SOLUTION. The system has three degrees of freedom, which may be taken as the Eulerian angles (φ, θ, ψ), and is conservative. The kinetic energy may be conveniently written in terms of the angular velocity components $\omega_1, \omega_2, \omega_3$ as follows:

$$T = \tfrac{1}{2}J_1\omega_1^2 + J_2\omega_2^2 + \tfrac{1}{2}J_3\omega_3^2$$

where $J_1 = J_2$ are the moments of inertia about the axes perpendicular to the axis of symmetry (spin axis) and J_3 is the moment of inertia about the spin axis. To apply Lagrange's equations, it is necessary to express $\omega_1, \omega_2, \omega_3$ (which are nonholonomic) in terms of the holonomic coordinates (φ, θ, ψ). This can be done by geometrical or analytical means which are somewhat lengthy and are omitted here. The result is

$$\omega_1 = \dot{\varphi} \sin \psi \sin \theta + \dot{\theta} \cos \psi$$
$$\omega_2 = \dot{\varphi} \cos \psi \sin \theta + \dot{\theta} \sin \psi$$
$$\omega_3 = \dot{\psi} + \dot{\varphi} \cos \theta$$

Substituting these values in the expression for T, gives

$$T = \tfrac{1}{2}J_1(\dot{\theta}^2 + \dot{\varphi}^2 \sin^2 \theta) + \tfrac{1}{2}J_3(\dot{\psi} + \dot{\varphi} \cos \theta)^2$$

The potential V is

$$V = mgz = mg\ell \cos \theta$$

Thus

$$L = T - V = \tfrac{1}{2}J_1(\dot{\theta}^2 + \dot{\varphi}^2 \sin^2 \theta) + \tfrac{1}{2}J_3(\dot{\psi} + \dot{\varphi} \cos \theta)^2 - mg\ell \cos \theta$$

The equations of motion may be written from L, but this is an unnecessary step since inspection shows that three integrals of motion are available. Two of these are obtained from the ignorable coordinates φ, ψ and the third from the conservation of energy. Thus, we write

(1) $\quad \dfrac{\partial L}{\partial \dot{\varphi}} = p_\varphi = \text{const.} = J_1 \dot{\varphi} \sin^2 \theta + J_3 (\dot{\psi} + \dot{\varphi} \cos \theta) \cos \theta$

(2) $\quad \dfrac{\partial L}{\partial \dot{\varphi}} = p_\psi = \text{const.} = J_3 (\dot{\psi} + \dot{\varphi} \cos \theta)$

(3) $\quad E = \text{const.} = \tfrac{1}{2} J_1 (\dot{\theta}^2 + \dot{\varphi}^2 \sin^2 \theta) + \tfrac{1}{2} J_3 (\dot{\psi} + \dot{\varphi} \cos \theta)^2 + mg\ell \cos \theta$

Substitution of (2) in (1) and (3) simplifies the equations as follows:

(4) $\qquad\qquad p_\phi = J_1 \dot{\varphi} \sin^2 \theta + p_\psi \cos \theta$

(5) $\qquad\qquad p_\psi = J_3 (\dot{\psi} + \dot{\varphi} \cos \theta)$

(6) $\qquad\qquad E = \tfrac{1}{2} J_1 (\dot{\theta}^2 + \dot{\varphi}^2 \sin^2 \theta) + \dfrac{p_\psi^2}{2J_3} + mg\ell \cos \theta$

These equations can be integrated by elimination of $\dot{\varphi}$ between (4) and (6), which leads to a quadrature in terms of θ. With θ known, both φ and ψ can be found.

This outlines the general procedure for obtaining the solution using ignorable coordinates and the conservation of energy.* The general motion of the spinning top consists of a steady precession or rotation of the spin axis about the vertical axis through the fixed point, plus nutation, or oscillation of the spin axis in a direction perpendicular to the direction of precession.

BIBLIOGRAPHY

1. A. A. Andronow and C. E. Chaikin, *Theory of Oscillations.* (Translated from the Russian, S. Lefshetz, ed.) Princeton: Princeton University Press, 1949.
2. R. E. D. Bishop and D. C. Johnson, *The Mechanics of Vibration.* Cambridge: Cambridge University Press, 1960.
3. E. B. Cole, *The Theory of Vibrations for Engineers*, Third Edition. New York: Macmillan, 1957.
4. E. U. Condon and H. Odishaw, eds., *Handbook of Physics.* New York: McGraw-Hill, 1958.
5. S. W. McCuskey, *An Introduction to Advanced Dynamics.* Reading, Mass.: Addison-Wesley, 1959.
6. N. Minorsky, *Nonlinear Oscillations.* Princeton: Van Nostrand, 1962.
7. E. C. Pestel and F. A. Leckie, *Matrix Methods in Elastomechanics.* New York: McGraw-Hill, 1963.
8. J. W. S. Rayleigh, *The Theory of Sound*, Vols. I and II. London: Macmillan, 1877.
9. A. Sommerfeld, *Mechanics.* New York: Academic Press, 1952.

---

* Details of the complete solution can be found in Ref. 9, pp. 196–200.

10. K. R. Symon, *Mechanics*, Second Edition. Reading, Mass.: Addison-Wesley, 1960.
11. F. S. Tse et al., *Mechanical Vibrations*. Boston: Allyn and Bacon, Inc., 1963.
12. M.I.T. Radiation Laboratory Series, Vol. 5, *Pulse Generators*. New York: McGraw-Hill, 1948. (See Chap. 6, Sects. 1 through 4, H. J. White.)
13. Y. Rocard, *General Dynamics of Vibrations*. New York: Frederick Ungar Publishing Co., 1960.
14. W. Seto, *Mechanical Vibrations*. New York: Schaum, 1964.
15. I. Crandall, *Theory of Vibrating Systems and Sound*. New York: Van Nostrand, 1927.
16. J. P. Den Hartog, *Mechanical Vibrations*, Fourth Edition. New York: McGraw-Hill, 1956.
17. S. Timoshenko and D. H. Young, *Vibration Problems in Engineering*, Third Edition. New York: Van Nostrand, 1956.
18. R. A. Fraser et al., *Elementary Matrices*. London: Cambridge University Press, 1947.
19. J. J. Stoker, *Nonlinear Vibrations in Mechanical and Electrical Systems*. New York: Interscience, 1950.
20. J. R. Barker, *Mechanical and Electrical Vibrations*. London: Methuen, 1964.
21. J. B. Vernon, *Linear Vibration Theory*. New York: Wiley, 1967.
22. T. E. Stern, *Nonlinear Networks and Systems*. Reading, Mass.: Addison-Wesley, 1965.
23. C. M. Harris and C. E. Crede, *Shock and Vibration Handbook*, 3 Vols. New York: McGraw-Hill, 1961.

12

FEEDBACK SYSTEMS

The primary objectives of this chapter are to introduce the reader to the broad field of feedback systems, and to show how the fundamental systems theory and methods covered in the earlier chapters of the book can be applied to the feedback area.

Following an introductory discussion, the chapter covers applications to feedback amplifiers, operational amplifiers, analog computers, and control systems. Specialized aspects of these fields are outside the scope of this book. However, detailed information on equipment, design methods, and the like, is readily available in textbooks and journals, of which a representative selection is given in the bibliography of this chapter. In this connection, it should be noted that, as in many other fields, design methods for control systems are in a state of flux owing to the impact of computer-aided methods, which are so rapidly rendering earlier methods obsolete.

A feedback system may be defined in rather broad terms as one in which a portion of the output is fed back into the input in such a manner as to influence the behavior or performance of the system. This is illustrated graphically in Figure 2 of the Introduction. Feedback may be incidental as, for instance, in high-gain amplifiers where stray coupling between the output and input may lead to uncontrolled parasitic oscillations. More commonly, however, feedback is intentionally introduced into a system in order to improve performance in terms of some stated or understood criteria or objectives. Examples of such purposeful feedback systems are legion, of which we may mention servomechanisms, cybernetic systems, management systems, automated production systems, and computer-controlled automobile traffic signals.

Many life processes are characterized by feedback phenomena, as for example posture control, eye movement, and homeostasis in man and animals. Most man-made systems incorporate feedback in one form or another. Business enterprises collect information on operations and conditions and feed back this information to the managers for the purpose of control and guidance toward certain objectives. Political systems are subject to feedback through elections and other activities of the populace. The servomechanisms and other automata of technology are based on fundamental concepts of feedback.

Despite the pervasiveness of feedback systems, the quantitative measure and analysis of such systems is only in the rudimentary stage for all but the automata of technology. In the latter area, however, theory and practice are approaching levels of sophistication characteristic of a true applied science, that is, with concepts and methods based primarily on fundamental science and rational mathematical processes rather than on empirical observation and subjective experience.

The feedback systems of technology are used mostly for feedback amplifiers, for analog computers and simulation, and for the automatic control of equipment and processes. For the latter, control may be effected through direct-coupled feedback loops, or through digital control actions. Automatic control systems based on the feedback principle go back many centuries. As an example, Singer (Ref. 1) refers to the invention of an automatic furnace by Drebbel in 1666 as "perhaps the first example of a feedback mechanism."

The first significant scientific paper on feedback mechanisms was published by Maxwell in 1868 (Ref. 2). In this paper he gave a mathematical analysis of the centrifugal governor invented by Huygens (1673) and used by Watt to control the speed of steam engines (Ref. 13, p. 606). It is to be noted that this control tends to oppose what the system is already doing and, hence, in modern terminology would be called negative feedback. In the middle of the nineteenth century practical operating difficulties arose with Watt's centrifugal governor because of "design improvements," and these led to a classic study and solution of the problem by a Russian engineer Vyshnegradskii in 1877 (Ref. 14). This was one of the first quantitative studies in the theory of machine control which answered questions raised by industrial practice.

Modern automatic feedback control stems largely from developments just before and during World War II (Ref. 11). In these developments, the theories and methods of servomechanisms were joined with those of communication systems networks and amplifiers to produce a fruitful, if not always entirely compatible, union.

Feedback systems currently enter into the design and operation of most equipment and processes above the elementary level. For example, feedback amplifiers are central to long distance telephony (Ref. 12) as well as to other communication systems. Automatic feedback controls are widely used to control voltage, frequency, power, and the like, in electrical systems. Automated production tools and systems such as high-speed precision rolling mills in steel plants, and automated process controllers in oil refineries are based on feedback and computer control. Space technology is fundamentally dependent on automatic feedback systems for control and guidance of missiles and space vehicles, tasks which are far beyond the capabilities of human operators.

Many fields of scientific research are being advanced through the utilization of sophisticated feedback controls, as for example in high energy physics, microchemistry, and life sciences. Research is under way in the application of feedback concepts to traffic flow problems in urban areas.

The basics of feedback control theory are contained within the systems framework given in earlier chapters of this book. Most feedback theory is concerned with linear systems, although nonlinear systems are receiving increasing attention. For continuously acting feedback systems, including both linear and nonlinear, the general theory of dynamical systems covered in Chap. 6 is applicable. The equations of motion for both linear and nonlinear continuously-acting systems can therefore be formulated and studied in accord with dynamical methods; even in those nonlinear cases where a general solution cannot be obtained, dynamical methods still serve to provide valuable information for analysis and design.

For many nonlinear systems interest centers on the effects of small disturbances from equilibrium. These can be effectively dealt with by linear approximations similar to the method used for small vibrations (cf. Chap. 11).

12.1 FEEDBACK AMPLIFIERS

Electronic amplifiers used in communication and control systems commonly incorporate feedback to improve performance and to achieve operating characteristics not otherwise possible. Positive feedback was used in regenerative detectors for radio receivers as early as 1912 (Ref. 19) to greatly increase sensitivity and range of reception. The more fundamental concept of negative feedback was introduced by Black in 1927 (Ref. 12) as a means of solving the repeater problem for the transcontinental (U.S.) telephone system where upwards of 100 amplifiers in tandem may be necessary for acceptable performance (Ref. 11).

The basic arrangement of the negative feedback amplifier is depicted in Figure 12.1. It consists of an ordinary amplifier of gain A, with a portion of the output fed back to the input. The input signal e_1 is combined with a portion β of the output e_2 and the difference signal fed to the amplifier. Output e_2 is related to input e_1 by the expression

$$e_2 = A(e_1 - \beta e_2)$$

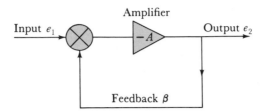

FIGURE 12.1 Schematic diagram of feedback amplifier.

or

$$\frac{e_2}{e_1} = \frac{A}{1 + \beta A} = \text{net gain of amplifier} \qquad (12.1)$$

Usually $\beta A >> 1$ and the net gain is closely approximated by

$$\frac{e_2}{e_1} \cong \frac{1}{\beta} \qquad (12.2)$$

For example, if $\beta = 10^{-2}$ and $A = 10^4$, then $\beta A = 10^2$ and

$$\frac{e_2}{e_1} \cong \frac{\beta}{1} = 100$$

The overall gain, although reduced below that without feedback, is virtually independent of the properties of the amplifier itself and depends only on β. Since β can be obtained from a resistance network of fixed characteristics, it can be made substantially independent of frequency, voltage fluctuations, changing characteristics of amplifier elements, and the many other variable factors which may adversely affect amplifier performance. Other major advantages of feedback amplifiers include greatly reduced amplitude distortion and improved signal-to-noise ratio (for noise introduced into the high-level parts of the amplifier, a major source in practice). Stabilized amplifiers of this type find innumerable applications in communication systems as already noted, as well as in instrumentation, in analog computers, and in many other fields.

Practical design of feedback amplifiers must take into account not only the gain but also such other factors as frequency response, phase shift, and the possibility of parasitic oscillations. Input signals will, in general, experience a phase shift in passing through the amplifier because of the presence of inductive and capacitive elements. This phase shift may cause the feedback signal to have a component tending to reinforce the input signal, i.e., the feedback will have a positive component, and oscillations may occur. These effects are closely related to the eigen-frequencies of the feedback amplifier system and can be analyzed in terms of the eigenvalues and eigenfrequencies (cf. Chap. 11).

Positive feedback may also be intentionally introduced into a system as, for example, the regenerative detector previously mentioned, or may be incorporated into certain types of oscillators. For positive feedback, β in Eq. (12.1) becomes negative, and very large gains (e.g., 10,000 or more for the regenerative detector) are possible by making $\beta A \rightarrow 1$. Positive feedback can be used to generate continuous oscillations as illustrated in the following analysis of the Wien-bridge oscillator. The basic principle involved is to counteract resistance losses by supplying energy from an external source through the medium of a feedback loop.

Example 1. Wien-bridge Oscillator. The well-known Wien-bridge oscillator is shown schematically in Figure 12.2. It consists of an amplifier of gain A and an R-C network which represents one-half of the bridge circuit. The remaining half (not shown) consists of a linear and nonlinear combination which has the sole function of automatically maintaining the amplifier gain at the precise value required to generate sinusoidal oscillations. Oscillators of this type are commonly used for the frequency range of 1 to 10^6 Hertz, i.e., where an extremely wide frequency range from subaudio to the megacycle band is needed.

Referring to the diagram of Figure 12.2, e_1 is related to e_1 by the equation

$$e_2 = ARi_2 = ARY_{21}e_1 = e_1 \tag{1}$$

where Y_{21} is the transfer function (defined in Chap. 8) from e_1 to i_2. From Eq. (8.12)

$$Y_{21} = \frac{\delta_{21}}{|Z|} \tag{2}$$

where Z is the impedance matrix and δ_{21} is the cofactor of Z_{12} in det $[Z]$. By inspection of Figure 12.2, matrix Z has the form

$$Z = \begin{bmatrix} R + \dfrac{2D^{-1}}{C} & -\dfrac{D^{-1}}{C} \\[2ex] -\dfrac{D^{-1}}{C} & R + \dfrac{D^{-1}}{C} \end{bmatrix} \tag{3}$$

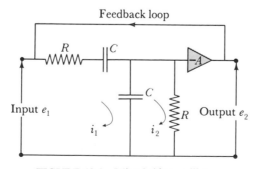

FIGURE 12.2 Wien-bridge oscillator.

To find the eigenvalues for the system, substitute Eqs. (2) and (3) in (1) and solve for e_2/e_1.

$$\frac{e_2}{e_1} = \frac{AR\,\delta_{21}}{|Z|} = \frac{AR\frac{1}{C}D^{-1}}{|Z|} = 1$$

Expanding det $[Z]$ and denoting RCD by the parameter λ leads to the equation

$$\lambda^2 + (3 + A)\lambda + 1 = 0$$

By choosing $-A = 3$, this reduces to

$$\lambda^2 + 1 = 0 \quad \text{or} \quad D = \pm \frac{i}{RC}$$

which corresponds to a normal vibration of angular frequency $\frac{1}{\sqrt{R^2C^2}} = \frac{1}{RC}$. Hence, the system will produce undamped oscillations in accord with

$$e_2 = E_0 \cos \frac{t}{RC} \tag{4}$$

12.2 APPLICATION OF FEEDBACK TO OPERATIONAL AMPLIFIERS AND ANALOG COMPUTERS

Analog computers (electronic analog simulators) are widely used to simulate systems which can be described by ordinary differential equations. Various mathematical operations such as integration, differentiation, addition, and multiplication are performed electronically by using special types of feedback amplifiers in combination with re-sistances, inductances, and other circuit elements. These amplifier arrangements are usually referred to as operational amplifiers. The basic circuit for a rather general type of operational amplifier is shown in Figure 12.3. Note that the amplifier reverses the sign of the input signal.

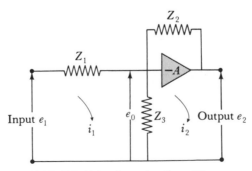

FIGURE 12.3 Operational amplifier.

From circuit theory (Chap. 8, Sect. 3), the relation between output e_2 and input e_1 is derived as follows

$$\dot{i}_1 - \dot{i}_2 = \frac{e_0}{Z_3} = \frac{e_1 - e_0}{Z_1} + \frac{e_2 - e_0}{Z_2} \tag{12.3}$$

Noting that $e_2 = -Ae_0$ and solving for e_2/e_1 gives

$$\frac{e_2}{e_1} = \frac{-\dfrac{Z_2}{Z_1}}{1 + \dfrac{1}{A}\left(\dfrac{Z_2}{Z_1} + \dfrac{Z_2}{Z_3} + 1\right)} \tag{12.4}$$

If the amplifier gain $A \gg 1$, the second term in the denominator becomes small and the transfer function can be approximated by

$$\frac{e_2}{e_1} \cong -\frac{Z_2}{Z_1}$$

Various mathematical operations may be simulated by proper choice of Z_1 and Z_2. Several of the most important types are discussed at this point.

(a) Integrating amplifier

Choosing $Z_2 = \dfrac{1}{C}D^{-1}$, $Z_1 = R$ leads to

$$e_2 \cong \frac{-D^{-1}e_1}{RC} = \frac{-1}{RC}\int e_1\,dt \tag{12.5}$$

so that the output e_2 is proportional to the integral of the input e_1. The operational amplifier therefore acts as an integrator circuit.

(b) Differentiation

Differentiation can be obtained by letting $Z_1 = R$ and $Z_2 = LD$, i.e.,

$$e_2 \cong -\frac{L}{R}De_1 \tag{12.6}$$

However, in practice it is difficult to construct high-Q inductors, i.e., L/R cannot be made much greater than unity. Moreover, differentiation tends to increase noise, which is a limiting factor in many applications.

(c) Multiplication

This is accomplished by letting $Z_1 = R_1$ and $Z_2 = R_2$, which gives

$$e_2 = -\frac{R_2}{R_1}e_1 \tag{12.7}$$

(d) Multiple inputs

It is frequently necessary to deal with inputs from several sources. This can be accomplished by using the modified circuit of Figure 12.4.

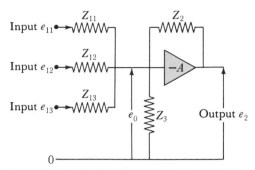

FIGURE 12.4   Multiple input operational amplifier.

For example, choosing $Z_{11} = R_1$, $Z_{12} = R_2$, $Z_{13} = R_3$ and $Z_2 = (CD)^{-1}$ leads to a multiple input integrator, i.e.,

$$e_2 = -\int \left(\frac{e_{11}}{R_1 C} + \frac{e_{12}}{R_2 C} + \frac{e_{13}}{R_3 C} \right) dt \qquad \textbf{(12.8)}$$

Analog computers designed to solve both linear and nonlinear differential equations are constructed by assembling operational amplifiers and function generators. For general purposes a large number of amplifiers (typically from 10 to several hundred) are used, with provisions for external interconnections to provide maximum flexibility in applications. Analog computers have several important advantages as compared to digital computers; they are much cheaper, easily programmed, the output can be displayed on an oscilloscope or on a continuous recorder in the form of a graph, and effects of parameter changes can be quickly and directly compared. On the other hand, analog computers are limited in accuracy (typically to about 0.1 percent) compared to the almost unlimited calculation accuracy of the digital computer. Analog computers also lack the versatility and high speed attainable with digital computers.

The general method for solving differential equations by analog computer will be made clear by the following example.

Example I. Analog Computer Solution of a Second Order Differential Equation. This example will serve to illustrate the procedure for solving ordinary differential equations on an analog computer. Let the equation be of the form

$$\frac{d^2\theta}{dt^2} + a\frac{d\theta}{dt} + b\theta = f(t)$$

where a and b are constants and $f(t)$ is a prescribed input function. The steps for finding the solution are as follows.

(1) Solve for $\dfrac{d^2\theta}{dt^2} = \ddot{\theta}$, i.e.,

$$\frac{d^2\theta}{dt^2} = f(t) - b\theta - a\frac{d\theta}{dt}$$

(2) Simulate $\ddot{\theta}$ on the analog
(see Fig. 12.5(a))

(3) Add simulator for $-a\dot{\theta}$ by an integrator and feed back to the input
(see Fig. 12.5(b))

(4) Add simulator for $-b\theta$ and feed back to the input
(see Fig. 12.5(c))

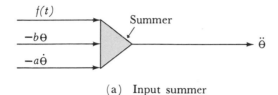

(a) Input summer

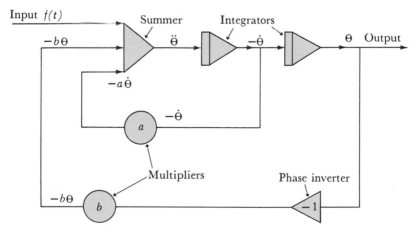

(b) Input summer plus simulator for $-a\dot{\theta}$

(c) Input summer plus simulators for $-a\dot{\theta}$ and $-b\theta$

FIGURE 12.5 Analog computer for solution of second order differential equation.

(5) The analog is operated as follows: An input voltage proportional to the forcing function $f(t)$ is applied to the computer; then the output voltage will be proportional to θ, i.e., to the system response. Commonly used forcing functions include the step function, the ramp function, and the parabolic function (see p. 413).

(6) Note that the analog can be simplified by combining the input summer and the first integrator, provided $\ddot{\theta}$ is not required explicitly.

12.3 SINGLE-VARIABLE FEEDBACK CONTROL SYSTEMS

Many feedback control systems can be described in terms of a single variable and a single feedback loop. More complex control systems, with multiple variables and multiple feedback loops, may be developed from the simpler one-variable structures.

The basic equation for one type of widely used single-loop control system can be derived by inspection from the schematic diagram of Figure 12.6a. The output y is related to the input x by the expression

$$y = G(x - Hy) = Gx - GHy$$

Solving for y gives

$$y = (I + GH)^{-1}Gx = Tx \qquad \textbf{(12.9)}$$

where G is called the forward transfer operator, H the feedback transfer operator, I the identity operator, and T the transfer operator. For continuous processes such as those represented, for example, by linear lumped parameter Lagrangian systems, the operators G and H may be reduced to polynomials and I to the number 1. More generally, the operators must be treated symbolically and identified according to the system. It is, of course, assumed that the inverse operator $(I + GH)^{-1}$ exists.

A somewhat more general type of single-loop control system uses a computer control as shown in Figure 12.6b. The computer may be a simple device capable of carrying out only such straightforward mathematical operations as integration, differentiation, and multiplication, or it may be a far more versatile general purpose digital computer. The latter type would be used, for example, for a city traffic control system where the average waiting time of all automobiles stopped by the traffic signal system is to be minimized. In deriving the basic equation for the computer control system, we shall assume that the computer operates on x and y to give

$$H(x, y) = H_1x - H_2y$$

where H_1 is an operator acting only on x and H_2 is an operator acting only

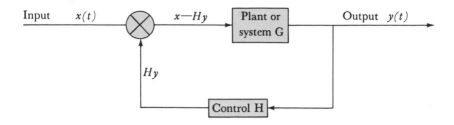

(a) Basic feedback loop

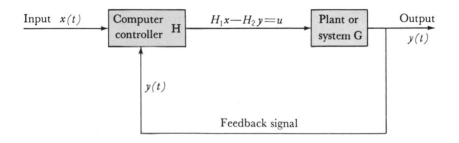

(b) Computer controller type

FIGURE 12.6 Single-loop control systems.

on y. Then, by inspection of the diagram of Figure 12.6b, it follows that

$$y = G(H_1 x - H_2 y) = GH_1 x - GH_2 y$$

Solving for y gives

$$y = (I + GH_2)^{-1} GH_1 x = Tx \qquad \textbf{(12.10)}$$

where again T is to be regarded as a transfer operator and it is assumed that the inverse operator $(I + GH_2)^{-1}$ exists.

It should be observed that the more general or computer type control system of Figure 12.6b reduces to the more restricted system of Figure 12.6a by making $H_1 = I$. This corresponds to applying x directly to the system G and therefore is more limited in scope than in cases where x is processed ahead of the system.

The input $x(t)$ in a physical control system may represent, for example, a coordinate such as a shaft position or the displacement of a point, or it may represent an electrical or thermal quantity.

In many applications it is desired to force the output $y(t)$ to behave in a prescribed manner with respect to the input $x(t)$. For instance, in an automatic frequency control system for a turbogenerator the objective is to hold the frequency at a given value (usually 60 Hertz in the U.S.). More generally, the target value x itself may be a prescribed function of time.

For linear Lagrangian systems, the transfer operator T may be identified as the transfer function and the system may then be analyzed in accord with the procedures of Chap. 8. It has long been conventional in control systems engineering to use the Laplace transform scheme, often supplemented by frequency analysis. However, the state variable method, eigenvalue and eigenvector concepts, and direct study of the differential equations themselves provide broader and often more valuable results.

In general, the transfer function for linear Lagrangian systems can be expressed as the ratio of two polynomials, i.e., as a rational function. Therefore, if T_{ij} represents the transfer function for an input x_j and an output y_i, it can be expressed in the form

$$T_{ij} = \frac{p_n s^n + p_{n-1} s^{n-1} + \cdots + p_1 s + p_0}{q_m s^m + q_{m-1} s^{m-1} + \cdots + q_1 s + q_0} \tag{12.11}$$

where s is the Laplace transform parameter.

Types of Control. The correction mechanism of a feedback control system may be designed to respond to the error signal in a variety of ways. One of the earliest and still most widely used methods is the *on-off* or *relay* type of control in which full power is applied as soon as the error signal is large enough to actuate the system. The thermostat type of temperature control system commonly used for home heating furnaces is an example of this type, as are some power drives for gun directors. More sophisticated versions of these bi-state systems, often referred to as *bang-bang control*, are considered in the modern theories of optimal control processes (see, for example, Refs. 21, 22, 29 and 34). Relay type control systems offer relatively simple construction and are useful for both moderate and high-performance controls.

Another extensively used type of control exerts corrective action continuously as a function of the error signal. *Continuous* control systems, as these are called, can be further classified according to the manner in which the error signal is used to generate the control action. The most commonly used types are as follows:

(1) Proportional control: the control action is proportional to the error signal, so that large errors tend to be corrected faster (limited of course by the maximum correction power available from the control system). Proportional control tends to improve system stability and performance.

(2) Derivative control: corrective action is made to depend on the time rate of change of the error signal. Thus, if the angular position θ of a shaft is being controlled, derivative control would be actuated by the angular speed $\dot\theta$; higher order derivative controls dependent on $\ddot\theta$, $\dddot\theta$, and so forth, follow by extension of the derivative control principle. Derivative control tends to anticipate corrective action and thereby to improve performance.

(3) Integral control: the control action depends on the time integral of the error signal. Again, first, second, and higher order integral controls can be introduced.

(4) Combination control: the several types of control signals may be combined in various ways to produce mixed or combination control methods. Such a combination control is illustrated in Ex. 2 of this section. In this example, the angular position of a shaft is controlled by both a proportional and an integral error signal. The integral control greatly increases the precision of control since the corrective action depends on the integrated value of the error signal rather than just on the instantaneous value of the error signal.

Step, Ramp, and Parabolic Driving Functions for Determining Time Response. Performance of an automatic feedback system is gauged in part by its response to specified disturbances or driving functions. In normal operation, the driving function is derived from the error signal and, hence, is not known in advance. Usually the error signal will have a strong random component. It is therefore necessary, for test purposes, to specify standard test inputs or driving functions. Three generally accepted types of such test inputs are as follows.

(1) Step function

$$f(t) = \begin{cases} \text{const.} = C & t > 0 \\ 0 & t < 0 \end{cases}$$

Note that $f(t)$ is discontinuous at $t = 0$.

(2) Ramp function

$$f(t) = \begin{cases} Ct & t > 0 \\ 0 & t < 0 \end{cases}$$

In this case $f(t)$ is continuous, but $f'(t)$ is discontinuous at $t = 0$.

(3) Parabolic function

$$f(t) = \begin{cases} Ct^2 & t > 0 \\ 0 & t < 0 \end{cases}$$

Here $f(t)$ and $f'(t)$ are continuous, but $f''(t)$ is discontinuous at $t = 0$.

Examples of Single-Loop Control Systems. As the first example, selected both for its current relevance and historical significance, we study the centrifugal governor used to control the operating speed of an engine. This type of control system is highly nonlinear, but its performance can be effectively studied by first finding the state of operating equilibrium and then studying performance and stability in terms of small deviations from equilibrium. It is thought-provoking to note that although the classic study of this problem (Ref. 14) was made by the Russian Vyshnegradskii in 1877 in terms of its application to the steam engine, the same methods and principles apply to the most modern devices and equipment. The basic method of analysis is not materially affected by substitution of other frequency-sensitive detection and control elements, so that the example still serves today as a prototype for a wide range of automatic feedback control systems.

The second example illustrates a combined proportional and integral control format as applied to an angular-position servomechanism.

Example I. Centrifugal Governor. The centrifugal governor is used to control the speed of engines and motors. In practice it is coupled to the shaft of the engine or motor, and to the control element as depicted in Figure 12.7. The steady-state speed of the engine is determined by the balance between power input and load torque. For piston engines a flywheel is used to smooth out the pulsations caused by the cyclic action of the pistons. In the steady-state condition the masses m of the centrifugal governor are at some angle θ_0 with respect to the rotating shaft of the governor, and the feedback linkage from the governor to the engine holds power input at a value just sufficient to maintain the speed at a constant value for a fixed load. If the load increases, the speed decreases and the flyballs drop somewhat, causing an increase in power and engine speed and thus counteracting the effect of the increased load.

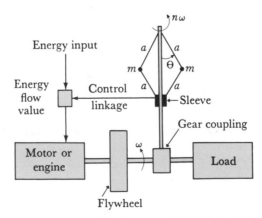

FIGURE 12.7 Centrifugal governor feedback control system.

A steady-state quantitative theory can be readily constructed to explain the regulating action. However, actual operation of the system is more complicated because the system is in fact dynamic, and under certain conditions the corrective action can overshoot, leading to a series of uncontrolled or wild oscillations. The possibility of this kind of instability is inherent in all feedback systems, and is one of the basic problems which must be investigated and solved by designers of automatic control systems.

Vyshnegradskii's study of the stability of the centrifugal governor is now summarized.* Referring to Figure 12.7, it is apparent that the system has two degrees of freedom, i.e., φ, the angular position of the engine shaft, and θ, which specifies the configuration of the centrifugal governor. The energy functions for the system may be written by reference to the diagram.

$$T = ma^2(\dot{\theta}^2 + n^2\dot{\varphi}^2 \sin^2 \theta) + \tfrac{1}{2}J\dot{\varphi}^2$$
$$V = 2mga(1 - \cos \theta)$$
$$F = \tfrac{1}{2}\mu\dot{\theta}^2$$
$$P = (\tau_s - \tau_\ell)\dot{\varphi}$$

where J is the moment of inertia of the flywheel and load, n is the transmission ratio of the gear coupling, μ is the effective viscous force coefficient of the governor mechanism (that the frictional force is viscous in nature is a simplifying assumption, but is sufficiently accurate for this analysis), τ_s is the driving torque supplied by the engine, and τ_ℓ is the load torque.

The equations of motion follow from Lagrange's equations (6.31).

$$\left.\begin{array}{l} 2ma^2\ddot{\theta} - 2ma^2n^2\dot{\varphi}^2 \sin \theta \cos \theta + 2mga \sin \theta + \mu\dot{\theta} = 0 \\ 2ma^2n^2\ddot{\varphi} \sin^2 \theta + 4ma^2n^2\dot{\varphi} \sin \theta \cos \theta\dot{\theta} + J\ddot{\varphi} - \tau_s + \tau_\ell = 0 \end{array}\right\} \quad (1)$$

Since the kinetic energy of the flywheel and load is far greater than that of the governor, the second equation of motion can be very closely approximated by

$$J\ddot{\varphi} - \tau_s + \tau_\ell = 0 \qquad (2)$$

Assuming that the control mechanism operates in such a manner that engine torque is proportional to the displacement of the sleeve on the governor, we have the relation

$$\tau_s = k \cos \theta \qquad k > 0 \qquad (3)$$

where k is the proportionality constant.

---

* Adapted from the analysis given by L. S. Pontryagin, *Ordinary Differential Equations*. Addison-Wesley, 1962. See p. 213.

To complete the analysis, it is convenient to reduce the system equations (1) to first order by means of the substitution $v = \dot{\theta}$, $\omega = \dot{\varphi}$, giving the set of state variable equations

$$
\left.
\begin{aligned}
\dot{\theta} &= v \\
\dot{v} &= \frac{n^2\omega^2}{2} \sin 2\theta - \frac{g}{a} \sin \theta - \frac{\mu}{2ma^2} v \\
\dot{\omega} &= \frac{k \cos \theta - \tau_\ell}{J}
\end{aligned}
\right\} \qquad (4)
$$

The conditions for steady-state operation are

$$
\dot{v} = v = \dot{\omega} = 0
$$

which, when substituted in (4), give

$$
k \cos \theta_0 = \tau_\ell
$$

$$
n^2\omega_0^2 = \frac{g}{a \cos \theta_0} = \frac{gk}{a\tau_\ell} \qquad (5)
$$

where ω_0 and θ_0 are the equilibrium values for the system.

Expanding set (4) in a Taylor's series about the point of equilibrium, and retaining first order terms, leads to the following equations.

$$
\Delta\dot{\theta} = \Delta v
$$

$$
\Delta\dot{v} = \dot{v} - \dot{v}_0 = \frac{\partial\dot{v}}{\partial\omega}\bigg]_0 \Delta\omega + \frac{\partial\dot{v}}{\partial\theta}\bigg]_0 \Delta\theta + \frac{\partial\dot{v}}{\partial v}\bigg]_0 \Delta v
$$

$$
= n^2\omega_0 \sin 2\theta_0 \, \Delta\omega + \left(n^2\omega_0^2 \cos 2\theta_0 - \frac{g}{a} \cos \theta_0 \right)\Delta\theta - \frac{\mu}{2ma^2} \Delta v
$$

$$
\Delta\dot{\omega} = -\frac{k}{J} \sin \theta_0 \Delta\theta
$$

where $\Delta\omega$, $\Delta\theta$, Δv represent incremental changes and may be taken as state variables for the system. Substitution of (5) in these equations reduces the set to the form

$$
D(\Delta\theta) = \Delta v
$$

$$
D(\Delta v) = -\frac{g \sin^2 \theta_0}{a \cos \theta_0} \Delta\theta - \frac{\mu}{2ma^2} \Delta v + \frac{2g \sin \theta_0}{a\omega_0} \Delta\omega
$$

$$
D(\Delta\omega) = -\frac{k}{J} \sin \theta_0 \, \Delta\theta
$$

where D is the differentiation operator d/dt. These equations may also be written in matrix form (6)

$$
\begin{bmatrix}
-D & 1 & 0 \\
-\dfrac{g \sin^2 \theta_0}{a \cos \theta_0} & -\left(D + \dfrac{\mu}{2ma^2}\right) & \dfrac{2g \sin \theta_0}{a\omega_0} \\
-\dfrac{k}{J}\sin \theta_0 & 0 & -D
\end{bmatrix}
\begin{bmatrix}
\Delta\theta \\
\Delta v \\
\Delta\omega
\end{bmatrix} = 0 \qquad (6)
$$

To find the eigenvalues for this set of homogeneous equations, replace D by λ, expand the determinant, and set it equal to zero. This gives the characteristic equation

$$
\lambda^3 + \frac{\mu}{2ma^2}\lambda^2 + \frac{g \sin^2 \theta_0}{a \cos \theta_0}\lambda + \frac{2gk \sin^2 \theta_0}{a\omega_0 J} = 0 \qquad (7)
$$

The Routh-Hurwitz criteria for stability (see Ref. 30, p. 405) require that

$$(A) \quad a_0 > 0 \qquad a_1 > 0 \qquad a_2 > 0 \qquad a_3 > 0$$

and

$$(B) \quad a_1 a_2 > a_0 a_3$$

where a_0, a_1, a_2, a_3 are the coefficients of the characteristic polynomial. Condition (A) clearly is satisfied since all the physical quantities are positive. Condition (B) is equivalent to

$$
\frac{\mu}{2ma^2} \cdot \frac{g \sin^2 \theta_0}{a \cos \theta_0} > \frac{2gk \sin^2 \theta_0}{a\omega_0 J}
$$

or to

$$
\frac{\mu J}{2ma^2} > \frac{2k \cos \theta_0}{\omega_0} = \frac{2\tau_\ell}{\omega_0} \qquad (8)
$$

This result may be further interpreted by noting from (5) that a change in load $d\tau_\ell$ leads to a change in speed $d\omega_0$ given by

$$
d(\omega_0^2 \tau_\ell) = 0
$$

or

$$
\frac{d\omega_0}{d\tau_\ell} = -\frac{\omega_0}{2\tau_\ell}
$$

Substituting in (8) gives

$$
\frac{\mu J}{2ma^2}\left|\frac{d\omega_0}{d\tau_\ell}\right| > 1 \qquad (9)
$$

Inspection of (9) shows that stability increases with viscous friction coefficient μ and with moment of inertia J, but decreases with the mass m of the flyballs. Furthermore, increasing the sensitivity of the governor tends to reduce the stability of the system.

In presenting this example, it is significant to note that Vyshnegradskii's investigation came about because of problems introduced by improvements and refinements (in the form of better machining methods, closer tolerances, and so forth) of the Watt governor. These supposed improvements in turn greatly reduced frictional losses and made a previously stable device quite unstable due to the tendency of the system to break into uncontrolled or undamped oscillations. Such unexpected and disconcerting outcomes of meritorious attempts to improve systems are very common in technology and in other fields as well.

*Example 2.   Angular Position Servomechanism.** Consider a servomechanism designed to maintain coincidence between the angle θ_2 of a load shaft and the angle θ_1 of a control shaft, as shown in Figure 12.8. The synchro-differential generator produces an error signal emf $E = k \, \Delta\theta$ where $\Delta\theta = \theta_2 - \theta_1$. This emf is amplified and integrated and the sum fed to the servomotor which in turn applies a torque τ to the load. The load (which includes the armature of the servomotor) has a moment of inertia J and a viscous friction coefficient μ.

It is desired to calculate the load response when the control shaft is suddenly started at an angular velocity ω_0, i.e.,

$$\theta_1 = \begin{cases} = 0 & t < 0 \\ = \omega_0 t & t > 0 \end{cases}$$

SOLUTION

The system design is such that the torque τ is related to the error signal by the equation

$$\tau = k_1 \, \Delta\theta + k_2 \int_0^t \Delta\theta \, dt$$

where, by inspection of Figure 12.8, k_1 and k_2 are positive constants given by $k_1 = ka_1 b$, $k_2 = ka_2 b$. Since the equation of motion for the servomotor and load is

$$\tau = J\ddot{\theta}_2 + \mu\dot{\theta}_2$$

we can write

$$J\ddot{\theta}_2 + \mu\dot{\theta}_2 = -k_1\Delta\theta - k_2 \int_0^t \Delta\theta \, dt$$

or, in operator form,

$$(JD^2 + \mu D)\theta_2 = -(k_1 + k_2 D^{-1})(\theta_2 - \theta_1)$$

Solving for θ_2 gives

$$\theta_2 = \frac{(k_1 + k_2 D^{-1})\theta_1}{JD^2 + \mu D + k_1 + k_2 D^{-1}}$$

---

* Adapted from Hazen (Ref. 4).

Input

θ_1

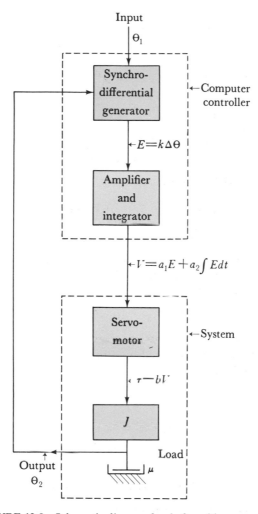

←Computer
controller

$E = k\Delta\theta$

$V = a_1 E + a_2 \int E\,dt$

←System

$\tau - bV$

J

Load

Output
θ_2

μ

FIGURE 12.8 Schematic diagram for shaft positioner control.

We solve this differential equation by the Laplace transform method. Since the initial conditions are all zero, the operator D can be replaced by the Laplace transform parameter s yielding

$$\bar{\theta}_2 = \frac{(k_1 s + k_2)\omega_0}{(Js^3 + \mu s^2 + k_1 s + k_2)s^2}$$

with

$$\bar{\theta}_1 = \frac{\omega_0}{s^2}$$

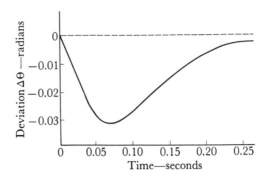

FIGURE 12.9   Time response of shaft position servomechanism.

Let s_1 and s_2, s_3 $(= -\alpha \pm i\beta)$ be the roots of

$$Js^3 + \mu s^2 + k_1 s + k_2 = 0$$

where s_1 is real and s_2, s_3 are complex conjugates (assuming only one real root). Expanding the expression for $\bar{\theta}_2$ in partial fractions* and taking the inverse transform gives

$$\theta_2 = C_1 e^{s_1 t} + C_2 e^{-\alpha t} \sin (\beta t + \varphi) + \omega_0 t$$

where C_1, C_2, and φ are constants of integration.

The result may be illustrated by a numerical example. Let $\omega_0 = 1$ radian per second, and let

$$J = 60$$
$$\mu = 4500$$
$$k_1 = 1.2 \times 10^5$$
$$k_2 = 1.1 \times 10^6$$

Then we find the roots

$$s_1 = -21.6 \qquad s_2, s_3 = -26.7 \pm 11.5i$$

Using these values and evaluating C_1, C_2, φ, gives the equation for $\Delta\theta$

$$\Delta\theta = \theta_2 - \theta_1 = -0.338e^{-21.6t} - 0.343e^{-26.7t} \sin (11.5t - 1.76)$$

which is plotted in Figure 12.9.

12.4 STABILITY OF FEEDBACK SYSTEMS

Feedback systems usually raise questions of stability of operation. This comes about because the feedback signal may tend to reinforce rather

---

* Note that in this expansion no term of the form $\dfrac{C}{s}$ appears and therefore there will be no constant term in the expression for θ_2.

than counteract disturbances or deviations of the system, with the result that the disturbance is magnified instead of diminished. If a given system is initially in a state of equilibrium and tends to return to equilibrium after being disturbed, the system is said to be stable. On the other hand, if the disturbance grows in magnitude, the system is said to be unstable and proper operation is impossible.

If the system is linear, the magnitude of the disturbance is unimportant because the nature of the system response in no way depends on the magnitude of the disturbance. But this is not the case for nonlinear systems. A nonlinear system may be stable for small disturbances and unstable for large disturbances. Thus, investigations of stability for nonlinear systems in general must include the effects of the magnitudes of the disturbances.

In practice, several different definitions of stability are used. The highest type of stability exists if the effect of the disturbance disappears with time, that is, if

$$\lim_{t \to \infty} |r_0 - y(t)| = 0$$

where $r_0 = $ constant is the desired output and $y(t)$ is the actual output. Such a system is said to be asymptotically stable. Lesser degrees of stability can be defined where the limit remains finite although not zero.

The stability of a linear system may be determined by examination of the eigenvalues of the system, i.e., the roots of the characteristic equation. If the eigenvalues all have negative real parts, the disturbances will decay and asymptotic stability results. On the other hand, if one or more eigenvalue has a positive real part, the disturbance will grow and the system is unstable. Thus, in Ex. 2 of this section all the eigenvalues have negative real parts and the control system is stable. Frequently it is tedious or impracticable to determine eigenvalues numerically. Various other schemes such as the Routh-Hurwitz, Nyquist, and Bode criteria (see Ref. 30 for details) are then used.

Energy concepts may be used to investigate the stability of nonlinear (and also linear) Lagrangian systems. In a stable system the total energy will decrease with time. On the other hand, in an unstable system the total energy will increase with time, coming from an outside source (such outside sources are always present for feedback systems). Stable equilibrium is associated with a minimum value of the potential energy (see Chap. 6). Liapunov (Ref. 32) generalized this concept and introduced other functions which have properties similar to potential functions. These are usually called Liapunov V-functions and include potential functions as a special case. Although techniques exist for finding suitable V-functions in some cases, no general technique is available, and this is a field of active research.

Returning now to linear systems of the type of Figure 12.6b, it is of interest to examine the effect of system parameters on stability and closeness of control. In the general case, the system G may be represented by a differential expression as follows:

$$u = G^{-1}y = (g_m D^m + g_{m-1}D^{m-1} + \cdots g_1 D + g_0) y \quad (12.12)$$

where y is the actual output, the g_i's are constants, D is the differentiation operator, and m is a positive integer.

The controller H may, in general, be chosen in accord with design requirements. To see how the design parameters may be selected, assume that the expression for the controller is of the form

$$h_{-1}\int_0^t (x - y) \, dt + (h_0 + h_1 D + \cdots + h_{m-1}D^{m-1}) y = u \quad (12.13)$$

For the closed-loop system, the two expressions (12.12) and (12.13) are equal. Differentiating both sides of the resulting equation and rearranging terms leads to the equation

$$g_m y^{(m+1)} + (g_{m-1} - h_{m-1}) y^{(m)} + \cdots + (g_1 - h_1)\ddot{y} + (g_0 - h_0)\dot{y}$$
$$+ h_{-1}y = h_{-1}r_0 \quad (12.14)$$

where we assume $x = 0$, $t < 0$, and $x = r_0$, $t > 0$. The eigenvalues of (12.14) are the roots of the characteristic equation (12.15).

$$g_m \lambda^{m+1} + (g_{m-1} - h_{m-1})\lambda^m + \cdots + (g_1 - h_1)\lambda^2$$
$$+ (g_0 - h_0)\lambda + h_{-1} = 0 \quad (12.15)$$

Stability of the system can be insured, since there are enough adjustable parameters h_{-1}, h_0, h_1, ..., h_{m-1} to control the location of all the eigenvalues in the complex λ-plane. With stability assured, it is evident that the steady-state solution reduces to

$$\lim_{t \to \infty} |r_0 - y| = 0$$

or y approaches r_0 asymptotically for large values of time t.

12.5 CONTROL OF INHERENTLY UNSTABLE SYSTEMS

In contrast to most earlier control systems such as the positioning servomechanism of Ex. 2, Sect. 12.3, many present day control problems involve the control of inherently unstable systems. For example, a satellite or missile moving in free space will tend to tumble because of the absence of frictional or damping forces. Similarly, a missile at takeoff is unstable because of its very low velocity. So-called vertical takeoff and landing craft are also inherently unstable and require control by means of reaction jets. The following two examples will bring out the principles and methods used in the control of inherently unstable systems. The underlying principle is to control the locations of the eigenvalues of the closed-loop system by proper selection of control parameters.

Example 1. Control of Inherently Unstable System. Determine the characteristics of a control device necessary to stabilize an inherently unstable system described by the relation

$$(D^2 - 4D + 3)\, y = u$$

where y is the system output and u the system input as depicted in Figure 12.6b. The system to be stabilized might, for example, be a linear positioning system with negative damping inherently present as in certain types of steering devices.

SOLUTION. The system can be stabilized by choosing u in accord with the conditions imposed by Eq. (12.15). Using the values of $m = 2$, $g_2 = 1$, $g_1 = -4$, $g_0 = 3$, the characteristic equation becomes

$$\lambda^3 + (-4 - h_1)\lambda^2 + (3 - h_0)\lambda + h_{-1} = 0$$

The Routh-Hurwitz stability criteria (Ref. 30) will be used. These are

(a) $\quad a_0 > 0 \qquad a_1 > 0 \qquad a_2 > 0 \qquad a_3 > 0$

and

(b) $\quad a_1 a_2 > a_0 a_3$

where a_0, a_1, a_2, a_3 are the coefficients of the characteristic polynomial. Because $a_0 = 1$, these conditions reduce to the following inequalities

$$a_1 = -4 - h_1 > 0 \quad \text{or} \quad h_1 < -4$$
$$a_2 = 3 - h_0 > 0 \quad \text{or} \quad h_0 < 3$$
$$a_3 = h_{-1} > 0$$
$$a_1 a_2 > a_3 \qquad \text{or} \quad (-4 - h_1)(3 - h_0) > h_{-1}$$

These conditions may be satisfied, for example, by choosing $h_{-1} = 1$, $h_0 = 2$, $h_1 = -6$, as may be seen by substituting in the characteristic equation. This gives

$$\lambda^3 + 2\lambda^2 + \lambda + 1 = 0$$

The eigenvalues are

$$\lambda_1 = -1.755 \qquad \lambda_2, \lambda_3 = -0.123 \pm 0.745i$$

and the system, therefore, is stable because all roots have negative real parts. The steady-state response clearly is given by

$$\lim_{t \to \infty} y = r_0$$

where r_0 is the input or control position (assumed constant after being set to this value). Thus, the output or controlled position y approaches r_0 asymptotically for each setting of r_0, assuming that the time between settings is long compared to the decay time constants $\dfrac{1}{0.123}$ and $\dfrac{1}{1.755}$

*Example 2. Stabilization of Inverted Pendulum.** There is given a dynamical system consisting of an inverted pendulum attached to a movable vehicle, as shown in Figure 12.10. It is obvious that this system is inherently unstable and that stability, if possible at all, will require a compensating feedback control device.

The first step toward determining the characteristics of the required control device is to find the dynamical equation for the system. By inspection, it is observed that there are two degrees of freedom.

$$[q_1, q_2] = [\theta, x]$$

Letting m_1 be the mass of the pendulum and m the mass of the vehicle, the Lagrangian is

$$L = T - V$$

where

$$T = \int \tfrac{1}{2}(\dot{x} + r\dot{\theta})^2 \, dm_1 + \tfrac{1}{2}m\dot{x}^2$$

$$= \tfrac{1}{2}m_1\dot{x}^2 + \dot{x}\dot{\theta}\int r \, dm_1 + \tfrac{1}{2}\dot{\theta}^2\int r^2 \, dm_1 + \tfrac{1}{2}m\dot{x}^2$$

$$= \tfrac{1}{2}(m_1 + m)\dot{x}^2 + m_1 R\dot{x}\dot{\theta} + \tfrac{1}{2}m_1\rho^2\dot{\theta}^2$$

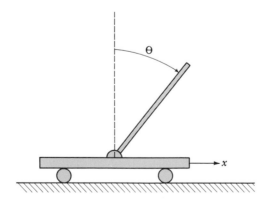

FIGURE 12.10   Inverted pendulum mounted on stabilizing vehicle.

* Suggested by Lynch and Truxal (Ref. 23).
See also "The Lever of Mahomet," by R. Courant and H. Robbins (Ref. 31, p. 412) for a problem of related interest.

and

$$V = m_1 g R_1 \cos \theta$$

$$\cong m_1 g R_1 (1 - \tfrac{1}{2}\theta^2)$$

for small angles. In these expressions R is the center of mass and ρ is the radius of gyration for the pendulum, both measured from the point of support of the pendulum.

Writing Lagrange's equations, we obtain

$$\frac{d}{dt}\left(\frac{\partial L}{\partial \dot\theta}\right) - \frac{\partial L}{\partial \theta} = m_1 R \ddot{x} + m_1 \rho^2 \ddot\theta - m_1 g R \theta = 0$$

and

$$\frac{d}{dt}\left(\frac{\partial L}{\partial \dot x}\right) - \frac{\partial L}{\partial x} = \frac{d}{dt}\left[(m_1 + m)\dot x + m_1 R \dot\theta\right] = 0$$

To control the motion of the vehicle, an electric motor drive is added such that the force applied to the vehicle is given by kE, where E is the control emf and k is a constant which relates E to the force. Also assume that a viscous force $-f\dot x$ acts on the vehicle, and that the mass of the pendulum is negligible compared to the mass of the vehicle. Then, the second Lagrangian equation above becomes

$$m\ddot x + f\dot x = kE$$

Lagrange's equations for the system may also be written in matrix form as follows

$$\begin{bmatrix} D^2 - \dfrac{Rg}{\rho^2} & \dfrac{R}{\rho^2}D^2 \\ \\ 0 & D\left(D + \dfrac{f}{m}\right) \end{bmatrix} \begin{bmatrix} \theta_1 \\ \\ x \end{bmatrix} = \begin{bmatrix} 0 \\ \\ \dfrac{k}{m}E \end{bmatrix}$$

The eigenfrequencies are given by the roots of

$$\det [A] = \left(\lambda^2 - \frac{Rg}{\rho^2}\right)\lambda\left(\lambda + \frac{f}{m_3}\right) = 0$$

where A is the matrix of the coefficients and λ is written for D. The roots are

$$\lambda_1, \lambda_2 = \pm\sqrt{\frac{Rg}{\rho^2}} \qquad \lambda_3 = 0 \qquad \lambda_4 = -\frac{f}{m}$$

By inspection, it is seen that there is one unstable root, λ_1.

In order to eliminate this unstable root, and to provide an appropriate control device to stabilize the pendulum in the vertical position, we introduce a control signal u as follows

$$u = \frac{k}{m} E = b_1 \dot\theta + b_2 \theta + b_3 \dot x$$

Substitution in the equation for the closed-loop control leads to the matrix

$$\begin{bmatrix} D^2 - \dfrac{Rg}{\rho^2} & \dfrac{R}{\rho^2} D^2 \\[3mm] -b_1 D - b_2 & D\left(D + \dfrac{f}{m} - b_3\right) \end{bmatrix}$$

The eigenfrequencies are given by the roots of the modified characteristic equation

$$\det[A] = \lambda\left[\lambda^3 + \left(\frac{f}{m} - b_3 + \frac{Ra_1}{\rho^2}\right)\lambda^2 + \frac{R}{\rho^2}(b_2 - g)\lambda + \frac{Rg}{\rho^2}\left(b_3 - \frac{f}{m}\right)\right] = 0$$

Using the Routh-Hurwitz criteria, the conditions for stability are as follows:

(1) $\qquad\qquad\qquad\qquad b_3 - \dfrac{f}{m} > 0 \quad\text{or}\quad b_3 > \dfrac{f}{m}$

(2) $\qquad\qquad\qquad\qquad b_2 - g > 0 \quad\text{or}\quad b_2 > g$

(3) $\qquad\qquad \dfrac{f}{m} - b_3 + \dfrac{R}{\rho^2} b_1 > 0 \quad\text{or}\quad \dfrac{R}{\rho^2} b_1 > b_3 - \dfrac{f}{m}$

(4) $\qquad\qquad \left(\dfrac{f}{m} - b_3 + \dfrac{R}{\rho^2} b_1\right)(b_2 - g) > g\left(b_3 - \dfrac{f}{m}\right)$

NUMERICAL EXAMPLE. Application of these theoretical results may be demonstrated by means of a numerical example. For this purpose, assume a simple pendulum of length 3 feet. Also assume the control vehicle with its associated electronic equipment has a net weight of 9.3 lbs. and a frictional constant of $f = 0.1$ slug ft.$^{-1}$. Using these values, the system constants are

$$m = \frac{9.3}{32.2} = 0.289 \text{ slug} \qquad f = 0.1 \text{ slug ft.}^{-1}$$

$$R = \rho = 3 \text{ ft.} \qquad \frac{f}{m} = 0.346 \qquad \frac{R}{\rho^2} = 1$$

The control parameters must be selected to insure stability, in accord with the Routh-Hurwitz inequalities. A suitable set is

$$b_3 = 2\frac{f}{m} \qquad b_2 = 2g \qquad \frac{R}{\rho^2} b_1 = 3\frac{f}{m}$$

as can be seen by direct substitution in the inequalities (1), (2), (3), and (4) above.

Substitution in the characteristic equation yields

$$\det A = \lambda(\lambda^3 + 0.69\lambda^2 + 10.7\lambda + 3.72) = 0$$

which has the roots

$$\lambda_1 = 0 \qquad \lambda_2 = -0.351 \qquad \lambda_3, \lambda_4 = -0.170 \pm 3.26i$$

The root $\lambda_1 = 0$ simply indicates that only the velocity $\dot{x}$ is being controlled, the value of x itself being unimportant. The remaining roots all have negative real parts and the system is therefore stable.

Requirements for the control device are set by the control signal u as follows:

$$u = b_1\dot{\theta} + b_2\theta + b_3\dot{x}$$

$$= 3\frac{f}{m} \cdot \frac{\rho^2}{R}\dot{\theta}_1 + 2g\theta + 2\frac{f}{m}\dot{x}$$

$$= 1.02\dot{\theta} + 62.4\theta + 0.69\dot{x} = \frac{kE}{m}$$

These values can be generated by combining an electric analog computer, a power amplifier, and an electric motor. The control vehicle force is kE, where E is the emf and k is an appropriate constant which relates E to the force applied to the vehicle.

*Example 3. Lateral Movement of a Vertical-Takeoff-and-Landing Aircraft. (VTOL).** In such an aircraft, exhausts from jet engines provide lift, and hovering is possible. The jets are mechanically coupled to the pilot's normal control stick and pedals, and serve to generate control moments about the pitch, roll, and yaw axes. However, it is essential to provide a supplementary automatic control system to enable the pilot to cope with the otherwise unmanageably complex operation.

The analysis here is restricted to a drastically simplified version of the system as indicated in Figure 12.11. To move laterally a certain distance x_0, the ship is rotated through a small angle θ which introduces a horizontal force component from the jets and causes the vehicle to be accelerated laterally. Upon approach to the position x_0 the vehicle is decelerated and hovers again.

Let y be the position of the control stick, and let the control moment be proportional to y. Then, we can write

$$\frac{d^2\theta}{dt^2} = \ddot{\theta} = k_1 y$$

where k_1 is a constant of the system.

---

* Adapted from Lynch and Truxal (Ref. 23).

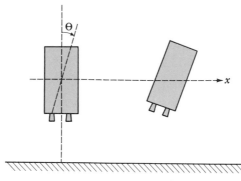

FIGURE 12.11   Lateral movement of VTOL craft.

Since F_x is proportional to $\sin \theta \simeq \theta$, we can write similarly that

$$\frac{d^2x}{dt^2} = \ddot{x} = k_2\theta$$

where k_2 is also a constant of the system.

Combining these two equations gives

$$k_1 y = D^2\theta = D^2 \left(\frac{1}{k_2} D^2x \right) = \frac{1}{k_2} D^4x$$

or

$$x = D^{-4}(k_1 k_2 y) = k_1 k_2 \iiiint y \, dt^4$$

That is, x is proportional to the fourth integral of y.

Conversely

$$y = \frac{1}{k_1 k_2} D^4 x$$

which means that the pilot's control stick motion must be based on the fourth derivative of the lateral motion.

Experience with man-machine systems shows that the human mind can differentiate a signal once (velocity) and even twice (acceleration), but that twice is the limit. Thus, a pilot could not possibly learn to fly the VTOL aircraft with this control system which requires differentiating a signal four times.

To cope with this flight problem, the dynamics of the system can be modified so that in effect only two differentiations or integrations by the pilot are required. This can be accomplished by introducing negative feedback operational amplifiers as shown in Figure 12.12, for which

$$x = \frac{k_1 k_2 y}{D^2(D^2 + \alpha D + \beta k_2)} = \frac{k_1 k_2 y}{f(D)}$$

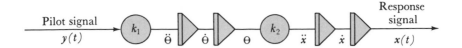

(a)   Original system with four integrations

$$x = \frac{k_1 k_2}{D^4} \, y$$

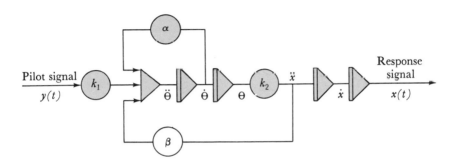

(b)   Modified system with two feedback loops

$$x = \frac{k_1 k_2}{D^2(D^2 + \alpha D + \beta k_2)} \quad y = \frac{k_1 k_2}{f(D)} \, y$$

FIGURE 12.12 Analog computer representation of VTOL lateral maneuver feedback control system.

The eigenvalues of $f(D)$ are the roots of $f(\lambda) = 0$, and are seen to be

$$\lambda_1 = \lambda_2 = 0 \qquad \lambda_3, \lambda_4 = \frac{-\alpha \pm \sqrt{\alpha^2 - 4\beta k_2}}{2} = -\frac{\alpha}{2} \pm \sqrt{\left(\frac{\alpha}{2}\right)^2 - \beta k_2}$$

If $\left(\dfrac{\alpha}{2}\right)^2 < \beta k_2$, the roots λ_3, λ_4 are complex, i.e.,

$$\lambda_3, \lambda_4 = -\frac{\alpha}{2} \pm i \sqrt{\beta k_2 - \left(\frac{\alpha}{2}\right)^2} = a \pm bi$$

These roots correspond to a damped sinusoid which, by proper choice of the parameters, can be made to attenuate rapidly.

As one maneuver strategy, let y be given a sudden positive displacement followed by an equal and opposite negative displacement in accord with the following equations.

$$y = y_0 \qquad 0 < t < \tfrac{1}{2}t_0$$

$$y = -y_0 \qquad \tfrac{1}{2}t_0 < t < t_0$$

where

$$x(t_0) = x_0 \qquad x(0) = \dot{x}(0) = \ddot{x}(0) = \dddot{x}(0) = 0$$

Then, using Laplace transform method, the transform $\bar{x}(s)$ for $0 \le t < \tfrac{1}{2}t_0$ is given by the equation

$$\bar{x}(s) = \frac{k_1 k_2 y_0}{s f(s)} = \frac{k_1 k_2 y_0}{s^3(s^2 + \alpha s + \beta k_2)}$$

Expanding the expression for $\bar{x}(s)$ in partial fractions gives

$$\bar{x}(s) = \left[\frac{A_1}{s^3} + \frac{A_2}{s^2} + \frac{A_3}{s} + \frac{2B_1(s - a) + 2B_2 b}{s^2 + \alpha s + \beta k_2} \right] k_1 k_2 y_0$$

where A_1, A_2, A_3, B_1, and B_2 are real constants which can be evaluated in terms of the parameters α and βk_2 by the standard methods of Laplace transform theory. Thus

$$A_1 = \left[\frac{s^3}{s^3(s^2 + \alpha s + \beta k_2)} \right]_{s=0} = \frac{1}{\beta k_2}$$

$$A_2 = \frac{d}{ds}\left[\frac{s^3}{s^3(s^2 + \alpha s + \beta k_2)} \right]_{s=0} = \frac{-\alpha}{\beta^2 k_2^2}$$

$$A_3 = \frac{d^2}{ds^2}\left[\frac{s^3}{s^3(s^2 + \alpha s + \beta k_2)} \right]_{s=0} = \frac{2\alpha^2}{\beta^3 k_2^3}$$

$$B_1 - B_2 i = \frac{1}{\dfrac{d}{ds}\,[s^3(s^2 + \alpha s + \beta k_2)]_{s=a+bi}}$$

The inverse transform of $x(s)$ can now be obtained by inspection, giving the solution

$$x(t) = \left[\frac{A_1 t^2}{2} + A_2 t + A_3 + 2e^{-\alpha t/2}(B_1 \cos bt + B_2 \sin bt) \right] k_1 k_2 y_0$$

Since $\dfrac{x(0)}{k_1 k_2 y_0} = A_3 + 2B_1 = 0$, it follows that

$$B_1 = \frac{-A_3}{2} = \frac{-\alpha^2}{\beta^3 k_2^2}$$

Similarly, from

$$\frac{\dot{x}(0)}{k_1 k_2 y_0} = A_2 - \alpha B_1 + 2b B_2 = 0$$

we find

$$B_2 = \frac{\alpha \left(\beta k_2 - \dfrac{\alpha^2}{2} \right)}{b \beta^3 k_2^3}$$

The solution above for $x(t)$ can be joined to the solution for the deceleration phase to obtain the complete solution.

Hardware for the feedback system will include a stable platform, a rate gyro to measure $\dot{\theta}$, and an accelerometer to measure $\ddot{x}$.

12.6 MULTIVARIABLE FEEDBACK CONTROL SYSTEMS

Control systems with several inputs and several outputs occur in practice and may be regarded as a generalization of the simpler one-variable structures studied earlier in this chapter. Thus the single-variable feedback system of Figure 12.6b may be generalized to the multiple-variable system of Figure 12.13. The input and output variables may be conveniently represented by the vectors

$$\mathbf{x} = [x_1, x_2, \dots, x_n] = \text{input or control vector}$$

and

$$\mathbf{y} = [y_1, y_2, \dots, y_n] = \text{output or controlled vector}$$

The plant input $\mathbf{u}$ is supplied to G by the controller H, where

$$\mathbf{u} = [u_1, u_2, \dots, u_r]$$
$$= H_1 \mathbf{x} - H_2 \mathbf{y}$$

Following the approach used for the single-variable system (Sect. 12.3), the basic equation for the multiple variable system is found from the relation

$$\mathbf{y} = G(H_1 \mathbf{x} - H_2 \mathbf{y}) = G H_1 \mathbf{x} - G H_2 \mathbf{y}$$

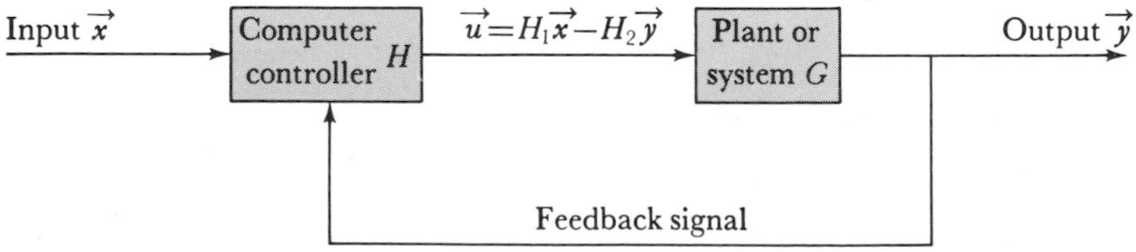

FIGURE 12.13 Multivariable feedback control system.

which when solved for **y** gives

$$\mathbf{y} = (I + GH_2)^{-1}GH_1\mathbf{x} = T\mathbf{x} \qquad (12.16)$$

Again, the quantities G, H_1, H_2 are operators which represent the system G and controller H, while I is the identity operator. For linear lumped parameter Lagrangian systems, the operators G, H_1, H_2 reduce to matrices and the operator I to the indentity matrix.

BIBLIOGRAPHY

1. C. Singer et al., *A History of Technology*, Vol. III. London: Oxford University Press, 1957. (See page 680.)
2. J. C. Maxwell, "On Governors." *Proceedings of the Royal Society of London, 16* 270, 1868.
3. N. Minorsky, "Directional Stability of Automatically Steered Bodies." *Journal of the American Society of Naval Engineers, 34* 280, 1922.
4. H. L. Hazen, "Theory of Servomechanisms." *Journal of the Franklin Institute, 218* 279, 1934.
5. H. Nyquist, "Regeneration Theory." *Bell System Technical Journal, XI* 126, 1932.
6. H. W. Bode, "Feedback Amplifier Design." *Bell System Technical Journal, XIX* 42, 1940.
7. R. Bellman and R. Kalaba, ed., *Selected Papers on Mathematical Trends in Control Theory*. New York: Dover, 1964.
8. H. J. James et al., ed., *Theory of Servomechanisms*. New York: McGraw-Hill, 1947. (Vol. 25, Massachusetts Institute of Technology Radiation Laboratory Series.)
9. G. J. Thaler and R. G. Brown, *Analysis and Design of Feedback Control Systems*, Second Edition. New York: McGraw-Hill, 1960.
10. J. Peschon, ed., *Disciplines and Techniques of Systems Control*. New York: Blaisdell, 1965.
11. H. W. Bode, "Feedback—The History of an Idea." Proceedings of Symposium on Active Networks and Feedback Systems, Polytechnic Institute of Brooklyn, Polytechnic Press, 1960.
12. H. S. Black, "Stabilized Feedback Amplifiers." *Bell System Technical Journal 13* 1, 1934.
13. H. Bateman, "The Control of an Elastic Fluid." *Bulletin of the American Mathematical Society, 51* 601, 1945.
14. I. A. Vyshnegradskii, "On Controllers of Direct Action." Ixv. *SPB Tekhnolog. Inst.*, 1877.
15. W. A. Lynch and J. G. Truxal, *Introductory System Analysis*. New York: McGraw-Hill, 1961.
16. N. Wiener, *Cybernetics*, Second Edition. Massachusetts Institute of Technology Press, 1961.
17. *Journal on Control*, published by Society for Industrial and Applied Mathematics (SIAM).
18. *IEEE Transactions on Automatic Control*, published by IEEE professional technical group on Automatic Control.
19. E. H. Armstrong, "Some Recent Developments of Regenerative Circuits." *Proceedings of the Institute of Radio Engineers, 10* 244, 1922. (Also see E. H. Armstrong, "The Regenerative Circuit." *Proceedings Radio Club of America*, April 1915.)
20. R. Bellman et al., "Some Aspects of the Mathematical Theory of Control Processes." The RAND Corporation, R-313, 1958.
21. J. P. LaSalle, "Time Optimal Control Systems." *Proceedings of the National Academy of Sciences, 45* 573, 1959.
22. V. Boltyanskii et al., "On the Theory of Optimal Processes." *Reports of the Academy of Sciences of the USSR, 110* 7, 1956. (Translated in Ref. 7, pp. 172–176.)

23. W. A. Lynch and J. G. Truxal, "Control System Engineering." *International Science and Technology*, p. 60, March 1966.
24. G. J. Thaler, "Cross Disciplinary Education in Control Engineering." *Journal of Engineering Education*, *53* 534, April 1963.
25. G. J. Thaler, "Advanced Automatic Control Theory." *Journal of Engineering Education*, *53* 777, June 1963.
26. E. S. Savas, *Computer Control of Industrial Processes*. New York: McGraw-Hill, 1965.
27. J. G. Truxal, ed., *Control Engineers Handbook*. New York: McGraw-Hill, 1958.
28. M. R. Hestenes, *Calculus of Variations and Optimal Control Theory*. New York: Wiley, 1966.
29. R. Oldenburger, ed., *Optimal and Self-Optimizing Control*. Cambridge: Massachusetts Institute of Technology Press, 1966. (A collection of 38 original papers on subject.)
30. W. Kaplan, *Operational Methods for Linear Systems*. Reading, Mass.: Addison-Wesley, 1962.
31. J. Newman, ed., *The World of Mathematics*. New York: Simon and Schuster, 1956.
32. J. P. LaSalle and S. Lefshetz, *Stability by Liapunov's Direct Method*. New York: Academic Press, 1961.
33. *Proceedings of the International Federation of Automatic Control, Moscow, 1960*, edited by J. F. Coles et al., Butterworths, 1961.
34. *Proceedings of the IEEE*, Special Issue on Computer-Aided Design, Nov. 1967, published by the Institute of Electrical and Electronics Engineers, Inc., 345 East 47th Street, New York, N.Y., 10017.
35. V. W. Everleigh, *Adaptive Control and Optimization Techniques*. New York: McGraw-Hill, 1967.
36. G. Leitman, *An Introduction to Optimal Control*. New York: McGraw-Hill, 1967.
37. K. Ogata, *State Space Analysis of Control Systems*. Englewood Cliffs, N.J.: Prentice-Hall, 1967.
38. P. de la Barriére, *Optimal Control Theory*. Philadelphia: Saunders, 1967.

13

EXTENSIONS OF SYSTEMS METHODS

Mastery of the philosophy, the mathematical and scientific foundations, and the general approach to application of systems methods, provides a strong basis for developing new applications as well as unifying existing ones. These applications can be at the conceptual, theoretical, or practical levels, and need not be hampered by the constraints which often limit progress within the framework of traditional fields.

Preceding chapters of this book have followed this theme by developing systems concepts, laying theoretical foundations for broadly applicable systems methods, and applying these methods to several major areas of physical and engineering science. By defining and treating a system abstractly, we avoid the undesirable situation of placing arbitrary and unnecessary limits on the number and kinds of fields to which systems theory and methods may be applied. It is shown that by utilizing this broad approach many apparently disparate fields and problems do, in fact, turn out to be closely related and can be efficiently treated by the same general methods. Such unification of knowledge and methods is of course the central goal of systems science.

With this background in mind, the purpose of this closing chapter is to give some vista of the breadth and richness of present and potential applications of systems methods. The list of systems applications is already impressive, and is growing steadily. In addition to physical and engineering areas, where the greatest use has been made thus far, systems applications are established in biology, economics, and business administration. Systems work is also developing in education, psychology, and sociology.

Some of the areas where the practical application of the systems approach has attained or is attaining major importance are industrial management, automation, electric power grids, military strategy and tactics, urban planning, medical and public health problems, transportation and traffic flow, air and water pollution problems, architectural planning and construction, information handling, and agriculture.

In the Introduction of this book a system is defined as a set of interacting elements. The analysis of a system is essentially quantitative and mathematical in character, and treats the interrelations between the elements as well as the overall functioning of the systems. Systems analysis is most advanced in the physical sciences and their associated technologies, because for these fields there exist precise general laws and well-developed mathematics for expressing the laws and coping with systems relations. In addition, generalization and the transfer of systems methods between the various physical science and engineering fields has aided greatly in the development of advanced systems analysis in these areas.

By comparison, systems methods in the biological sciences, although offering tremendous potential, are still in the fairly early stages of development. The barriers to development have been the lack of general theories, insufficiently developed mathematical methods, and a dearth of quantitatively trained scientists in biological fields. This latter deficiency is being overcome in part by the influx of workers and the transfer of methods, from physics, chemistry, mathematics, and engineering to biological areas.

The situation with regard to systems methods in most behavioral fields somewhat parallels that which exists in the biological sciences. Theories and laws in these fields have been largely descriptive or verbal in nature, and often seem to be tentative and uncertain. In addition, a complicating factor is that elements of these systems are people, and the systems themselves are the organizations evolved and devised by people; compared to the physical and biological sciences, which are essentially objective and impersonal, the behavioral fields deal with elements whose behavior is subjective and often random and conflicting.

Systems in the areas of economics and management tend to be better structured than in other behavioral and social fields and are, therefore, more amenable to quantitative systems approaches. Additionally, there is greater incentive for systems application in industry and business because the pay-off is in dollars rather than in abstract social values.

Evidence of the rising interest in systems and systems application is seen in the growing number of undergraduate and graduate systems programs which are appearing in educational institutions. Some of these are, happily, interdisciplinary as the term *systems* itself would appear to require. Additional evidence of interest in systems is found in the expanding literature and in the amount of professional society activity.

13.1 PHYSICAL AND ENGINEERING SCIENCES

Many examples of systems in physical science developed during the nineteenth century could be given, but here it will suffice to mention the great advance and simplification brought about in chemistry by Mendeleev's development of the periodic table of the chemical elements, and in physics by the formulation of Maxwell's equations of electromagnetic theory.

As brought out in Chapter 5, twentieth century development of systems, especially large-scale complex technological systems, may be regarded as having its origin and impetus in the massive scientific and technological efforts of the 1940's. The consequent explosion of systems knowledge and applications triggered by these efforts continues at an ever-increasing pace.

The range of application of systems methods to physics, chemistry, and engineering fields will be made evident by the areas and examples given in the following list.

Areas and Examples of Systems Applications in the Engineering and Physical Sciences:

A. Engineering fields

1. Electrical
2. Mechanical
3. Civil and structural
4. Industrial
5. Chemical
6. Aerospace
7. Transportation and traffic
8. Systems engineering

B. Process and equipment areas

1. Communications
2. Feedback and optimal control systems
3. Automation
4. On-line process control (oil refineries, paper mills, etc.)
5. Digital and analog computers
6. Process dynamics
7. Electric networks
8. Mechanical systems
9. Electromechanical systems
10. Generalized networks and dynamic analogies

B. Process and equipment areas (*contd.*)

11. Computer simulation
12. Inertial guidance
13. Automated analysis and design
14. Electric power systems
    (a) Power flow analysis
    (b) Transient stability
    (c) Economic loading and planning
15. Design optimization
16. Industrial process modeling (steel mills, etc.)
17. Chemical process control

C. Other areas

1. Applied mathematics
2. Chemical kinetics
3. Crystal structure
4. Organic chemistry
5. Matrix methods in geometrical optics
6. System stability problems
7. Operations research

The applications already covered in this book give a sufficient cross section to enable the student to apply the same principles and methods to solving problems in other physical and engineering science fields. In addition, the references given in this and preceding chapters will prove helpful in extending the scope of applications.

13.2 BIOLOGICAL SCIENCES

Systems methods are being adapted increasingly for use in the biological sciences, a trend to be expected in view of the complex, interconnected entities typically dealt with in these fields. Waterman and Morowitz, in a recent compendium of theoretical and mathematical biology (Ref. 11), give many examples of interacting entities in biology such as networks of neurons, cycles of metabolites, pools of genes, individuals in a group, and species in a population. They observe that the notion of the whole being greater than the mere sum of its parts is a familiar concept to biologists, but that the problem of predicting the properties of the whole from the rules of interaction between its components is difficult and uncertain because neither general theories nor general mathematical methods for such analyses have been developed.

Another sort of difficulty arises from the dichotomy which exists in biology itself between the qualitative and quantitative approaches to the field. As a result of this dichotomy, and of the dearth of quantitatively trained biologists, the quantitative phase seems to have been largely taken over by biophysicists, biochemists, biomathematicians, and more recently by bioengineers, that is, by scientists whose mother fields have a long history of dedication to quantitative systems methods.

Morowitz (Ref. 11, Chap. 2), lists formal theory, physical theory, and systems theory as comprising the three general areas of theoretical biology. In this line-up, formal theory is based on a set of postulates put forth as mathematical statements from which predictions can be deduced and compared with experimental results. Physical theory attempts to apply to biological areas the established concepts of physics and chemistry, as for example thermodynamics, molecular structure, and molecular interactions. This approach is closely related to molecular biology. Systems theory is regarded by Morowitz as a special case of formal theory derived from the modern development of information theory, operations research, network theory, servosystem theory, and computer development.

The lack of general laws in biology (laws comparable, for example, with Newton's laws of motion and Maxwell's equations of electromagnetic theory in physics) greatly complicates the task of establishing useful mathematical models for biological systems. Thus, systems models used in biology are of necessity based on limited theory, and depend a great

deal on empirical data.* In effect, each problem must be treated separately. However, the influx of workers with experience and training in the
physical sciences, mathematics, and engineering is advancing the productive use of systems in biology.

The impact of electronic computers on the biological sciences
promises to be as great, if not greater than that experienced in the physical
sciences. Biological systems characteristically involve large numbers of
interrelated variables; studies of these systems require processing of huge
masses of data, work which could not be done without computers. Ledley,
in a review of the scope of computer applications in biological fields
(Ref. 11, Chap. 10), emphasizes the point that in biology computers are
far more than mere rapid calculators. Their real significance is that they
open the way for fundamental advances through new kinds of experiments
and the development and verification of new theories which would
otherwise be quite impossible. Ledley points out that systems simulation
by computer frequently is the only practical approach to the study of
highly involved biological phenomena.

Typically, computer simulations are necessary to study effectively
the behavior of systems where a great deal is known about the local
component aspects and where it is desired to study how these many
complicated parts interact to produce the behavior of the system as a
whole. Some typical examples are the systems treatment of the overall
functioning of the eye, the application of systems methods to the evaluation
of new drugs, and studies of the correlation of diseases with various possible
etiologies.

Another major application referred to by Ledley is the solution of
rate equations† for biomedical processes as, for example, flow equations
for the heart and circulatory system. In this study the heart and circulatory system are represented by 22 simultaneous differential equations,
which is about the smallest number possible to adequately represent even
the more elementary aspects of the interrelations between the heart and
the rest of the circulatory system. These equations, supplemented by
certain experimental measurements , are used to study and predict various
complicated effects which occur in the circulatory system.

Cybernetics. In the 1940's the mathematician Norbert Wiener
coined the term *cybernetics* for the science of control and communication
in the animal and in the machine (Ref. 12). Wiener's objective was to
unify those areas of engineering, mathematics, and physiology which
center about problems of control and communication. In this effort he
and his associates were fully cognizant of the resistance of academic

---

* One view, not infrequently expressed in the literature, is that mathematical theory
in biology may be regarded as scarcely more than a sophisticated form of curve fitting.

† This is essentially the state variable approach referred to in Chap. 4, p. 155 of
this text.

monopolies to intrusions, and of the hazards of invading the no man's land between disciplines. But he also recognized that it is these boundary regions of science which offer the richest opportunities. Parenthetically, it may be noted that Wiener is only one of many world-famous mathematicians and physicists who have had a deep interest in the life sciences. For example, consider the work of the eminent mathematician von Neumann on the parallels between living organisms and automata (Ref. 13), and of the theoretical physicist Schrödinger on the relation of quantum mechanics and statistical mechanics to biological phenomena (Ref. 14).

While the ultimate direction of cybernetics is not certain, there is no doubt that the application of feedback control theory to biological systems is a field of increasing interest and significance. Many homeostatic or regulatory control systems met with in biology can be usefully analyzed and studied using the general theory of feedback control systems developed for technological applications. Some examples listed in a recent text on the subject (Ref. 15) are as follows:

1. Body temperature
2. Cardiac output
3. Water and electrolyte balance
4. Respiration
5. Blood chemistry; sugar level; red and white cells
6. Audition
7. Control of posture and locomotion
8. Oculomotor system-tracking and stabilization
9. Pupillary reflex system

All these systems comprise several organs, have feedback channels, and have as their function the regulation of physical variables or actions of the body. Moreover, similar to engineering control systems, they can be satisfactorily modeled by differential equations, and treated by familiar engineering and mathematical methods.

Several scientific and technical societies publish journals which point up the interdisciplinary nature of cybernetics. For example, The Institute of Electrical and Electronics Engineers has a Systems Science and Cybernetics Group with its own publication (Ref. 29). The word *cybernetics* has somewhat varied meanings. The Russian usage of the term, for example, is much broader than in this country. The major Russian publication in the field, *Engineering Cybernetics* (bimonthly; available in English translation), covers such topics as "Systems Theory and Control Problems," "Automatic Control," "Information Processing," and "Automata and Computers."

Bioengineering, Bionics, Biotechnology. Several systems areas of growing interest have emerged in the years since about 1955 as a result of the interactions between certain engineering and life science groups (Ref. 26). These newer fields of interdisciplinary endeavor, which combine concepts from biology and engineering, are variously called bioengineering, biomedical engineering, bionics, and biotechnology, according to the aspect emphasized.

Bioengineering and biomedical engineering essentially apply engineering principles and methods to biological and medical problems for the purpose of advancing the latter fields. One of the major focuses of the field is the adaptation of systems engineering ideas and methods to physiological functions. Bionics may be thought of as the converse of bioengineering, that is, the use of knowledge and properties of living systems for the purpose of advancing engineering and technology. A more formal definition of bionics (Ref. 26, p. 3) is the "science of systems which function after the manner of, or in the manner of, or resembling, living systems." Trends in bionics research include adaptive learning, methods for simulating some of the higher-level functions of the human brain, and computer systems having visual pattern recognition capability. These latter areas are sometimes characteristized as "machine learning" and "artificial intelligence" (Ref. 27).

Biotechnology refers more specifically to the man-machine relationship (Ref. 28). It is also an interdisciplinary subject and uses engineering and mathematical tools to describe the functioning of the human operator in man-machine systems. Interest in biotechnology centers in the aerospace industry and aerospace systems, where the limitations of the human operator are perhaps most evident.

Those wishing to explore quantitative system studies in the biological sciences are referred to the rapidly growing literature in this field, a cross section of which is given in the appended bibliography. Also, it is of interest to note the emergence of graduate-level programs in biological systems analysis (Ref. 16).

13.3 ECONOMICS, INDUSTRY, AND MANAGEMENT

The current role of systems methods in economics, industry, and management may be gauged in terms of existing applications of quantitative methods and systems concepts to these fields.

Simple arithmetic is, of course, the most common mathematical tool, and suffices for coping with the quantitative phases of many business problems and systems. Present-day business data-processing systems may be regarded as an amplified form of arithmetic and can extend through the medium of electronic computers, previous arithmetical capabilities a million-fold or more, but do not introduce any new mathematical principles. In addition, statistics and graphs have long been accepted aids in economics and business enterprises.

However, it is only since about 1950 that more advanced mathematical methods, paced by the successes of linear programming and operations research transferred at that time from military problems to the civilian arena, have entered the repertory of economics and management on a significant scale. As a part of this trend, calculus, differential equations, linear algebra, and other branches of mathematics are now being applied to the study of economic models and to many management and business problems.

The branch of economics concerned with the mathematical and statistical aspects of the subject is called econometrics or mathematical economics. Included in this context are such areas or topics as production theory, consumption theory, dynamic theory of growth, and welfare economics. Both static and dynamic models are used in these studies.

The complexities inherent in the management and operation of modern large-scale business enterprises have led to greatly increased interest in the use of more sophisticated mathematics in these fields. Mathematical programming methods and operations research (treated in Sect. 4 of this chapter) were developed for the express purpose of coping with organization and operations problems. Other mathematical methods developed for dealing with these problems include game theory, value theory, queueing theory, and decision theory. There are also various scheduling techniques, among which the best-known are PERT (program evaluation and review technique) and CPM (critical path method).

Management Systems. Management may be thought of as the decision-making or problem-solving segment of an organization. In this context, organization has been defined as a "purposeful system of cooperation among individuals for increasing personal welfare" (Ref. 49, p. 22). The system aspects are similar to those indicated schematically in Figure 1 of the Introduction of this text; the system is the organization, the input is human resources, and the output human welfare. Traditionally, the managerial function has been considered an art, relying on the personal experience, intuition, and judgment qualities of individual managers rather than on any scientific basis. However, as organizations increase in size and complexity, the need for more scientific management becomes more urgent. The parallel with engineering in this respect is very evident: the trend of engineering toward an applied science base, which began several decades ago under the pressures of coping with larger and more complex engineering systems, may well be duplicated in coming years in the management field.*

In addition to the usefulness of systems concepts and methods in the overall design of business organizations, they are also equally useful in solving strategic and tactical problems which occur in the operation

---

* References 39 and 45 give a modern view of the emergence of management as an applied and quantitative science.

of these organizations. Trial and error methods tend to be costly and can be disastrous if the stakes are high. Fiat methods based on experience, although workable in conventional situations, are clearly limited where new conditions must be dealt with.

The trend toward a new and more scientific approach to solving business problems began in the early 1950's and stems from the wartime operations research methods of the 1940's. Since these pioneering efforts, there has been a remarkable growth of decision-making methods and techniques developed primarily for business and industry. Even though foreign to the traditional intuitive and personal judgment approach to solving business problems, these newer methods, based on scientific analysis, are receiving substantial attention and significant acceptance in business circles. Although at first there seemed to be something like a mathematical and computer barrier to acceptance of the newer methods, this appears to be gradually disappearing under the impact of the obvious successes of these methods and the entrance of a new generation, quantitatively oriented and trained, into the business field.

By far the most widely employed form of management, used not only in business but also in educational, political, and other organizations, is the hierarchal or bureaucratic type. The major feature of the bureaucratic scheme is the allocation of authority within an organization. The classic method of depicting an authority allocation pattern is by means of the familiar organization chart which, in effect, is a block diagram of the authority system.

Another form of representation may be made using the branch of mathematics known as graph theory (Refs. 52 and 78). In this theory, a *graph* is a finite set of objects, called *points*, with connections between some of the points, called *lines*. One use of graph theory is made in organization theory, which deals with the structure of groups of human beings working together. The basic relationships of a group may be portrayed in a graph-theoretic model, as for example in Figure 13.1 which represents the authority structure for a small organization and is self-explanatory.* Using the abstract results of graph theory, it is possible to represent and derive many of the characteristics of even very complex organizations in a relatively simple way. The utility of this method is that, among other things, it makes it possible to discover pathological conditions in an organization, such as communication gaps, conflicts of authority, or instabilities due to the presence of unsuspected feedback loops.

Authority in a bureaucratic structure flows downward from the

---

* Graph-theoretic diagrams of this type are met with in many fields, usually under different names, e.g., circuit diagrams in engineering and physics, sociograms in psychology, simplexes in topology, PERT diagrams, as well as the organization structure use of Figure 13.1. One of the earliest applications of graph theory was to the Königsberg bridge problem (Euler, 1736).

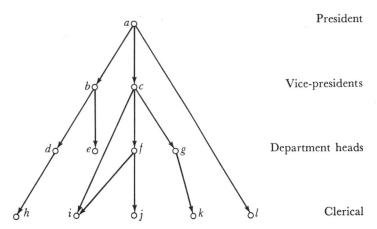

FIGURE 13.1 Graph-theoretic organization chart (after Kemeny and Snell, Ref. 52).

president or chief executive officer to successive lower management levels. Decision-making in the bureaucratic system usually remains a largely unplanned activity. A certain amount of authority is delegated to each manager who then, within the limits of that authority, takes the steps which in his judgment appear to be necessary to achieve his objectives. If the individualism of a submanager conflicts or does not coincide with organizational policies and plans, poor decisions are likely to result, with consequences which the top managers will have to correct. These deficiencies, together with the widespread use of the computer and other analytical tools now available to management, are leading to reappraisal of the traditional bureaucratic format.

There has been some tendency to confuse the newer tools and techniques which serve as aids to management with the management system itself. For example, the procedures for collecting, processing, and retrieving information, usually with computers as a central element, are frequently and incorrectly called management systems. Without denying the utility of such information systems, it is apparent that such usage of the term management systems is merely arrogation.

Appending various modern managerial tools (such as the computer information system mentioned above) in a piecemeal fashion to existing management structures undoubtedly has appeal and some advantages, but it is too limited a concept to serve as a basis for long-term improvement. According to one viewpoint (Ref. 49) it is now becoming feasible to develop total, integrated management systems based on the systems analysis methods used so successfully for physical systems and for the complex communications, military, and space programs.

The reader will recall, from preceding chapters of this book, that in the analysis of physical systems the basic approach is to divide the system with which one is concerned into increasingly smaller subsystems until the basic components of the system are identified. The functioning of each basic component is then described mathematically. Then, reversing this process by tying all the components together to form the original system, and then solving the mathematical expressions which represent the system and its inputs, gives the overall performance or output of the system. This approach has, of course, proved highly successful in dealing with many physical and engineering systems. The same approach, extended to infinite processes, underlies the infinitesimal calculus of mathematics.

Young (Ref. 49) discusses the extension of these systems ideas to the management process. In this extension he uses analogous steps in constructing (or, more generally, designing) a model and simulating the operation of the management system. The model might be a block diagram or a mathematical representation of some sort. Simulation of the operating properties of the management system, as represented by the model, might be done on a digital or analog computer. By a sequence of such model constructions and simulations, and by testing each model to determine its strengths and shortcomings, it becomes possible to arrive at improved, if not optimal, systems. This sequential design-improvement method is often used in designing engineering structures and systems.

A further analogy between management and the control component in a physical system may be made in terms of feedback control function. In this context, the systems aspects of an organization as a whole are similar to those indicated schematically in Chap. 12, Figure 6 of this book. The output is human welfare. The actual output is compared with the expected output (as determined by earlier experience, by extrapolation of this, and by comparative experience in similar organizations). Differences between expected and actual welfare levels comprise an error signal for the system, to which the management process is (or should be) responsive by formulating solutions and decisions in such a way as to correct for the error signal. In addition to the error signal, the organization and the control function must cope with disturbances and "noise" in the form of environmental inputs. Thus, the overall function of management is to provide the control and self-regulation feature of an organization. This control is not only adaptive but can also, at least within limits, impose new or changed patterns of behavior on the organization.

Although adaptation of the systems analysis methods of science and technology to management is still in the early stage, nevertheless, productive work, as indicated by Young, is being done and case studies can be cited.

Decision Theory and Processes. Decision theory is not a single theory of how to make decisions, but instead includes a group of techniques for weighing the many decision-related factors in a logical fashion. There are several major categories of decision situations which may be described as follows:

1. Those in which all the factors are known or predictable, but the complexity is too great for the human mind to grasp;

2. Those in which probability factors are present and involve risks;

3. Those where uncertainties (as distinguished from risks), such as the behavior of a competitor, must be taken into account.

As one example of decision theory, many companies, using mathematical techniques and modern computers, are simulating the possible consequences of important business actions, such as building new plants, installing new inventory systems, or raising prices. By these methods many alternatives can be investigated in a short time, and better decisions made.

Decision processes can, in essence, be reduced to a mathematical basis. A decision problem exists when several courses of action and several outcomes are possible. An individual or an organization has a set of goals or preferences; there are several courses of action possible, not all of which are equally good in terms of the goals or preferences; the decision to be made involves choosing the course of action that best satisfies the goals or objectives. The decision process consists of several steps. First, there usually will be at least some conflict of purpose among the individuals comprising an organization; these differences must be reconciled by a balancing of interests, by discussion and negotiation, or by some other social means. Second, the alternative courses of action in terms of the established objectives must be determined and the outcome of each course of action predicted and evaluated. Finally, the best solution must be chosen.

The major difficulty which commonly arises in carrying out the second and third steps, even where most of the factors are known, is the prediction or determination of the outcome of each of the various possible courses of action. If the number of outcomes is small, one might think that trials on the actual system would present the simplest approach, but actually this is seldom, if ever, true, because of the disruption which would be caused by this kind of experimentation. And, contrary to what might be expected at first glance, investigation of various possible courses of action by simple enumeration on a computer can lead to an astronomical number of calculations (millions or more) for even relatively simple cases.

The more practical approach is to develop a model for the system and then investigate the outcomes of various courses of action in terms of the model. For systems having only a small number of courses of action the model may be purely mental and the investigations made by intuitive means. But for more complex situations, quantitative or mathematical models are necessary so that the best solution can be arrived at by taking advantage of the many highly efficient mathematical procedures which are available.

The systems approach with its strong mathematical orientation provides the basis for the battery of new decision processes which have evolved in recent years. Among the more prominent of these processes are:

1. Operations research (see Sect. 13.4);

2. Mathematical programming: linear programming, nonlinear programming, dynamic programming, etc.

3. Econometric methods;

4. Scheduling methods: PERT, CPM, etc.;

5. Queueing theory and Markov processes;

6. Computer simulation;

7. Industrial dynamics (application of engineering control theory to industrial problems);

8. Engineering-economic systems methods (a blend of engineering and economics methods, particularly as applied to industries based on science and advanced technology, e.g., aerospace, communications, automation, computers).

Even this brief look at the background and status of quantitative methods in economics, industry, and management reveals a promising development which, although not comparable to the quantification of physics and some branches of engineering, nevertheless does provide a strong base for further advances in the application of quantitative and systems methods.

13.4 OPERATIONS RESEARCH AND ITS APPLICATIONS

In about 1940, under the stress of a world war, operations research as a recognized and significant activity was created in Britain and the United States by scientists who were searching for methods to effect major improvements in complex military operations. An account of this early work, together with some of its more general aspects, is given by Morse and Kimball in their pioneering book on the subject, published in unclassified form in 1951 (Ref. 59). They define operations research as: "a scientific method of providing executive departments with a quantitative basis for decisions regarding the operations under their control." Expanding on this definition, they note that operations research "is an *applied science* utilizing all known scientific techniques as tools in solving a specific problem, in this case providing a basis for decision by an executive department."

The term *operations research* seems to have been coined by military people to distinguish this research, which is concerned with operations, from the more familiar kind of research carried on in laboratories. Operations research is not a branch of engineering because, unlike engineering, it is concerned with the *use* of equipment, rather than with its design and production. (Operations research is, however, usefully applicable to certain engineering problems).

Creation in a short time of a new field of applied science in an area of human activity with little or no pre-existing pure science base undoubtedly was possible only under the conditions of a world war where the stakes were high and the potential benefits enormous, and where scientists of high ability were available. In creating the new field, operational data and, in some cases, operational experiments were used to determine facts; theories to explain the facts were elaborated; and finally, the facts and theories were used and tested in the prediction of future operations. The aim, as in any other science, was to translate empirical data into generalized theories which could be treated mathematically to obtain new results.

In this way the quantitative predictive modes, characteristic of pure science, were established. The applied science phases were implemented by formulating quantitative measures of effectiveness for the various operations and then improving or, in the ideal case, optimizing effectiveness based on these criteria to obtain maximum pay-off.

Much of the strength and value of the pioneer work on operations research was derived from the ability of the scientists to strip away details and work effectively with very approximate mathematical models which concentrate on the really important parameters. This methods of working, which is very characteristic of research physicists, led to many successful applications of operations research where factors of improvement of from 300 to 1000 percent were achieved (Ref. 59, p. 38).

In the late 1940's, industry became interested in operations research and activity began, especially in some of the larger industrial corporations where operational and organizational problems were becoming matters of major concern. Professional interests of the military and industrial operations researchers were amalgamated with the formation in 1952 of the *Operations Research Society of America*. Similar amalgamations occurred in Britain and in several other European countries.

Currently, operations research is active in the military, industrial, academic, and consulting fields. Distinctions between operations research, the more modern aspects of industrial engineering, and what has become known as management science, now seem to have become blurred to the point where the three terms are used interchangeably (Ref. 60). It is evident that activities in all these areas fit profitably into the framework of systems in that the same mathematical methods (principally linear algebra, mathematical programming, probability, statistics, and numerical calculus) are used. Reliance is placed on electronic computers for the practical handling of large problems, and scientific attitudes and methods are stressed.

As noted by Churchman et al. (Ref. 61), operations research is an example of a *systems* approach, since *system* implies an interconnected complex of functionally-related components.

Because of the magnitude and the level of complexity of many military applications, the mathematical methods required have become highly sophisticated, and comprise a specialized field of study in themselves (Ref. 62). However, at the present stage of development the majority of industrial problems tend to be amenable to less complex mathematical methods (although it is of course possible to find some examples of less complex military, and of more complex industrial, problems). In addition to military and industrial applications, which are now well established, there are now emerging significant developments in the area of urban planning (see Ex. 3, p. 454).

Operations research problems are commonly concerned with such areas as allocations, inventory, distribution and transportation, queueing, sequencing and priority assignments, market research, and competitive strategies.

In the context of this book, a few relatively simple examples will suffice to demonstrate the general applicability of systems methods in operations research. These examples involve elements of linear programming which is described briefly in Chap. 2. For detailed treatments of the many complex problems which arise in operations research and which are amenable to the systems approach, the reader is referred to the extensive literature now available on this subject, a cross section of which is given in the appended bibliography (Refs. 59 through 67).

Example I.   Military Problem*

The strategic bomber command receives instructions to interrupt the enemy's tank production. The enemy has four key plants located in separate cities, and destruction of any one plant will effectively halt the production of tanks. There is an acute shortage of fuel, which limits the supply to 48,000 gallons for this particular mission. Any bomber sent to any particular city must have at least enough fuel for the round trip plus a reserve of 100 gallons.

The number of bombers available to the commander and their descriptions are listed in the following table.

Bomber type	Description	Miles per gallon	Number available
1	Heavy	2	48
2	Medium	3	32

---

* After Sasieni, Yaspan, and Friedman, Ref. 63, p. 242.

Information about the location of the plants and their vulnerability to attack by a medium bomber and a heavy bomber is given in the following table.

Plant	Distance from base, miles	Probability of destruction by	
		a heavy bomber	a medium bomber
1	450	0.10	0.08
2	480	0.20	0.16
3	540	0.15	0.12
4	600	0.25	0.20

How many of each type of bomber should be dispatched, and how should they be allocated among the four targets, in order to maximize the probability of success? (Assume that no damage is inflicted on a plant by a bomber that fails to destroy it.)

SOLUTION

Let x_{ij} be the number of bombers of type i sent to city j:

$$i = 1, 2 \qquad j = 1, 2, 3, 4$$

We wish to maximize the possibility of destroying at least one plant, and this is equivalent to minimizing the probability of not destroying any plant. We will use Q to denote this probability. So

$$Q = (1 - 0.10)^{x_{11}}(1 - 0.20)^{x_{12}}(1 - 0.15)^{x_{13}}(1 - 0.25)^{x_{14}}(1 - 0.08)^{x_{21}}$$

$$\times (1 - 0.16)^{x_{22}}(1 - 0.12)^{x_{23}}(1 - 0.20)^{x_{24}}$$

We now have to impose the restrictions on the x_{ij} due to availability of fuel and aircraft.

Fuel:

$$\frac{2 \times 450}{2} x_{11} + \frac{2 \times 480}{2} x_{12} + \frac{2 \times 540}{2} x_{13} + \frac{2 \times 600}{2} x_{14}$$

$$+ \frac{2 \times 450}{3} x_{21} + \frac{2 \times 480}{2} x_{22} + \frac{2 \times 540}{2} x_{23} + \frac{2 \times 600}{3} x_{24}$$

$$+ 100(x_{11} + x_{12} + x_{13} + x_{21} + x_{22} + x_{23} + x_{24}) \leq 48,000$$

Thus

$$550x_{11} + 580x_{12} + 640x_{13} + 700x_{14} + 400x_{21}$$

$$+ 420x_{22} + 460x_{23} + 500x_{21} \leq 48,000 \qquad \textbf{(1)}$$

Aircraft:

$$x_{11} + x_{12} + x_{13} + x_{14} \le 32 \tag{2}$$

$$x_{21} + x_{22} + x_{23} + x_{24} \le 48 \tag{3}$$

Now, although the restrictions on the x_{ij} are linear, the function Q is not linear. However, minimizing Q is equivalent to minimizing log Q, and log Q is linear in the x_{ij}. Moreover, minimizing log Q is equivalent to maximizing $-\log Q = \log (1/Q)$. Any convenient base of logarithms will do, and we will use a base 10, as $\log_{10}$ is readily available in tables. Thus, we wish to maximize

$$\log \frac{1}{Q} = 0.0457x_{11} + 0.0969x_{12} + 0.07041x_{13} + 0.12483x_{14}$$

$$+ 0.03623x_{21} + 0.06558x_{22} + 0.05538x_{23} + 0.09691x_{24}$$

subject to the restrictions (1) through (3).

Example 2.   Production Scheduling Problem*

Chemical Products, Inc., has orders for 16,000 gallons of Product 1 and 8000 gallons of Product 2. These two products will be made during the next week in a department that has the equipment shown in Figure 13.2. Product 1 can be made by Process A or B and Product 2 can be made by Process C or D. We assume negligible hang-over from one lot to another and also assume that the processing of one product through one machine after another does not result in the second being contaminated. Thus, there is no reason to avoid lot splitting. Machine T is such that a very expensive procedure is required if it is not kept active at all times. We will require that the production program keep it fully loaded or else it will not be used at all (the latter being impossible this week, as a study of the processes will reveal). The plant operates around the clock, so there will be 168 hours available time on each machine. With the data given as follows, determine the optimum production schedule in terms of

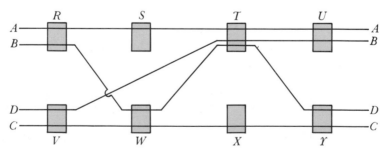

FIGURE 13.2   Equipment diagram for production scheduling problem.

---

* After Llewellyn, Ref. 64, p. 360.

minimum cost

Process	R	S	T	U	V	W	X	Y
A	100	200	150	400	—	—	—	—
B	120	—	140	—	—	80	—	175
C	—	—	—	—	200	350	140	280
D	—	—	125	250	185	—	—	—
Cost per hour	$4.50	$8.50	$6.90	$10.40	$6.70	$8.50	$5.25	$8.10

SOLUTION. This problem could, presumably, be set up in a variety of ways, but since there are only four basic elements to choose from—the four production processes—it will be possible to set up the problem with only four structural variables.

The first two constraints specify the amounts of the two products to be made. Product 1 can be made by either of processes A and B; hence, the first constraint is

$$x_1 + x_2 = 16,000$$

Similarly, the second constraint is found to be

$$x_3 + x_4 = 8000$$

These can be conveniently scaled to the following:

$$x_1 + x_2 = 16$$
$$x_3 + x_4 = 8$$

The second set of constraints arises from the limitations on the capacities of the machines. The capacity of each of the processes is limited by the least efficient step in the process. The four constraints are

$$x_1 \leq 168,000$$
$$x_2 \leq 13,440$$
$$x_3 \leq 23,500$$
$$x_4 \leq 21,000$$

Comparison with the first set of constraints shows that all but the second are redundant. The remaining one, scaled as before, is

$$x_2 \leq 13.44$$

The third set of constraints arises from the condition that, while the processes may follow each other in the various machines, they may not overlap or process more than one batch at a time. It cannot be stated, at present, whether it will be possible to sequence the processes so that the machines will be operating at capacity whenever they are running.

Consequently, it will do for the present to assume that they will. The eight constraints are the following:

$$1.2x_1 + x_2 \leq 20{,}800$$
$$x_1 \leq 33{,}600$$
$$x_1 + 1.072x_2 + 1.2x_4 \leq 25{,}200$$
$$x_1 + 1.6x_4 \leq 67{,}200$$
$$x_3 + 1.082x_4 \leq 33{,}600$$
$$4.375x_2 + x_3 \leq 58{,}800$$
$$x_3 \leq 23{,}500$$
$$1.6x_2 + x_3 \leq 47{,}000$$

The first two of these, and the fifth and seventh, are redundant with respect to the first set of constraints. The fourth is redundant with respect to the one immediately preceding it. Hence, there remain the following three.

$$x_1 + 1.072x_2 + 1.2x_4 \leq 25.2$$
$$4.375x_2 + x_3 \leq 58.8$$
$$1.6x_2 + x_3 \leq 47$$

It is to be expected that the third set of constraints contains the second set, since it contains information about all possible processes going through a given machine, instead of just one. Since this is, in fact, the case, the constraint

$$x_2 \leq 13.44$$

may be eliminated from the problem, being redundant with respect to the second of the remaining constraints in the third set.

The last constraint arises from the fact that machine T is to be operated at capacity at all times. This is expressed in

$$x_1 + 1.072x_2 + 1.2x_4 = 25.2$$

The entire set of constraints is therefore the following:

$$x_1 + x_2 = 16$$
$$x_3 + x_4 = 8$$
$$4.375x_2 + x_3 \leq 58.8$$
$$x_1 + 1.072x_2 + 1.2x_4 \leq 25.2$$

Furthermore, it can be seen that the fourth of these constraints is redundant with respect to the first two, since the first two give, in part

$$x_2 + x_3 \leq 24$$

Finally, the costs of operation of the various machines arc summed for each process to give the objective function

$$z = 159.5x_1 + 239x_2 + 124.2x_3 + 133x_4$$

To find the program of minimum cost, the initial matrix is constructed as follows:

$$\begin{bmatrix} 25.2 & 1 & 1.072 & 0 & 1.2 & 0 \\ 16 & 1 & 1 & 0 & 0 & 0 \\ 8 & 0 & 0 & 1 & 1 & 0 \\ 58.8 & 0 & 4.375 & 1 & 0 & 1 \end{bmatrix}$$

This matrix, since the necessary artificial variables have not been added, does not represent a basic feasible solution. It is possible, however, to eliminate the need for artificial variables by employing a linear transformation. In this case, the transformation is made by inspection. Typically, however, the transformation is made in the first phase of a two-phase simplex method, using a modified objective function to drive out the artificial variables. The initial basic feasible solution is displayed below.

$$\begin{array}{ccccc} -159.5 & -239 & -124.2 & -133 \\ \end{array}$$

$$\begin{array}{c} -133 \\ -159.5 \\ -124.2 \\ 58.5 \end{array} \begin{bmatrix} 7.66 & 0 & 0.06 & 0 & 1 & 0 \\ 16 & 1 & 1 & 0 & 0 & 0 \\ 0.34 & 0 & 0.06 & 1 & 0 & 0 \\ 58.5 & 0 & 4.44 & 0 & 0 & 1 \end{bmatrix}$$

It is apparent at this point that this matrix represents the optimum solution, provided that the program's elements can be sequenced in such a way that no conflict between the programmed processes arises.

Provided that the program can be properly sequenced, it will be the following:

A	16,000 gal
B	none
C	340 gal
D	7660 gal

The table of data may now be reconstructed to give the following elapsed times for the program

	R	S	T	U	V	W	X	Y
A	160	80	106.7	40	—	—	—	—
C	—	—	—	—	1.7	0.971	2.43	1.214
D	—	—	61.3	30.6	41.4	—	—	—

For each process, the following sequence of machines is used:

$$A \quad R—S—T—U$$
$$B \quad V—W—X—Y$$
$$D \quad V—T—U$$

It is not likely that process C will create a problem, since it shares one machine with process D only, with plenty of time to spare. The major conflict is likely to occur at machine T and between processes A and B. In fact, if process A is run through machine T prior to process D, it will be impossible either to fully utilize the machine or to complete process D. On the other hand, if process D is begun first on machine V, followed by process C, and run directly through machine T, it will be possible to clear the machine in 61.3 hours. Process A will clear machine S with 8 hours to spare. Furthermore, through proper staging, it will be possible to operate all the machines at full capacity. Therefore, the costs per gallon for each process will be as stated in the objective function. The cost per thousand gallons of product 2 will be \$132.58, and for Product 1 it will be \$159.50. The total cost will be \$3610.

Example 3.   City Planning.*

City planning tends to be more complex than industrial planning not only because of the many variables involved but, more importantly, because the decision-making and goal-setting functions are divided among many people whose objectives are diverse and often conflicting.

For many purposes, a city can be viewed as an operating system characterized by land, buildings, people, jobs, amenities, costs, and revenues. Although analysis of an overall urban system, even a simplified one, does not seem practical at present, it is still possible to treat in a useful manner certain subsystems connected with cities. In the present example, urban renewal is chosen as a subsystem and a mathematical model developed for the purpose of evaluating the potential impact of renewal decisions. It is to be noted that the systems approach to urban planning differs markedly from and is vastly superior to earlier treatments based on the master plan concept because the latter are static in nature and therefore become obsolete in a short time. The systems approach on the other hand contains built-in provisions for taking account of time-dependent factors.

The problem posed (as stated by Herrmann) is to develop a quantitative simulation model which would be useful in formulating housing renewal policies and programs, and which would help answer such vital questions as the effects of zoning changes on the pattern of housing

---

* After Herrmann, Ref. 65, who describes the research leading to the first "workable simulation model . . . built for the purpose of forward planning in a city" (San Francisco).

construction, and the effect of rigidly enforced building codes on housing quality vis-a-vis increased costs.

SIMULATION MODEL AND RESULTS.   Following is a summary of the steps used in developing a model, with a discussion of its application and some of the results obtained.

Building the Model: This set of tasks was divided into five steps:

(1) The programmers grouped together all those decision makers in the San Francisco population who behaved in a similar manner. They were able, by using 1960 census data and special studies of their own, to divide the population into 114 such groups, each group having, generally, the same income, family size, race, age, and housing preference.

(2) The groups were classified as living in 106 neighborhoods, defined, again, on the basis of significant similarities.

(3) All dwelling units were then grouped into 27 categories, each of which existed in a variety of physical conditions, and then were allocated among the neighborhoods.

(4) It was assumed that investors would be willing to build new units or to rehabilitate existing ones whenever it became profitable for them to do so.

(5) The "rule" of profitability just described was tied to the variables of yield, price, and the cost of the construction or rehabilitation.

Unit of Measurement: The fundamental unit of housing used in the computer was the dwelling unit. In order to keep inventories of dwelling units commensurable among different housing types, and to keep the inventory of space in the city within the memory capacity of a large computer (in this case an IBM 7090), the programmers invented, for purposes of the model, a land unit they called a "fract."

In the Community Renewal Program (CRP) model a fract was considered equal to two acres. Thus, there were 4980 fracts of residential land in San Francisco, including vacant land available for residential development.

Each fract was located in one of the 106 neighborhoods; the total number of fracts in each of them did not change throughout a run. Each fract was homogeneous; it could not contain mixed land uses. Although it had to lie within the boundaries of a specific neighborhood, it was not necessarily a contiguous group of parcels. Instead, it was a grouping of parcels in a neighborhood which had common characteristics of housing type and condition. Transitions in use of space and housing, such as new construction, conversions and mergers, or changes in the condition of the stock, occurred a complete fract at a time.

Six binary "flags" were used. Each of these flags had a different meaning and could be turned "on" or "off" for each fract. For example, one flag assumed that, because two "projects" could not be undertaken on the same property within a two-year cycle period, the fract would not be available for another transition in the time period. Another flag indicated that the fract was owned by a public agency or was otherwise inaccessible to the private market. A third flag showed whether a fract was available for occupancy by residential users in spite of ownership by a public body. And so on with the other three flags.

Through similarly rigid control of all the major inputs, programmers were able to use the model to simulate the interactions and effects on residential housing of public policies, programs, and actions, the investment behavior of the private market, and the location decisions of households. All decision-making groups in the model were programmed to interact as they normally would, that is, the anticipated effect of various land-use actions on the decisions of buyers, sellers, builders, and others was made as realistic as possible. Existing space (land and buildings) could be matched, therefore, with the needs of previously determined potential users. This process was carried out with the help of the following adjustments:

If the existing supply of space proved inadequate for the needs of the various users, the computer made the necessary changes in the "space stock" to allow the need to be met with the financial resources available.

Financial feasibility was determined by comparing the rent-paying ability of the prospective user with the cost of making the change in the building and with the yield which could be anticipated from the change. When the comparison indicated profitable circumstances, the computer added the appropriate number of new units to the housing inventory.

To complete the analysis, the computer was asked to consider what the effects of the additional housing would be on such matters as shifts in rent levels, new market values, modifications in the tax base, neighborhood amenities, and a range of public actions.

Impact of Public Actions: The great strength of a model like that used in the CRP project lies in its ability to suggest the relative impact of public actions on the housing market. In order to maximize this potentiality in the San Francisco analysis, it was necessary to make two runs of the model through the computer so that comparative simulations could be made of the market (1) without public actions and (2) with public actions:

For purposes of the first simulation, or the base run, programmers assumed a continually increasing population slightly in excess of the target population presented in the CRP. They also assumed a somewhat higher income distribution. The only public actions included in the run were those already in force or approved, such as a zoning change or a renewal project.

The public-action run was identical to the base run except that it included a simulation of the type of code enforcement, zoning, clearance, and public-housing actions called for in the CRP.

The same populations were used in both runs to test the capacity of the market to carry an excess load, that is, the ability of the private sector to increase the quantity and improve the quality of San Francisco's housing stock to meet the needs of the projected population.

The two computer runs provided a great deal of significant information for analysis. Thus, the base run indicated that with the projected 16 percent increase in population between 1960 and 1978, there would be a corresponding increase in housing units of 14 percent. During the 18-year period, the private market would increase the housing stock by 21 percent (gross); the additions would, of course, be new units.

Of especially great interest was the projected improvement in substandard housing. When code enforcement was added, in the public-action run, the number of substandard units decreased by 9 percent while the total number of units was increasing. When code enforcement was not simulated, however, low-quality units increased by 3 percent.

The effect of the CRP, therefore, was to encourage the upgrading of some 5000 dwelling units from substandard to standard condition. This meant a 40 percent higher rate of investment by the private market than was the case in the absence of code enforcement.

Another important finding concerned the price or rent effects that would accompany improvement in housing quality. If the number of substandard units decreased, the price on the remaining substandard units would go up. By 1979, rents on these units would be increasing at a rate double that for higher quality dwellings. Clearly, there would be a need for expansion in the supply of standard, low-cost housing, as well as a concerted effect to increase the wage-earning capacity of the low-income segment of the population.

The pioneering interdisciplinary results achieved in this example show the great potential which exists for systems research and quantified studies of the social apparatus.

BIBLIOGRAPHY

Physical and Engineering

1. D. M. Eakin and S. P. Davis, "An Application of Matrix Optics." *Journal of Applied Physics*, p. 758, Sept. 1966.
2. F. F. Kuo, "Network Analysis by Digital Computer." *Proceedings of IEEE*, *54*, 1966.
3. L. A. Schmit, "Automated Design." *International Science and Technology*, p. 63, June 1966.
4. R. F. Harrington, "Matrix Methods for Field Problems." *Proceedings of IEEE*, *55* 136, 1967.

5. G. E. Schafer, "A Systems Concept of Electromagnetic Measurements in the U.S.A." *Proceedings of IEEE*, *55* 775, 1967.
6. T. J. Williams, *Systems Engineering for the Process Industries.* New York: McGraw-Hill, 1961.
7. E. C. Pestel and F. A. Lechie, *Matrix Methods in Elastomechanics.* New York: McGraw-Hill, 1963. (Gives the systems approach to the dynamics and statics of complex elastic structures which occur in engineering practice; stresses the use of linear algebra and computer methods in solving otherwise intractable problems.)
8. W. C. Hurty and M. F. Rubinstein, *Dynamics of Structures.* Englewood Cliffs, N.J.: Prentice-Hall, 1964. (A unified systems treatment of structural dynamics in the context of matrix and energy methods.)
9. L. A. Pipes, *Matrix Methods for Engineering.* Englewood Cliffs, N.J.: Prentice-Hall, 1963. (Gives applications to engineering problems in electric circuit theory, dynamics, elasticity, and vibrations.)
10. D. Foster, *Modern Automation.* London: Pitman, 1963.

Biological Sciences

11. T. H. Waterman and H. J. Morowitz, eds., *Theoretical and Mathematical Biology.* New York: Blaisdell, 1965.
12. N. Wiener, *Cybernetics of Control and Communication in the Animal and the Machine*, Second Edition. New York: Wiley, 1961.
13. J. von Neumann, *The Computer and the Brain.* New Haven: Yale University Press, 1958.
14. E. Schrödinger, *What is Life?* London: Cambridge University Press, 1944.
15. J. H. Milsum, *Biological Control Systems Analysis.* New York: McGraw-Hill, 1966.
16. See *Notices of American Mathematical Society*, Feb. 1967, p. 217.
17. A. J. Lotka, *Elements of Physical Biology.* Baltimore: Williams and Wilkins, 1925.
18. R. W. Jones and J. S. Gray, "System Theory and Physiological Processes." *Science*, *140* 461, 1963.
19. R. S. Ledley and L. B. Lusted, "Medical Diagnosis and Modern Decision Making." Proceedings of Symposia in Applied Mathematics, *14* 117, 1962 (American Mathematical Society).
20. Bulletin of Mathematical Biophysics.
21. "Mathematical Theories of Biological Phenomena." *Annals of the New York Academy of Sciences*, *96* Art. 4, 895, March 2, 1962.
22. R. Bernhard, "Interdisciplinary Discord—A Debate on the Physiology of Color Vision." *Scientific Research*, May, 1967. (A report on a controversial debate on "How to apply systems engineering properly to biological systems," at the IEEE New York Convention, March 1967.)
23. R. N. Linebarger and R. D. Brennan "Digital Simulation for Bio-Medical System Studies." *Proceedings of the 18th Annual Conference on Engineering in Medicine and Biology*, 1965.
24. T. H. Waterman, "Revolution for Biology." *American Scientist*, *50* 548, 1962. (A general article, appraising the status and progress of biology from a descriptive to a quantitative science.)
25. K. E. F. Watt, "Problems in Population Input-Output Dynamics." *General Systems*, *IX* 163, 1964.
26. V. W. Bolie, "Trends and Developments in Bionics and Bioengineering." Presented to First Annual Rocky Mountain Bioengineering Symposium, May 1964, U.S. Air Force Academy, Colorado.
27. M. L. Minsky, "Artificial Intelligence." *Scientific American*, *215* 246, 1966.
28. L. J. Fogel, *Biotechnology: Concepts and Applications.* Englewood Cliffs, N.J.: Prentice-Hall, 1963. (Text on man-machine relation; uses mathematical tools to describe the functioning of the human operator.)
29. *IEEE Transactions on Systems Science and Cybernetics.* Published quarterly by the Institute of Electrical and Electronics Engineers, New York.
30. R. W. Jones and J. S. Gray, "System Theory and Physiological Processes." *Science*, *140* 461, 1963.
31. N. T. J. Bailey, *The Mathematical Approach to Biology and Medicine.* New York: Wiley, 1967. (States the case for the importance of mathematics in biology. Even more,

this book investigates and advocates the logical use of computers and operational research methods, i.e., systems methods, in biology.)

32. F. M. Patterson and D. M. Levy, "An Example of State-Variable Analysis from Physiology and Medicine." *IEEE Transactions on Education*, *E-10* 100, 1967.

33. K. E. F. Watt, *Systems Analysis in Ecology*. New York: Academic Press, 1967.

Economics, Industry, and Management

34. E. C. Bursk and J. F. Chapman, *New Decision-Making Tools for Managers*. Cambridge, Mass.: Harvard University Press, 1963.

35. S. Karlin, *Mathematical Methods and Theory in Games, Programming, and Economics*. Reading, Mass.: Addison-Wesley, 1959.

36. B. V. Dean et al., *Mathematics for Modern Management*. New York: Wiley, 1963.

37. M. E. Stern, *Mathematics for Management*. Englewood Cliffs, N.J.: Prentice-Hall, 1963.

38. J. W. Forrester, *Industrial Dynamics*. New York: Wiley, 1961.

39. C. McMillan and R. F. Gonzalez, *Systems Analysis, A Computer Approach to Decision Models*. Homewood, Ill.: Richard C. Irwin, 1965.

40. H. E. Koenig and T. J. Manetsch, "From Physical to Socio-Economic Systems." *Engineering Education*, *57* 704, 1967.

41. W. K. Linvill, "The Engineering-Economic Systems Program at Stanford." *IEEE Spectrum* p. 99, April 1966.

42. G. E. Burress, "Input-Output Tables: A Tool for Management." *Data Processor*, *IX* 9, 1966.

43. "Systems Education in the United States." Study conducted by the Education Committee of Systems and Procedures Association, Cleveland, Ohio: 1966.

44. R. A. Johnson et al., *The Theory and Management of Systems*. New York: McGraw-Hill, 1963.

45. S. L. Optner, *Systems Analysis for Business Management*. Englewood Cliffs, N.J.: Prentice-Hall, 1960.

46. J. W. Forrester, "Common Foundations Underlying Engineering and Management." *IEEE Spectrum* p. 66, Sept. 1964.

47. *Systems and Procedures Journal*, published bimonthly for the Systems and Procedures Association, Cleveland, Ohio.

48. H. Thiel et al., *Operations Research and Quantitative Economics*. New York: McGraw-Hill, 1965.

49. S. Young, *Management: A Systems Analysis*. Glenview, Illinois: Scott, Foresman, 1966.

50. A. Charnes and W. Cooper, *Management Models and Industrial Applications of Linear Programming*, Vols. I and II. New York: Wiley, 1961.

51. *Management Science*, published quarterly by The Institute of Management Sciences, Baltimore, Md.

52. J. G. Kemeny and J. L. Snell, *Mathematical Models in the Social Sciences*. Boston: Ginn, 1962.

53. M. Greenberger, "The Uses of Computers in Organizations." *Scientific American*, *215* 192, 1966.

54. A. Tustin, *The Mechanism of Economic Systems*. London: Heinemann, 1963.

55. F. A. Koomanoff and J. A. Bontadelli, "Computer Simulation of Railroad Freight Transport Operations." *Battelle Technical Review*, *16* 15, 1967.

56. *PERT Guide for Management Use*. (Prepared by U.S. Government PERT Coordinating Group June 1963. Available from Superintendent of Documents, U.S. Government Printing Office, Washington, D.C.)

57. K. J. Arrow et al., *Studies in the Mathematical Theory of Inventory and Production*. Stanford, Calif.: Stanford University Press, 1958.

58. Van Court Hare, Jr., *Systems Analysis: A Diagnostic Approach*. New York: Harcourt, Brace, and World, 1967.

Operations Research

59. P. M. Morse and G. E. Kimball, *Methods of Operations Research*. New York: Wiley, 1951.

60. N. N. Barish, "Operations Research and Industrial Engineering: The Applied Science and Its Engineering." *Operations Research, 11* 387, 1963.
61. C. W. Churchman et al., *Introduction to Operations Research*. New York: Wiley, 1957.
62. *Operations Research*, Journal of the Operations Research Society of America, published bi-monthly. (This journal contains numerous papers relating to military applications, as well as to more general aspects of the field).
63. M. Sasieni et al., *Operations Research—Methods and Problems*. New York: Wiley, 1959.
64. R. W. Llewellyn, *Linear Programming*. New York: Holt, Rinehart and Winston, 1964.
65. C. C. Herrmann, "Systems Approach to City Planning." *Harvard Business Review, 44* 71, 1966.
66. *International Abstracts in Operations Research*, Published bi-monthly for the International Federation of Operational Research Societies by the Operations Research Society of America.
67. H. Thiel et al., *Operations Research and Quantitative Economics*. New York: McGraw-Hill, 1965.

General

68. J. P. Eberhard, "Technology for the City." *International Science and Technology*, p. 18, September 1966.
69. J. M. Buchnan and G. Tullock, *The Calculus of Consent*. Ann Arbor: University of Michigan Press, 1962.
70. D. N. Chorafas, *Systems and Simulations*. New York: Academic Press, 1965.
71. J. A. Logan, "The Potentialities of Systems Approaches To Environmental Health." Presented at National Congress on Environmental Health Management, New York City, April 26, 1967.
72. H. M. Adelman, "Systems Analysis and Planning for Public Health Care in New York City." Presented at National Congress on Environmental Health Management, New York City, April 26, 1967.
73. W. Buckley, *Sociology and Modern Systems Theory*. Englewood Cliffs, New Jersey: Prentice-Hall, 1967.
74. F. H. Leimkuhler, "Systems Analysis in University Libraries." Presented at American Society for Engineering Education Annual Meeting, June, 1965.
75. I. G. Wilson and M. E. Wilson, *Information, Computers and Systems Design*. New York: Wiley, 1965. (An overview of the system engineering process; covers information processing on computers in relation to design.)
76. R. D. Luce, ed., *Developments in Mathematical Psychology*. The Free Press of Glencoe, Illinois, 1960. (A study of behavioral models in learning theory, information theory, and manual tracking. Essentially a research review of progress in these fields. Also of interest in showing systems trends.)
77. H. Soloman, ed., *Mathematical Thinking in the Measurement of Behavior*. The Free Press of Glencoe, Illinois, 1960. (Covers small group processes, utility theory, and mental ability. Complements Ref. 77.)
78. C. Berge, *The Theory of Graphs and its Applications*. London: Methuen, 1962. (Translation of the 1958 French edition. A fundamental treatment of graph theory which preserves mathematical integrity while covering applications to transport networks, game theory, electric networks, optimal route problems, and the like.)
79. D. Meister and G. F. Rabideau, *Human Factors in System Development*. New York: Wiley, 1965. (A new technology which represents a fusion of the elements of one of the newest sciences [psychology] and those of one of the oldest [engineering], a field less than 20 years old.)
80. J. S. Coleman, *Introduction to Mathematical Sociology*. London: The Free Press of Glencoe, Collier–Macmillan Limited, 1964. (This fundamental book is an initial effort to construct a mathematical framework adequate for the behavioral sciences; it is also a plea for the quantitative and systems approach to social and psychological -processes.)
81. R. Boguslaw, *The New Utopians: A Study of System Design and Change*. Englewood Cliffs, N.J.: Prentice-Hall, 1965. (Essentially a social commentary on the "systems world," i.e., the world being created by scientists and engineers whose touchstone is efficiency, reliability, and efficiency rather than the individual.)
82. P. Suppes, "The Uses of Computers in Education." *Scientific American, 215* 206, 1966.

PROBLEMS

Chapter 1

1. Given $\mathbf{x} = [1, 0, 0]$, $\mathbf{y} = [-\frac{1}{2}, 0, \frac{2}{3}]$, $\mathbf{z} = [0, \frac{1}{4}, 2]$

Find (i) $\mathbf{x} + 2\mathbf{y} + \mathbf{z}$

    (ii) $3(\mathbf{x} + \mathbf{y}) - 2(\mathbf{y} - \mathbf{z})$

    (iii) Assuming that unit masses are at the end points of $\mathbf{x}$, $\mathbf{y}$, $\mathbf{z}$, find the center of mass.

    (iv) Solve the vector equation $6\mathbf{y} + 5\mathbf{w} = \mathbf{x}$.

2. Given $\mathbf{x} = [1, 1 + i, 0]$, $\mathbf{y} = [1 + 2i, 0, i]$

Find (i) $2\mathbf{x} + \mathbf{y}$

    (ii) $3\mathbf{x} - 2\mathbf{y}$

3. Test the following sets of vectors for linear independence

    (i) $[1, 0, 1]$, $[0, 2, 2]$, $[3, 7, 1]$

    (ii) $[1, i, 1 + i]$, $[i, -1, 2 - i]$, $[0, 0, 3]$

4. Find the scalar products of the following pairs of vectors:

(a) $\mathbf{x} = [1, 4, 3, 6]$ $\mathbf{y} = [-2, 1, -1, 3]$

(b) $\mathbf{x} = [1 + i, 2, 1 + 2i]$ $\mathbf{y} = [1 - i, 2i, -1 + i]$

5. Find the norms and the lengths of the vectors of problem 4.

6. Find the angles between the following pairs of vectors:

(a) $\mathbf{x} = [2, 1, 3, 2]$ $\mathbf{y} = [1, 2, -2, 1]$

(b) $\mathbf{x} = [1, 2, 2, 3]$ $\mathbf{y} = [3, 1, 5, 1]$

7. Use the following data to perform the indicated numerical calculations:

$$\mathbf{x} = (2, 0, 1)$$
$$\mathbf{y} = (1, -1, 0)$$
$$\mathbf{z} = (3, 2, 1)$$
$$a = 2$$
$$b = 1$$

(i) $\mathbf{x}(\mathbf{y} \cdot \mathbf{z})$

(ii) $(\mathbf{x} \cdot \mathbf{y})\mathbf{z}$

(iii) $a\mathbf{x} + b\mathbf{y}$

    Verify the following:

$$a(\mathbf{x} + \mathbf{y}) = a\mathbf{x} + a\mathbf{y}$$
$$(a + b)\mathbf{x} = a\mathbf{x} + b\mathbf{x}$$

8. Can the following set of vectors serve as a basis in V_3?

$$\mathbf{x}_1 = [1, 1, 0], \quad \mathbf{x}_2 = [1, 0, 1], \quad \mathbf{x}_3 = [0, 1, 1]$$

461

9. Which pairs of the following vectors are orthogonal?

$$\mathbf{w} = (1, 2, 0, 1)$$
$$\mathbf{x} = (3, -1, 2, 0)$$
$$\mathbf{y} = (-2, 3, 1, -4)$$
$$\mathbf{z} = (4, 0, 2, 1)$$

10. Given the vectors

$$\mathbf{A} = (1, 0, 2, -1)$$
$$\mathbf{B} = (0, 1, 4, 2)$$
$$\mathbf{C} = (0, 5, -1, 4)$$
$$\mathbf{D} = (1, 1, -1, -1)$$

(i) Are they linearly independent?

(ii) Find an orthonormal basis for these vectors.

11. Verify the triangle inequality and Schwarz's inequality and find the angle between each of the following pairs of vectors. Using the second pair, verify whether dot multiplication in Hermitian space is commutative.

$$\mathbf{A} = (3, 1, -2, 7)$$
$$\mathbf{B} = (-1, 2, 2, -6)$$
$$\mathbf{C} = (1 + i, 2 - 3i, 4 - i)$$
$$\mathbf{D} = (-2 - i, 3 - 2i, -5 + 3i)$$

12. Express the Hermitian inner product of the complex vectors $(\mathbf{a} + \mathbf{b}i)$ and $(\mathbf{c} + \mathbf{d}i)$. Explain, in consequence, why the relation is or is not commutative and why the Hermitian norm is real. Explain the distinction between the Euclidean inner product and the Hermitian inner product, and between the Euclidean and Hermitian vector space.

13. Given

$$A = \begin{bmatrix} 1 & 2 \\ 2 & 5 \end{bmatrix} \qquad B = \begin{bmatrix} 3 & 4 \\ -2 & -1 \end{bmatrix}$$

Find AB, BA, A^{-1}

14. Find the current vector $\mathbf{i} = [i_1, i_2]$ for the resistance network in the following figure:

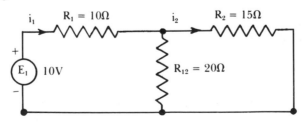

15. Compute the following product by partitioning the matrices as indicated.

$$
\left[
\begin{array}{ccccccc}
1 & 0 & 2 & -1 & -4 & 0 & 0 \\
0 & 1 & 1 & 3 & 5 & 0 & 0 \\
-2 & 3 & 0 & 0 & 0 & 5 & -1 \\
1 & 0 & 1 & 0 & 0 & -1 & 2 \\
0 & -1 & 0 & 1 & 0 & 3 & -4 \\
1 & 0 & 0 & 0 & 1 & 0 & 1
\end{array}
\right]
\left[
\begin{array}{ccccc}
3 & -4 & 0 & 0 & 0 \\
-1 & 2 & 0 & 0 & 0 \\
1 & 0 & 0 & 5 & -1 \\
0 & 1 & 0 & -3 & 2 \\
0 & 0 & 1 & -1 & -6 \\
0 & 1 & 0 & 3 & 1 \\
1 & 0 & 1 & -1 & -4
\end{array}
\right]
$$

16. Find the normal form and rank of the following matrix:

$$
\begin{bmatrix}
1 & 2 & 1 & 0 \\
2 & 0 & 0 & 2 \\
3 & 1 & 3 & 5 \\
4 & 4 & 2 & 2 \\
2 & 5 & 0 & -3
\end{bmatrix}
$$

17. Find the inverse of the following matrix:

$$
\begin{bmatrix}
1 & 2 & 3 \\
2 & 3 & 2 \\
1 & 2 & 2
\end{bmatrix}
$$

18. Find the transpose, adjoint, determinant, and inverse of the following matrix:

$$
\begin{bmatrix}
1 & 0 & 0 & 0 \\
1 & 1 & 0 & 0 \\
1 & 2 & 1 & 0 \\
1 & 3 & 3 & 1
\end{bmatrix}
$$

19. Find the eigenvalues and normalized eigenvectors of the matrix

$$
A = \begin{bmatrix}
1 & 2 \\
5 & 4
\end{bmatrix}
$$

20. Use a matrix solution to find the currents in the branches of the following network:

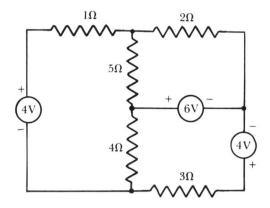

21. Find the eigenvalues and eigenvectors of the following matrix:

$$\begin{bmatrix} 3 & 2 & 2 & -4 \\ 2 & 3 & 2 & -1 \\ 1 & 1 & 2 & -1 \\ 2 & 2 & 2 & -1 \end{bmatrix}$$

22. Find a matrix P such that $P^{-1}AP = D$, where D is a diagonal matrix

$$A = \begin{bmatrix} 2 & 2 & 1 \\ 1 & 3 & 1 \\ 1 & 2 & 2 \end{bmatrix}$$

23. Find A^{-1} of problem 22, using the Cayley-Hamilton theorem.

24. The characteristic polynomial of B is

$$\varphi(\lambda) = \lambda^3 - 3\lambda^2 + 2\lambda = 0$$

Show that B^{-1} does not exist.

25. Recall: Every real symmetric matrix is orthogonally similar to a diagonal matrix whose elements are the eigenvalues of the matrix.
   Show: For an n-square real symmetric matrix, A

(a) A has n independent eigenvectors.
(b) The eigenvectors are orthogonal.
(c) The eigenvalues of A^{-1}, provided it exists, are the reciprocals of those of A, and the eigenvectors are the same as those of A.

26. Transform the quadratic form $\mathbf{x}'A\mathbf{x}$ to canonical form, using Lagrange's reduction.

$$A = \begin{bmatrix} 2 & 0 & -1 \\ 0 & 2 & 0 \\ -1 & 0 & 2 \end{bmatrix}$$

27. In the foregoing exercise, $\mathbf{x}$ is transformed into $\mathbf{y}$ by the inverse of the transformation $\mathbf{x} = P\mathbf{y}$. Find P, and show that $P'AP = D$, where D is the matrix of the canonical form.

28. Transform $\mathbf{x}'A\mathbf{x}$ into canonical form by the method of eigenvalues. Show that the matrix Q which gives the transformation

$$Q'AQ = \begin{bmatrix} \lambda_1 & 0 & 0 \\ 0 & \lambda_2 & 0 \\ 0 & 0 & \lambda_3 \end{bmatrix}$$

is orthogonal.

29. A system has a set of inputs b_1, b_2, b_3, to which the outputs are given by the transformation

$$A\mathbf{x}_i = \mathbf{b}_3'; \quad i = 1, 2, 3$$

Let $\mathbf{b}_1' = (2, -1, 0)$, $\mathbf{b}_2' = (1, 0, -1)$, $\mathbf{b}_3' = (1, 1, 1)$. Show that the inputs can be written as a matrix B, giving the matrix solution

$$A^{-1}B = X$$

Find X, where A is given in problem 22.

30. An electric network has resistance elements which we denote by R_1, $R_2, \ldots, R_5$. We can represent the currents and emf's for the set of resistors by the vectors

$$\mathbf{I} = [2, 3, 1, 4, 1] \qquad \mathbf{E} = [1, 4, 2, 1, 2]$$

What physical quantity does the scalar product $\mathbf{E} \cdot \mathbf{I}$ represent? Calculate the value of this quantity.

31. Reduce to canonical form

$$Q = \mathbf{x} \begin{bmatrix} 1 & 2 & 3 & 4 \\ 2 & 1 & 5 & 6 \\ 3 & 5 & 2 & 3 \\ 4 & 6 & 3 & 4 \end{bmatrix} \mathbf{x}'$$

32. Reduce to canonical form

$$Q = \mathbf{x} \begin{bmatrix} 0 & 1 & 1 & 1 \\ 1 & 0 & 0 & 0 \\ 1 & 0 & 0 & 0 \\ 1 & 0 & 0 & 0 \end{bmatrix} \mathbf{x}'$$

33. Reduce to canonical form

$$Q = x^2 + 4y^2 + z^2 - 4xy + 2yz + 2zx$$

34. Given the quadratic form $ax^2 + 2bxy + cy^2$. When is it positive definite?

35. Same question for $ax^2 + by^2 + cz^2 + 2dxy + 2eyz + 2fzx$.

Chapter 2

1. Study the variation of the function

$$y = \frac{3x^2 - 4}{(x - 2)^2(x + 1)}$$

In particular study its stationary points. If $-2 \leq x \leq 3$, $-4 \leq y \leq 1$, find the new maxima and minima introduced by this limitation.

2. Study the variation of the function

$$y = \frac{x^2 - 3x + 1}{x^2 - 5x + 1} + \frac{x^2 + 5x + 1}{x^2 + 3x + 1}$$

by studying each fraction alone and adding. In particular study the asymptote. Find the additional maxima and minima introduced by the condition $0 \leq x \leq 5$, $0 \leq y \leq 10$.

3. Same question for $y = x\sqrt{(x - a)(x - b)}$, $0 < a < b$

$$-1 \leq x \leq +1$$

4. Same question for $y = \sqrt[5]{x^4(x - 1)}$; $-1 \leq x \leq +1$

5. Same question for $y = \dfrac{\sin 2x}{1 + \sin x}$; $-\pi \leq y \leq \pi$

6. Given an idealized Pelton wheel water turbine for which it is assumed that the buckets completely reverse the relative velocity of the water stream impinging on them, and that all energy losses are negligible, find:

(a) The values of the bucket velocity v for maximum and minimum power outputs. What are the absolute maximum and absolute minimum values of P?
(b) The absolute maximum of the reaction force of the water stream on the buckets.

Note: Let A = cross-sectional area of water stream

ρ = density of water

F = force on bucket due to change of momentum of water stream

$\quad = \rho A v_0 [(v_0 - v) - (v - v_0)]$

P = power = Fv

7. Find the area of the largest rectangle which can be inscribed in the ellipse

$$(x^2/a^2) + (y^2/b^2) = 1$$

Use method of the Lagrangian multiplier.

8. A dealer buys at most 6000 units of a certain commodity per year from 3 factories A, B, C. By contract he must buy at least 1000 units from A and B together and at least 1000 from C. If the cost is \$100 per unit from A, \$200 per unit from B, and \$300 per unit from C, find the maximum and minimum amounts of money possible the dealer pays to A, B, and C together. Use a graphical 3 dimensional method and call x, y, z the number of units bought from A, B, and C.

9. Find the stationary values and extrema of

$$f(x, y) = x^2 + 2y^2 - x$$

10. Find the extrema points of the function $f(x, y) = x^2 + y^2 + 2$ subject to the constraint $F(x, y) = x - y + 1 = 0$. Use the Lagrange multiplier method.

11. Find the maxima and minima of the energy function for the total electric charge q_0 stored on two capacitors C_1 and C_2. (Use the Lagrange multiplier method.)

12. Given $f(x, y, z) = x^3 + 2y^3 + 3x^3 + 5x^2 - 6y^2 + 7z^2$, find all maxima and minima for $x \geq 0$, $y \geq 0$, $z \geq 0$. Study the behavior on the boundaries defined by $0 \leq x \leq 5$; $0 < y \leq 6$; $0 \leq z \leq 1$, and thus for the given region find the greatest and smallest value of f for the given region.

13. Find the absolute extrema values of

$$f(x, y) = \sqrt{(6 - x)(6 - y)(x + y - 6)}$$

on the triangle bounded by $x = 6$, $y = 6$, $x + y = 6$.

14. (a) Show that the maximum of $f(x, y, z) = x^2 y^2 z^2$ on the sphere $x^2 + y^2 + z^2 = R^2$ is $R^6/27$.

(b) Show that the minimum of $g(x, y, z) = x^2 + y^2 + z^2$ on the surface $x^2 y^2 z^2 = R^6/27$ is R^2.

15. Find the extrema of the function $f(x, y) = y^4 + x^2 - 2xy$.

16. Find the maximum and minimum of

$$f(x, y) = 2x^2 - 3y^2 + 2 \qquad \text{for} \qquad x^2 + y^2 \leq 1$$

17. Find the maxima and minima of $f(x, y) = 4x^2 - 12xy + 9y^2$.

18. According to Einstein's theory of specific heats the total heat energy of a crystalline substance is

$$Q = 3Nk\theta/(e^{\theta/T} - 1)$$

where N is the number of atoms k, θ are constants. Find the specific heat dQ/dT of the substance and its maximum for $T \leq \alpha\theta$, where α is a given positive constant.

19. Find the maximum of $x^2 + y^2 + z^2$ if $x + y + z \leq 4$, $x + 4y + z < 8$, $4x + y + z \leq 8$, $x + y + 4z \leq 8$.

20. Given the two points $A(1, 0)$, $B(-1, 0)$, find the P such that $\overline{PA^2} + \overline{PB^2}$ is either maximum or minimum, with the condition that P is on the circle $x^2 + y^2 - 2x - 2y + 1 = 0$.

Chapter 3

1. Find the equation of the curve which passes through the points $(0, 0)$ and $(2, 3)$ and which makes the integral I stationary, where

$$I = \int_0^2 (x + y + y'^2)\, dx$$

2. Find the arc $y = y(x)$ that gives the shortest distance between the points $(2, 1)$ and $(5, 4)$.

3. A cable of length L and of uniform weight w per unit length hangs between two points (a, c) and (b, d) in a vertical plane. Find the equation of the curve it takes if the potential energy of the cable

$$V = \int_a^b yw\, ds$$

is minimal.

4. Minimize the integral $\int_{t_0}^{t_1} (m^2 x'^2 + n^2 x^2)\, dt$

5. Given the sphere $x = R \cos \varphi \cos \theta$, $y = R \cos \varphi \sin \theta$, $z = R \sin \varphi$. Consider the curves on the surface of the sphere defined by $\theta = \theta(\varphi)$ and show that the length of the curve is given by

$$I = R \int_{\varphi_0}^{\varphi_1} \left[1 + \cos^2 \varphi \left(\frac{d\theta}{d\varphi} \right)^2 \right]^{1/2} d\varphi$$

By minimizing I show that the shortest distance of two points on a sphere is a great circle.

6. Show that the shortest distance between two points on a circular cylinder is a helix.

7. Given the two parallel planes $ax + by + dz + d_1 = 0$ and

$$ax + by + cz + d_2 = 0$$

and $P_1(x_1, y_1, z_1)$ a point of the first plane. Find the shortest distance of P_1 to the second plane.

8. The cross section of a channel is an isoceles trapezium (cf. Figure). The frictional effect of liquid flowing through the channel is proportional to the wet part of the perimeter $(2a + 2t)$. What form should the cross section have if

for constant area the wet perimeter is to be minimum? (Ans. $\frac{1}{2}$ of regular hexagon.)

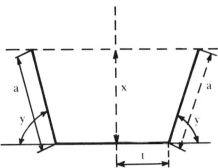

9. Find the triangle of largest area that can be inscribed into a given circle of radius R.

10. Currents I_1, I_2, I_3 are fed to load resistors R_1, R_2, R_3 respectively. Find the values of the currents under the assumption that the power P is minimum and that $I_1 + I_2 + I_3 = I_0$ (constant).

11. Find a curve of length L between two points A and B in a vertical plane whose center of gravity is the lowest possible.

12. Find a curve that passes through A and B in the (x, y)-plane and that by rotation about OX generates a surface S whose surface area A is constant and such the volume enclosed is the smallest. (The problem leads to an elliptic integral.)

13. By finding the maximum for the integral

$$I = \int_0^\pi y^2 \, dx$$

with the condition

$$\int_0^\pi y'^2 \, dx = 1,$$

show that

$$\int_0^\pi (y^2 - y'^2) \, dx \leq 0 \qquad \text{for all } y \text{ such that}$$

$y(0) = u(\pi) = 0$.

14. Find the extremal conditions for

$$I = \int_{(r)} P(x, y) \, dx + Q(x, y) \, dy = \int_a^b [P + Qy'] \, dx$$

15. By studying the Euler-Lagrange equation of the two problems show that between two points
(i) the problem of encompassing the largest area for a given length of curve (problem of Dido)

(ii) the problem of finding a curve encompassing a given area for the smallest possible length are equivalent.

Find the Euler-Lagrange equations of the following problems:

16. $\int_0^1 y'^2 \, dx = \text{minimum}.$

$$\int_0^1 y^2 \, dx = 1, \quad y(0) = 0, \quad y(1) = 0$$

17. $\int_0^1 xy'^2 \, dx = \text{minimum}$

$$\int_0^1 xy^2 \, dx = 1, |y(0)| < \infty, \quad y(1) = 0$$

18. $\int_{-a}^{+a} (1 - x^2)y'^2 \, dx = \text{minimum}$

$$\int_{-a}^{+a} y^2 \, dx = b, |y(-a)| < \infty, \quad |y(a)| < \infty$$

19. Prove that the Euler-Lagrange equation for the following integral

$$I = \int_a^b f(x, y, y', y'') \, dx$$

is $\dfrac{\partial f}{\partial y} - \dfrac{d}{dx}\left(\dfrac{\partial f}{\partial y'} \right) + \dfrac{d^2}{dx^2}\left(\dfrac{\partial f}{\partial y''} \right) = 0$

20. Generalize the result of exercise 19 for the integral

$$I = \int_a^b f(x, y, y', y'', y''', \dots, y^{(m)}) \, dx$$

Chapter 4

Write in vector form the following differential equations, or systems of differential equations. Specify in every case the components of the vectors introduced.

1. $D^4x + 3(D^3x)x + 5(D^2x)(Dx)x + 7(Dx)x^2 + t(t - 1)x^4 = e^t$

2. $Dx + D^2y + D^3z = e^t$
   $D^3x + Dy + D^2z = e^t \cos t$
   $D^2x + D^3y + Dz = e^{2t} \sin t + t$

3. $Dx = x + 2y + 3z$
   $Dy = 2x + 3y + 4z$
   $Dz = 3x + 4y + 5z$

4. $t\, D^3x + t^2\, D^2x + t^3\, Dx + t^4x = t^5$

5. Prove that if X is a matrix with variable elements such that $X(s + t) = X(s)X(t)$, then $dX/ds = MX$, where M is a constant matrix, provided all elements of X are continuous.

6. Given the differential equation $(Dx)^3 - t^3x^3 = 0$, separate the three branches and write it in vector form.

7. Show that the linear autonomous system of problem 3 reduces for x to the form

$$D^3x + \alpha\, D^2x + \beta\, Dx + \gamma x = 0$$

8. Given the system of differential equations with constant coefficients

$$\mathbf{Dx} = A\mathbf{x}$$

$$\mathbf{x} = [x, y]$$

$$A = \begin{bmatrix} p & q \\ r & s \end{bmatrix}$$

Discuss the matrix of its solutions in functions of A.
    In the discussion of linear autonomous systems of Section 4.4, show that

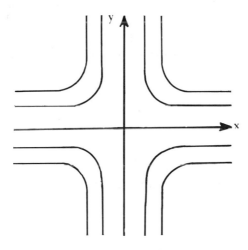

9. If $\lambda_1 < 0 < \lambda_2$ the origin is a saddle point (see sketch).

10. If $\lambda_1 = \lambda_2$ is real, A is of rank zero, then the origin is a degenerate nodal point (see sketch) (distinguish between $\lambda > 0$ and $\lambda < 0$).

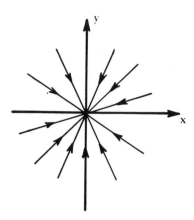

11. If $\lambda_1 = \lambda_2$ is real, A is of rank 1, then the origin is a nodal point.

12. If the λ's are purely imaginary, we have a vertex point.

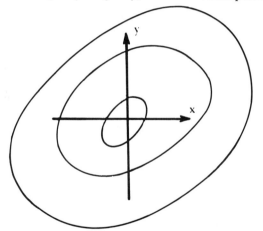

13. If $\lambda_1 = \bar{\lambda}_2$, then we have spiral points.

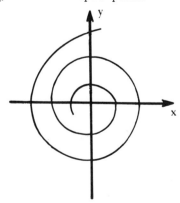

Give a phase plane interpretation for the following differential equations:

14. $D^2x + x + \dfrac{1}{x - 1} = 0$

15. $D^2x + Dx/|Dx| = 0$

16. $D^2x + Dx\,|Dx| + \sin x = 0$

17. Consider Eq. (4.26) general solution of (4.13). Show that if $\lambda_j \to \lambda_k$; i.e., $\lambda_k = \lambda_j + \Delta\lambda$ and $\Delta\lambda \to 0$, then $te^{\lambda_j t}$ is a particular solution of the differential equation.

18. Generalize the result of problem 17 and obtain the solution for a p times repeated λ.

Chapter 6

1. For the transformation from Cartesian to spherical coordinates, determine the following quantities:
(a) the transformation matrix
(b) the transformation vectors
(c) the line element ds
(d) the Jacobian and hence the singular points for the transformation.

2. The same problem for the transformation from Cartesian (x, y) to parabolic (f, h) coordinates.

3. Given the system shown in the sketch. Set up the expressions for the kinetic and potential energies, and from these write Lagrange's equations of motion for the system.

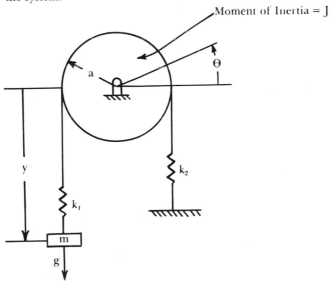

4. Solve the equations of motion for problem 3, subject to the initial conditions

$$y(0) = y_0, \quad \dot{y}(0) = \dot{y}_0, \quad \theta(0) = \theta_0, \quad \dot{\theta}(0) = \dot{\theta}_0.$$

5. Set up Lagrange's equations of motion in curvilinear orthogonal coordinates (u, v, w) for a particle of mass m subject to a force whose components are Q_u, Q_v, Q_w.

6. Apply the results of problem 5 to the case of cylindrical polar coordinates.

7. A simple pendulum of mass m_2 is suspended from a mass m_1 which can move freely on a horizontal line lying in the plane of the pendulum. Find Lagrange's equations of motion for the system.

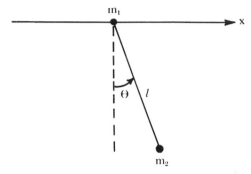

8. A train has a maximum acceleration α, maximum deceleration β, and a maximum cruising velocity v_0. Determine the minimum time required for the train to travel between two station stops a distance S_0 apart.

9. A rhombus $ABCD$ is formed of four rods of negligible mass joined together. Equal and opposite forces F are applied inward at A and C, and a flexible cable connects B and D. Find:
(a) The equation for the tension T in the cable in terms of θ and F
(b) The values of T for $F = 1000$ lbs, $\theta = 30°, 60°, 90°, 120°, 150°$.

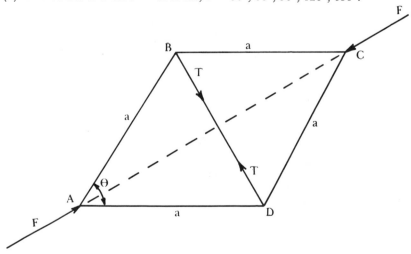

10. Study the motion of the mass m constrained to move along a smooth rigid rod which rotates at constant angular velocity ω about a fixed vertical axis as shown in the diagram.

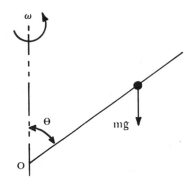

11. Prove that Lagrange's equations are not affected by the addition of the total time derivative of any function of coordinates and time to the Lagrangian.

12. Given the system which consists of a pendulum suspended from a fixed cylinder as shown in the sketch. Find the Lagrangian and Lagrange's equations of motion. How is this system related to a simple pendulum?

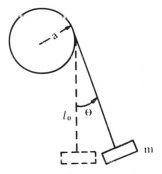

13. Derive the equations of motion for a simple pendulum by the direct use of Hamilton's principle.

14. An electron (charge $-e$) moves in a plane under the action of a fixed positive nucleus of charge $+e$. Find the equations of motion by the direct application of Hamilton's principle.

15. Given a system with a Lagrangian

$$L = \tfrac{1}{2}\dot{\mathbf{q}}'M(\mathbf{q})\dot{\mathbf{q}} - V(\mathbf{q})$$

show that the Hamiltonian is

$$H = \tfrac{1}{2}\mathbf{p}'M^{-1}\mathbf{p} - V$$

Using this result, find H for the case

$$L = \tfrac{1}{2}(\dot{q}_1^2 + \dot{q}_1\dot{q}_2 + \dot{q}_2) - V(\mathbf{q})$$

16. The Lagrangian for a particle moving in space is

$$L = \tfrac{1}{2}m(\dot{x}_1^2 + \dot{x}_2^2 + \dot{x}_3^2) - V + a_1\dot{x}_1 + a_2\dot{x}_2 + a_3\dot{x}_3$$

where V, a_1, a_2, a_3 are given functions of x_1, x_2, x_3. Show that the equations of motion have the form

$$m\ddot{x}_1 = -\frac{\partial V}{\partial x_1} + \dot{x}_2\left(\frac{\partial a_2}{\partial x_1} - \frac{\partial a_1}{\partial x_2}\right) - \dot{x}_3\left(\frac{\partial a_1}{\partial x_3} - \frac{\partial a_3}{\partial x_1}\right)$$

17. Derive the mathematical statement of d'Alembert's principle for a system of particles, i.e.,

$$\sum_{j=1}^{n} (\mathbf{F}_j - m_j\ddot{\mathbf{r}}_j) \cdot \delta\mathbf{r}_j = 0$$

starting from Hamilton's principle as a basic postulate.

18. Two particles of mass m_1 and m_2, connected by a rigid rod of negligible mass and length l, rest under gravity on the interior of a smooth spherical surface of diameter $2r > l$. Find the angle of repose θ between the bar and a horizontal plane.

19. Same as problem 18, but for three particle masses constrained at the vertices of a triangle.

20. Euler's equations of motion for a rigid body rotating about a fixed point are

$$J_1\dot{\omega}_1 = (J_2 - J_3)\omega_2\omega_3 + M_1$$
$$J_2\dot{\omega}_2 = (J_3 - J_1)\omega_3\omega_1 + M_2$$
$$J_3\dot{\omega}_3 = (J_1 - J_2)\omega_1\omega_2 + M_3$$

where J_1, J_2, J_3 are the principal moments of inertia of the body, ω_1, ω_2, ω_3 are the angular velocities about the principal axes, and M_1, M_2, M_3 are the applied or external moments about the principal axes. Derive these equations by expressing the kinetic energy T in terms of Euler's angles (see Ex. 4, Sect. 11.4) and substituting in Lagrange's equations. Note that this system has three state variables $(\omega_1, \omega_2, \omega_3)$ and that these are nonholonomic.

21. Find Hamilton's canonical equations for a system whose Hamiltonian is

$$H = \frac{1}{2}\left[\frac{p_1^2 + p_2^2}{q_1^2 + q_2^2}\right] + \frac{1}{q_1^2 + q_2^2}$$

Show that this represents the orbit of a particle in the (q_1, q_2)-plane and that this orbit is a conic section.

22. Prove that the transformation

$$q' = \ln\left(\frac{1}{q}\sin p\right), \qquad p' = q\cot p$$

is a contact or canonical transformation and write Hamilton's canonical equations in terms of (q', p').

23. Given a system of n particles of masses $m_1, m_2, \ldots, m_n$ located at points $(x_1, y_1, z_1), (x_2, y_2, z_2), \ldots, (x_n, y_n, z_n)$. Find the axis about which the moment of inertia of the system of particles is a minimum.

24. A body of mass m_1 is placed on a horizontal rotating table and is connected by a cord to a second body of mass m_2 as shown in the sketch.

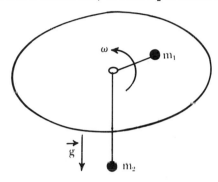

Assuming that m_1 is constrained to move radially on the rotating table and that friction is negligible, find the condition for stable equilibrium if such a position exists. Also derive the equation of motion.

Chapter 7

1. A conducting sphere of radius a is given a charge q. Determine the electrostatic energy associated with the charged sphere, assuming that the sphere is isolated from other bodies.

2. Suppose that by a virtual process the radius of the sphere shrinks to infinitesimal magnitude, the charge remaining unchanged. How does this affect the electrostatic energy? What is the explanation?

3. Find the energy of the electrostatic field produced by a volume charge density $\rho = \alpha/r^2$ in the space between two concentric spheres of radii a and b, $b > a$.

4. Given a parallel plate capacitor partially submerged with its plates vertical in an insulating liquid of density τ and permittivity ε. Determine the distance y to which the liquid will rise in the capacitor (plate separation x_0 and potential difference V_0).

5. Consider two conductors having self-capacitances of C_1 and C_2 at potentials of φ_1 and φ_2. Calculate the force between the two conductors, assuming that the separation of the conductors is much greater than their linear dimensions.

6. Similar to problem 5, except prove that the coefficients of capacitance of the system are given by the matrix

$$C = \begin{bmatrix} C_1\left(1 + \dfrac{C_1 C_2}{r^2}\right) & -\dfrac{C_1 C_2}{r} \\[2ex] -\dfrac{C_1 C_2}{r} & C_2\left(1 + \dfrac{C_1 C_2}{r^2}\right) \end{bmatrix}$$

where r is the distance between the conductors.

7. Find the vector potential and therefrom the magnetic field produced by a current I flowing in a long straight wire of radius b.

8. Calculate the force per unit length between the two wires of an electric power transmission line carrying a direct current of 100 amperes and separated by a distance of 1 meter.

9. Prove, by making use of Laplace's equation, that the potential function for an electrostatic field cannot have an extremum at any point in space which is not occupied by an electric charge.

10. Derive the expression for the capacitance of a system of capacitors $C_1, C_2, \ldots, C_n$ connected in parallel.

11. Same as problem 10 except for the capacitors connected in series.

12. Given a system of charged conductors in which the conductors are maintained at constant potentials by being connected to external batteries. Prove by using the principle of virtual work that the interconductor forces are given by

$$\mathbf{F}_e = \nabla(\tfrac{1}{2}\varphi C \varphi')$$

Contrast this with the result of Eq. (7.26) and explain the difference in sign.

13. A sinusoidal electric field $E_0 \sin \omega t$ is established in a homogeneous material having a relative dielectric constant k and conductivity σ. Determine the frequency ω for which the amplitudes of the conduction current density and the displacement current density are equal. What is the frequency for copper, distilled water, and polystyrene?

Chapter 8

1. A length $4a$ of uniform wire is bent into the form of a square, and the opposite vertices are joined with straight pieces of the same wire, which are in contact at their point of intersection. A given current enters at the intersection of the diagonals and leaves at a vertex. Find the current strength in the various parts of the network, and show that its whole resistance is equal to a length $\dfrac{a\sqrt{2}}{2\sqrt{2}+1}$ of the wire.

2. A generator produces an emf of $\varepsilon(t)$ and has an internal resistance R and internal inductance L. Determine the nature of the load such that the greatest energy is supplied to the load over a given time interval τ.

3. Find the mesh currents for the circuit below, where the switch is closed at time $t = 0$ and the initial stored energy in the circuit is zero.

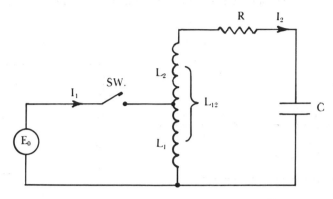

4. An emf $\varepsilon(t)$ is applied to the circuit shown in the diagram. It is desired to design the circuit so that it acts as a pure resistance. Is this possible, and if so what are the conditions which must be fulfilled?

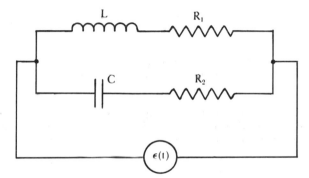

5. Calculate the mesh and branch currents for the resistance network shown in the following electric circuit.

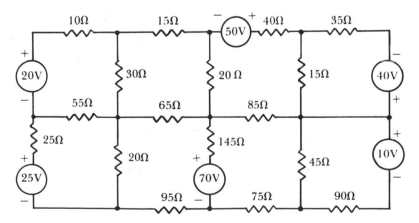

6. The electrical circuit shown in the diagram is used in connection with certain types of radar pulse generation devices for the purpose of quenching pulse "overshoot." Immediately following a pulse, the initial current i_0 in L and the

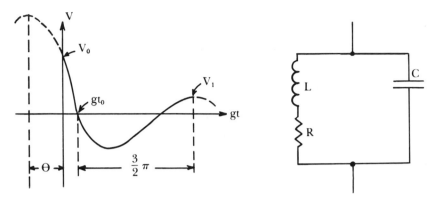

initial charge q_0 on C are given by

$$i_0 = \frac{V_0}{R}[1 - e^{-(R/L)\,\tau}]$$

$$q_0 = CV_0$$

It is desired to bring the capacitor voltage V to zero as fast as possible, i.e., to minimize t_0 subject to the constraint that V_1 (see diagram) is held constant. Assume that $V_1 = 0.2V_0$.

7. For the following system, find the currents which give minimum heat loss in terms of the total current I, using the method of Lagrangian multipliers. Show that the result is the same as that given by Kirchhoff's laws.

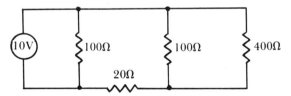

8. The following circuit is a bridge containing a null detector whose impedance is Z_D. The unknowns R_x and L_x are to be determined by balancing the bridge.

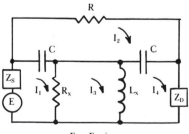

$$E = E_0 \sin \omega t$$

(a) Set up the impedance matrix A, using the indicated mesh currents.
(b) Show why setting $I_4 = 0$ reduces the problem to a homogeneous linear system.
(c) Set the cofactor of a_{14} equal to zero and find R_x and L_x in terms of R, C and ω.
(d) Explain, in terms of part b, why these values satisfy the given conditions.

9. The following circuit is inductively coupled with a mutual inductance of -10.6 mh between coils, and driven by a simple harmonic source $E = E_0 \sin 120 \pi t$.
(a) Set up the impedance matrix.
(b) Construct an equivalent circuit which has no magnetic coupling.

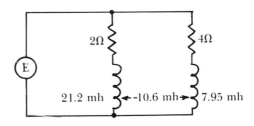

10. The following circuit is driven by sources $E_1 = 10 \sin 120 \pi t$ and $E_2 = 10 \cos 120 \pi t$.
(a) Set up the impedance matrix.
(b) Find the steady state currents, including the phase angles.

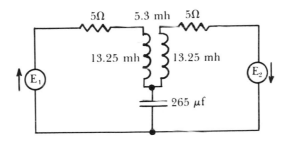

11. For the following circuit, analyze the transient and the steady state response to an input $E = \cos 120 \pi t$. Assume zero initial stored energy.

$$31.6\Omega \qquad 0.1325\text{h} \qquad 53\mu\text{f}$$

12. For the following circuit, the applied emf is of the illustrated form. The Fourier series representation is

$$E = \frac{8}{\pi^2} \sum_{m=0}^{\infty} \frac{(-1)^m}{(2m+1)^2} \sin (2m+1)\omega t$$

(a) Choose a typical term of E and find the associated impedance matrix.

(b) Find the steady state current response to that term.
(c) Using the superposition theorem, find the total response

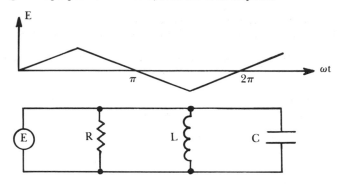

13. Continuation of problem 12.

(a) Taking the complete waveform, use Laplace transforms to find the current. Let the initial current in the coil be $I_2(0)$.

(b) Compare part a with part c of problem 12. Find the initial vector $\mathbf{I}(0)$ and $\mathbf{q}(0)$.

(c) How many terms would it take to give 10 percent accuracy for
  (1) the input?
  (2) the output?

14. Transform the following dynamical equation to state variable form

$$\left\{ \begin{bmatrix} 2 & 1 \\ 1 & 3 \end{bmatrix} D^2 + \begin{bmatrix} 0.5 & 0 \\ 0 & 0.2 \end{bmatrix} D + \begin{bmatrix} 1 & 0 \\ 0 & 1 \end{bmatrix} \right\} \mathbf{q} = \begin{bmatrix} 1 \\ 0 \end{bmatrix}$$

$$\mathbf{x}_0 = \begin{bmatrix} \mathbf{q}_0 \\ \dot{\mathbf{q}}_0 \end{bmatrix} = \begin{bmatrix} 2 \\ 0 \\ 1 \\ 0 \end{bmatrix}$$

15. In problem 14 find the eigenvalues of the system matrix B (Eq. 8.30), the corresponding transformation matrix P (Eq. 8.32), and hence the solution for $\mathbf{x}$ (Eq. 8.39).

16. Complete the solution of problem 14 using the Laplace transform method and thereby check the answer to problem 15.

17. Apply Kron's mesh analysis method to solve the three-mesh network of Figure 8.7.

18. Confirm the solution of the inductively coupled a-c circuit of Figure 8.10 using Kron's mesh analysis method.

19. Analyze the continuous structure shown by the displacement method for the applied forces. Modulus of elasticity (E) and moment of inertia (I) are constant throughout the structure.

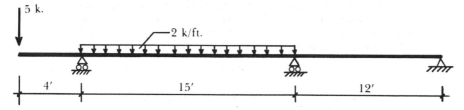

20. Find the displacements (rotation and translation) for the rectangular plane frame shown due to applied loads. Also compute the end moments of all members. Assume $E = 29,000$ k/in.^{2}

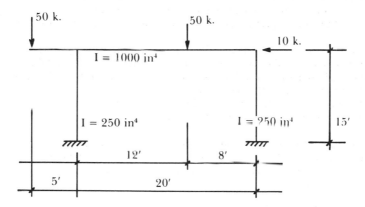

21. For the given loading, find the bar forces of the truss shown utilizing the displacement method of analysis. Hint: Note that in the mathematical model of the truss, bar forces are treated similar to the end moments of frame members, and loads applied to joints similar to fixed end moments. Numbers in circles indicate cross sectional areas in (in.^{2}).

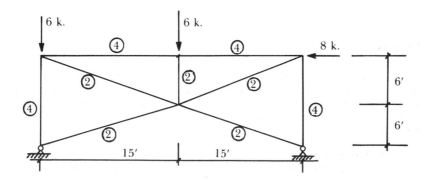

Chapter 9

1. A satellite is launched at an altitude of 200 miles. Determine the initial velocity (magnitude and direction) necessary for a circular orbit. Also discuss the motions which result from launching velocities smaller and greater than that necessary for a circular orbit.

2. Determine the escape velocities from the Sun, Jupiter, and our Moon.

3. The satellite spacecraft Explorer XI was launched into an Earth orbit with a perigee altitude of 304 miles and apogee altitude of 1113 miles. Find the velocities at these points, the eccentricity of the orbit, and the orbital period.

4. A satellite is launched into an elliptical orbit of eccentricity 0.50, a perigee distance of 500 miles, and the orbital plane is oriented at an angle of 45° with respect to the equator. Assuming that the perigee point is in the equatorial plane of the Earth, find the equation or equations which give the apparent path of the satellite with respect to the Earth taking into account the Earth's rotation.

5. Assume that a 200 kg.-satellite moves in a circular orbit with a period of exactly 3 hours. Another satellite of twice the mass of the first satellite is to be placed in orbit alongside the original satellite. Determine the following quantities:
(a) The escape velocity of satellite 2 from the Earth.
(b) The magnitude of the velocity of the second satellite.
(c) The relative velocities of the two satellites.

6. Assume that a space probe of mass 100-kg. is shot from the Earth with an initial velocity of 12 km./sec. It is desired to know whether the probe will orbit the Earth, or whether it will escape the Earth's gravitational field. Neglect any atmospheric effects.

7. An Earth satellite has a mass of 100 kg. and an orbital period of 117 minutes. Find the major axis of the orbit, the velocity at perigee and at apogee, the total energy of the satellite, and the eccentricity of the orbit.

8. The satellite of problem 7 fires rockets in a direction tangential to the orbit. It is observed sometime later that the period of the satellite has increased to 150 minutes. Find the total energy imparted to the satellite to effect the transfer to the new orbit.

Chapter 10

1. A shunt-wound d-c motor operating from a 200-volt power source has an armature resistance of 0.5 ohm, a no-load speed of 1200 rpm, a no-load armature current of 2.0 amps., and a full-load armature current of 30 amps. Find the steady-state speed of the motor and the torque developed under full-load conditions.

2. The motor of problem 1 is operated with an external resistance of 1 ohm in series with the armature and the mechanical load on the motor adjusted

until the armature current is again 30 amps. Determine the speed and the torque developed under the latter conditions. Find an equivalent electric network which represents the motor and load system, assuming steady-state conditions.

3. Continuation of problem 2. Find an equivalent electric network which represents the motor and load system, assuming that the load torque is proportional to speed, the armature inductance is 0.02 henry, and the moment of inertia of the armature and load is 1.0 slug-ft^2.

4. Given a shunt-wound d-c generator rated at 500 volts, 40 amps., 1500 rpm. The armature inductance is 0.01 henry. Find the equations of motion for the system, assuming that the generator is driven by a prime mover of maximum torque equal to that necessary for full-load output of the generator and that the load resistance matches the full-load output of the generator. Solve the equations of motion under the condition that the driving torque is applied in accord with the law $\tau = \tau_0(1 - e^{-0.2t})$, where τ_0 is the full load torque and t is the time in seconds, $t \geq 0$.

5. The system equations for commutator machines are derived in Sect. 10.2 for the case where the commutator brushes are aligned with the fixed axes. Derive these equations for the case where the commutator brushes are displaced from the fixed axes by an angle α.

6. Derive the system equations for the three-phase induction motor (cf. Example 3 for the two-phase induction motor).

7. Given a rotating magnetic system consisting of one stator and one armature derived from Figure 10.2 by deleting coils 2 and 4. Voltages $V_1(t)$ and $V_3(t)$ are applied to the system. Find the expressions for the current vector $\mathbf{I}$ and the torque τ_m for the system.

8. A two-phase, 60 cps, synchronous machine (cf. Figure 10.5) is operating in the steady-state as a generator. The field winding is energized by a constant d-c voltage such that the field current is 3 amps. Parameters of the machine are $L_a = 0.12$ henry, $R_a = 1.0$ ohm, $L_{as} = 0.10$ henry. Find the generated voltages in the rotor. What is the load voltage for an inductive load $12 + 15i$ ohms connected to each rotor? Determine the power output of the machine and the phase angle γ when the inductive load $12 + 15i$ ohms to each rotor?

9. Derive the system equations for the device known as an "amplidyne" (see Ref. 5, p. 345), starting with the results of Sect. 10.1 of this chapter.

10. Find the torque equation for a series wound d-c motor assuming constant inductances for both the stator and armature windings.

Chapter 11

1. A mass m is connected to three linear springs on a horizontal plane as shown. Find the position of equilibrium and determine the eigenfrequencies of the system.

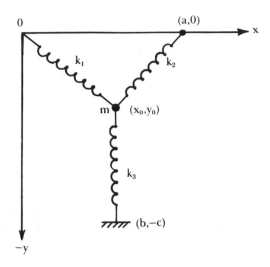

2. The system of three equal masses joined by three identical springs shown in the accompanying figure may be regarded as a mechanical model of the O_3 molecule.* Determine the eigenfrequencies of the system and show the normal modes graphically.

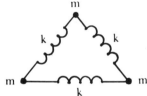

3. A conservative system is free to oscillate in one dimension, and has a potential function $V = \frac{1}{2}kx^2$ and kinetic energy $T = \frac{1}{2}m\dot{x}^2$, where k and m are constant. The total energy is given by $E = \frac{1}{2}kx_0^2$, and the second boundary condition is $t(0) = t_0$.
(a) Write the solution in integral form.
(b) Use a change of variable or other appropriate method to evaluate the integral.
(c) Write x as a function of t.
For problems 4, 5, and 6 find:
(a) the kinetic energy function
(b) the potential energy function
(c) the equilibrium position
(d) the stiffness matrix K for small vibrations
(e) the dynamic matrix $M^{-1}K$
(f) the eigenvalues and associated eigenvectors of the dynamic matrix
(g) the general solution

4. Two pendulums have equal lengths, a, bobs of equal mass m, and are coupled with a smooth hinge. They are constrained to motion in the vertical plane in a constant gravitational field.
(h) Show that the dynamic matrix is not symmetric.

---

* This problem is discussed in the following reference: A. Nussbaum, "Group Theory and Normal Modes." *American Journal of Physics, 36* p. 529, June, 1968.

5. Three pendulums have equal lengths, $2a$, and bobs of equal mass, m. They are suspended from smooth hinges with equal spacing, a, and coupled midway between the hinge and bob by springs of constant k and a relaxed length of zero.

(h) Find the normal modes of vibration.

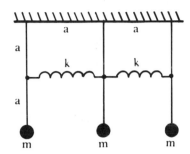

6. Two equal masses are constrained to horizontal motion by slots, and coupled by springs of constant k and zero relaxed length

(h) Find the normal modes of vibration.

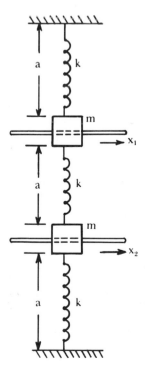

7. Determine an electric network analog for the small vibrations of the coupled pendulum system of Figure 11.6.

8. Develop the equation of motion for the analytical balance shown in the sketch, assuming small viscous damping due to the drag of the atmosphere on the pans of the balance. Find the expression for the sensitivity of the balance.

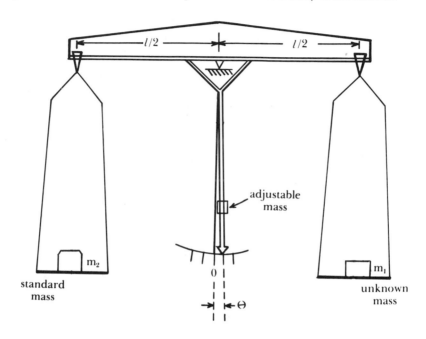

9. Calculate the frequency of vibration of a wooden cylinder of uniform cross-section floating in a fluid of density ρ_0.

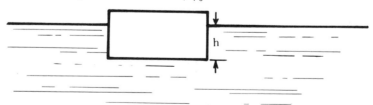

10. The electromechanical device shown in the sketch operates by displacing the capacitor plate from the equilibrium position and releasing it, no external forces being applied thereafter. Establish the equation of motion and calculate the energy transferred to the load, represented by the linear resistor R_L. Let C_0 be the capacitance at equilibrium and neglect edge effects during the motion. Investigate the stability conditions for the system.

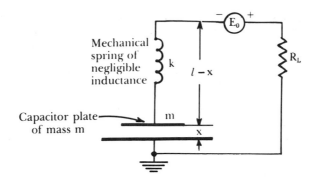

11. Inverted pendulum. A plane pendulum can be made to execute stable vibrations about the inverted position ($\theta = \pi$ in Figure 11.1c) by causing the point of support to oscillate vertically according to the law $y = a \cos \omega t$, where

$$\omega \gg \sqrt{\frac{g}{l}}$$

Show that the condition for $\theta = \pi$ to be a position of stable equilibrium is

$$a\omega > \sqrt{2gl}$$

Discuss the vibrations which can occur about $\theta = \pi$ under this condition.

12. Given a system of n degrees of freedom consisting of n identical one-dimensional systems of eigenfrequency λ_0 coupled together in such a way that the Lagrangian is

$$L = \tfrac{1}{2}(\dot{q}_1^2 + \dot{q}_2^2 + \cdots + \dot{q}_n^2) - \tfrac{1}{2}\lambda_0^2(q_1^2 + q_2^2 + \cdots + q_n^2)$$
$$+ a(q_1 q_2 + q_2 q_3 + \cdots + q_{n-1} q_n)$$

(a) Find the characteristic equation.
(b) Calculate the eigenfrequencies for the case $\lambda_0^2 = 1$, $a = 0.01$.
(c) Find mechanical and electrical analogs for the system and specify the physical parameters corresponding to conditions stated under part b.

13. Electromechanical parametric oscillator. The prongs of a tuning fork form one electrode of a variable capacitor which in turn is a part of an electrical resonant circuit as shown in the accompanying sketch. Hartley (*Physical Review*

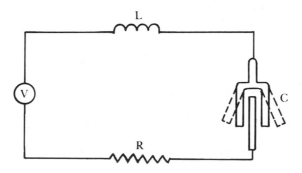

15 289, 1929) showed theoretically that both the electrical and mechanical circuits can exhibit negative resistance at ω_1 and ω_2 when V is an a-c emf source of angular frequency $\omega_1 + \omega_2$. More recently it has been shown that this parametric oscillator is an interesting parallel or analog to the stimulated Raman laser. (B. Salzberg, Airborne Instruments Laboratory) Investigate the characteristics of the parametric oscillator system.

14. **Vibrations of continuous structures.** The small vibrations of periodic structures are examined in Ex. 12 of Sect. 11.1 of this chapter. Therein is derived the equations of motion for a system of n particles of equal mass m mounted at equal distances on a taut elastic string (Fig. 11.7). This system serves as an analog for many other physical systems. It is of interest to extend the results of Ex. 12 to the vibrations of a continuous string and its various analogs.

By allowing the number of particles n to increase without limit, while keeping the total mass of the particles per unit length of string constant, the system approaches a continuous distribution of mass, i.e., a continuous string of mass ρ per unit length.

Find:

(a) The solution $q(x, t)$, where q is the displacement of the string at the distance x from one of the fixed ends of the string and t is the time, by considering the limit of the solution of the form of Eq. (11.16).
(b) The equation of motion for the continuous string by using Hamilton's principle with appropriate expressions for T and V. Note that this leads to a partial differential equation.

Chapter 12

1. Give specific examples of feedback systems from technology, life processes, management, transportation, and the space sciences. Show the main structural features of the examples cited by means of block diagrams. Discuss the overall characteristics of at least one of the examples.

2. Find the equation relating the output signal e_2 to the input signal e_1 (cf. Figure 12.1) for an amplifier with two feedback loops in parallel.

3. Determine the output-input equation for two feedback amplifiers in tandem as shown in the following diagram.

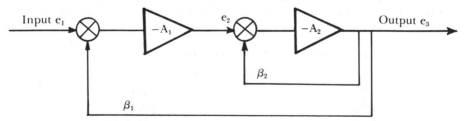

4. Devise feedback operational amplifiers using resistor and capacitor elements which will (a) integrate and (b) differentiate input signals.

5. Derive the equations which apply to problem 4 and discuss their properties.

6. Devise an analog computer system to solve third order differential equations of the form

$$\frac{d^3\theta}{dt^3} + a\frac{d^2\theta}{dt^2} + b\frac{d\theta}{dt} + c\theta = f(t)$$

where a, b, c are constants and $f(t)$ is a prescribed input function (cf. Ex. 1 of Sect. 12.2).

7. Similar to problem 6, but for a fourth order differential equation.

8. Show how the small vibrations of the coupled pendulum system of Figure 11.6 can be simulated on an analog computer.

9. Derive the expression for a two-loop servomechanism system shown in the accompanying diagram where s is the Laplace transform parameter.

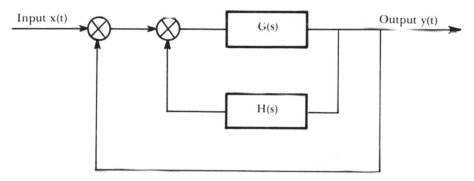

10. Compensation networks are often introduced into feedback systems to alter performance characteristics. The block diagram for one such arrangement is shown in the sketch; $A(s)$ represents the compensation network and K_1 and K_2 are constants. Find the transfer function for the system in general form and then apply this to a specific example (covered in many textbooks on the subject, as for example B. C. Kuo, *Automatic Control Systems*, Englewood Cliffs, New Jersey, Prentice-Hall, 1962).

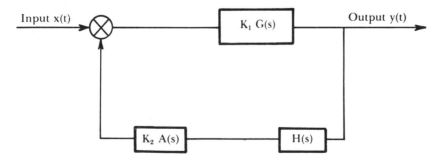

11. Given a system described by the relation

$$(D^2 - 6D + 4)y = u$$

Find the characteristic of a control device necessary to stabilize the above system (cf. Ex. 1 of Sect. 12.5).

12. Same as problem 11 except the system is described by the relation

$$(D^3 + 2D^2 - 4D + 3)y = u$$

13. Determine the characteristics of a control device designed to stabilize a dynamic system similar to that shown in Figure 12.10 except that there are two inverted pendulums. Complete the solution by assigning numerical values to the parameters and carrying out the resulting numerical calculations.

14. Using literature references, find three specific examples of multi-variable feedback control systems (cf. Ref. 10 of this chapter).

INDEX

493